CONTRIBUTIONS TO MECHANICS

PROFESSOR MARKUS REINER, Dr. Tech.

Gerard Swope Professor of Mechanics,
Technion—Israel Institute of Technology, Haifa.

Member of The Israel Academy of the Sciences
and the Humanities.

Born 5 January 1886.

CONTRIBUTIONS TO MECHANICS

Markus Reiner Eightieth Anniversary Volume

Edited by

DAVID ABIR

Technion – Israel Institute of Technology, Haifa

THE QUEEN'S AWARD TO INDUSTRY 1966

PERGAMON PRESS

OXFORD · LONDON · EDINBURGH · NEW YORK

TORONTO · SYDNEY · PARIS · BRAUNSCHWEIG

Pergamon Press Ltd., Headington Hill Hall, Oxford
4 & 5 Fitzroy Square, London W.1
Pergamon Press (Scotland) Ltd., 2 & 3 Teviot Place, Edinburgh 1
Pergamon Press Inc., Maxwell House, Fairview Park, Elmsford, New York 10523
Pergamon of Canada Ltd., 207 Queen's Quay West, Toronto 1
Pergamon Press (Aust.) Pty. Ltd., 19a Boundary Street, Rushcutters Bay, N.S.W. 2011, Australia
Pergamon Press S.A.R.L., 24 rue des Écoles, Paris 5e
Vieweg & Sohn GmbH, Burgplatz, 1, Braunschweig

First edition 1969

Library of Congress Catalog Card No. 68-26907

Printed in Hungary

08 012822 X

CONTENTS

DYNAMICS

MATERIAL PROPERTIES

RHEOLOGY

PREFACE

To THE numerous friends of Professor Markus Reiner it may come as a surprise to learn that he is in his eighties now (born 5 January 1886). His vitality and activity defy his age. Of the dozens of research projects conducted by Reiner, the two most important were published at the ages of 59 and 62.† In 1947, being 61 years old (the retirement age for many people from active work), he opened a new and flourishing chapter in his life as Professor of Mechanics at the Technion—Israel Institute of Technology, Haifa, thus becoming an academic after almost 40 years of work as a practising civil engineer. At the age of 70 Reiner exhibited his centripetal pump, which caused a stir and keen discussion among professional circles.

Reiner continues to display vivid interest and energy in many facets of activity, both intellectual and physical; he still publishes papers and swims almost daily in the sea or pool.

May he carry on his energetic and fruitful activity for many more years!

The prolific and diverse contribution of Professor Reiner in the field of Mechanics in general, and Rheology in particular, is well known. It stands out not only in quality, but also in quantity. Though he spent the first four decades of his adult life as a full-time engineer, doing research as a hobby in his spare time only (apart from two years devoted solely to research), he has published to date an impressive number of publications—172 papers and notes. The more significant of these, sixty-eight in number, are given in the selected list of his works, which follows this Preface. The list also includes Reiner's three fundamental books (one of them written in collaboration with Dr. G.W. Scott Blair), which were translated into several languages.

Reiner's work has been recognized by numerous academic and professional institutions, both in his own country, Israel (where he has been living close to half a century), and in a number of other countries. The list of awards, prizes, honorary memberships, etc., bestowed upon him by many learned societies is both lengthy and noteworthy.

† (1) "A mathematical theory of dilatancy", *American Journal of Mathematics* **67,** 350–62 (1945). (2) "Elasticity Beyond the Elastic Limit", *ibid.* **70,** 433–46 (1948).

The present book, too, is a manifestation of esteem and affection to Professor Reiner, expressed by thirty-three authors from various countries, who eagerly contributed original works. The number of contributors might easily have been much larger; but in order to avoid too voluminous a book, invitations for papers were confined to personal friends of Markus Reiner in the narrow professional field. Even so, it is quite likely that a good number of people who might fall within this category of friends have been left out. To these, sincerest apologies are hereby offered; any such omission was unintentional and a result of unawareness.

The papers in this volume are conveniently grouped in parts, according to their contents. This grouping, however, should not be considered as rigid; neither should it be taken as a measure of relative merit or significance. In each part, the papers are arranged in alphabetical order of the authors.

The list of contributions opens with a biographical survey of Reiner's life, written by Dr. Scott Blair—his close friend for almost four decades.

In editing the volume no attempt was made to unify the style of writing of the various authors; it was thought that the personal way of presentation of the individual contributors should be preserved. Some editorial changes, however, have been introduced in the manuscripts, either for the sake of clarity of the subject-matter or in accordance with English language requirements. By the same token, the spelling of words which are written differently in Great Britain and America has been left intact according to preference of the individual authors; however, in order to avoid misunderstandings and confusion, uniformity has been adopted throughout the book in designating the serial numbers of References by square brackets, and of equations by parentheses.

Finally, thanks are expressed to the authorities of the Technion–Israel Institute of Technology, Haifa, for their encouragement in the preparation of this volume; and sincere gratitude and appreciation are due to Eng. E. Goldberg, who contributed significantly as language editor for part of the material.

DAVID ABIR

SELECTED LIST OF PUBLICATIONS BY MARKUS REINER

BOOKS

1943: *Ten Lectures on Theoretical Rheology*, R. Mass, Jerusalem. (Translated into Russian, Moscow, 1947.)

1949: *Deformation and Flow*, H. K. Lewis & Co., London. (Translated into Japanese, Tokyo, 1957.)

1949: *Twelve Lectures on Theoretical Rheology* (revised and enlarged edition of 1943 book), North Holland Publishing Co., Amsterdam. (Translated into French, Paris, 1955; and Polish, Warsaw, 1958.)

1957: *Agricultural Rheology* (in collaboration with G. W. Scott Blair), Routledge & Kegan Paul, London. (Translated into Japanese, Tokyo, 1960.)

1960: *Lectures on Theoretical Rheology* (revised and enlarged edition of 1949 book), North Holland Publishing Co., Amsterdam.

1960: *Deformation, Strain and Flow* (revised and enlarged edition of 1949 book), H. K. Lewis & Co., London. (Translated into Russian, Moscow, 1963; and German, Munich, 1967.)

SELECTED PAPERS

1911: Deformationswiderstand, tangentiale Kohaesion and Reibungswiderstand von Fluessigkeiten, *Zeitschrift des oesterreichischen Ingenieur- u. Architekten Vereins*, **63**, 803–7.

1914: Ueber die Torsionsbeanspruchung von Wellen, *Zeitschr. d. oesterr. Ing. u. Arch. Ver.* **18,** 1–6.

1915: Ueber das Auftreten von Normalspannungen bei der Torsion prismatischer Staebe, *Oesterr. Wochenschrift fuer den oeffentlichen Baudienst.*

1917: Ueber die Berechnung der Auflagereibung von Tragwerken, *Zeitschr. d. oesterr. Ing. u. Arch. Ver.* **69,** 574–6, 584–7.

1924: Structural steelwork reinforced with concrete, *Concrete and Constr. Eng.* **19,** 445.

1925: Ueber die Torsion prismatischer Staebe durch Kraefte, die auf den

Mantel einwirken. *Zeitschrift f. angewandte Mathematik u. Mechanik*, **5,** 409–17.

1925: Safety of retaining walls against sliding, *Concr. and Constr. Eng.* **20,** 183.

1926: Ueber die Stroemung einer elastischen Fluessigkeit durch eine Kapillare, *Kolloid-Zeitschrift* **39,** 80–87.

1926: Zur Theorie der Strukturturbulenz, *Kolloid-Zeitschr.* **39,** 314–15.

1927: (in collaboration with R. RIWLIN) Ueber die Stroemung einer elastischen Fluessigkeit im Couette Apparat, *Kolloid-Zeitschr.* **43,** 1–5.

1929: The general law of flow of matter, *J. Rheology* **1,** 11–20.

1930: In search of a general law of flow of matter, *Rheology* **1,** 250–60.

1930: Zur Hydrodynamik der Kolloide, *Zeitschr. f. angew. Math. u. Mech.* **10,** 400–13.

1931: Slippage in a non-Newtonian liquid, *Rheology* **2,** 337–50.

1932: Outline of a systematic survey of rheological theories, *Rheology* **3,** 245–56.

1932: Die Berechnung des Einflusses einer festen Wand auf den Aggregatzustand einer Fluessigkeit aus Viskositaetsmessungen, *Physikalische Zeitschrift* **33,** 499–502.

1933: Ueber den Gebrauch der Potenzfunktion zur Darstellung einer naturgesetzlichen Beziehung, *Die Naturwissenschaften* **21,** 294–9.

1933: A contribution to the theory of the straight beam, *Quarterly J. of Mathematics* (Oxford), **4,** 12–20.

1933: (in collaboration with E. C. BINGHAM) The rheological properties of cement and cement mortar stone, *Physics* **4,** 88–96.

1933: (in collaboration with R. SCHOENFELD–REINER) Viskometrische Untersuchungen hochmolekularer Naturstoffe: Kautschuk in Toluol, *Kolloid Zeitschr.* **65,** 44–62.

1934: The theory of non-Newtonian liquids, *Physics* **5,** 321–41.

1934: Viscometric studies of rubber solutions, *Physics* **5,** 342–9.

1938: (in collaboration with A. FREUDENTHAL) Failure of a material showing creep (a dynamical theory of strength), *Proc. 5th Intern. Congr. Appl. Mechanics.*

1939: (in collaboration with K. WEISSENBERG) A thermodynamical theory of the strength of materials, *Rheology Leaflet* No. 10, pp. 12–20.

1945: A mathematical theory of dilatancy, *Am. J. Math.* **67,** 350–62.

1945: (in collaboration with A. ARNSTEIN) Creep of cement and cement mortar, *Civil Engineering* **40,** 198–202.

1946: The coefficient of viscous traction, *Am. J. Math.* **68,** 672–80.

1946: A simple instrument for measuring blast, *J. Scientific Instruments* **23,** No. 12.

1948: Research on the workhardening of polycrystalline metals. Palestine Board for Scientific & Industrial Research, Jerusalem.

1948: Research on the rheological properties of bitumen and asphalt. Palestine Board for Scientific & Industrial Research, Jerusalem.

1948: Elasticity beyond the elastic limit, *Am. J. Math.* **70,** 433–46.

1949: On volume or isotropic flow, *Applied Scientific Research* **A. 1,** 471–88.

1949: (in collaboration with G. W. SCOTT BLAIR and G. MOCQUOT) Sur un aspect de méchanisme de formation des lainures dans le fromage de Gruyère compte, *Le Lait* **29,** 351–7.

1949: (in collaboration with G. W. SCOTT BLAIR and H. B. HAWLEY) The Weissenberg effect in sweetened condensed milk, *J. Soc. Chem. Ind.* **68,** 327–8.

1950: The volume flow of asphalt, *J. Soc. Chem. Ind.* **69,** 257–60.

1950: (in collaboration with A. ARNAN and M. TEINOWITZ) Research on loading tests of reinforced concrete floor structures. Research Council of Israel, Jerusalem.

1951: An investigation into the rheological properties of bitumen: Maxwell body and elastic dispersions. *Bull. Res. Counc. Israel* **1,** No. 3, 5–25.

1952: (in collaboration with I. BRAUN) Problems of cross-viscosity, *Quart. J. Mech. Appl. Math.* **5,** 42–53.

1952: (in collaboration with I. BRAUN) Note on dimensions in tensor analysis, *Bull. Res. Counc. Israel* **1,** 81–82.

1952: A possible cross-viscosity effect in air, *Bull. Res. Counc. Israel* **2,** 65.

1952: A tea-pot effect, *Bull. Res. Counc. Israel* **2,** 265.

1953: On volume viscosity, *Proc. 8th Intern. Congr. Appl. Mechanics*, Istanbul, p. 273.

1954: Volume rheology, *Proc. 2nd Intern. Congr. Rheology*, 1953, Butterworths, London, pp. 310–15.

1955: The complete elasticity law for some metals according to Poynting's observations. *Appl. Sci. Res.*, **A. 5,** 281–95.

1956: The tea-pot effect, *Physics Today* **9,** 16–20.

1956: (in collaboration with M. HANIN) On isotropic tensor functions and the measure of deformation *ZAMP* **7,** 377–93.

1956: Phenomenological macrorheology, *Rheology 1* (ed. F. EIRICH), pp. 9–62, Academic Press, New York.

1956: (in collaboration with B. POPPER) The application of the centripetal effect in air to the design of a pump, *Brit. J. Appl. Phys.* **7,** 452–3.

1957: A centripetal pump effect in air, *Proc. Roy. Soc.* **A. 240,** 173–89.

1958: (in collaboration with L. RINTEL) The complete stress–strain law in infinitesimal elasticity, *Bull. Res. Counc. Israel* **6C,** 113–26.

1958: Rheology, *Encyclopaedia of Physics* **6** (ed. S. FLUEGGE), pp. 434–550, Springer-Verlag, Berlin.

1958: The centripetal-pump effect in a vacuum pump, *Proc. Roy. Soc.* **A. 247,** 152–67.

1959: The flow of matter, *Scientific American* **201,** 122–38.

1960: The rheology of concrete, *Rheology 3* (ed. F. EIRICH), pp. 341–64, Academic Press, New York.

1960: Cross-stresses in the laminar flow of liquids, *Physics of Fluids* **3,** 427–32.

1960: The stress–strain relation of elasticity and the measure of strain, *ZAMM* **40,** 415–20.

1960: Plastic yielding in anelasticity, *J. Mech. Phys. Solids* **8,** 255–61.

1962: (in collaboration with A. FOUX) Cross-stresses in the flow of air at reduced pressures, IUTAM Intern. Symposium, *Second-order Effects in Elasticity, Plasticity and Fluid Dynamics,* Haifa, 1962 (ed. M. REINER and D. ABIR), Pergamon Press, Oxford, 1964.

1962: (in collaboration with Z. KARNI) The general measure of deformation, IUTAM Symposium, Haifa, *ibid.*

1964: (in collaboration with A. FOUX) Cross-stresses in the flow of rarefied air, *3rd Intern. Congress in the Aeronautical Sciences,* Stockholm, 1962, Spartan Books, Washington, D.C.

1965: (in collaboration with A. FOUX) Cross-stresses in the flow of different gases, *Topics in Applied Mechanics*—Schwerin Memorial Volume (ed. D. ABIR, F. OLLENDORFF and M. REINER), Elsevier Co., Amsterdam.

1965: (in collaboration with D. ABIR and H. MANOR) The solid viscosity of mild steel, *ibid.*

1965: Anomalous viscosity, *Research Frontiers in Fluid Dynamics* (ed. SEEGER and TEMPLE), pp. 171–91, Interscience Publishers, New York.

1965: Second order effects, *ibid.*, pp. 193–211.

1965: Second order stresses in the flow of gases, *Proc. Intern. Congress Rheology,* Providence, 1963, pp. 267–79.

1966: Finite simple torsion of a circular cylinder, *Israel J. Technology* **4,** 187–92.

1966: Deformation, *Encyclopedia of Polymer Science and Technology* **4,** 620–47, John Wiley & Sons, New York.

1967: (in collaboration with Z. KARNI) The complete tensor expressions of the Green and Almansi measures of deformation, *ZAMP* **18,** 131–2.

PROFESSOR MARKUS REINER: A BIOGRAPHICAL SKETCH

G. W. SCOTT BLAIR

National Institute for Research in Dairying, University of Reading, Shinfield, Reading, England

I FEEL very much honoured to have been invited to write a brief biographical sketch for the Volume commemorating the eightieth birthday of my old friend Markus Reiner. Although we live in different continents, we have worked together on rheology over a period of nearly 40 years and are the closest of friends.

Markus Reiner was born on 5 January 1886, of Jewish parents, in Czernowitz, at that time the capital of the Province of Bukowina which was part of the Austro-Hungarian Empire. The population of Bukowina was mixed: Ukrainians in the north, Rumanians in the south, and about 25% of Jews, who spoke either Yiddish or German, but did not think of themselves as Germans.

His father was, in turn, wine-merchant, restaurateur and proprietor of a delicatessen shop. He and his wife had to work hard to support themselves and their six children. Markus was the youngest child and he was much helped by an older sister, who, among other things, taught him to play the piano, a hobby which he has enjoyed all his life.

He was at Primary and Secondary Schools in Czernowitz and then went to study Civil Engineering at the famous "Technische Hochschule" at Vienna. Among his teachers was J. Finger, whose work covered Second-order Elasticity. Reiner had no idea at this time that he would some day himself work in this field. In 1909 he obtained the degree of "Ingenieur" and, finding better prospects for a good position in Berlin, he moved there as an employee in an office of structural design. (In those days, one was free to move from any one country to another to find a suitable post—excepting only Russia.) He also worked for a time in Essen but later returned to Czernowitz to be near his par ents, having found a position in the Structural Design Office of the State Railways Directorate. At this time he was much influenced by the works of Tolstoy and the early writings of Martin Buber, later to become a close friend. The Jews in Bukowina were mostly in trade, and the young Markus felt the call of the soil, especially in the land of his ancestors, "Erets Israel". He became a Zionist but the First World War started before he could implement his plans to go to Palestine. Just before the war started he was awarded the degree of Doctor of Technology at Vienna. His thesis dealt with problems of torsion and he had already published a paper on Lubrication Theory.

Early in the war the Russians occupied Czernowitz, and Reiner found himself a lieutenant in the Engineering Corps of the Austrian Army, supervising such projects as the reconstruction of blown-up bridges in Serbia and in Italy. After the war he returned for a time to his position in Czernowitz, then part of Rumania; but in 1922 he emigrated to Palestine.

He was greatly attracted by the experiments in communal living embodied in the Kibbutz Movement and he started to work as a labourer at an orange plantation. But, with his age and experience (he was then 36), his comrades decided that he could do better service in a post under the British Mandate Government in the Department of Public Works in Jerusalem, where he stayed for 25 years, until the British left Palestine and the State of Israel came into existence.

As Chief Civil and Structural Engineer of the Department of Public Works, he had many interesting assignments, including structural support for the Church of the Holy Sepulchre and the restoration of irrigation channels in Jericho, originally constructed by Herod the Great.

At this time Reiner frequently co-operated with the Government Chief Architect, A. St. B. Harrison, doing the structural design work for many beautiful buildings designed by him. They have remained lifelong friends.

Until 1926 Reiner had only occasional time to spare for strictly scientific work. At about this time, however, preparations were made for the realization of the old plan for a Hebrew university in Jerusalem. The first Professor to arrive was Professor A. Fodor, a Colloid Chemist. Reiner was then living in the same house with Fodor and they often talked about subjects of common interest. One day Fodor showed him an article in the *Physikalische Zeitschrift* by Freundlich *et al.* on the problem of flow of an elastic liquid through a tube. Though knowing nothing about colloid or physical chemistry, Reiner saw at once that the solution of the problem given was incorrect and worked out his own solution. This Fodor sent to Wo. Ostwald who published it in the *Kolloid-Zeitschrift*. The paper raised great interest. Hatschek wrote from London, congratulating Reiner; and Herschel also wrote from Washington, drawing attention to the fact that Reiner's equation had already been published by Buckingham. It became known as the Buckingham–Reiner equation. Herschel wrote to Eugene C. Bingham at Lafayette College and Bingham invited Reiner to visit him there. This he did for a couple of months in the summer of 1928. Some years previously Bingham had published his famous finding that "paint is not a liquid, but a soft plastic solid" and had become interested in plasticity. He arranged for a Plasticity Symposium at Lafayette College in which both chemists and physicists participated; and he became convinced that it was necessary to find for them a common ground. When Reiner came to Easton, Bingham said to him, "Here am I, a chemist, and you are an engineer working on similar problems. There must be a field in physics to which these problems belong and in which we both can work." To which Reiner replied, "There is such a field; it is called the Mechanics of Continuous Media." "That will not do," said Bingham, "this highbrow designation will frighten away the chemists." He consulted his Professor of Classical Languages who suggested to him "Rheology" in accordance with Heraclitus' saying: *panta rhei.* A Society of Rheology was founded and first met in Washington, D.C., in December 1929, a meeting I well remember: thus rheology was born. This provided Reiner with a field of scientific activities lasting for the subsequent 40 years. However, he continued to work as an engineer and science was merely his hobby until 1947 when he was appointed Professor of Mechanics at the Technion, now the Israel Institute of Technology in Haifa.

He spent 2 years (1932–4) in America with Bingham, acting as an Assistant Editor for the *Journal of Rheology* which Bingham had founded, and doing certain experimental work in which he was assisted by his second wife, Dr. Rebecca Schoenfeld, who was a chemist. Reiner now came into contact with many scientists working in this field, either personally or by correspondence, among the latter, myself. I sent him an inquiry about the date of a paper of his in order to establish priority with a paper of mine. Such questions of priority have in many cases led to antagonisms, the most famous being that between Newton and Leibnitz and their followers about priority in the discovery of the calculus. In this case, however, it led to an intimate friendship which has lasted ever since.

In the years 1929–30 I had the good fortune to hold a Rockefeller Fellowship in America and, during vacations, to visit as many rheologists as I could find, including, of course, Professor Bingham. Wherever I went, I heard of "Dr. Reiner" who had been in America just before me. In view of this and having already corresponded with him, it was a great pleasure to my wife and myself when the Reiners, passing through England on their way to America in 1931, came and visited us. This was the first time that we met.

The first Mrs. Reiner (*née* Grete Obernik) died while comparatively young, leaving a son and a daughter. His second wife, Rivka (the Hebrew form of her name by which she was always known), was a Colloid Chemist. Her death a few years ago, after a short illness, was a great tragedy for her husband. She had been his constant helper, companion and scientific colleague for some 33 years. She is also greatly missed by her many friends, one of the closest of whom was my wife. Her two daughters, as well as Reiner's two children by his first marriage, are all happily married and have all taken to the Kibbutz way of life, thus fulfilling, in the second generation, their father's own ambitions as a young man. There are also now numerous grandchildren growing up as "sabras" (native-born Israelis) in the four Kibbutzim where their parents work.

Markus Reiner has never been a political party man; but, with his late friends Magnes, Buber, and others, he sponsored a group, in Mandate times, to work for Jewish–Arab unity. He must indeed be pleased that all his four children are members of a movement which stresses the importance of making every attempt, however difficult, to end the enmity with Israel's Arab neighbours (while not ignoring the need for the defence of Israel in the meantime). The Movement also urges that everything should be done to give Israel Arabs their fullest rights as citizens, a cause dear to Markus Reiner's heart.

Markus Reiner is a citizen of the world. Not only has he travelled widely, but he has encouraged his children so far as has been possible to do the same: one daughter was at school for a time in Reading, England, and another recently spent 2 years in London. He himself comes each year to England and, through his initiative, I have been able to work with him on three occasions in Israel, a country which I came to love as much as Reiner loves England; so we have kept continually in touch since the end of the Second World War. Just before its start, in 1938, Markus, Rivka and his two small daughters came for a few months to England and lived at a farm near my home. No one knows how the Hebrew-speaking children so quickly made contact with the farmer's family! It was during this visit that I discovered that Reiner was not only a fine research worker, but an excellent "tutor". Trained as a chemist, I had never learned tensor analysis, which a rheologist should know something about. A little private coaching in the farm-house, by the light of an oil lamp, was a very great help to me.

Reiner has published several books, including a modern version of a course of lectures (*Lectures on Theoretical Rheology*), which he originally gave at the University of Jerusalem and his *Deformation, Strain and Flow* which has been a help to rheologists in many lands. His books have been translated into many languages. One book, which we wrote together, on "Agricultural Rheology" has, apparently, been more appreciated in Japanese than in its original English version!

For many years, he lectured to students on Strength of Materials; now, his work is mainly research, sponsored by the U.S. Navy, Air Force, and Department of Agriculture. Always popular as a lecturer, he is, perhaps, even better as an after-dinner speaker, having, like most Israelis, so many good stories. At a time when he was obviously in great distress of mind (1963), he spoke at Providence, R.I., U.S.A., at the Fourth International Congress on Rheology. At this after-dinner talk he introduced the non-dimensional "Deborah Number" which he based on the saying of the Prophetess Deborah—"The Mountains Flowed Before the Lord". Unfortunately bad health prevented my being there, but a number of colleagues who were present afterwards told me that they regarded his speech as one of the highlights of the Congress.

It is difficult to summarize a lifetime of research within a few short paragraphs. I can only try to highlight the principal achievements of Reiner's massive contribution to rheology. He started, as we have seen, with the independent calculation (in 1926) of Buckingham's equation (1921) for Bingham systems flowing in capillary tubes. Soon afterwards, with the late Miss R. Riwlin (not to be confused with her nephew, Dr. R. S. Rivlin), he derived a similar equation for coaxial cylinder viscometry. He was also considerably influenced at about this time by the work of Wo. Ostwald and wrote several papers on Structural Turbulence and what used to be called "Structural Viscosity" as well as the Newtonian portion of the flow-curve which Ostwald called the "Laminarast". He also made a critical study of Ostwald's use of power equations. Around 1930 equations were often sought for non-Newtonian systems not following Bingham's equation; and Taylor, Maclaurin and other series were tried. Reiner was in the forefront of attempts to find general equations for these complex systems. He also studied slip at the capillary wall and proposed a classification of rheological properties which may be regarded as a precursor to the Tables of Deformations drawn up by the (then) British Rheologists' Club† and by the Dutch and other Societies of Rheology many years later.

† Now the British Society of Rheology.

Starting with what are now called Maxwell, Kelvin and Saint Venant Bodies, he was responsible for building and collating many of the model systems used by other workers.

Around 1938–9 he extended his engineering work on strength of materials, publishing several rheological papers on this subject; and immediately after the war, he turned his attention to dilatancy and work-hardening as well as publishing further work on rheological classification.

He was among the first to predict the importance of "secondary normal stresses" and foretold the existence of what are now called "Weissenberg Effects" before anybody had seen such phenomena.

Reiner was always ready to turn his skill as an engineer to unlikely uses. A French colleague (M. Mocquot) and I once consulted him about the causes of some characteristic cracks in Gruyère cheese and he at once found an analogy with cracks which he had often observed in concrete buildings! Later, we were able to show the Weissenberg Effect, which he had predicted, to be present in certain aged condensed milks.

Meanwhile, he had time to help substantially in such diverse fields as his wife's work on rheology of flour doughs and my own studies on bovine cervical mucus.

But perhaps his most exciting and controversial findings came when he had reached an age when many people have retired. Led by observations in the rheology of bitumen, he started studies on the behaviour of gases and, later, liquids, between two very carefully aligned concentric disks, very close together, one of which is rotated at high speeds. He used this for the construction of a centripetal pump. Some aerodynamic experts maintain that the effect is due simply to the imperfections in the alignment of the apparatus and the whole question still remains *sub judice*.

Meanwhile, Reiner had not forgotten metals and has studied second-order effects in their elastic response; and the changes in length which take place when metal rods are twisted (Poynting Effects).

With Dr. David Abir, he edited the *Proceedings of an International Symposium*, arranged by IUTAM *on Second-order Effects in Elasticity, Plasticity and Fluid Dynamics*, held in Haifa in 1962. It may be mentioned that in his introductory lecture, Professor Truesdell, the well-known authority, said that "direction was given to this field by two classical papers of Reiner, published in 1945 and 1948" in the *Americal Journal of Mathematics*. Few rheologists—and indeed few research workers in any field—have worked over so wide a range of subjects and materials. Prophets may go without honour in their own countries but with rheologists this is not always so. Markus Reiner's work has been very fully appreciated in Israel.

He has received many honours, including the Israel Prize, the Rothschild Prize and the Weizmann Prize of the City of Tel-Aviv, and he is a member of the Israel Academy of the Sciences and the Humanities.

Long ago, it was true indeed that "The days of our years are three score years and ten; and if by reason of strength they be fourscore years, yet is their strength labour and sorrow". Fortunately, today, modern medicine and surgery have extended this span. Markus Reiner has had his share of labour and sorrow, but I am sure that he feels that he has had much happiness too and that he has still many more fruits to contribute to rheology (every time I see him, he passes me *some* useful suggestion). We dare to hope that many more years lie ahead of him, as is indeed very probable in this new age in which we live. So, on his eightieth birthday we may well wish him "Many happy returns of the day",

וכן ירבו בארץ-ישראל

(And may such men multiply in the Land of Israel.)

FUNDAMENTALS

A PHENOMENOLOGICAL APPROACH TO QUANTUM THEORY

J. M. BURGERS

Institute for Fluid Dynamics and Applied Mathematics, University of Maryland, U.S.A.

ABSTRACT

Starting from an analysis of what is involved in an observation and in the evaluation of its possible results, a new approach is sought for the interpretation of certain features commonly accepted in quantum theory on the basis of its mathematical formalism, viz. the spread in the results of an observation and the occurrence of eigenvalues and eigenstates. This requires a discussion of how to ascertain that in a series of similar observations one is always dealing with the "same" object and whether the observations are sufficiently decisive in order that they may throw the object into an "eigenstate" for that type of observation without leaving traces of the effects of previous observations.

The primary outcome of physical observations is seen as furnishing sets of eigenvalues and probability distributions of eigenvalues. An observation (A) can be defined as "decisive," when in the next observation (B) the probability ν_{nm} of obtaining an eigenvalue b_m belonging to B is uniquely dependent upon the eigenvalue a_n which had been the result of A. One can introduce mathematical functions (first conceived as abstract entities) to be used as representatives of eigenstates and assume that functions belonging to one type of observation can be expressed linearly in terms of the functions belonging to another observation. The representations become useful when the probability numbers ν_{nm} can be related to the coefficients occurring in the developments, and they obtain a definite form when the sets of functions can be chosen so that they are eigensolutions of differential equations to be associated with the observations, with the condition that the eigenvalues of each equation shall correspond to the eigenvalues which are the results of the associated observation.

Further points discussed are non-commutability of observations, conjugate observables and observables which do not have a conjugate. The construction of a dynamical theory to account for the change of probability distributions with increasing intervals of time between successive observations, is not considered in the paper.

NOTATION

a_n, b_m sets of eigenvalues coming forward from observations A, B, respectively;

ϕ_n, ψ_m eigenstates belonging to observations A, B, respectively;

a_{nm} coefficients occurring in the development of ϕ_n in terms of the ψ_m;

ν_{nm} probability that observation B shall give the eigenvalue b_m when the preceding observation A had given the eigenvalue a_n.

The following lines contain some thoughts on a fresh approach to certain features of quantum theory. The starting-point is a discussion of what can be observed and of the interpretation to be given to the results of observations. As a consequence of the limited length permitted for this article the discussion must be restricted to a brief sketch; also, various points cannot be elaborated as yet and must be left for further investigation. Nevertheless, it is hoped that the considerations presented here may evoke some interest.

I start from the very general assumption that we can make observations; we can observe phenomena, situations, things, people, etc. The origin of science is the desire to find order and rules in the results of our observations. It must be granted that we can never state the result of an observation in a form which shall have meaning for other people, if there is not already available some method of interpretation, some theory; there are no "bare facts" which can be communicated without the assumption of a method of description. However, although we cannot make ourselves free from the entire complex of ideas amidst which we are living, we have the ability to shift our point of view and this can often be helpful as a means for arriving at new insights and at a new form of understanding.

Without going into a general analysis of the notion of observation, I take as a sufficiently typical case that there is an object to be observed and an observational apparatus, even if the latter would be, for instance, our fingers touching something. When object and apparatus come to interaction, there is some effect upon the apparatus which we call the *result of the observation*. There may also be some effect upon the object; to this we come later. A first point in collecting results of observations is the problem whether these results can be considered as *reproducible*. When we attempt to approach the questions involved here from an empirical standpoint, we must first try to ascertain whether our object and whether our observational apparatus always are in the same state; if this would not be the case we have no ground to expect much consistency in the observational procedure. Whether our object and our apparatus each time can be brought into the same state requires preliminary observations, and these can require still further preparatory work; moreover, we must reckon with the possibility that a rigorous decision concerning the state of the object cannot be obtained. We may be in danger here of falling into unending regressions. However, we can find guidance in the empirical result that in certain domains of investigation, in particular in the domain of physics, a high degree of reproducibility can be reached without an impossible amount of labor. The restriction to physics contains both a restriction to certain classes of objects and to certain classes of observational apparatus and procedure. Evidently there are various domains of research—for instance, a study of the behavior of the entities called nations—where reproducibility both of objects and of methods of observation is wellnigh unattainable. I mention this to stress the selective nature of research in physics.

I now introduce an important point of view which in fact contains a rejection of a hypothesis accepted in what we call "classical physics": I pretend that careful scrutiny of the results of physical observations, carried out with the utmost refinement attainable, demonstrates the impossibility of ensuring absolute reproducibility. *There always remains a certain spread in the results*, a spread which seems to have grounds inherent in the nature of things and which cannot be diminished *ad libitum*. Thus our first conclusion is: Observations carried out with an apparatus which to the best of our knowledge has always been prepared in the same way, on an object or set of objects which likewise, to the best of our knowledge, have been prepared in the same way, in general give results which are spread over a certain range of values. We call these possible results the *eigenvalues* connected with the observation, or, in the usual terminology, the *eigenvalues of the observable which is measured in the set-up*. The eigenvalues sometimes cover a continuous range; in other cases they may form a set of discrete values (for convenience we keep to the latter assumption in the following lines).

The reader will notice that what I take here as an observational result is that which in treatises on quantum theory is presented as a part of the formalism of the theory. The theory, however, is made to explain observational results and everybody expects that experimental results are in conformity with this theory. Thus I can take for granted that the results of certain sets of observations present the behavior here described and I may use this as an empirical datum for the construction of an interpretation.

Having encountered dispersion in the results of observations we shall be interested in the statistical distribution of the results and the question will turn up whether this distribution is always the same. Summing up the experimental work that has been done in physics, the conclusion seems warranted that, if our objects are non-living, we can expect that for a given way of preparing the object and a given way of preparing the observational apparatus (which likewise should not be subject to unforeseen actions or whims of living beings) *fixed frequency distributions are indeed obtained.* Hence we can state that the first result of physical observation is to become acquainted with sets of eigenvalues and with frequency distributions of these eigenvalues. Together these results are characteristic for the object, the way in which it has been prepared, the apparatus, and the way in which the apparatus has been prepared.

A point to be noted is that all activities mentioned, the preparation of the object and the apparatus, the recording of the results and even processing of results and comparison of frequency distributions can be carried out mechanically by appropriate machinery and computing apparatus. It is true that eventually the result must become part of human knowledge as otherwise no theory can be made. But it is not necessary that the human mind is directly involved in any of the observations; the question whether an observation should terminate through a form of conceptual activity in an understanding mind does not arise here.

Returning to the description of observations we take account of the fact that there exist objects which can be subjected to different forms of preparatory treatment before they are brought into contact with the observational apparatus. We can expect that these objects then will have been brought into states different from the state into which they had been brought previously. Quite generally, the question can arise whether we still may consider the objects as identical with the original ones, now that they have been treated differently. This is not a question which can be answered *in abstracto* once for all; it requires detailed consideration in every particular case. But in physics it is customary to speak of electrons as being the same type of objects even if they find themselves at different locations or if they have different velocities. In such cases it has been noted that when the "same" objects have been brought into a different state and then are subjected to the same type of observation (the nature and the state of the observational apparatus not having been changed), *the same eigenvalues make their appearance;* only their statistical distribution has changed. To generalize this result we invert the statement and *take it as a definition of the "sameness" of an object that the set of eigenvalues has not changed.* An alteration of the frequency distribution of the eigenvalues is then an indication of an "altered state" of the (same) object.

We can check whether there is a fixed relation between the various states into which an object can be brought before an observation, and the frequency distributions of eigenvalues coming out of the observation. If this appears to be the case, we consider this as an indication that our object belongs to the general class of things investigated in physics and that it does not give a sign of life or of other seemingly arbitrary changes. Trivial as this remark may seem to be, it is not altogether gratuitous when we take the word observation in a very general sense; again it is useful to stress that physics refers to the behavior of a carefully selected group of objects and of observations.

The next question which turns up is whether we can find something behind the sets of eigenvalues and their statistical distributions? To approach this problem we put forward the hypothesis that the observation influences the state of the object and that each eigenvalue is indicative of a particular way of influencing and thus of a particular state (an *eigenstate*) into which the object has been brought by the observation. The idea that the observation influences the state of the object is a natural one, and the hypothesis of the specificity of the influence finds support in the fact that immediate repetition of an observation on the same object with the same apparatus gives the same eigenvalue as the original observation; this suggests that state and eigenvalue are connected with one another. A further argument comes forward from a more lengthy consideration.

Our assumption amounts to the idea that observations producing different eigenvalues can be considered as providing a method for bringing the object into different states, every state having an eigenvalue as its label. We suppose that we can collect the objects after the observation into groups, each group containing objects all in the same state. We then submit these groups to a further observation. We call the first observation to be of type A, giving eigenvalues $a_1, a_2, \ldots$, with which we associate eigenstates $\phi_1, \phi_2, \ldots$ (the letters ϕ_n here are no more than names for the eigenstates). The second observation will be called to be of type B; it gives eigenvalues $b_1, b_2, \ldots$ and we use letters $\psi_1, \psi_2, \ldots$ as names for the corresponding eigenstates. In conformity with what has been stated before we suppose that the eigenvalues b_m occur whatever the state ϕ_n into which the object had been brought by observation A. This must be checked, of course, and if it is found to be the case it ensures that we always have to do with the same object. If now it appears that each state ϕ_n is associated with a particular distribution of the eigenvalues b_m as result of observation B, we can take this as proof that our supposition of different states has a real ground. In checking a precaution must be taken: the interval of time between observation A and observation B should be the *same* in all experiments, as it often appears that frequency distributions change when this time interval is changed. This is an important empirical result, which forces us to accept that the state of a system changes with time. For the moment we leave this point aside; we shall return to it at the end of the paper.

The results obtained so far induce us to introduce sets of frequency numbers ν_{nm}, with $\sum_m \nu_{nm} = 1$ (for every n), referring to the frequency of appearance of the various eigenvalues b_m as a result of observation B carried out on the object when it has been brought into state ϕ_n by observation A.

The proposed notation suggests that the observed frequency distribution is dependent exclusively upon the state into which the object had been brought by observation A; it does not give attention to that which may have happened with the object before observation A was carried out. Can we be certain that this is right, so that the successive observations may be compared to what is assumed for the transitions in a Markoff chain? A little thinking will show that this cannot be generally correct: observation A might have been of a very superficial type and formally it can dwindle into nothing; if this would be the case it cannot destroy or eliminate all effects of the previous history of our objects. Hence we must introduce a condition on our observations, and we shall say that physically useful, or for short, *decisive observations are such that eliminate the effects of the preceding history*, so that they throw the object into one of their own eigenstates which are fully characterized by the eigenvalue coming forward from the observation. Here we have an example that a relation which in the usual treatment of quantum theory is presented as a kind of postulate without precisely telling what an observation or a measurement entails, in reality involves a discussion of observational technique. That the matter is not simple will be understood when we note that increase of sharpness and of penetrating power of an observation, needed to prevent it from being too superficial, must not go so far that it destroys the object, a possibility which cannot be excluded *a priori*.

Mathematics can now be called to help in the following way. For each type of observation we attempt to find or construct an equation of such nature that it has eigensolutions, corresponding to a set of eigenvalues of a parameter occurring in that equation. If we succeed in finding an equation which gives precisely the eigenvalues which came forward from the observation, an important step has been made. Of course, in general it is extremely difficult to construct such an equation from scratch, but here we may refer to the accumulated mathematical knowledge which has developed in connection with quantum theory and from this stock of knowledge we can select that which seems to fit the case. At the moment it is sufficient to consider the fitting equation simply as a reasonable guess. Having succeeded in making this step, it is natural then to associate the eigenfunctions of the equation with the corresponding eigenstates of the object, according to the labeling by the eigenvalues. In this way we arrive at a principle which gives mathematical content to the symbols ϕ_n, etc.

Two different observations in general will need different equations and consequently involve two different systems of eigenfunctions. Now with many equations of the type here in use the systems of eigenfunctions satisfy a condition of completeness which makes it possible that an eigenfunction ϕ_n of observation A can be expressed linearly in terms of the eigenfunctions ψ_m of observation B:

$$\phi_n = \sum_m a_{nm}\psi_m.$$

The hypothesis is now introduced that *the coefficients a_{nm} appearing in such an expression bear some relation to the frequency numbers ν_{nm}*. As is well known, this hypothesis is confirmed by the relations found in those atomic systems to which quantum theory applies, provided complex functions are used and provided the frequency numbers are related to the squares of the absolute values of the coefficients a_{nm} occurring in the expressions. This again can be tested empirically.

An interesting consequence is connected with the scheme described. We mentioned that the decisive observation A throws the object into one of its eigenstates, for instance into state ϕ_n. When observation B likewise is a decisive one, it will throw the object into one of its own eigenstates, say ψ_m; according to what has been said it thereby obliterates all reminiscences of the previous state ϕ_n. Hence, when now observation A is repeated it will not necessarily reproduce the eigenvalue a_n: on the contrary, we must expect that again a whole spectrum of eigenvalues a_l will make its appearance, with a frequency distribution ν_{ml} that can be calculated from the coefficients $\hat{a}_{ml}$ in the development

$$\psi_m = \sum_l \hat{a}_{ml}\phi_l.$$

When this is found to be case, it is said that *the decisive observations A and B interfere with one another and that they do not commute*. A conclusion which may be drawn from this is that it will not be possible to devise some method of observation which would give the results of observations A and B simultaneously. So far there is no empirical result which contradicts this expectation.

The introduction of a definition of interference and non-commutability suggests that there may exist cases where observations do not interfere with one another in the manner described above for the observations *A* and *B*. Thus far we left the possibility of such cases out of sight in order to concentrate upon the more striking and more general case, but it is important to give attention to this matter. A trivial case is the one where the quantity measured in one observation is a simple function of the quantity measured in the other observation; e.g. one observation might measure a length and the other observation the square of that same length. In such a case the eigenvalues of the second observation are the squares of the eigenvalues of the first observation, and both observations have the same system of eigenfunctions. However, there are cases of a different nature; for instance, in the case of an electron measurements of its three coordinates do not interfere with one another; they do commute and in principle can be carried out simultaneously in a single observation. On the contrary the measurement of a coordinate and that of the corresponding momentum do interfere with one another.

This situation requires a considerable extension of the experimental program and of the necessary processing of the results. Given a certain object, one must perform all kinds of observations upon it, or in a slightly different wording, one must measure all kinds of observables connected with that object. It will be found that many observables can be considered as functions of other observables and we must investigate which is the minimum set of independent observables with which the object is endowed. Amongst these we then must find out which ones do not interfere with one another's measurement and which do. In any practical case this involves a discussion of what is the most convenient choice of a set of independent observables.

Here again we may refer to the accumulated knowledge, which makes us expect that in general the independent observables can be classified in three groups, one group being called coordinates, a second group being called generalized or canonical momenta, and a third group which seem to play the part of constant parameters. The entities of the last group are not called "observables" in the terminology of standard quantum theory; from the experimental point of view, however, they have full right to be considered as such. The "momenta" make their appearance in our experimental scheme as observables each one of which does not commute with one of the coordinates, but their dynamical nature will not be immediately evident; they are not yet entities connected with velocities and mass. Complications can arise; for example, in the case of an electron we distinguish three coordinates (which can be measured with respect to an orthogonal Cartesian system); three momenta conjugated to these coordinates, while in this group we must also include the spin of the electron for which there is no corresponding coordinate; and finally we have mass and electric charge as parameters.

As mentioned, measurements of the three coordinates do not interfere with one another. It is possible to imagine a combined procedure of observation in which the three coordinates are measured simultaneously. An observation which provides a simultaneous measurement of the independent coordinates and of the spin, is called a *maximal* or a *complete observation*. To give full scope to the theory developed in the preceding pages, the observations considered there should be complete observations. Each such observation at one stroke will give a set of eigenvalues, one eigenvalue for each entity measured. We may express this differently by assuming that what we have called an "eigenvalue" in reality stands for a set of numbers, each one referring to an observable. With each such set there is associated an eigenfunction that is obtained as the product of the eigenfunctions associated with the separate observables.

The investigation of the questions which turn up here requires a large amount of research and several points must be clarified for which the description given above does not yet give a procedure. Various features can be deduced from a re-discussion of the work done in physical research on atomic systems, but this re-discussion must start from the problem how best to classify experimental results. It is impossible to work this out in the extent of the present article and moreover we are facing questions for which no answer as yet is available. The difference between the point of view chosen here and the customary treatment is that until now we have not introduced the Hamiltonian function for any object, so that we cannot make use of the formalisms connected with that function. For instance, our discussion naturally leads to the question whether there is an intrinsic dispersion in the magnitudes of mass and electric charge, a problem which is ignored in the usual treatment where these entities are taken as absolute constants. Another difficulty is that there are observables as angular momentum and spin, which we must consider as belonging to the class of momenta, but with which we cannot unequivocally associate observable coordinates, so that there is an imbalance between the group of coordinates and that of the momenta.

It is probable that a full solution of these difficulties cannot come forward from the type of mathematical research discussed above, and that it can only be reached via an analysis of a further problem, viz. *the change with time of the state of an object* in a period in which it is not subjected to deliberate observation. That such changes exist is demonstrated by experiments of the type discussed before, when the interval of time between two successive observations A and B is changed: in general this will lead to a change of the statistical distribution of the eigenvalues coming out of the second observation. Even the repetition of the same observation A after an increasing interval of time can show a change of the statistical distribution of the eigenvalues with increasing time. Hence we must conclude that each object can suffer changes of state, which changes may be due either to features in the object itself, or to interactions with other objects not belonging to the observational apparatus. As even the properties of space and time are dependent upon the presence of matter, no object can be said to be ever alone in the universe and every object can suffer influences and changes from its environment.

The study of these changes forms the most important part of physical research and *must provide the basis for a dynamical theory of the objects under investigation*, which is the ultimate goal of most physical research. This topic is here approached from the changes observed in the statistical distribution of eigenvalues. When mathematics is called to help the idea will present itself that the eigenfunctions must be considered as functions of the time, and that, instead of an equation given eigenfunctions and eigenvalues only at a certain instant, we shall need an equation giving time-dependent eigenfunctions. When such time-dependent eigenfunctions are expressed in terms of the eigenfunctions corresponding to a time-independent equation connected with a certain method of observation, the coefficients of the development will become functions of the time, and, if we have guessed well, the absolute values of these coefficients should predict the change of the frequency numbers with time.

It can be expected that in an attempt to construct such time-dependent equations attention must be given to the parts played by momenta and coordinates together. Indeed, one would have to build up a theory which in a sense is an analog of the reasoning which has been used in the development of Hamiltonian mechanics from the beginnings made by Galileo and Newton, plus the transition from there to quantum mechanics, but the argumentation and the assumptions should now be based upon the behavior of the observables as revealed in the experiments described above. This is a subject which will need the elaboration of new chains of thought that may be of great interest. Apparently even the first steps will require appropriate inventiveness.

Notwithstanding the incompleteness of the discussion presented in these pages, it may have become evident that there is a promise for the evolution of an interesting theory. It will be a theory which starts from observational results, introduces mathematical functions to describe these results, and then attempts to extend the mathematical system in such a way that it will embrace the development in time. Schrödinger's, Heisenberg's and Dirac's equations should come at the end of the theory, instead of being introduced as postulates at the beginning. In principle the mathematical scheme must serve as a means for the calculation and prediction of observational results; it does not purport to furnish a description of nature. But once equations are obtained, they can be compared with other equations describing phenomena with which we feel more or less acquainted; analogies found in equations then may point to analogies in the properties of the entities to which our analyses refer.

FINAL REMARK

Although it has been stated on p. 14 above that the results of a set of observations—the eigenvalues and their frequency distribution—are characteristic "for the object, the way in which it has been prepared, the apparatus, and the way in which the apparatus has been prepared" all taken together, the effects of the apparatus have somewhat got out of sight in the rest of the article. It is necessary to remember that along with the states of the object, also the states of the apparatus are a factor of importance. The many observational techniques developed in experimental physics give a partial indication concerning ways for separating apparatus effects from effects connected with object states. Nevertheless, the ultimate decision in each case is dependent upon a choice made by the observer, or by another physicist who attempts to interpret the experimental results. Thus in observing energy levels of an atom by means of spectroscopic observations decisions must be made in order to separate from one another apparatus effects, Doppler and Stark effects, ideally sharp energy levels, effects due to the finite lengths of wave trains, and perhaps still others. When the observable is the electronic charge, a similar decision must be made. At present it is an article of faith that this charge has a sharply defined value and that the dispersion noticed in results of measurements must be due to apparatus states. When this view is correct, it would mean that the charge has only a single eigenvalue (in the case of the system proton–neutron modern considerations seem to indicate that the charge has two eigenvalues, viz. 0 and e). However, since Dirac has ventured the hypothesis that the constant of gravitation may be a function of the time, there is nothing which would forbid us to consider also e, m, c, h as variable quantities. From the point of view of cosmogony one can even advance the hypothesis that the apparent constancy of these entities is a result of some development which took place much longer ago than the present "age of the universe" of the order of 20 billion years, and that originally these entities had fluctuating magnitudes. Perhaps they still suffer very small fluctuations. All such possibilities must be taken into consideration when one is building up a theory concerning physical phenomena from observational data in order to explain the eigenvalues.

October 1965.

MECHANICS OF MICROPOLAR CONTINUA†

A. CEMAL ERINGEN‡

School of Aeronautical and Engineering Sciences, Purdue University, Lafayette, Indiana, U.S.A.

ABSTRACT

The recent theory of micromorphic materials developed by Eringen and his co-workers is presented in a simpler form suitable for mathematical and experimental work. Microfluids and microelastic solids are explained. The concept of micropolar fluids and micropolar solids as special classes of micromorphic materials are presented. The thermodynamical restrictions on elasticity and viscosity coefficients are discussed. The field equations and boundary conditions are derived for both micropolar fluids and solid media. The solution is given for a channel flow of micropolar fluids which bring out some new and interesting physical phenomena not encountered in the classical theory of viscous fluids.

1. INTRODUCTION

The classical theories of continuum mechanics deal with the deformations and motions of materials that possess continuous mass densities, under the effect of external loads. The theory of continuous media is built upon the premise that any volume element ΔV in a body can be taken to its limit dV without affecting the distribution of mass. By this hypothesis, in the limit, the identity of the material points in a volume element is lost and their individual motions coincide with that of the center of mass. Atomistic models of the materials have shown that the mass density can fluctuate violently with the size of ΔV especially when it is below a certain limit. For many materials that possess coarse structure and fibers, the continuum theory is expected to break down in a macroscopic range of ΔV.

Colloidal fluids, liquid crystals, and fluids containing additives are known to fall beyond the domain of application of the classical theory of Stokesian fluids. For such media we need a new theory which incorpo-

† An invited lecture partially based on this paper was presented at the 9th Midwestern Mechanics Conference on Solid Mechanics, 16–18 August 1965, University of Wisconsin, Madison, Wisconsin. The work contained in this paper was sponsored by the Office of Naval Research, U.S.A.

‡ Present address: Department of Aerospace and Mechanical Sciences, Princeton University, Priceton, N.J., U.S.A.

rates the micromotions of the particles contained in a material volume element, ΔV.

Eringen and Suhubi [1, 2] and Eringen [3–6] have introduced theories of microelastic solids and microfluids in which the "micromotions" of the material points contained in ΔV with respect to its centroid are taken into account in an average sense. Materials affected by such micromotions and "microdeformations" are called *micromorphic materials* [4]. Both microelastic solids and microfluids are studied. According to these theories the micromorphic media can support distributed stress moments and body couples and their behavior is influenced by the spin inertia. These theories, however, even in the linear case, are too complicated for engineering applications since the underlying mathematical problem is not easily amenable to the solution.

A subclass of micromorphic materials is the *micropolar media* which exhibit "microrotational" effects and microrotational inertia. Such media can be subjected to surface and body couples only and the material points in a volume element can undergo only rigid rotational motions about their centers of mass. This class of media possesses certain elegance in its mathematical formulation and simple enough to attract the attention of mathematicians and engineering scientists. Physically, micropolar media may represent the materials that are made up of dipole atoms or dumbbell molecules. Thus, for example, certain fibrous solids and liquid crystals can be represented adequately with this model. In fact, animal blood happens to fall into this category.

Recent experiments with fluids [7, 8] containing minute amounts of additives indicate that the skin friction near a rigid body is considerably lower (up to 30–35%) than the same fluids without additives. The classical Navier–Stokes theory is incapable of predicting this new physical phenomenon. Recent theoretical work on pipe [5] and channel flow presented in Section 6 indicate that the theory of micropolar fluids can be used to explain this phenomenon. While we must reserve the final judgement until conclusive experiments are carried out using the present viewpoint, we can at least state that the theory does affirm the existence of such phenomena.

In Sections 2, 3 and 4 we discuss the linear theory of micromorphic materials. In Section 5 the theory of micropolar elastic solids and micropolar fluids is presented. Section 6 is devoted to the solution of the problem of channel flow of micropolar fluids.

We believe the theory of micropolar media opens up a very worthwhile branch of continuum mechanics. It should find important applications dealing with a variety of solids and fluids. Interesting theoretical and experimental studies in this field are awaiting future workers.

2. DEFORMATION AND MOTION

Consider a small volume element ΔV consisting of material particles in an undeformed body B. Let the rectangular coordinates X_K, $(K = 1, 2, 3)$ or $\mathbf{X}$ denote the position vector of the center of gravity of ΔV, Fig. 1. Any other point in ΔV will have the position $\mathbf{X}'$ given by

$$\mathbf{X}' = \mathbf{X} + \boldsymbol{\Xi}. \tag{2.1}$$

If ΔV is allowed to move and deform under external effects, the new position $\mathbf{x}'$ of $\mathbf{X}'$ will be

$$\mathbf{x}' = \mathbf{x}(\mathbf{X}, t) + \boldsymbol{\xi}(\mathbf{X}, \boldsymbol{\Xi}, t), \tag{2.2}$$

where $\mathbf{x}$ is the new position vector of the center of mass of ΔV and $\boldsymbol{\xi}$ is the

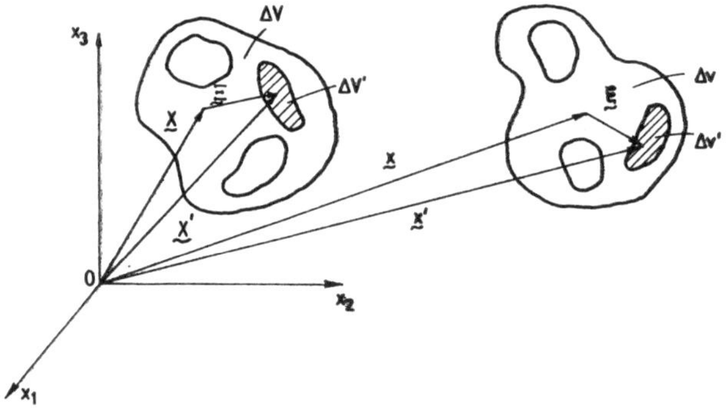

FIG. 1. Motion of microelements in a volume element.

new relative position of the point $\mathbf{X}'$ from $\mathbf{x}$. We assume that the material points in ΔV undergo a homogeneous deformation about $\mathbf{x}$, so that

$$\boldsymbol{\xi} = \boldsymbol{\chi}_K(\mathbf{X}, t)\Xi_K, \tag{2.3}$$

where summation over the repeated index K is understood. Here the three vectors $\boldsymbol{\chi}_K$, $(K = 1, 2, 3)$ are the basic unknowns of the theory, as $\mathbf{x}$, so that the otion of $\mathbf{X}$ cannot be determined by the knowledge of the centroidmal motion $\mathbf{x}(\mathbf{X}, t)$ alone, as in classical continuum mechanics. The rectangular coordinates of $\mathbf{x}$ and $\boldsymbol{\xi}$ are respectively denoted by x_k and ξ_k. Through (2.3) we write

$$\xi_k = \chi_{kK}(\mathbf{X}, t)\Xi_K. \tag{2.4}$$

Here and throughout, summation over the repeated indices is understood in the range (1, 2, 3). Also capital latin indices refer to the material coordinates X_K and small indices to the spatial coordinates x_k.

Let χ_K^{-1} be the reciprocal three vector to χ_K so that

$$\chi_{lk}\chi_{Kk}^{-1} = \delta_{kl}, \qquad \chi_{kK}\chi_{Lk}^{-1} = \delta_{KL}, \tag{2.5}$$

where δ_{kl} and δ_{KL} are the usual Kronecker deltas. It is well known that the solution of $(2.5)_1$ or $(2.5)_2$ is

$$\chi_{Kk}^{-1} = \frac{\text{cofactor } \chi_{kK}}{\det(\chi_{kK})} \tag{2.6}$$

provided that

$$\det(\chi_{kK}) \neq 0, \tag{2.7}$$

where det abbreviates determinant.

With the aid of (2.5), from (2.4) we can solve for Ξ_K.

$$\Xi_K = \chi_{Kk}^{-1}\xi_k. \tag{2.8}$$

The velocity and acceleration vectors are obtained by differentiating (2.2) with respect to time t with fixed $\mathbf{X}$ and Ξ, i.e.

$$\mathbf{v}' = \dot{\mathbf{x}}+\dot{\boldsymbol{\xi}}$$
$$\mathbf{a}' = \dot{\mathbf{v}}'.$$

Upon using (2.3) and (2.8) these can be written as

$$\mathbf{v}' = \mathbf{v}+\nu_k\xi_k \tag{2.9}$$
$$\mathbf{a}' = \mathbf{a}+(\dot{\nu}_l+\nu_k\nu_{kl})\xi_l, \tag{2.10}$$

where

$$\mathbf{v} \equiv \dot{\mathbf{x}}, \quad \mathbf{a} \equiv \dot{\mathbf{v}}. \tag{2.11}$$
$$\boldsymbol{\nu}_l \equiv \dot{\chi}_K\chi_{Kl}^{-1}.$$

Here ν_{kl} is the kth component of $\boldsymbol{\nu}_l$ which we name the gyration tensor.† The gyration tensor is the physical expression of the rotational and extensional velocities of any material point in the deformed element Δv with respect to its center of mass. We thus see that in the mechanics of micromorphic materials in addition to the velocity and accelerations of the center of mass, $\mathbf{x}$, of Δv we need the velocities and accelerations of particles in Δv about the center of mass. This situation arises from the fact that the element of volume is deformable. Since all materials to some degree are granular or fibrous the volume element dv can be approximated more realistically by a deformable one.

† ν_{kl} used here corresponds to ν_{kl} of our paper [1] but ν_{lk} of [3–6].

Without repeating the proofs presented in [3] we can easily show that: *A necessary and sufficient condition for the motion to be microrigid is vanishing*

$$d_{kl} \equiv \tfrac{1}{2}(v_{k,l}+v_{l,k}), \tag{2.12}$$

$$b_{kl} \equiv v_{l,k}+\nu_{kl} \tag{2.13}$$

when these vanish, these also vanish

$$a_{klm} \equiv \nu_{kl,m}, \tag{2.14}$$

where an index following a comma indicates partial differentiation with respect to the spatial coordinates, e.g.

$$v_{k,l} \equiv \partial v_k/\partial x_l.$$

Tensors **d**, **b** and **a** play central roles in the theory of micromechanics. Of course **d** is well known to us from classical continuum mechanics. It is the deformation rate tensor. We name **b** and **a** the *microdeformation rate tensors* of the second and third order. These tensors possess a very important property, namely that they transform like absolute tensors when the spatial frame of reference **x** is rotated rigidly. The quantities having this property are called *objective* [9].

The infinitesimal change in angles and the square of arc length can be calculated when the following tensors are known:†

$$2e_{kl} \equiv u_{k,l}+u_{l,k} \tag{2.15}$$

$$\varepsilon_{kl} \equiv \varphi_{kl}+u_{l,k} \tag{2.16}$$

$$\gamma_{klm} \equiv -\varphi_{kl,m}. \tag{2.17}$$

Here u_k is the deformation vector associated with the center of mass of Δv and $\boldsymbol{\varphi}$ is the microdeformation of any point in Δv relative to its center of mass. They are related to the deformation gradients $X_{K,k}$ and χ_{Kk}^{-1} by

$$X_{K,k} = (\delta_{lk}-u_{l,k})\delta_{Kl}$$

$$\chi_{Kk}^{-1} = (\delta_{lk}-\varphi_{lk})\delta_{Kl}, \tag{2.18}$$

where δ_{Kl} is also a Kronecker delta which takes the value 1 when $K = l$ and zero otherwise. The ultimate determination of the deformation in a solid microcontinuum requires the knowledge of u_k and φ_{kl} and in a fluid microcontinuum v_k and ν_{kl}. In addition, of course, the mass density, temperature and microrotatory changes will have to be determined. To accomplish this we must establish the laws of motion and the character of the continuum, i.e. the constitutive equations.

† For the large deformations see ref. [2], eq. (3.4).

3. LAWS OF MOTION

The laws of motion of classical continuum mechanics are not adequate for the determination of the field variables of microcontinua. In our previous work [1] and [3] we introduced additional laws through a simple statistical means. We assume that the principles of conservation of mass, balance of momentum, balance of moment of momentum and conservation of energy are valid for each "microelement" of the macrovolume element Δv. Thus

$$\frac{\partial \varrho'}{\partial t}+(\varrho v'_k),_k = 0, \tag{3.1}$$

$$t'_{kl,k}+\varrho(f'_l-\dot{v}'_l) = 0, \tag{3.2}$$

$$t'_{kl} = t'_{lk}, \tag{3.3}$$

$$\varrho'\dot{\varepsilon}' = t'_{kl}v'_{l,k}+q'_{k,k}+\varrho' h', \tag{3.4}$$

where ϱ', v'_k, t'_{kl}, f'_l, ε', q'_k and h' are, respectively, the mass density, the velocity vector, the stress tensor, the body force per unit mass, the internal energy density per unit mass, the heat vector (directed outward from the surface of the body) and the heat source of any microvolume element in Δv. By taking the first three statistical moments of (3.1), the first two statistical moments of (3.2) (the average, and the first moment) and the average of (3.4) about the centroid of Δv and using (3.3) we obtain the following set of equations which constitute the basic laws of microcontinua.†

Conservation of mass

$$\frac{\partial \varrho}{\partial t}+(\varrho v_k),_k = 0 \quad \text{in} \quad \mathcal{V}. \tag{3.5}$$

Conservation of microinertia

$$\frac{\partial i_{km}}{\partial t}+i_{km,r}\, v_r - i_{rm}\nu_{kr} - i_{kr}\nu_{mr} = 0 \quad \text{in} \quad \mathcal{V}. \tag{3.6}$$

Balance of momentum

$$t_{kl,k}+\varrho(f_l-\dot{v}_l) = 0 \quad \text{in} \quad \mathcal{V}. \tag{3.7}$$

Balance of first stress moments

$$t_{ml}-s_{ml}+\lambda_{klm,k}+\varrho(l_{lm}-\dot{\sigma}_{lm}) = 0 \quad \text{in} \quad \mathcal{V}. \tag{3.8}$$

Conservation of energy

$$\varrho\dot{\varepsilon} = t_{kl}v_{l,k}+(s_{kl}-t_{kl})\nu_{lk}+\lambda_{klm}\nu_{lm,k}+q_{k,k}+\varrho h \quad \text{in} \quad \mathcal{V}. \tag{3.9}$$

† The analysis is lengthy and involved. The reader is referred to ref. [1–2] for the derivations.

In addition we have the principle of entropy formulated as an inequality

$$\varrho\Gamma \equiv \varrho\dot{\eta} - \left(\frac{q_k}{\theta}\right)_{,k} - \frac{\varrho h}{\theta} \geqslant 0 \quad \text{in} \quad \mathcal{V} \tag{3.10}$$

valid for all independent processes. In these equations

ϱ ≡ mass density
i_{km} ≡ microinertia moments $= i_{mk}$
t_{kl} ≡ the stress tensor
f_l ≡ the body force per unit mass
s_{kl} ≡ the microstress average $= s_{lk}$
λ_{klm} ≡ the first stress moments
l_{lm} ≡ the body moments
$\dot{\sigma}_{lm}$ ≡ inertial spin
ε ≡ internal energy density per unit mass
q_k ≡ heat vector directed outward from the surface of body
h ≡ heat source per unit mass
η ≡ entropy density per unit mass
θ ≡ absolute temperature

The inertial spin $\dot{\sigma}_{kl}$ is related to ν_{kl} by

$$\dot{\sigma}_{kl} = i_{ml}(\dot{\nu}_{km} + \nu_{kn}\nu_{nm}). \tag{3.11}$$

The quantities ϱ, t_{kl}, ε, q_k and h are well known to us from the classical continuum theory. Among the newly introduced quantities, roughly speaking, i_{km} is a kind of rotatory microinertia of the micro mass elements in Δv about its center of mass; the tensor $\boldsymbol{\lambda}$ with twenty-seven components represents a statistical moment of the stress about the centroid of an area element; while $\mathbf{l}$ is the statistical moment of forces acting in Δv about its center of mass. The spin inertia $\dot{\boldsymbol{\sigma}}$ is the inertia of micromotions about the center of mass of Δv.

On the surface $\mathcal{S}$ of the body we have the following boundary conditions, [1],

$$t_{kl}n_k = t_l \quad \text{on} \quad \mathcal{S} \tag{3.12}$$

$$\lambda_{klm}n_k = \lambda_{lm} \quad \text{on} \quad \mathcal{S}, \tag{3.13}$$

where t_l and λ_{lm} are, respectively, the surface tractions and surface stress moments acting on the surface of the body.

We note that while equations (3.5) and (3.7) are well known from the classical theory, the six equations (3.6) are entirely new and the nine equations (3.8) and the energy equation (3.9) are modified and extended forms of

(3.3) and (3.4). In fact the symmetric part of (3.8) is certainly new. When $\mathbf{i} = \boldsymbol{\lambda} = \mathbf{l} = \mathbf{s} = \boldsymbol{\nu} = \mathbf{0}$ the above equations reduce to those of the classical continua.

If we exclude the phenomena of heat conduction in the present theory, the determination of the motion requires the knowledge of the following nineteen unknowns,

$$\varrho(\mathbf{x}, t), \quad i_{kl}(\mathbf{x}, t), \quad v_k(\mathbf{x}, t), \quad \nu_{kl}(\mathbf{x}, t) \tag{3.14}$$

as against the four unknowns v_k and ϱ of the classical theory.

4. CONSTITUTIVE EQUATIONS

The character of the continuum is reflected through the constitutive equations. For a microelastic medium the constitutive equations relate (t_{kl}, s_{kl}, λ_{klm}, ε, q_k, η) to (e_{kl}, ε_{kl}, γ_{klm}, i_{km}, $\theta_{,k}$, θ) and for a microfluid to (d_{kl}, b_{kl}, a_{klm}, i_{km}, $\theta_{,k}$, θ). Thus we may write symbolically for a *microelastic solid:*

$$(\mathbf{t}, \mathbf{s}, \boldsymbol{\lambda}, \varepsilon, \mathbf{q}, \eta) = \mathbf{F}(\mathbf{e}, \boldsymbol{\varepsilon}, \boldsymbol{\gamma}, \mathbf{i}, \nabla\theta, \theta), \tag{4.1}$$

and for a *microfluid:*

$$(\mathbf{t}, \mathbf{s}, \boldsymbol{\lambda}, \varepsilon, \mathbf{q}, \eta) = \mathbf{F}(\mathbf{d}, \mathbf{b}, \mathbf{a}, \mathbf{i}, \nabla\theta, \theta). \tag{4.2}$$

Of course, $\mathbf{F}$ is different type of tensor function for each of $\mathbf{t}$, $\mathbf{s}$, $\boldsymbol{\lambda}$, ε and η. It is a second-order tensor function for $\mathbf{t}$ and $\mathbf{s}$, a third-order tensor function for $\boldsymbol{\lambda}$, a vector function for $\mathbf{q}$ and scalar point functions for ε and η. Such general equations and, more generally, equations of microviscoelasticity were studied and their reductions were made in [1]–[4], by use of the axioms of constitutive theory [9]. Here we are concerned only with the linear theory characterizing media of a more limited type. The following assumptions are made:

(a) the constitutive equations are linear,
(b) that medium is microisotropic,
(c) heat conduction is negligible,
(d) the present theories must reduce to classical theories when $\boldsymbol{\chi}_K = \mathbf{0}$.

The assumption (a) states that tensor equations (4.1) and (4.2) are linear in their independent variables. Assumption (b) requires that $i_{kl} = i(x, t)\,\delta_{kl}$. Hence through (3.6) we get

$$i = \text{const} = j/2 \quad \text{on material lines} \tag{4.3}$$

and therefore (3.11) reduces to

$$\dot{\sigma}_{kl} = \frac{j}{2}(\dot{\nu}_{kl}+\nu_{kn}\nu_{nl}) \tag{4.4}$$

and the dependence of **F** on **i** disappears. The assumption (c) eliminates heat conduction, and (d) forces the final constitutive equations to go to those of the classical theory when the microdeformation is negligible. Below we give the constitutive equations for **t**, **s** and $\boldsymbol{\lambda}$ in the case of a microelastic solid:

$$\mathbf{t} = [(\lambda_1+\tau)e_{kk}+\eta_1\varepsilon_{kk}]\mathbf{I}+2(\mu_1+\sigma_1)\mathbf{e}+\varkappa_1\boldsymbol{\varepsilon}+\nu_1\boldsymbol{\varepsilon}^T, \tag{4.5}$$

$$\mathbf{s} = [(\lambda_1+2\tau)e_{kk}+(2\eta_1-\tau)\varepsilon_{kk}]\mathbf{I}+2(\mu_1+2\sigma_1)\mathbf{e}+(\nu_1+\varkappa_1-\sigma_1)(\boldsymbol{\varepsilon}+\boldsymbol{\varepsilon}^T), \tag{4.6}$$

$$\begin{aligned}\lambda_{klm} = {} & \tau_1(\varphi_{kr,r}\,\delta_{ml}+\varphi_{rr,l}\,\delta_{mk})+\tau_2(\varphi_{rk,r}\,\delta_{ml}+\varphi_{rr,m}\,\delta_{kl})\\ & +\tau_3\varphi_{rr,k}\,\delta_{ml}+\tau_4\varphi_{lr,r}\,\delta_{mk}+\tau_5(\varphi_{rl,r}\,\delta_{mk}+\varphi_{mr,r}\,\delta_{kl})\\ & +\tau_6\varphi_{rm,r}\,\delta_{kl}+\tau_7\varphi_{lm,k}+\tau_8(\varphi_{mk,l}+\varphi_{kl,m})\\ & +\tau_9\varphi_{lk,m}+\tau_{10}\varphi_{ml,k}+\tau_{11}\varphi_{km,l},\end{aligned} \tag{4.7}$$

where $\lambda_1, \tau, \eta_1, \mu_1, \sigma_1, \nu_1, \varkappa_1$ and τ_1 to τ_{11} are microelastic modulii of which λ_1 and μ_1 replace the classical Lamé constants. A superposed T represents the transpose of the matrix, and **I** is the unit tensor, i.e.

$$\mathbf{I} = \begin{bmatrix} 1 & 0 & 0 \\ 0 & 1 & 0 \\ 0 & 0 & 1 \end{bmatrix}, \quad \varepsilon^T_{kl} = \varepsilon_{lk}.$$

The constitutive equations for microfluids are similar to (4.5) except that **e**, $\boldsymbol{\varepsilon}$, and $\varphi_{kl,m}$ are now respectively replaced by **d**, **b** and $\nu_{kl,m}$ (cf. [3]) and a term $(-\pi\mathbf{I})$ is added to the right sides of (4.5) and (4.6), where π is the pressure.

A thermodynamic discussion can be made showing that for both solids and fluids

$$\mathbf{q} = 0, \qquad \eta = -\frac{\partial\psi}{\partial\theta}, \tag{4.8}$$

where $\psi = \varepsilon-\theta\eta$ is the free energy which depends on ϱ^{-1} and θ for fluids and θ and strain measures for solids. Alternatively

$$\mathbf{q} = \mathbf{0}, \qquad \theta = \frac{\partial\varepsilon}{\partial\eta} \tag{4.9}$$

when ε is considered to be a function of the same arguments as that of ψ.

In the case of fluids the pressure π is also obtained to be†

$$\pi = \left.\frac{\partial \psi}{\partial \varrho^{-1}}\right|_{\eta} . \tag{4.10}$$

Upon substitution of (4.5) to (4.7) into (3.7) and (3.8) together with (3.5) we obtain thirteen equations for the thirteen unknowns ϱ, u_k and φ_{kl} since $\mathbf{e}$, $\boldsymbol{\varepsilon}$, $\mathbf{v}$ and $\dot{\boldsymbol{\sigma}}$ are expressible in terms of these quantities. The theory of microsolids and microfluids is therefore complete at least in form. Under appropriate boundary and initial conditions the solution of the thirteen partial differential equations should predict the behavior of micromedia.

The theory outlined above is still too complicated for application. For fluid not only do we have to deal with convective terms emanating from $\dot{v}_k$ and $\dot{\nu}_{kl}$ but a large number of viscosity terms containing second-order partial derivatives coupled through thirteen equations. These partial differential equations are not so readily accessible to the mathematical treatment of nontrivial boundary and initial value problems. We therefore present the concept of micropolar media next.

5. THEORY OF MICROPOLAR MEDIA

Definition. A body is called micropolar if for all motions

$$\begin{aligned} \lambda_{klm} &= -\lambda_{kml} \quad \varphi_{kl} = -\varphi_{lk} \quad \text{(solids)} \\ \nu_{kl} &= -\nu_{lk} \quad \text{(fluids)}. \end{aligned} \tag{5.1}$$

For the linear theory through (2.5) and $(2.18)_2$ we can show that when $\varphi_{kl} = -\varphi_{lk}$ so is $\nu_{kl} = -\nu_{lk}$. For in this case $(2.11)_3$ gives

$$\nu_{kl} = \dot{\varphi}_{kl} \quad \text{(linear theory)}. \tag{5.2}$$

The notion of micropolar media appears to have important implications. Such materials are physically realistic. They exhibit only microrotational effects and can support body and surface couples. Materials consisting of dipoles are examples of such media. Diatomic gases, fluids containing dumbbell molecules, e.g. liquid crystals and certain types of crystalline solids that are made up of two spherical masses tied together with a rigid bar are found in nature. Animal blood is a liquid that can be represented by this model.

† For these and other interesting results we refer the reader to refs. [1] and [3], see also refs. [5] and [6].

By use of the skew symmetric properties of $\boldsymbol{\lambda}$ and $\boldsymbol{\varphi}$ the basic equations of such continua can be simplified a great deal. We first define the *couple stress tensor* m_{kr}, *microrotation vector* φ_k, *microangular velocity* ν_k, *microrotation inertia* $\dot{\sigma}_k$ and body couple l_k by

$$\begin{aligned} m_{kr} &\equiv -\varepsilon_{rlm}\,\lambda_{klm}, \\ \varphi_r &\equiv -\tfrac{1}{2}\varepsilon_{rkl}\,\varphi_{kl}, \\ \nu_r &\equiv -\tfrac{1}{2}\,\varepsilon_{rkl}\,\nu_{kl}, \\ \dot{\sigma}_r &\equiv -\varepsilon_{rki}\,\dot{\sigma}_{kl}, \\ l_r &\equiv -\varepsilon_{rkl}\,l_{kl}. \end{aligned} \tag{5.3}$$

The couple stress tensor m_{kr} has the same sign convention as the stress tensor t_{kl}, Fig. 2.

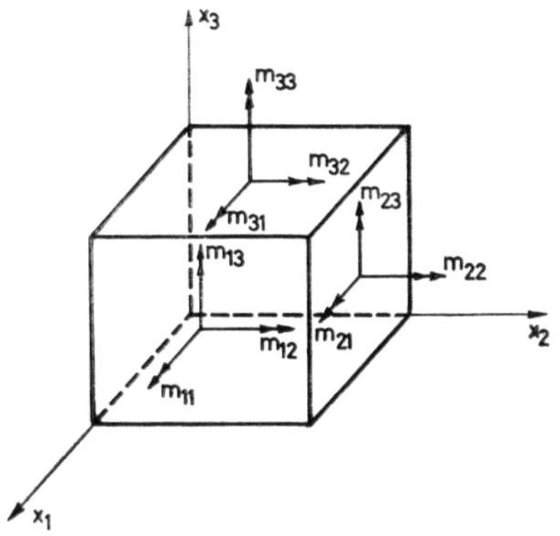

FIG. 2. Positive couple stress components.

Upon using (5.3) in (3.8) and (3.9) these equations can be expressed respectively as [6]

$$m_{rk,r}+\varepsilon_{klr}t_{lr}+\varrho(l_k-\dot{\sigma}_k) = 0, \tag{5.4}$$

$$\varrho\dot{\varepsilon} = t_{kl}(v_{l,k}-\varepsilon_{klr}\nu_r)+m_{kl}\nu_{l,k}+q_{k,k}+\varrho h. \tag{5.5}$$

The boundary conditions (3.13) can be written as

$$m_{rk}n_r = m_k \quad \text{on} \quad \mathcal{S}, \tag{5.6}$$

where m_k is the surface couple acting on $\mathcal{S}$.

Similarly the constitutive equations for $\mathbf{t}$ and $\boldsymbol{\lambda}$ can be reduced to simpler forms for this case [5], [6]. Below we give the results of these manipulations:

A. *Micropolar elasticity*

$$t_{kl} = \lambda u_{r,r}\,\delta_{kl}+\mu(u_{k,l}+u_{l,k})+\varkappa(u_{l,k}-\varepsilon_{klr}\,\varphi_r), \tag{5.7}$$

$$m_{kl} = \alpha\varphi_{r,r}\,\delta_{kl}+\beta\varphi_{k,l}+\gamma\varphi_{l,k}, \tag{5.8}$$

where λ, μ, $\varkappa$, α, β and γ are the new elastic constants of which λ and μ are the Lamé constants of the classical theory of elasticity.

B. *Micropolar fluids*

$$t_{kl} = (-\pi+\lambda_v v_{r,r})\,\delta_{kl}+\mu_v(v_{k,l}+v_{l,k})+\varkappa_v(v_{l,k}-\varepsilon_{klr}\,\nu_r), \tag{5.9}$$

$$m_{kl} = \alpha_v\,\nu_{r,r}\,\delta_{kl}+\beta_v\,\nu_{k,l}+\gamma_v\,\nu_{l,k}, \tag{5.10}$$

where λ_v, μ_v, $\varkappa_v$, α_v, β_v and γ_v are the coefficients of viscosity of which λ_v and μ_v are the classical viscosities of the Navier–Stokes theory.
In this theory the stress average s_{kl} disappears from the equations. Finally the expression of $\dot{\sigma}_r$ follows from (5.3), and (4.4), i.e.

$$\dot{\sigma}_r = j\ddot{\varphi}_r = j\dot{\nu}_r. \tag{5.11}$$

Upon substituting (5.7), (5.8) and (5.11) into (3.7) and (5.4) we obtain the field equations

Micropolar elasticity

$$(\lambda+\mu)u_{l,lk}+(\mu+\varkappa)\,u_{k,ll}+\varkappa\varepsilon_{klm}\varphi_{m,l}+\varrho\left(f_k-\frac{\partial^2 u_k}{\partial t^2}\right) = 0, \tag{5.12}$$

$$(\alpha+\beta)\,\varphi_{l,lk}+\gamma\varphi_{k,ll}+\varkappa\varepsilon_{klm}u_{m,l}-2\varkappa\varphi_k+\varrho\left(l_k-j\,\frac{\partial^2\varphi_k}{\partial t^2}\right) = 0. \tag{5.13}$$

Here we have six partial differential equations for the two unknown vector fields u_k and φ_k.

Micropolar fluids

$$-\pi_{,k}+(\lambda_v+\mu_v)\,v_{l,lk}+(\mu_v+\varkappa_v)\,v_{k,ll}+\varkappa_v\,\varepsilon_{klm}\,\nu_{m,l}+\varrho(f_k-\dot{v}_k) = 0, \tag{5.14}$$

$$(\alpha_v+\beta_v)\,\nu_{l,lk}+\gamma_v\nu_{k,ll}+\varkappa_v\,\varepsilon_{klm}\,v_{m,\,l}-2\varkappa\nu_k+\varrho(l_k-j\dot{\nu}_k) = 0. \tag{5.15}$$

In these six equations the unknown vector fields are the velocity field v_k and microrotation field ν_k. These equations must be supplemented by the equation of continuity (3.5). Note that $\dot{v}_k$ and $\dot{\nu}_k$ appearing in (5.14) and (5.15) are the material derivatives of v_k and ν_k so that their expressions contain convective terms, i.e.

$$\dot{v}_k = \frac{\partial v_k}{\partial t}+v_{k,l}\,v_l, \qquad \dot{\nu}_k = \frac{\partial \nu_k}{\partial t}+\nu_{k,l}\,v_l. \tag{5.16}$$

We observe that the constitutive coefficients $\varkappa$ and $\varkappa_v$ play an important role in this theory since for $\varkappa = 0$ the equations (5.12) and (5.13) are uncoupled and for $\varkappa_v = 0$, (5.14) and (5.15) are uncoupled. We also note that when $\boldsymbol{\varphi} = 0$ or $\boldsymbol{\nu} = 0$, (5.12) and (5.14), respectively, reduce to the Navier's equation and Navier–Stokes equations of the classical theories with μ and μ_v of classical theories respectively replaced by $\mu+\varkappa$ and $\mu_v+\varkappa_v$ while (5.13) and (5.14) lead to $\boldsymbol{l} = 0$.

In the case of fluids the classical Stokes conditions $3\lambda_v+2\mu_v = 0$ is now replaced by

$$3\lambda_v+2\mu_v+\varkappa_v = 0. \tag{5.17}$$

For *incompressible fluids* $\varrho =$ const and

$$v_{k,k} = 0. \tag{5.18}$$

In this case the thermodynamic pressure π must be replaced by an unknown pressure p to be determined through the solution of each problem.

The above field equations are subject to certain boundary and initial conditions. Since the ultimate decision on these conditions requires the proof of existence and uniqueness theorems we only cite here a few natural conditions.†

Boundary conditions on tractions (both media)

$$\left.\begin{aligned} t_{kl}\,n_k &= t_l \\ m_{kl}n_k &= m_l \end{aligned}\right\} \quad \text{on} \quad \mathcal{S}. \tag{5.19}$$

Displacement and velocity conditions (solids)

$$\left.\begin{aligned} u_k(x, 0) &= \overset{\circ}{u}_k(x), &\quad \dot{u}_k(\mathbf{x}, 0) &= \overset{\circ}{v}_k(\mathbf{x}) \\ \varphi_k(\mathbf{x}, 0) &= \overset{\circ}{\varphi}_k(\mathbf{x}), &\quad \dot{\varphi}_k(\mathbf{x}, 0) &= \overset{\circ}{\nu}_k(\mathbf{x}) \end{aligned}\right\} \text{in } \mathcal{V}. \tag{5.20}$$

Mixed types of conditions specifying (5.17) on a part of $\mathcal{S}$, $\mathcal{S}_t$ and (5.18) on the remainder $\mathcal{S}_u \equiv \mathcal{S}-\mathcal{S}_t$ are possible. Still other mixing (such as specifying t_1, t_2 and u_3) can be made. We do not, however, delve into these matters further since they require the proof of the uniqueness theorem.

Velocity conditions (fluids)

$$\left.\begin{aligned} v_k(\mathbf{x}, 0) &= \overset{\circ}{v}_k(\mathbf{x}) \\ \nu_k(\mathbf{x}, 0) &= \overset{\circ}{\nu}_k(\mathbf{x}) \end{aligned}\right\}, \quad \text{in } \mathcal{V}. \tag{5.21}$$

These conditions are the expression of adherence of the fluid to a solid boundary when $\mathbf{v}^\circ$ and $\boldsymbol{\nu}^\circ$ are the velocity and microrotations of the solid boundary. For a rigid stationary boundary $\mathbf{v}^\circ = \boldsymbol{\nu}^\circ = \mathbf{0}$. We do not intend

† For the uniqueness theorems for micropolar elasticity see Eringen [6].

to open the controversial question of adherence of viscous fluids to a solid boundary here.

We close this section by a brief discussion of the restrictions emanating from the second law of thermodynamics. This is stated as

THEOREM. *The Clausius–Duhem inequality* (3.10) *for micropolar fluids is satisfied identically for all independent motions if and only if*

$$\begin{aligned} &0 \leqslant 3\lambda_v+2\mu_v+\varkappa_v, \quad 0 \leqslant 2\mu_v\varkappa_v, \quad 0 \leqslant \varkappa_v \\ &0 \leqslant 3\alpha_v+\beta_v+\gamma_v, \quad -\gamma_v \leqslant \beta_v \leqslant \gamma_v, \quad 0 \leqslant \gamma_v. \end{aligned} \tag{5.22}$$

For the proof of this theorem we refer the reader to our work [5].

Finally a similar procedure for the nonnegative internal energy can be shown to give for the micropolar elastic solid

$$\begin{aligned} &0 \leqslant 3\lambda+2\mu+\varkappa, \quad 0 \leqslant 2\mu\varkappa, \quad 0 \leqslant \varkappa \\ &0 \leqslant 3\alpha+\beta+\gamma, \quad -\gamma \leqslant \beta \leqslant \gamma, \quad 0 \leqslant \gamma. \end{aligned} \tag{5.23}$$

For proof see ref. [6].

6. MICROPOLAR CHANNEL FLOW

Here we give an example of a simple solution of the field equations (5.14) to (5.15) for a steady channel flow confined between two parallel walls. We take the right-handed rectangular coordinates $x_1 \equiv x$, $x_2 \equiv y$, $x_3 \equiv z$ with the x-axis along the axis of the channel, y perpendicular to the walls and z being perpendicular to the (x, y)-plane. For a steady channel flow we seek to determine the velocity and microrotation components

$$\begin{aligned} &v_x \equiv v(y), \quad v_y = v_z = 0 \\ &\nu_z \equiv \nu(y), \quad \nu_x = \nu_y = 0 \end{aligned} \tag{6.1}$$

The equation of continuity (3.5) is satisfied for $\varrho =$ const and (5.14) and (5.15) with $\boldsymbol{f} = \boldsymbol{l} = 0$ give $p_{,y} = p_{,z} = 0$ and

$$-p_{,x}+(\mu_v+\varkappa_v)v''+\varkappa_v\nu' = 0, \tag{6.2}$$

$$\gamma_v\nu''-\varkappa_v v'-2\varkappa_v\nu = 0, \tag{6.3}$$

where a superposed prime indicates differentiation with respect to y. We also used p to denote hydrostatic pressure in place of π.

The general solution of these two linear equations for v and ν is

$$\begin{aligned} v &= A \cos ky+B \sinh ky+mk^{-2}y^2+Ck^{-2}y+Dk^{-2}, \\ \nu &= -(\mu_v+\varkappa_v)\varkappa_v^{-1}[Ak \sinh ky+Bk \cosh ky+\gamma_v m(2\mu_v+\varkappa_v)^{-1}y] \\ &\qquad -\tfrac{1}{2}Ck^{-2}, \end{aligned} \tag{6.4}$$

where A, B, C and D are constants of integration and

$$k \equiv \left(\frac{2\mu_v+\varkappa_v}{\mu_v+\varkappa_v}\times\frac{\varkappa_v}{\gamma_v}\right)^{\frac{1}{2}}, \qquad m \equiv \frac{\varkappa_v p_{,x}}{\gamma_v(\mu_v+\varkappa_v)}. \tag{6.5}$$

We assume that the fluid sticks to the boundaries $y = \pm h$. Hence

$$v(h) = v(-h) = 0, \qquad \nu(h) = \nu(-h) = 0 \tag{6.6}$$

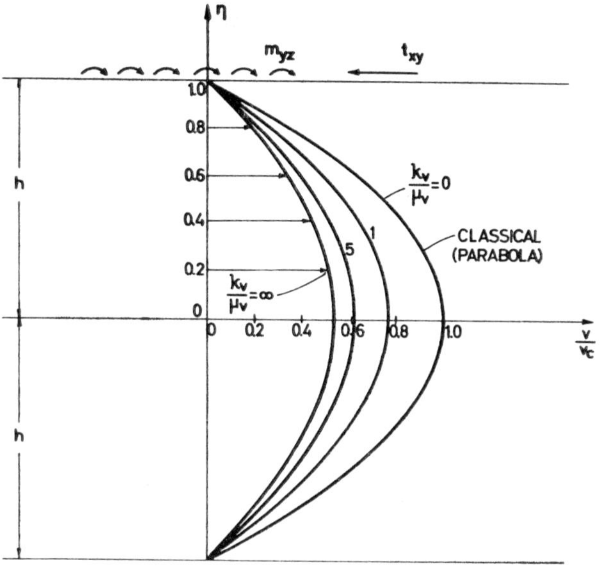

FIG. 3. Velocity profile.

Using these four boundary conditions we determine A, B, C and D. The resulting solution is

$$\frac{v}{v_c} = 1-\eta^2+\frac{\varkappa_v}{\mu_v+\varkappa_v K}\frac{\cosh K}{\sinh K}\left(\frac{\cosh K\eta}{\cosh K}-1\right) \tag{6.7}$$

$$\frac{\nu h}{v_c} = \eta-\frac{\sinh K\eta}{\sinh K}, \tag{6.8}$$

where

$$\begin{aligned} v_c &= -p_{,x}h^2/(2\mu_v+\varkappa_v) \\ \eta &\equiv y/h \\ K &\equiv kh = \left(\frac{2\mu_v+\varkappa_v}{\mu_v+\varkappa_v}\times\frac{\varkappa_v}{\gamma_v}\right)^{\frac{1}{2}} h. \end{aligned} \tag{6.9}$$

The solutions (6.7) goes to the classical channel flow for $\varkappa_v = 0$, and for this case $\nu = 0$. Here v_c is the maximum velocity in the classical channel flow which occurs at $y = 0$.

For various values of K we give on Fig. 4 plots of the velocity difference from the classical channel flow and on Fig. 5 $\nu h/v_c$. We see that for micropolar fluid the velocity profile is no longer parabolic. Moreover,

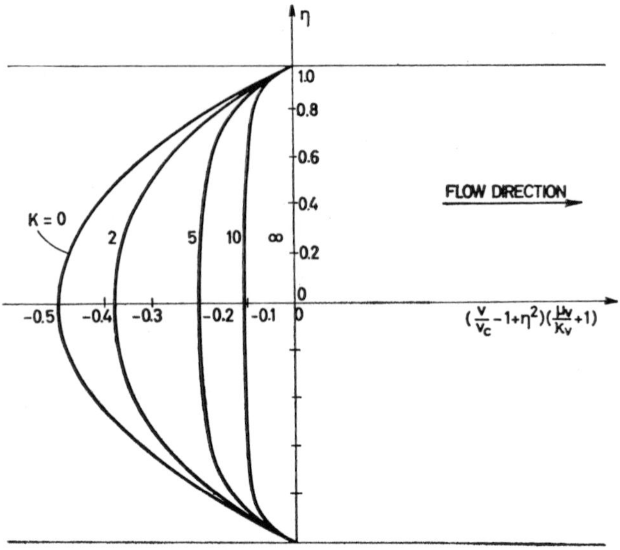

FIG. 4. Adverse microflow.

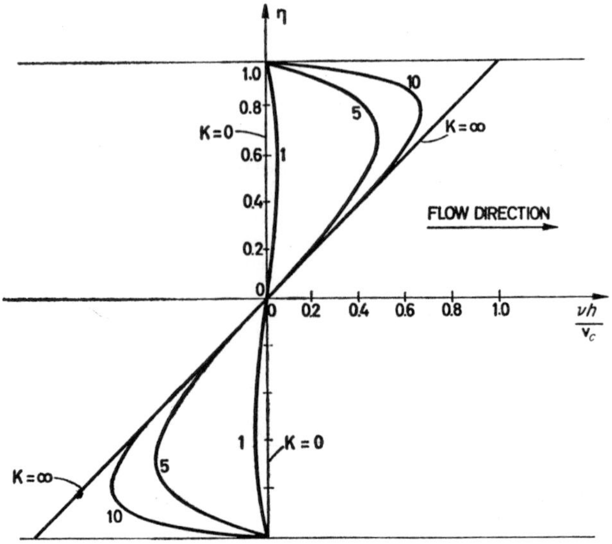

FIG. 5. Microrotation.

the velocity here is smaller than that of the classical Navier–Stokes fluids. Of course, the microrotation ν is totally missing in the Navier–Stokes theory.

The components of the stress and couple stress are tensors calculated through (5.9) and (5.10). Hence

$$\begin{aligned} t_{xx} &= t_{yy} = t_{zz} = -p, \\ t_{xy} &= p_{,x}\, h\left(\eta - \frac{\varkappa_v}{\mu_v+\varkappa_v}\,\frac{\sinh K\eta}{\sinh K,}\right), \\ t_{yx} &= p_{,x} h\eta, \end{aligned} \tag{6.10}$$

and all other $t_{kl} = 0$,

$$\begin{aligned} m_{yz} &= p_{,x}\frac{\gamma_v}{2\mu_v+\varkappa_v}\left(\frac{K\cosh K\eta}{\sinh K} - 1\right), \\ m_{zy} &= p_{,x}\frac{\beta_v}{2\mu_v+\varkappa_v}\left(\frac{K\cosh K\eta}{\sinh K} - 1\right) \end{aligned} \tag{6.11}$$

and all other $m_{kl} = 0$. We note that $t_{xy} \neq t_{yx}$ whenever $\varkappa_v \neq 0$.

On Fig. 3 are shown the surface tractions and couples on the fluid surface adjacent to the wall at $\eta = 1$ for $p_{,x} < 0$, $\beta_v < 0$ and, of course, $\gamma_v > 0$. Accordingly, the net effect of m_{yz} will be a reduction in the shearing stress t_{yz} of the classical theory near the wall. If the walls consist of thin plates not resistant to couple then $m_{yz} < 0$ will make the plates move

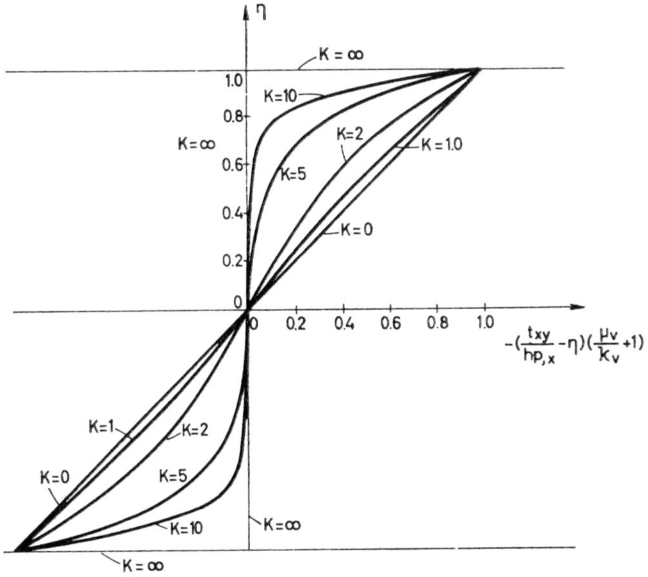

FIG. 6. Shear stress difference.

away from each other in the direction of the flow. The couple m_{yz} depending on K may vanish at a plane $\eta = \eta_0$ determined by

$$K \cosh K\eta_0 - \sinh K = 0. \tag{6.12}$$

The effect of m_{zy} is felt at a section $z =$ const and it varies with η. Near the rigid walls it tends to turn a line parallel to x-axis inwards in the direction of flow and at $y = 0$ the x-axis outwards. For $\beta_v > 0$ the effect of m_{zy} is reserved. These actions are, however, eliminated by the adjacent layers of the fluids except, of course, at $z = \pm \infty$.

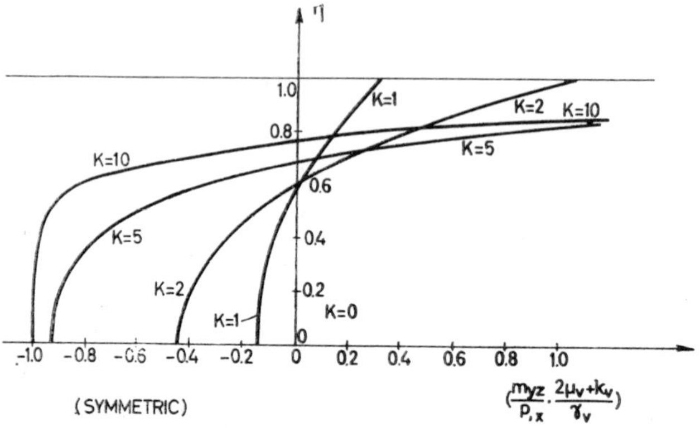

FIG. 7. Couple stress.

REFERENCES

1. A. C. ERINGEN and E. S. SUHUBI, "Nonlinear theory of simple micro-elastic solids—I", *Int. J. Engng. Sci.* **2**, 189 (1964).
2. E. S. SUHUBI and A. C. ERINGEN, "Nonlinear theory of simple micro-elastic solids—II", *Int. J. Engng. Sci.* **2**, 389 (1964).
3. A. C. ERINGEN, "Simple microfluids", *Int. J. Engng. Sci.* **2**, 205 (1964).
4. A. C. ERINGEN, "Mechanics of Micromorphic Materials", *Proceedings XI Intern. Congress of Applied Mech.*, Springer-Verlag (1965).
5. A. C. ERINGEN, *Theory of Micropolar Fluids*, ONR Report (1965).
6. A. C. ERINGEN, Theory of micropolar elastic solids, scheduled for publication in *J. Math. and Mech.*
7. J. W. HOYT and A. G. FABULA, *The Effect of Additives on Fluid Friction*, U.S. Naval Ordnance Test Station Report (1964).
8. W. M. VOGEL and A. M. PATTERSON, *An Experimental Investigation of the Effect of Additives Injected into the Boundary Layer of an Underwater Body*, Pacific Naval Lab. of the Defense Research Board of Canada, Report 64–2.
9. A. C. ERINGEN, *Nonlinear Theory of Continuous Media*, McGraw-Hill (1962).

APPLICATIONS OF SINGULAR PERTURBATIONS AND BOUNDARY-LAYER THEORY TO SIMPLE ORDINARY DIFFERENTIAL EQUATIONS

SYDNEY GOLDSTEIN†

Division of Engineering and Applied Physics, Harvard University, Mass., U.S.A.

1. Singular perturbation methods for finding approximations to solutions of differential equations seem to have had their origin in Prandtl's boundary-layer theory for considering laminar motions of a viscous fluid, i.e. for studying the Navier–Stokes equations of motion. These equations are non-linear partial differential equations. The considerable development of the ideas by several investigators (in particular by Lagerstrom and Kaplun and others of their colleagues at the California Institute of Technology) [1, 2] also began largely with the same equations, and, in addition, the method was adapted to the consideration of equations describing other physical phenomena. Several writers have recently considered the application of the ideas to simpler ordinary differential equations, partly to introduce discussions that are mathematically more rigorous, and partly for pedagogical purposes. In what follows some of the simplest possible applications are set out—applications to linear ordinary differential equations. This was done in the first place for pedagogical purposes, but it seemed to acquire a certain interest of its own, elementary as it all was. It is, however, still set out in as simple a manner as possible, and rigorous mathematical formulations and proofs are not included.

Moreover, it soon appeared that even for singular perturbations of ordinary linear differential equations there is a large number of possible

† I take this opportunity to pay tribute to Prof. Markus Reiner. Although we had met previously at international meetings, I came to know Prof. Reiner more intimately when I went to the Technion, Haifa, in 1950, and rapidly came to feel for him both great affection and the deepest respect. So I am glad to be allowed to contribute to this volume for Prof. Reiner's eightieth anniversary, even though my subject has, perforce, only a tenuous connection with Prof. Reiner's major interests and with the subjects considered in other papers in the volume, and even though considerations of space allowed me to go not much further than an introduction to the matters dealt with.

types of behavior, and, although a more nearly complete classification of the possible types would have some interest, no attempt at such a classification is made here. Limitations of space restrict this account to a comparatively few examples.

As usually set out, the method of singular perturbations deals with approximations for a small value of a parameter. It is both convenient and instructive to begin with a comparison of this method, as applied to second-order ordinary linear differential equations, with the Liouville–Green method for finding approximations for large values of a parameter.

2. Let

$$f(x, y, y', y,'' \ldots, y^{(n)};\ \varepsilon) = 0 \tag{1}$$

(with primes denoting derivatives) be an ordinary differential equation of order n, in which ε is a small parameter which enters in such a way that when it is put equal to zero the order of the resulting equation, which will be called the reduced equation, is less than n. So the number of boundary conditions that can be satisfied is also reduced. Boundary conditions for the full equation (1) are assumed specified so that the system has a solution. The cases in which we are interested are those in which this solution has a limit as $\varepsilon \to 0$, and this limit is a solution of the reduced equation with a smaller number of appropriate boundary conditions. How do we know the appropriate boundary conditions for the reduced equation? How do we restore the "lost" boundary conditions, and improve the "approximation" to the solution of the full system represented by the solution of the reduced system?

The simplest case is that in which ε enters simply as a multiplier of the highest derivative and Eq. (1) is of the form

$$\varepsilon y^{(n)} + \phi(x, y, y', y'', \ldots, y^{(n-1)}) = 0, \tag{2}$$

but many other cases are of course possible. The reduced equation may not be of order $n-1$, but of a lower order, and then ε may enter in many different ways in the terms that vanish when $\varepsilon = 0$.

For a linear equation, (2) is of the form

$$\varepsilon y^{(n)} + A_{n-1}(x) y^{(n-1)} + \ldots + A_1(x) y' + A_0(x) y = C(x). \tag{3}$$

3. For a second-order equation write

$$\varepsilon y'' + A(x) y' + B(x) y = C(x). \tag{4}$$

Now, from the usual theory of asymptotic approximations for a large parameter to solutions of a second-order linear ordinary equation, asymptotic approximations when h is large to two independent solutions of

$$v'' - [h^2 \chi_0(x) + h \chi_1(x) + \chi_2(x)] v = 0 \tag{5}$$

are, away from a zero of $\chi_0(x)$, given by

$$v \sim |\chi_0|^{-\frac{1}{4}} \exp\left[\pm \int^x \left(h\chi_0^{\frac{1}{2}} + \frac{\chi_1}{2\chi_0^{\frac{1}{2}}}\right) dx\right]\left[1 + \frac{f_1(x)}{h} + \frac{f_2(x)}{h^2} + \dots\right], \qquad (6)$$

where $f_1, f_2, \dots$ are functions still to be determined. The integral may be taken from any constant value of x; changing this constant simply multiplies each of the approximations by a constant factor. (Jeffreys and Jeffreys, *Methods of Mathematical Physics* [3], § 17.13. See also [4].) The approximate solutions are of exponential type when χ_0 is positive and oscillatory when χ_0 is negative.

The method fails near a zero of χ_0. If there is such a zero at $x = 0$, say, approximate solutions can be found in a transition zone near $x = 0$. These approximate solutions are found in terms of a stretched variable, $X = h^s x (s > 0)$, and can be matched, on each side of the transition zone, to solutions of the form (6) or linear combinations of such solutions. For a zero of χ_0 of order r, $s = 2/(r+2)$. For a simple zero ($r = 1$) the approximate solutions (6) are of exponential type on one side of the zero and oscillatory on the other, and in the transition zone the approximate solutions are Airy integrals of $[\chi_0'(0)]^{1/3}X$.

If we put (4) into normal form by writing

$$y = v \exp\left(-\frac{M}{2\varepsilon}\right), \quad M = \int_c^x A(x)\, dx \qquad (7)$$

(where c is an arbitrary constant at this stage), then

$$v'' - v\left(\frac{A^2}{4\varepsilon^2} + \frac{A' - 2B}{2\varepsilon}\right) = \frac{C}{\varepsilon} \exp\left(\frac{M}{2\varepsilon}\right), \qquad (8)$$

which for $C = 0$ is of the form (5) with

$$h = 1/(2\varepsilon), \quad \chi_0 = A^2, \quad \chi_1 = A' - 2B, \quad \chi_2 = 0. \qquad (9)$$

It follows from Eq. (6) that, away from any zero of A, two approximate complementary functions of (4) are

$$y_1 = \frac{1}{E} \quad \text{and} \quad y_2 = \frac{E}{A} \exp\left(-\frac{M}{\varepsilon}\right), \quad \text{where } E = \exp \int_c^x \frac{B}{A}\, dx, \qquad (10)$$

and M is given in Eq. (7).

There is also a particular integral to which an approximation P is given by

$$P = \frac{1}{E} \int_c^x \frac{EC}{A}\, dx\,. \qquad (11)$$

y_1 and P are slowly varying, and are, respectively, a complementary function and a particular integral of the reduced equation

$$A(x)y' + B(x)y = C(x). \tag{12}$$

y_2 is an approximation to a rapidly varying complementary function. It may be found directly from Eq. (4) by first substituting

$$y = u \exp(w/\varepsilon) \tag{13}$$

in (4) with $C = 0$, and equating to zero the coefficients of ε^{-1}, ε^0, ε after division by $\exp(w/\varepsilon)$. The expression for w comes directly from the coefficient of ε^{-1}, and the approximation for u from the coefficient of ε^0 if u is considered expanded in a power series in $\varepsilon\,(u = u_0+\varepsilon u_1+\varepsilon^2 u_2+\ldots)$ and only the first term retained.

We here interrupt the discussion of the second-order linear equation to remark that an exactly similar method may be applied to the nth-order linear equation (3). There is a slowly varying particular integral and $n-1$ slowly varying complementary functions, approximations to which are a particular integral and the complementary functions of the reduced equation. There is also a rapidly varying complementary function, an approximation to which is found by the substitution (13) to be

$$y_n = \left[A_{n-1}(x)\right]^{-(n-1)} \exp\left[\int \frac{A_{n-2}(x)}{A_{n-1}(x)}\,dx\right] \exp\left[-\frac{1}{\varepsilon}\int A_{n-1}(x)\,dx\right] \tag{14}$$

away from any zero of A_{n-1}.

4. We now return to the second-order equation (4) and consider problems with given boundary conditions. Without loss of generality we may suppose that one or two boundary conditions are given at $x = 0$, and take $c = 0$ as the lower limit of the integrals in M, E, and P in Eqs. (7), (10), and (11). We may either suppose that two boundary conditions are given at $x = 0$, say the values of $y(0)$ and $y'(0)$, or that the boundary conditions are given at two different values of x, say the values of $y(0)$ and $y(1)$. Of course, we are assuming that $A(0) \neq 0$, that $A(1) \neq 0$ in the second case above, and that A has no zero in the relevant range of x, so that there is no internal transition zone. Unless otherwise stated, we suppose $x \geqslant 0$ in the relevant range.

An approximation to a general solution of Eq. (4) for small ε is

$$y = P+K_1y_1+K_2y_2, \tag{15}$$

where K_1, K_2 are constants (i.e. functions of ε but not of x) to be found from the boundary conditions.

With $y(0)$, $y'(0)$ given and both $O(1)$, K_1 is $O(1)$, K_2 is $O(\varepsilon)$, and K_1, K_2/ε are found with errors of order ε:

$$K_1 = y(0), \quad \frac{K_2}{\varepsilon} = -\left[\frac{B(0)}{A(0)}y(0)+y'(0)-\frac{C(0)}{A(0)}\right]. \tag{16}$$

When $M = O(1)$, y_2 is exponentially small for small ε for $M > 0$, i.e. for $A > 0$. But near $x = 0$, M is approximately $A(0)x$, and when $x = O(\varepsilon)$, $y_2 = O(1)$ and is, in fact, approximately $[1/A(0)] \exp[-A(0)X]$, where $X = x/\varepsilon$.

If we keep x fixed and let $\varepsilon \to 0$, the approximate solution (15) tends to $P+y(0)E^{-1}$, which is a solution of the reduced equation (12) and satisfies the boundary condition $y = y(0)$ at $x = 0$ but does not satisfy the boundary condition $y' = y'(0)$ at $x = 0$.

For $A > 0$, the terms K_2y_2 and K_2y_2' in y and y' from Eq. (15) are exponentially small in $x > 0$ except in a "boundary layer" near the boundary $x = 0$ whose thickness is of order ε, where $x = O(\varepsilon)$. (This is true for $x < 0$ if $A < 0$.) There is a boundary layer for positive x if $A > 0$ and for negative x if $A < 0$. In the boundary layer, when $x = O(\varepsilon)$ the term K_2y_2 in y is $O(\varepsilon)$, but the term K_2y_2' in y' is $O(1)$, and the inclusion of K_2y_2 in Eq. (15) enables the second boundary condition to be satisfied. ($y_2'/y_2 = O(\varepsilon^{-1})$; this is what we mean by a rapidly varying solution.)

Although K_2y_2 is $O(\varepsilon)$ at most, Eq. (15) does not give y correct to order ε. For this we have to find P, y_1, and K_1 to order ε. It is a simple enough matter to do this if required. The value of y in Eq. (15) makes its value at $x = 0$ equal to $y(0)+K_2/A(0)$, the second term being of order ε, and the additional terms in Eq. (15) must be chosen just to cancel this second term and to satisfy

$$A(x)y'+B(x)y = -\varepsilon\frac{d^2}{dx^2}(P+K_1y_1).$$

If we now keep $X = x/\varepsilon$ fixed and expand Eq. (15) in powers of ε up to ε^1, this extra term just contributes $-K_2/A(0)$, and the result is

$$y = P(0)+K_1y_1(0)+\varepsilon X[P'(0)+K_1y_1'(0)]-\frac{K_2}{A(0)}[1-e^{-A(0)X}]$$

$$= y(0)+\varepsilon X\left[\frac{C(0)}{A(0)}-\frac{B(0)}{A(0)}y(0)\right]+\frac{\varepsilon}{A(0)}\left[\frac{B(0)}{A(0)}y(0)+y'(0)-\frac{C(0)}{A(0)}\right][1-e^{-A(0)X}] \tag{17}$$

The boundary-layer approach to the problem is simpler. The method consists of assuming the existence of a boundary layer and testing if the work goes through with an exponentially small error in the asymptotic expansions. We assume "outer" and "inner" expansions. The outer expansion will in this case be in powers of ε, with x used as independent variable, and with the assumption that y'/y, y''/y', etc., stay bounded as $\varepsilon \to 0$ with x in the range considered. This cuts out the rapidly varying complementary function and cannot be complete. For the inner expansion (in the boundary layer) the variable used is the stretched variable $X = x/\varepsilon^\nu$, with $\nu\,(> 0)$ chosen to make the term $\varepsilon y''$ in the differential equation of the same order as the second term when X-derivatives are all of the same order as the dependent variable. Here clearly $\nu = 1$, so

$$X = x/\varepsilon. \tag{18}$$

Then in this case the inner expansion is also in powers of ε, with X as independent variable. It is assumed that in the boundary layer derivatives with respect to X are bounded as $\varepsilon \to 0$.

For the expansions write

$$y_{\text{outer}} = g_0(x)+\varepsilon g_1(x)+\varepsilon^2 g_2(x)+\ldots \tag{19}$$

$$y_{\text{inner}} = f_0(X)+\varepsilon f_1(X)+\varepsilon^2 f_2(X)+\ldots \tag{20}$$

Substitute (19) into (4), and equate coefficients of powers of ε. Then

$$A(x)g_0'+B(x)g_0 = C(x), \tag{21}$$

$$A(x)g_n'+B(x)g_n = -g_{n-1}''\ (n \geqslant 1). \tag{22}$$

The differential equations for the g are first-order equations and one boundary condition is needed for each. The equation for g_0 is the reduced equation (12).

In terms of X, Eq. (4) becomes

$$\frac{d^2y}{dX^2}+A(\varepsilon X)\frac{dy}{dX}+\varepsilon B(\varepsilon X)y = \varepsilon C(\varepsilon X),$$

i.e. if A, B, and C are analytic,

$$\frac{d^2y}{dX^2}+[A(0)+\varepsilon XA'(0)+\ldots]\frac{dy}{dX}+\varepsilon[B(0)+\varepsilon XB'(0)+\ldots]y$$
$$= \varepsilon[C(0)+\varepsilon XC'(0)+\ldots]. \tag{23}$$

Substitute (20) into (23) and equate coefficients of powers of ε. The equations for f_0 and f_1 are

$$\frac{d^2f_0}{dX^2}+A(0)\frac{df_0}{dX} = 0, \tag{24}$$

$$\frac{d^2f_1}{dX^2}+A(0)\frac{df_1}{dX} = C(0)-B(0)f_0-XA'(0)f_0'. \tag{25}$$

The boundary conditions are given at $x = 0$, i.e. at $X = 0$. Since $dy/dx = (1/\varepsilon)(dy/dX)$, at $X = 0$, $y = y(0)$ and $dy/dX = \varepsilon y'(0)$. Thus the boundary conditions on the f are

$$f_0(0) = y(0), \quad f_0'(0) = 0, \quad f_1(0) = 0, \quad f_1'(0) = y'(0), \quad f_n(0) = f_n'(0) = 0 \quad \text{for } n \geqslant 2. \tag{26}$$

The differential equations and the boundary conditions at $X = 0$ determine the f completely. The boundary conditions for the determination of the g must come from the matching conditions. If we substitute $x = \varepsilon X$ in the outer solution and expand in powers of ε, or if we expand the outer solution in powers of x and then substitute $x = \varepsilon X$, and rearrange in powers of ε, we have

$$y_{\text{outer}} = g_0(0) + \varepsilon[Xg_0'(0) + g_1(0)] + \ldots \tag{27}$$

The matching conditions are then that, as $X \to \infty$ (i.e. $\varepsilon \to 0$, $x \neq 0$)

$$f_0 \sim g_0(0), \quad f_1 \sim Xg_0'(0) + g_1(0). \tag{28}$$

The values of $g_0(0)$ and $g_1(0)$ are then found from the asymptotic expansions of f_0 and f_1, respectively. The term $Xg_0'(0)$ in the asymptotic expansion of f_1 must be automatically correct.

The solutions for f_0 and f_1 are

$$f_0 = y(0) \tag{29}$$

$$f_1(X) = X\left[\frac{C(0)}{A(0)} - \frac{B(0)}{A(0)}y(0)\right] + \frac{1}{A(0)}\left[\frac{B(0)}{A(0)}y(0) + y'(0) - \frac{C(0)}{A(0)}\right][1 - e^{-A(0)X}]. \tag{30}$$

Hence the boundary condition for g_0 is $g_0(0) = y(0)$, and the solution for g_0 is

$$g_0 = P + K_1 y_1 \tag{31}$$

in the previous notation.

This makes

$$g_0'(0) = P'(0) + Ky_1'(0) = \frac{C(0)}{A(0)} - \frac{B(0)}{A(0)}y(0) \tag{32}$$

and the term $Xg_0'(0)$ in the asymptotic expansion of f_1 is, in fact, correct.

If now $A(0) > 0$, the second matching condition is satisfied if

$$g_1(0) = \frac{1}{A(0)}\left[\frac{B(0)}{A(0)}y(0) + y'(0) - \frac{C(0)}{A(0)}\right] = -\frac{K_2}{\varepsilon}\frac{1}{A(0)}. \tag{33}$$

These are the results at which we had previously arrived.

If $A(0) < 0$, the matching conditions cannot be satisfied (and there is not a boundary layer) for X and x positive. We assume that $A(0) > 0$.

Note that the boundary conditions for f_1 are lost if $\exp[-A(0)X]$ is neglected, and this term is exponentially small unless $X = O(1)$, $x = O(\varepsilon)$.

On the other hand, the inner solution joins on to the outer solution when X is large.

In general the series for the inner and outer solutions are each asymptotic, and not convergent. The inner-solution series is valid when $X = x/\varepsilon$ is bounded. The validity of the outer-solution series requires that the expansion of exponential terms such as $\exp(-A(0)x/\varepsilon)$ should be the zero expansion, so that this series is valid only for $\delta(\varepsilon) \leqslant x$, where δ is > 0 and δ/ε is unbounded as $\varepsilon \to 0$. There is no region of overlap of the two series. Nevertheless, the two series do "touch", so that matching is possible. This is the most usual case. In rare cases in less elementary problems the two series may overlap. As we shall see it is also possible to construct examples in which the series do not "touch" and matching is impossible.

Moreover, a discussion of "intermediate" solutions, overlapping at one end with the inner solution and at the other with the outer solution for cases such as that discussed above, will be found in the publications on the subject. We shall not discuss these "intermediate" solutions here. However, it is worth-while briefly to describe the "composite" solution, as it is called in the published literature.

If we substitute $x = X/\varepsilon$ in the inner expansion and expand in powers of ε with x fixed, we arrive at the outer expansion of the inner expansion, $y_{\text{inner}}^{\text{outer}}$. Since the expansion of $\exp[-A(0)x/\varepsilon]$ is the zero expansion, this is the same as dropping the exponential term in y_{inner}, and $y_{\text{inner}} - y_{\text{inner}}^{\text{outer}}$ is just the exponential term in y_{inner}.

If we substitute $x = \varepsilon X$ in the outer expansion, and expand in powers of ε with X fixed, we obtain the inner expansion of the outer expansion, $y_{\text{outer}}^{\text{inner}}$. But the matching condition is simply $y_{\text{outer}}^{\text{inner}} = y_{\text{inner}}^{\text{outer}}$.

The composite expansion is then taken as

$$y_{\text{outer}} + y_{\text{inner}} - y_{\text{inner}}^{\text{outer}} = y_{\text{inner}} + y_{\text{outer}} - y_{\text{outer}}^{\text{inner}}. \tag{34}$$

For the equation and boundary conditions considered above the term that must be added to y_{outer} to find the composite expansion is just $[K_2/A(0)] \exp(-A(0)x/\varepsilon)$ as far as $O(\varepsilon)$, and the composite expansion is

$$g_0 + \varepsilon g_1 + \frac{K_2}{A(0)} \exp[-A(0)x/\varepsilon]. \tag{35}$$

Now $\exp[-A(0)x/\varepsilon]$ is just the first approximation to $\exp[-M(x)/\varepsilon]$ in the boundary layer; in fact, with X fixed, $\exp[-M/\varepsilon] = \exp[-A(0)X] \times [1 + O(\varepsilon)]$. Also, in the previous notation, $y_2 = (E/A)\exp(-M/\varepsilon)$, and $E(0)/A(0) = 1/A(0)$, so the composite expansion (35) agrees as far as we have gone with the previous result (15). But if this is as far as we are going, and if the exponential term has any numerical importance at all outside the boundary layer, the previous approximation, $K_2 y_2$, is preferable in general, since it allows for the variation in E, A, and M—particularly M—with x outside the boundary layer.

The composite expansion is a general asymptotic expansion, not a Poincaré expansion. Such expansions are not unique, and some prescription is necessary to determine the terms one by one. Such a prescription has been given above by using inner and outer expansions with the matching conditions.

5. We now consider the same differential equation (4), but with the boundary conditions

$$y(0) = a \quad \text{and} \quad y(1) = b. \tag{36}$$

We take $A > 0$ in $0 \leqslant x \leqslant 1$.

The lower limit of integration, c, in the integrals in M, E, and P in Eqs. (7), (10) and (11) is still taken as zero.

First, we consider the approximate solution in the form (15), with y_1, y_2, and P given by Eqs. (10) and (11). If we put in the new boundary conditions and solve for K_1 and K_2, we find that

$$K_1\left\{1-\frac{A(0)}{A(1)}\,[E(1)]^2 e^{-M(1)/\varepsilon}\right\} = E(1)[b-P(1)]-(a-1)\,\frac{A(0)}{A(1)}\,[E(1)]^2 e^{-M(1)/\varepsilon}, \tag{37}$$

$$K_2\left\{1-\frac{A(0)}{A(1)}\,[E(1)]^2 e^{-M(1)/\varepsilon}\right\} = A(0)\{a-1-E(1)[b-P(1)]\}. \tag{38}$$

$M > 0$ since $A > 0$. If $\exp\,[-M(1)/\varepsilon]$ is neglected,

$$K_1 = E(1)[b-P(1)], \tag{39}$$

$$K_2 = A(0)\{a-1-E(1)[b-P(1)]\}. \tag{40}$$

Now the solution, $P+K_1 y_1$, of the reduced equation (12) satisfies $y(1) = b$ with $\exp\,[-M(1)/\varepsilon]$ neglected, but does not satisfy $y(0) = a$.

Near $x = 0$, $M = A(0)x+\ldots$, and, since $E(0) = 1$, the term $K_2 y_2$ is

$$\frac{K_2}{A(0)}\exp\,[-A(0)x/\varepsilon+\ldots]. \tag{41}$$

Near $x = 1$, $M = M(1)-A(1)\,[1-x]+\ldots$, and the term $K_2 y_2$ is

$$K_2\frac{E(1)}{A(1)}\exp\,[-M(1)/\varepsilon]\,\exp\,\{A(1)[1-x]/\varepsilon+\ldots\}, \tag{42}$$

which contains an exponentially small multiplier and a *positive* exponential.

With $A > 0$, there is a boundary layer at $x = 0$ but not at $x = 1$. If we had $A < 0$, there would be a boundary layer at $x = 1$ but not at $x = 0$.

With $A > 0$, the discussion by the boundary-layer method proceeds as follows. We could convince ourselves that there cannot be a boundary layer near $x = 1$ by using the same kind of procedure as we previously used near $x = 0$. We should begin by taking $\xi = 1-x$ in the differential equation (4), "stretching" ξ in the same way as we stretched x in (18) (i.e. $X = \xi/\varepsilon$), and assuming an inner expansion of the type (20). We see at once that, with $A > 0$, positive exponentials occur in the solutions for the functions in the inner expansion, and it is impossible to carry out a matching procedure with an outer solution. In fact, of course, we knew already that that would be the case, and that the positive exponential occurs with a multiplier which, as $\varepsilon \to 0$, is asymptotically equal to zero (see Eq. (42)).

Now, still with $A > 0$, we assume there is a boundary layer at $x = 0$ but not at $x = 1$. We again assume outer and inner expansions of the types (19) and (20), respectively, with X defined by Eq. (18) as before. The differential equations satisfied by the g and the f are the same as before, namely Eqs. (21), (22), (24), (25), etc. But now the outer expansion must satisfy the boundary condition at $x = 1$, i.e.

$$g_0(1) = b, \qquad g_n(1) = 0 \quad \text{for} \quad n \geqslant 1. \tag{43}$$

The g are thus defined. The inner expansion must satisfy the boundary condition at $x = 0$, i.e.

$$f_0(0) = a, \quad f_n(0) = 0 \quad \text{for} \quad n \geqslant 1. \tag{44}$$

There is thus only one boundary condition for each f. The other condition for each f must come from the matching conditions, which are still given by (28), and can indeed be satisfied with $A > 0$.

The solution for g_0 with the given boundary condition is

$$g_0 = P + K_1 y_1, \tag{45}$$

with K_1 given by Eq. (37), or Eq. (39) if $\exp[-M(1)/\varepsilon]$ is neglected. The solution for f_0 with the boundary condition in (44) and the asymptotic value given by Eq. (28) is

$$\begin{aligned} f_0(X) &= g_0(0) + [a - g_0(0)]e^{-A(0)X} \\ &= g_0(0) + \frac{K_2}{A(0)} e^{-A(0)X}, \end{aligned} \tag{46}$$

since

$$g_0(0) = 1 + K_1 = 1 + E(1)[b - P(1)]. \tag{47}$$

We shall not continue the discussion of the inner and outer expansions to a further stage here. To this stage the composite expansion is

$$P + K_1 y_1 + \frac{K_2}{A(0)} \exp\,[-A(0)x/\varepsilon], \tag{48}$$

and may again be compared with the approximate solution obtained by the previous method.

We may remark in passing that the composite expansion may be found by starting from an assumed mixed expansion of the form

$$y = g_0(x) + h_0(X) + \varepsilon[g_1(x) + h_1(X)] + \ldots, \tag{49}$$

where the g are the same as before and $x = \varepsilon X$. The details are fairly easily worked out.

We may summarize the results for Eq. (4) as follows. One boundary condition is required for the functions in the outer expansion, and two for the functions in the inner expansion. If $y(0)$ and $y'(0)$ are given there is a boundary layer near $x = 0$ for $x > 0$, but not for $x < 0$, if $A > 0$, and conversely if $A < 0$; the inner expansion is fixed by the boundary conditions, and the necessary conditions for the outer expansion come from the matching conditions. If $y(0)$ and $y(1)$ are given there is a boundary layer at $x = 0$ but not at $x = 1$ if $A > 0$, and conversely if $A < 0$; the outer expansion is fixed by the boundary condition where there is no boundary layer; one boundary condition for each of the functions in the inner expansion is given at the boundary where there is a boundary layer, and a second condition for each of these functions is found from the matching conditions. In all cases all other requirements of the matching conditions are satisfied automatically.

Many special examples may now be worked out for special values of A, B and C. It is interesting and instructive to consider the simple case $A = 1$, $B = 0$. This has been rather fully and rigorously discussed by Erdelyi [5] with $y(0)$ and $y(1)$ given, and the discussion with $y(0)$ and $y'(0)$ given may, of course, be carried through similarly. Erdelyi remarks that this example is "virtually one that has been used by Lagerstrom for expository purposes". The full outer, inner, and composite expansions may be found, and the exact solution may be obtained explicitly in closed form. For expository purposes it is even worth while to take special simple expressions for C, for example $C = p + qe^{\alpha x}$ (p, q, α constants), so the equation discussed is

$$\varepsilon y'' + y' = p + qe^{\alpha x}. \tag{50}$$

The discussion is elementary, and need not be reproduced here. In general the inner and outer expansions are asymptotic but not convergent; in the simple case they converge (though not to the exact solution) and are also asymptotic.

It may also be pointed out that if we find the composite expansion by assuming a series of the type (49), then we may try to improve the composite expansion by applying the boundary conditions for the h at $x = 1$ exactly, without neglecting exponentially small terms. In general this improvement of the approximation is poor, but this is not so in the special case when A is a constant and B is zero. In fact in the simple case discussed here we recover the exact solution completely by finding the composite expansion in this way and inserting the boundary conditions correctly.

6. The results on the existence and location of boundary layers are easily generalized to the Eq. (3) of the nth degree. In this case n boundary conditions are required and they may be given at any number of points not exceeding n. All the boundary conditions may be given, for example, at $x = 0$, or some (or one) at $x = 0$ and the rest at $x = 1$ (say), or some (or one) at $x = 0$ and some (or one) at $x = 1$ and the rest at any number of points, not exceeding $n-2$, between $x = 0$ and $x = 1$.

The reduced equation, obtained by putting $\varepsilon = 0$ in Eq. (3), has a particular integral P and $n-1$ independent complementary functions, y_1, $y_2, \ldots, y_{n-1}$. In general these cannot be found explicitly in closed form. They are approximations to a particular integral and $n-1$ independent complementary functions of Eq. (3). In addition an approximation to the rapidly varying complementary function, y_n', of Eq. (3) is given by Eq. (14). In Eq. (14) take zero as the lower limit of integration in the integrals. An approximation to a general solution of Eq. (3) is given by

$$y = P+K_1y_1+K_2y_2+K_3y_3+\ldots+K_{n-1}y_{n-1}+K_ny_n. \tag{51}$$

If the values of y and its first $n-1$ derivatives are given at $x = 0$, then $K_1, K_2, \ldots, K_{n-1}$ are $O(1)$ and K_n is $O(\varepsilon^{n-1})$. When $A_{n-1} > 0$ there is a boundary layer at $x = 0$ for $x > 0$ but not for $x < 0$, and conversely when $A_{n-1} < 0$.

When $A_{n-1} > 0$, and boundary conditions are given partly at $x = 0$ and partly at other, positive, values of x, there is a boundary layer at $x = 0$ but not at any other, positive, value of x at which a boundary condition is given. If a boundary condition is given at $x = c > 0$, and we change the definitions of y_n and K_n by taking c as the lower limit of integration in the integrals in (14), then for $x < c$, y_n contains as a factor a positive exponential of an expression with ε^{-1} as a factor, but the new multiplier K_n now contains a negative exponential of a multiple of ε^{-1}. In boundary-layer parlance, if we look for an inner solution for x near to and less than c, it would involve a positive exponential of $A_{n-1}(c)$ $[c-x]/\varepsilon$, but we know that the multiplier would be exponentially small, and asymptotically equal to zero (cf. Eq. (42)). Of course, for $A_{n-1} < 0$, there would be a boundary layer at the largest value of x at which a boundary condition is given, and not elsewhere.

It is a straightforward matter, though rather tedious, to work out the functions f in the inner, or boundary-layer, solution, and the matching conditions of the inner and outer expansions. With $X = x/\varepsilon$, as in Eq. (18), expansions of the same forms, (19) and (20), are assumed. The equation for f_0 is

$$\frac{d^n f_0}{dX^n} + A_{n-1}(0)\frac{d^{n-1} f_0}{dX^{n-1}} = 0, \tag{52}$$

and the equations for the following f have the same operator on the left-hand side and increasingly complicated non-zero terms on the right. The matching conditions far f_0 and f_1 are the same as before (Eq. (28)). The solution for f_0 depends on the boundary conditions at $x = 0$, but must in any case be of the form constant+(constant) $\exp[-A_{n-1}(0)X]$; and so on. (It has, of course, been assumed that the boundary conditions do not depend on ε.)

7. We now return to the second-order equation (4), and consider cases in which A, while not identically zero, has a zero at $x = 0$, so that $x = 0$ lies in a transition zone. If a boundary condition is given at $x = 0$, there may be a layer that is both a boundary layer and part of a transition zone.

B may or may not be zero at $x = 0$. Let A and B be analytic, and, for some range of values of x near $x = 0$,

$$A = a_n x^n + a_{n+1}x^{n+1} + \dots \qquad (a_n \neq 0), \tag{53}$$

$$B = b_m x^m + b_{m+1}x^{m+1} + \dots \qquad (b_m \neq 0), \tag{54}$$

where n is a positive integer and m a positive integer or zero.

The study made here is to some extent connected with the study of solutions of Eq. (5) in a transition zone. Since $\chi_0 = A^2$ (Eq. (9)), χ_0 has a zero of even order at $x = 0$. When χ_0 has a zero of any order, and $\chi_1 = 0$, the first approximations in a transition zone and the connections across a transition zone have been studied [6], the analysis involving Bessel functions. When $\chi_1 = 0$, $A' = 2B$ (Eq. (9)), so $m-n+1 = 0$. The connection between the Bessel functions in the work cited and the confluent hypergeometric functions used here is clear. When χ_0 has a single zero the presence of χ_1 does not affect the method, which was given by Jeffreys and uses Airy functions; but when χ_0 has a double zero the result is modified if $\chi_1 \neq 0$, and parabolic cylinder functions may be used [7].

With A and B as in Eqs. (53) and (54), we shall consider three cases; (i) $m-n+1 > 0$, (ii) $m-n+1 = 0$, (iii) $m-n+1 < 0$. The last case is sufficiently illustrated by $n = m+2$, and in fact by $n = 2$, $m = 0$. The first case may be included in the second by putting b_m, and as many more of the b as necessary, equal to zero, but it is much simpler than the second case, and it is convenient to consider it separately, with b_m definitely not zero.

We shall be content to study the complementary functions ($C = 0$ in Eq. (4)) in general, leaving particular integrals to examples.

We begin by writing down approximate expressions near $x = 0$ for M, E, y_1, and y_2, as given by Eqs. (7) and (10).

$$M = \frac{a_n x^{n+1}}{n+1} + \frac{a_{n+1} x^{n+2}}{n+2} + \ldots, \tag{55}$$

with $M(0) = 0$. With

$$\nu = \frac{b_m}{a_n}, \qquad C_1 = \frac{1}{a_n}\left(b_{m+1} - \frac{a_{n+1}}{a_n} b_m\right), \tag{56}$$

$$\frac{B}{A} = \nu x^{m-n} + C_1 x^{m-n+1} + \ldots. \tag{57}$$

When $m-n+1$ is a positive integer

$$E = \exp \int (B/A)\, dx = 1 + \nu \frac{x^{m-n+1}}{m-n+1} + \ldots \qquad (\text{with } E(0) = 1), \tag{58}$$

$$y_1 = E^{-1} = 1 - \nu \frac{x^{m-n+1}}{m-n+1} + \ldots, \tag{59}$$

$$y_2 = \frac{E}{A} \exp\left(-\frac{M}{\varepsilon}\right) = \frac{1}{a_n x^n}(1+\ldots)\exp\left[-\frac{1}{\varepsilon}\left(\frac{a_n x^{n+1}}{n+1} + \ldots\right)\right]. \tag{60}$$

With

$$X = x/\varepsilon^{1/(n+1)}, \tag{61}$$

$$y_1 = 1 - \varepsilon^{\frac{m-n+1}{n+1}} \nu \frac{X^{m-n+1}}{m-n+1} + \ldots,$$

$$y_2 = \frac{\varepsilon^{-n/(n+1)}}{a_n X^n} \exp\left(-\frac{a_n}{n+1} X^{n+1}\right) [1 + O(\varepsilon^{1/(n+1)})]. \tag{62}$$

When $m-n+1 = 0$, we may take

$$E = x^{\nu}(1+C_1x+\ldots), \tag{63}$$

$$y_1 = x^{-\nu}(1-C_1x+\ldots), \quad y_2 = \frac{x^{\nu-n}}{a_n}(1+\ldots)\exp\left[-\frac{1}{\varepsilon}\left(\frac{a_nx^{n+1}}{n+1}+\ldots\right)\right]. \tag{64}$$

With X as in Eq. (61),

$$y_1 = \varepsilon^{-\nu/(n+1)}X^{-\nu}[1+O(\varepsilon^{1/(n+1)})],$$

$$y_2 = \frac{\varepsilon^{(\nu-n)/(n+1)}}{a_n}X^{\nu-n}\exp\left[-\frac{a_n}{n+1}X^{n+1}\right][1+O(\varepsilon^{1/(n+1)})]. \tag{65}$$

When $m-n+1 = -1$, $n = m+2$, we may take

$$E = e^{-\nu/x}x^{C_1}[1+(\text{const})\,x+\ldots], \tag{66}$$

$$y_1 = e^{\nu/x}x^{-C_1}[1-\ldots], \tag{67}$$

$$y_2 = e^{-\nu/x}\frac{x^{C_1-n}}{a_n}[1+\ldots]\exp\left[-\frac{1}{\varepsilon}\frac{a_nx^{n+1}}{n+1}+\ldots\right]. \tag{68}$$

The results for $n = m+3$, $m+4$, etc., may be written down in the same way. The approximations will be valid in a range of x that will depend on ε.

Now consider Eq. (4), with $C = 0$, for small x.

$$\varepsilon y''+(a_nx^n+a_{n+1}x^{n+1}+\ldots)y'+(b_mx^m+b_{m+1}y^{m+1}+\ldots)y = 0. \tag{69}$$

Now use a stretching factor $1/\delta$, and put $X = x/\delta$. The equation becomes

$$\frac{d^2y}{dX^2}+\frac{\delta^{n+1}}{\varepsilon}(a_nX^n+\delta a_{n+1}X^{n+1}+\ldots)\frac{dy}{dX}$$
$$+\frac{\delta^{m+2}}{\varepsilon}(b_mX^m+\delta b_{m+1}X^{m+1}+\ldots)y = 0. \tag{70}$$

Unless $m-n+1 = 0$, $n+1 = m+2$, the multipliers of the last two terms are of different orders of magnitude. In accordance with the spirit of "boundary-layer theory", we must take the larger multiplier to be $O(1)$.

Hence, for $m-n+1 > 0$, $m+2 > n+1$, we take $\delta = \varepsilon^{1/(n+1}$ and

$$X = x/\varepsilon^{1/(n+1)}. \tag{71}$$

For the inner expansion we now assume

$$y = f_0(X)+\delta f_1(X)+\ldots. \tag{72}$$

The equation for f_0 comes from the first two terms only and is

$$f_0''(X)+a_nX^nf_0'(X) = 0. \tag{73}$$

The multiplier of the last term in Eq. (70) is δ^{m-n+1}, and the equation for f_1 depends on whether $m-n+1 \geqslant 1$, $m \geqslant n$. If $m = n$ the equation for f_1 is

$$f_1''(X)+a_nX^nf_1'(X) = -a_{n+1}X^{n+1}f_0'-b_nX^nf_0. \tag{74}$$

If $m > n$, $m-n+1 > 1$, the last term is absent.

When a_n is positive, we may take as the two independent solutions for f_0

$$f_{01} = 1, \quad f_{02} = \int_X^{\infty} \exp\left(-\frac{a_nw^{n+1}}{n+1}\right) dw$$
$$= \frac{1}{a_nX^n} \exp\left(-\frac{a_nX^{n+1}}{n+1}\right)\left[1+O\left(\frac{1}{X^{n+1}}\right)\right] \tag{75}$$

for large positive X. f_{01} and f_{02} are also complementary functions for f_1. When $f_0 = f_{01} = 1$, a particular integral for f_1 is $-(b_n/a_n)X$, with $m = n$, as above.

We now change our previous notation for convenience, and suppose we are given that

$$y(0) = \alpha, \qquad y'(0) = \beta. \tag{76}$$

Then the boundary conditions for f_0 and f_1 are

$$f_0(0) = \alpha, \quad f_0'(0) = 0, \quad f_1(0) = 0, \quad f_1'(0) = \beta. \tag{77}$$

Now

$$f_{02}(0) = a_n^{-1/(n+1)}(n+1)^{-n/(n+1)}\Gamma\left(\frac{1}{n+1}\right), \quad f_{02}'(0) = -1, \tag{78}$$

and the solutions for f_0 and f_1 are

$$f_0 = \alpha, \quad f_1 = -\alpha\frac{b_n}{a_n}X+\left(\alpha\frac{b_n}{a_n}+\beta\right)[f_{02}(0)-f_{02}(X)]. \tag{79}$$

Hence, in "the boundary layer",

$$y = \alpha\left(1-\delta\frac{b_n}{a_n}X\right)+\delta\left(\alpha\frac{b_n}{a_n}+\beta\right)[f_{02}(0)-f_{02}(X)]$$
$$\sim\alpha\left(1-\delta\frac{b_n}{a_n}X\right)+\delta\left(\alpha\frac{b_n}{a_n}+\beta\right)\left[f_{02}(0)-\frac{1}{a_nX^n}\exp\left(-\frac{a_nX^{n+1}}{n+1}\right)\right] \tag{80}$$

for large X, which matches with

$$[\alpha+\delta f_{02}(0)(\alpha b_n/a_n+\beta)]y_1-\delta(\alpha b_n/a_n+\beta)\varepsilon^{n/(n+1)}y_2$$
$$= [\alpha+\varepsilon^{1/(n+1)}f_{02}(0)(\alpha b_n/a+\beta)]y_1-\varepsilon(\alpha b_n/a_n+\beta)y_2, \tag{81}$$

where y_1 and y_2 are given by Eqs. (59) and (60), or (62), and then away from $x = 0$ by Eq. (10). This is for $m = n$. If $m > n$, $m-n+1 > 1$, the results are the same as if we simply put $b_n = 0$ above.

All this is very similar to the case $m = 0$, $n = 0$, where A and B have no zeros at $x = 0$. There, with $n = 0$, $m = 0$, both the inner and outer expansions are in powers of ε. Here the expansions are in powers of δ, although in the outer expansion in powers of δ all the terms up to and including that in δ^n are multiples of the first term, which is here y_1.

Our primary interest here is in boundary layers, with boundary conditions given at $x = 0$, but before we leave this case we may note that the results for f_{01} and f_{02} may also be used to find connections between the approximations on the two sides of the transition zone near $x = 0$. Note that for large $|X|$, y_1/y_2 is either exponentially small or exponentially large. It is exponentially small for $X > 0$, and also $X < 0$ if $n+1$ is even, n odd; it is exponentially large if $X < 0$ and n even. So there is necessarily indeterminacy in the connections. The only accurate solutions which are uniquely determined by the asymptotic approximations are those approximated by y_2 for $X > 0$, or for $X < 0$, n odd, and by y_1 for $X < 0$, n even.

Also it may be shown that when X is large and negative and n odd,

$$f_{02}(X) = 2f_{02}(0) - f_{02}(-X) \sim 2f_{02}(0) + \frac{1}{a_n X^n} \exp\left(-\frac{a_n X^{n+1}}{n+1}\right), \tag{82}$$

but when X is large and negative and n even, f_{02} is exponentially large and

$$f_{02}(X) - f_{02}(0) \sim \frac{1}{a_n X^n} \exp\left(-\frac{a_n X^{n+1}}{n+1}\right). \tag{83}$$

It follows that if we start from the side of x positive, then, whether n is odd or even, y_1 is matched with $f_{01}(X)$ in the transition layer and with y_1 for x negative outside the transition layer; y_2 is matched with $\varepsilon^{-n/(n+1)} f_{02}(X)$ in the transition layer and, when n is odd, with $2\varepsilon^{-n/(n+1)} f_{02}(0) y_1 + y_2$ for x negative, but when n is even it is matched with $\varepsilon^{-n(n+1)} f_{02}(0) y_1 + y_2$ for x negative; in the last case y_2/y_1 is exponentially large just outside the transition region for x negative, and the multiple of y_1 is not relevant.

If we start from the side of x negative, and consider only the solutions that are, relatively, small just outside the transition layer, then we see that for n even, y_1 is matched with f_{01} in the transition layer and with y_1 for x positive; for n odd, y_2 is matched with $\varepsilon^{-n/(n+1)}[f_{02}(X) - 2f_{02}(0)]$ in the transition layer and with $y_2 - 2\varepsilon^{-n/(n+1)} f_{02}(0) y_1$ for x positive.

Then connections are for $a_n > 0$; the connections when $a_n < 0$ can be discussed similarly.

8. Next, consider the case $m - n + 1 = 0$, so that $n+1 = m+2$ and the second and the third terms in Eq. (70) are of the same order of magnitude.

All three terms must therefore be of the same order; again we must take $\delta = \varepsilon^{1/(n+1)}$ and $X = x/\varepsilon^{1/(n+1)}$, as in Eq. (71). For the inner expansion we assume a series of the same form, Eq. (72), as before, but the equations for f_0 and f_1 are now

$$f_0''(X)+a_nX^nf_0'(X)+b_{n-1}X^{n-1}f_0(X) = 0, \tag{84}$$

$$f_1''(X)+a_nX^nf_1'(X)+b_{n-1}X^{n-1}f_1(X) = -(a_{n+1}X^{n+1}f_0'+b_nX^nf_0). \tag{85}$$

Two independent solutions for f_0, which are also independent complementary functions for f_1, etc., are easily found as hypergeometric series. They are

$$f_{01} = {}_1F_1\left\{\frac{b_{n-1}}{(n+1)a_n};\frac{n}{n+1};-\frac{a_nx^{n+1}}{n+1}\right\}$$

and

$$f_{02} = x_1F_1\left\{\frac{a_n+b_{n-1}}{(n+1)a_n};\frac{n+2}{n+1};-\frac{a_nx^{n+1}}{n+1}\right\}, \tag{86}$$

where

$${}_1F_1(a;\, b;\, x) = 1+\frac{a}{1!b}x+\frac{a(a+1)}{2!b(b+1)}x^2+\dots. \tag{87}$$

(If we put $b_{n-1} = 0$ we recover the previous case, with $m = n$.) The analysis may be continued along these lines. I have considered fully only the case $n = 1$, $m = 0$, when

$$X = x/\varepsilon^{\frac{1}{2}}, \tag{88}$$

and

$$f_0''(X)+a_1Xf_0'(X)+b_0f_0(X) = 0, \tag{89}$$

$$f_1''(X)+a_1Xf_1'(X)+b_0f_1(X) = -(a_2X^2f_0'+b_1Xf_0). \tag{90}$$

It is convenient to work in terms of Weber's parabolic cylinder functions $D_\nu(x)$. (See Whittaker and Watson's *Modern Analysis* [8], §§ 16.5 et seq.) Write

$$\nu = b_0/a_1 \tag{91}$$

as in Eq. (56) (ν may be positive or negative or, in a degenerate case, zero). *When a_1 is positive* write

$$\xi^2 = a_1X^2 = a_1x^2/\varepsilon. \tag{92}$$

Then

$$\frac{d^2f_0}{d\xi^2}+\xi\frac{df_0}{d\xi}+\nu f_0 = 0, \tag{93}$$

$$\frac{d^2f_1}{d\xi^2}+\xi\frac{df_1}{d\xi}+\nu f_1 = -a_1^{-3/2}\left(a_2\xi^2\frac{df_0}{d\xi}+b_1\xi f_0\right). \tag{94}$$

Two independent solutions of Eq. (93) for f_0, which are also complementary functions for f_1, are $e^{-\frac{1}{4}\xi^2} D_{\nu-1}(\xi)$ and $e^{-\frac{1}{4}\xi^2} D_{-\nu}(i\xi)$. Also $e^{-\frac{1}{4}\xi^2} D_\nu(-\xi)$ is a solution. But $D_{-\nu}(i\xi)$ is complex (neither purely real nor purely imaginary) when ξ is real, unless ν is zero or a negative integer, and $D_\nu(-\xi) = (-1)^\nu D_\nu(\xi)$ when ν is a positive integer or zero. Therefore we take a new standard second solution, $e^{-\frac{1}{4}\xi^2} F_\nu(\xi)$, where

$$F_\nu(\xi) = e^{\frac{1}{2}\nu\pi i} D_{-\nu}(i\xi) - i\left(\frac{\pi}{2}\right)^{\frac{1}{2}} \frac{D_{\nu-1}(\xi)}{\Gamma(\nu)}$$

$$= \tfrac{1}{2}\left[e^{\frac{1}{2}\nu\pi i} D_{-\nu}(i\xi) + e^{-\frac{1}{2}\nu\pi i} D_{-\nu}(-i\xi)\right]$$

$$= e^{-\frac{1}{2}\nu\pi i} \cos \nu\pi D_{-\nu}(i\xi) + \left(\frac{\pi}{2}\right)^{\frac{1}{2}} \frac{e^{-(\nu-\frac{1}{2})\pi i}}{\Gamma(\nu)} D_{\nu-1}(-\xi). \qquad (95)$$

The second expression on the right in Eq. (95) follows from the first since

$$e^{\frac{1}{2}\pi i} D_{\nu-1}(\xi) = \frac{\Gamma(\nu)}{(2\pi)^{1/2}} \left[e^{\frac{1}{2}\nu\pi i} D_{-\nu}(i\xi) - e^{-\frac{1}{2}\nu\pi i} D_{-\nu}(-i\xi)\right]. \qquad (96)$$

It follows from the second expression that $F_\nu(\xi)$ is real for any real ξ and ν. Also when ν is zero or a negative integer, $F_\nu(\xi)$ reduces to $e^{\frac{1}{2}\nu\pi i} D_{-\nu}(i\xi)$. Moreover,

$$D_{\nu-1}(0) \equiv \frac{2^{\frac{1}{2}\nu-\frac{1}{2}}\pi^{\frac{1}{2}}}{\Gamma(1-\frac{1}{2}\nu)}, \quad F_\nu(0) = \frac{\pi^{\frac{1}{2}}}{2^{\frac{1}{2}\nu}\,\Gamma(\frac{1}{2}+\frac{1}{2}\nu)} \cos \tfrac{1}{2}\nu\pi, \qquad (97)$$

and from the recurrence formulae for the D,

$$\frac{d}{d\xi}\left[e^{-\frac{1}{4}\xi^2} D_{\nu-1}(\xi)\right] = -e^{-\frac{1}{4}\xi^2} D_\nu(\xi);$$

$$\frac{d}{d\xi}\left[e^{-\frac{1}{4}\xi^2} F_\nu(\xi)\right] = -\nu e^{-\frac{1}{4}\xi^2} F_{\nu+1}(\xi). \qquad (98)$$

Hence if we write

$$f_{01} = e^{-\frac{1}{4}\xi^2} D_{\nu-1}(\xi), \quad f_{02} = e^{-\frac{1}{4}\xi^2} F\nu(\xi) \qquad (99)$$

for the two independent solutions of Eq. (93) or complementary functions of Eq. (94),

$$\left.\begin{aligned} f_{01}(0) &= \frac{2^{\frac{1}{2}\nu-\frac{1}{2}}\pi^{\frac{1}{2}}}{\Gamma(1-\frac{1}{2}\nu)}, \quad f_{01}'(0) = -\frac{2^{\frac{1}{2}\nu}\pi^{\frac{1}{2}}}{\Gamma(\frac{1}{2}-\frac{1}{2}\nu)}, \\ f_{02}(0) &= \frac{\pi^{\frac{1}{2}}}{2^{\frac{1}{2}\nu}\Gamma(\frac{1}{2}+\frac{1}{2}\nu)}\cos\tfrac{1}{2}\nu\pi, \quad f_{02}'(0) = \frac{\pi^{\frac{1}{2}}}{2^{\frac{1}{2}\nu-\frac{1}{2}}\Gamma(\frac{1}{2}\nu)}\sin\tfrac{1}{2}\nu\pi, \end{aligned}\right\} \tag{100}$$

so that

$$f_{01}(0)f_{02}'(0)-f_{01}'(0)f_{02}(0) = \sin^2\tfrac{1}{2}\nu\pi+\cos^2\tfrac{1}{2}\nu\pi = 1. \tag{101}$$

If now the values of y and dy/dx at $x = 0$ are α and β, as before, since $\xi = (a_1/\varepsilon)^{\frac{1}{2}}\, x$, the boundary conditions for f_0 and f_1 are

$$f_0(0) = \alpha, \quad f_0'(0) = 0, \quad f_1(0) = 0, \quad f_1'(0) = \beta/a_1^{1/2}. \tag{102}$$

Hence the solution for f_0 is

$$f_0 = \alpha[f_{02}'(0)f_{01}(\xi)-f_{01}'(0)f_{02}(\xi)]. \tag{103}$$

To find a particular integral for f_1 in Eq. (94) note that

$$\xi D_\nu(\xi) = D_{\nu+1}(\xi)+\nu D_{\nu-1}(\xi), \tag{104}$$

$$\xi^2 D_\nu(\xi) = D_{\nu+2}(\xi)+(2\nu+1)D_\nu(\xi)+\nu(\nu-1)D_{\nu-2}(\xi), \tag{105}$$

and it may be shown that

$$\xi F_\nu(\xi) = \nu F_{\nu+1}(\xi)+F_{\nu-1}(\xi), \tag{106}$$

$$\xi^2 F_\nu(\xi) = \nu(\nu+1)F_{\nu+2}(\xi)+(2\nu-1)F_\nu(\xi)+F_{\nu-2}(\xi). \tag{107}$$

Also

$$\left[\frac{d^2}{d\xi^2}+\xi\frac{d}{d\xi}+\nu\right]\left[e^{-\frac{1}{4}\xi^2}D_{\mu-1}(\xi)\right] = (\nu-\mu)\,e^{-\frac{1}{4}\xi^2}\,D_{\mu-1}(\xi), \tag{108}$$

$$\left[\frac{d^2}{d\xi^2}+\xi\frac{d}{d\xi}+\nu\right]\left[e^{-\frac{1}{4}\xi^2}F_\mu(\xi)\right] = (\nu-\mu)\,e^{-\frac{1}{4}\xi^2}\,F_\mu(\xi). \tag{109}$$

Thus if we take $f_0 = f_{01}$ (Eq. (99)) a particular integral for f_1 in Eq. (94) would be

$$\begin{aligned} P_{11}(\xi) = -\frac{1}{a_1^{3/2}}e^{-\frac{1}{4}\xi^2}\Big\{&\frac{a_2}{3}D_{\nu+2}(\xi)+[(2\nu+1)a_2-b_1]D_\nu(\xi) \\ &+[\nu-1][b_1-\nu a_2]D_{\nu-2}(\xi)\Big\}, \end{aligned} \tag{110}$$

and with $f_0 = f_{02}$ (Eq. (99)) a particular integral for f_1 would be

$$P_{12}(\xi) = -\frac{1}{a_1^{3/2}} e^{-\frac{1}{4}\xi^2} \Big\{ \nu(\nu+1)(\nu+2)\frac{a_2}{3} F_{\nu+3}(\xi) + \nu[(2\nu+1)a_2 - b_1]F_{\nu+1}(\xi) + [b_1 - \nu a_2]F_{\nu-1}(\xi) \Big\}. \tag{111}$$

For the value of f_0 in Eq. (103) a particular integral for f_1 is therefore

$$P(\xi) = \alpha[f'_{02}(0)P_{11}(\xi) - f'_{01}(0)P_{12}(\xi)], \tag{112}$$

and with the boundary conditions for f_1 in Eq. (102) the solution for f_1 is

$$f_1(\xi) = P(\xi) - P(0)\{f'_{02}(0)f_{01}(\xi) - f'_{01}(0)f_{02}(\xi)\} + \left\{\frac{\beta}{a_1^{1/2}} - P'(0)\right\}\{f_{02}(0)f_{01}(\xi) - f_{01}(0)f_{02}(\xi)\}. \tag{113}$$

Note that for ξ large and positive

$$f_{01}(\xi) = e^{-\frac{1}{4}\xi^2} D_{\nu-1}(\xi) \sim e^{-\frac{1}{2}\xi^2}\xi^{\nu-1}\left\{1 - \frac{(\nu-1)(\nu-2)}{2\xi^2} + \dots\right\}$$

$$= e^{-\frac{1}{2}a_1X^2} a_1^{\frac{1}{2}(\nu-1)} X^{\nu-1}\left\{1 - \frac{(\nu-1)(\nu-2)}{2a_1X^2} + \dots\right\}, \tag{114}$$

and

$$f_{02}(\xi) = e^{-\frac{1}{4}\xi^2} F_\nu(\xi) \sim \xi^{-\nu}\left\{1 + \frac{\nu(\nu+1)}{2\xi^2} + \dots\right\}$$

$$= a_1^{-\frac{1}{2}\nu} X^{-\nu}\left\{1 + \frac{\nu(\nu+1)}{2a_1X^2} + \dots\right\}, \tag{115}$$

where exponentially small terms in f_{02} have been neglected. We may now consider the matching with an outer expansion. Here, unlike the previous case, the exponentially small terms in f_0 are not identically zero, and we neglect exponentially small terms in matching with the solution outside the boundary layer, so we neglect y_2, which is exponentially small compared with y_1, just outside the boundary layer. (y_2 and y_1 are given by Eqs. (64) and (65) with $n = 1$.) In fact we adopt the usual boundary-layer procedure, and assume an outer expansion. If g_0 is the leading term in the outer expansion, g_0 is simply a multiple of y_1. Now if exponentially small terms are neglected

$$f_0 \sim -\alpha f'_{01}(0)\, a^{-\frac{1}{2}\nu} X^{-\nu}\left\{1 + \frac{\nu(\nu+1)}{2a_1X^2} + \dots\right\} \tag{116}$$

(from Eqs. (103), (114) and (115)), so we should take $-\alpha f'_{01}(0)a^{-\frac{1}{2}\nu}\varepsilon^{\frac{1}{2}\nu}X^{-\nu}$ as the leading term in g_0. Thus for the outer expansion it is better to change the notation, and consider the expansion of $\varepsilon^{-\frac{1}{2}\nu} y_{\text{outer}}$. The leading term will be $-\alpha f'_{01}(0) a_1^{-\frac{1}{2}\nu} y_1$, but at the next approximation the multiplier of y_1 will be altered by a term of order $\varepsilon^{\frac{1}{2}}$. Thus we may write

$$\varepsilon^{-\frac{1}{2}\nu} y_{\text{outer}} = \left[-\nu f'_{01}(0)\, a_1^{-\frac{1}{2}\nu} + O\left(\varepsilon^{\frac{1}{2}}\right)\right] y_1 + \varepsilon g_1 + \dots \quad (117)$$

Note the term $-C_1 x^{-\nu+1}$ in y_1.

We may consider how the matching would proceed without going through all the computations. The term of order $\varepsilon^{\frac{1}{2}}$ in the multiplier of y_1 in Eq. (117) will lead to a multiple of $\varepsilon^{\frac{1}{2}} X^{-\nu}$ and the term $-C_1 x^{-\nu+1}$ in y_1 to a multiple of $\varepsilon^{\frac{1}{2}} X^{-\nu+1}$, in fact to $\varepsilon^{\frac{1}{2}} \alpha f'_{01}(0) a_1^{-\frac{1}{2}\nu} C_1 X^{-\nu+1}$, and both of these must come from the asymptotic expansion of f_1 for large X. If x is $O(1)$, to begin with, g_1 gives a term of order $\varepsilon^{\frac{1}{2}\nu+1}$ in y. So as regards algebraic terms in the asymptotic expansions of f_0 and f_1 for large X, the matching will be with terms in $X^{-\nu-2}$, or $\varepsilon^{\frac{1}{2}} X^{-\nu-1}$, or $\varepsilon X^{-\nu}$, and so on. The first would match with a term in the asymptotic expansion of f_0, the second with a term in f_1, the third with a term in f_2, and so on. Consideration of f_2 would be needed for the third.

The term in $X^{-\nu-2}$ in the asymptotic expansion of f_0 is, in fact, shown in Eq. (116), and this must match with a term in εg_1 in Eq. (117). Consider the asymptotic expansion of f_1. (See Eq. (113).) $f_{01}(\xi)$ and $P_{11}(\xi)$ are both exponentially small, and are neglected. From Eqs. (111) and (115) we see that $P_{12}(\xi)$ leads to terms in $X^{-\nu+1}$, $X^{-\nu-1}$, $X^{-\nu-3}$, and so on, while $f_{02}(\xi)$ leads to terms in $X^{-\nu}$, $X^{-\nu-2}$, and so on. The first of the terms in $\varepsilon^{\frac{1}{2}} f_1$ from P_{12} should give correctly the contribution from $-C_1 x^{-\nu+1}$ in y_1, as given above. The second and third of the terms in $\varepsilon^{\frac{1}{2}} f_1$ from P_{12} provide terms that match with terms from εg_1 and $\varepsilon^2 g_2$, respectively. The first of the terms in $\varepsilon^{\frac{1}{2}} f_1$ from $f_{02}(\xi)$ must match with the term of order $\varepsilon^{\frac{1}{2}}$ in the multiplier of y_1 in Eq. (117); the second of the terms from $f_{02}(\xi)$ alters the multiplier of g_1 (by a term of order $\varepsilon^{\frac{3}{2}}$ in Eq. (117)), and so on. So far the "bookkeeping" works out correctly. Moreover, if we insert the leading terms of the asymptotic expansions of $P_{12}(\xi)$ and $f_{02}(\xi)$ we find that

$$f_1 = \alpha f'_{01}(0) a^{-\frac{1}{2}\nu-1} (b_1 - \nu a_2) X^{-\nu+1}$$
$$+ a_1^{-\frac{1}{2}\nu} \left\{ P(0) f'_{01}(0) + \left[P'(0) - \frac{\beta}{a_1^{1/2}}\right] f_{01}(0) \right\} X^{-\nu} + O(X^{-\nu-1}). \quad (118)$$

The first term is just $\alpha f'_{01}(0) a_1^{-\frac{1}{2}\nu} C_1 X^{-\nu+1}$, with C_1 given by Eq. (56) with

$n = 1, m = 0$, as it should be. The second term gives the value of the term of order $\varepsilon^{\frac{1}{2}}$ in the multiplier of y_1 in Eq. (117).

If we wish to consider connections across the transition layer near $x = 0$ instead of a boundary-value problem, this is easily done, since the asymptotic expansions of $f_{01}(\xi)$ and $f_{02}(\xi)$ for large negative ξ are known. The asymptotic expansion of $f_{01}(\xi)$ has different forms according as ν is, or is not, a positive integer, while that of $f_{02}(\xi)$ has different forms according as $\nu - \frac{1}{2}$ is, or is not, an integer. It turns out that the first approximation to the solution when ξ is large and negative is

$$\frac{(2\pi)^{\frac{1}{2}}}{\Gamma(1-\nu)} f_{02} - f_{01} \cos \nu\pi$$

but we shall not consider any details here.

9. If we still take $n = 1$, $m = 0$, but a_1 is negative, it is advisable to start again and define ξ by

$$\xi = -a_1 X^2 = -a_1 x^2/\varepsilon \tag{119}$$

instead of Eq. (92). The equation for f_0 is

$$\frac{d^2 f_0}{d\xi^2} - \xi \frac{d f_0}{d\xi} - \nu f_0 = 0, \tag{120}$$

with two independent solutions

$$f_{01}(\xi) = e^{\frac{1}{4}\xi^2} D_{-\nu}(\xi), \quad f_{02}(\xi) = e^{\frac{1}{4}\xi^2} F_{-\nu+1}(\xi). \tag{121}$$

For further analysis we should need the formulae

$$\frac{d}{d\xi}\left[e^{\frac{1}{4}\xi^2} D_{-\nu}(\xi)\right] = -\nu e^{\frac{1}{4}\xi^2} D_{-\nu+1}(\xi), \tag{122}$$

$$\frac{d}{d\xi}\left[e^{\frac{1}{4}\xi^2} F_{-\nu+1}(\xi)\right] = e^{\frac{1}{4}\xi^2} F_{-\nu}(\xi). \tag{123}$$

However, for $x > 0$, y_2/y_1 is now exponentially large, and if $y(0)$, $y'(0)$ are given, the solution for $x > 0$ away from $x = 0$ will, in general, involve the rapidly varying, exponentially increasing function y_2, and the usual method of finding outer expansions fails completely. It no longer makes sense to talk about a boundary layer at all for $x > 0$. However, there is still a transition layer, and we could still consider the connections across this layer. There is another interesting question that can be asked. We can assume that the solution in which we are interested is the slowly varying solution away from $x = 0$, so that in fact there can be a boundary layer, and we can ask for (say) the values of the solution and its derivative at $x = 0$. This can also be done with a non-homogeneous equation, with a particular integral involved. If $A(x)$, $B(x)$, and $C(x)$ are of the correct form, the solution outside the transition layer is fixed (i.e. the "outer" solution is fixed) by its asymptotic expression at infinity. It is necessary that the equations for the functions g_1, g_2, etc., in an outer expansion should have solutions of a smaller order at $x = \infty$ than the one selected for g_0. Examples of this must, however, be left to another publication, and I shall conclude the paper with a preliminary consideration of a case in which $m-n+1 < 0$, namely $n = 2$, $m = 0$, which is characteristic of this class, for which the usual method of matching an inner and an outer solution fails completely.

Before proceeding I should first like to make another remark about the functions F_ν introduced in the preceding section.

10. The connection with the parabolic cylinder functions, D_{-n}, of the repeated integrals of $e^{-\frac{1}{2}x^2}$ and of the complementary error function is presumably well known. If $L_0(x) = e^{-\frac{1}{2}x^2}$ and

$$L_n(x) = \int_x^\infty L_{n-1}(w)dw \quad \text{for} \quad n \geqslant 1 \tag{124}$$

(n integral), then

$$L_n(x) = e^{-\frac{1}{4}x^2} D_{-n}(x). \tag{125}$$

The purpose of this paragraph is to point out that the functions $F_n(x)$ previously introduced are connected with the repeated integrals of $e^{\frac{1}{2}x^2}$. The proof is simple, and will not be given here. The result is as follows:

If $M_0(x) = e^{\frac{1}{2}x^2}$ and

$$M_n(x) = \int_0^x M_{n-1}(w)dw \quad \text{for} \quad n \geqslant 1, \tag{126}$$

then

$$\left.\begin{aligned} M_0(x) &= e^{\frac{1}{4}x^2} F_0(x), \quad M_1(x) = e^{\frac{1}{4}x^2} F_1(x), \\ M_2(x) &= e^{\frac{1}{4}x^2} F_2(x)+1, \quad M_3(x) = e^{\frac{1}{4}x^2} F_3(x)+x, \end{aligned}\right\} \tag{127}$$

and, in general,

$$M_{2n-1}(x) = e^{\frac{1}{4}x^2}F_{2n-1}(x)+\frac{x^{2n-3}}{(2n-3)!}-\frac{1}{1.3.}\frac{x^{2n-5}}{(2n-5)!}+\frac{1}{1.3.5.}\frac{x^{2n-7}}{(2n-7)!}-\ldots+\frac{(-1)^n}{1.3.5\cdots(2n-3)}x$$

$$M_{2n}(x) = e^{\frac{1}{4}x^2}F_{2n}(x)+\frac{x^{2n-2}}{(2n-2)!}-\frac{1}{1.3.}\frac{x^{2n-4}}{(2n-4)!}+\frac{1}{1.3.5.}\frac{x^{2n-6}}{(2n-6)!}-\ldots+\frac{(-1)^n}{1.3.5.\cdots(2n-3)}\frac{x^2}{2!}+\frac{(-1)^{n+1}}{1.3.5\cdots(2n-1)} \tag{128}$$

for $n \geqslant 2$.

11. When $m-n+1 < 0$, the method of using a stretched independent variable for an inner expansion and matching with an outer expansion in terms of the unstretched independent variable breaks down completely.

The circumstances are sufficiently illustrated for the present by taking $n = 2$, $m = 0$—in fact by putting $A(x) = x^2$, $B(x) = 1$, $C(x) = 0$ in equation (4) and considering

$$\varepsilon y''+x^2y'+y = 0. \tag{129}$$

Here $A > 0$, and let us begin by assuming that we are given that

$$y(0) = \alpha, \quad y'(0) = \beta, \tag{130}$$

and seek for an inner expansion. If we assume a stretching factor δ, and change to an independent variable $X = x/\delta$, the equation becomes

$$\frac{d^2y}{dX^2}+\frac{\delta^3}{\varepsilon}X^2\frac{dy}{dX}+\frac{\delta^2}{\varepsilon}y = 0. \tag{131}$$

The third term is of a higher order of magnitude than the second, so with the usual procedure we must take $\delta = \varepsilon^{\frac{1}{2}}$, write

$$X_2 = x/\varepsilon^{\frac{1}{2}}, \tag{132}$$

and

$$\frac{d^2y}{dX_2^2}+y+\varepsilon^{\frac{1}{2}}X_2^2\frac{dy}{dX_2} = 0. \tag{133}$$

If we now assume an inner expansion

$$y_{\text{inner}} = f_0(X_2)+\varepsilon^{\frac{1}{2}}f_1(X_2)+\varepsilon f_2(X_2)+\ldots, \tag{134}$$

then

$$\frac{d^2f_0}{dX_2^2}+f_0 = 0, \quad \frac{d^2f_n}{dX_2^2}+f_n = -X_2^2\frac{df_{n-1}}{dX_2}, \tag{135}$$

and the boundary conditions are

$$f_0(0) = \alpha,\quad f_0'(0) = 0,\quad f_1(0) = 0,\quad f_1'(0) = \beta,\quad f_n(0) = f_n'(0) = 0 \quad \text{for}\quad n \geqslant 2. \tag{136}$$

The solutions for f_0 and f_1 are easily found, and to order $\varepsilon^{\frac{1}{2}}$ (with X_2 of order unity)

$$y_{\text{inner}} = \Re\left[e^{iX_2}\left\{\alpha\left[1-\varepsilon^{\frac{1}{2}}\left(\frac{X_2^3}{6}+\frac{iX_2^2}{4}-\frac{X_2}{4}\right)\right]-i\varepsilon^{\frac{1}{2}}\left(\beta-\frac{\alpha}{4}\right)\right\}\right], \tag{137}$$

where $\Re$ denotes "the real part of". (137) is a valid asymptotic approximation for small ε if X_2 is bounded as $\varepsilon \to 0$, and it clearly cannot be valid when $\varepsilon^{\frac{1}{2}}\ X_2^3$ is not bounded, i.e. when $x/\varepsilon^{\frac{1}{3}}$ is not bounded; it cannot be valid when $x = O(\varepsilon^m)$ if $m < \frac{1}{3}$.

Note next that away from $x = 0$ the approximate solutions y_1 and y_2 (Eq. (10)) are

$$y_1 = \exp(1/x),\quad y_2 = x^{-2}\exp(-x^3/3\varepsilon - 1/x) \tag{138}$$

(cf. Egs. (67) and (68); note $n = 2$, $m = 0$, $\nu = 1$, $C_1 = 0$.) y_1 is the slowly varying approximate solution. y_2 is exponentially small for positive x unless x is of order $\varepsilon^{\frac{1}{3}}$ or smaller.

If we assume an outer expansion

$$y_{\text{outer}} = K(\varepsilon)[g_0(x)+\varepsilon g_1(x)+\varepsilon^2 g_2(x)+\ldots], \tag{139}$$

then

$$\left.\begin{aligned} g_0 = y_1 = \exp(1/x)_1 \quad & xg_n'+g_n = -\varepsilon g_{n-1}'' \quad \text{for}\quad n \geqslant 1,\\ e^{-1/x}g_1 = \frac{1}{5x^5}+\frac{2}{2x^4},\quad & e^{-1/x}g_2 = \frac{1}{50}\frac{1}{x^{10}}+\frac{29}{90}\frac{1}{x^9}+\frac{11}{8}\frac{1}{x^8}+\frac{10}{7}\frac{1}{x^7},\end{aligned}\right\} \tag{140}$$

and so on. The lowest power of x in $e^{-1/x}g_n$ is x^{-5n}, and the highest is $x^{-(3n+1)}$.

Equation (139) is a valid approximation if x is independent of ε. It cannot be valid when $x \to 0$ as $\varepsilon \to 0$ unless x is at least of an order as large as $\varepsilon^{\frac{1}{5}}$.

Therefore there can be no matching between these crude inner and outer expansions. In fact, if $x = \delta X$, $g_0 = y_0 = \exp(1/\delta X)$ has an essential singularity in δ at $\delta = 0$, and cannot be expanded in any way.

Usually, although there is no overlap in the domains of validity of the inner and outer expansions, the least order of magnitude of x for the possible validity of the outer expansion is at any rate the same as the largest order of magnitude of x for the possible validity of the inner expansion. (Here they are $\varepsilon^{\frac{1}{5}}$ and $\varepsilon^{\frac{1}{3}}$.) In fact, in the simplest cases the matching is done simply by discarding terms like $e^{-x/\delta}$ in the outer expansion and e^{-X} in the inner expansion. In more complicated cases the statement above has remained true when matching was possible.

Therefore the boundary-layer thickness cannot be said to be of order $\varepsilon^{\frac{1}{2}}$. The following questions may be asked. (1) What may be said to be the order of magnitude of the boundary-layer thickness? (2) What is the structure of the solution in the boundary-layer? (3) With the given boundary conditions (130) what is the multiple, $K(\varepsilon)$, of y_1 or g_0 to which the solution approximates outside the boundary layer? The last question, being strictly quantitative, is probably the hardest to answer.

The answers to these questions must be left to another publication. This will also contain the examples mentioned at the end of Section 9. Equations such as

$$\varepsilon^2 y'' + \varepsilon A(x) y' + B(x) y = 0$$

and

$$\varepsilon y'' + \varepsilon A(x) y' + B(x) y = 0$$

will also be discussed, with examples and remarks about generalizations to equations of the nth order, and an application of the method of singular perturbations to at least one non-linear ordinary differential equation will be set out.

The work reported here was begun when I was a Visiting Research Fellow at the Australian National University in Canberra. Its revision and the preparation of the paper were supported by the U.S. Office of Naval Research.

REFERENCES

1. S. Kaplun and P. A. Lagerstrom, "Asymptotic expansions of Navier–Stokes solutions for small Reynolds numbers"; S. Kaplun, "Low Reynolds number flow past a circular cylinder"; P. A. Lagerstrom, "Note on the preceding two papers", *J. Math. Mech.* **6,** 585–606 (1957).
2. P. A. Lagerstrom and J. D. Cole, "Examples illustrating expansion procedures for the Navier–Stokes equations", *J. Rat. Mech. Anal.* **4,** 817–82 (1955).
3. H. Jeffreys and B. S. Jeffreys, *Methods of Mathematical Physics*, Cambridge University Press (1946).
4. H. Jeffreys, *Asymptotic Approximations*, Clarendon Press (1962).
5. A. Erdelyi, "An expansion procedure for singular perturbations", *Atti Accad. delle Scienza di Torino*, **95** (1), 651–72 (1960–1).
6. S. Goldstein, "A note on certain approximate solutions of linear differential equations of the second order, with an application to the Mathieu equation", *Proc. London Math. Soc.* **(2) 28,** 81–90 (1928).
7. S. Goldstein, "A note on certain approximate solutions of linear differential equations of the second order" (2), *Proc. London Math. Soc.* **(2) 33,** 246–52 (1931).
8. E. T. Whittaker and G. N. Watson, *Modern Analysis*, Cambridge University Press (1948).

THE GENERAL STRAIN TENSOR

Z. KARNI

Department of Mechanics, Technion—Israel Institute of Technology, Haifa

ABSTRACT

Based on the most general isotropy between two second-order asymmetric tensors, shown to be reducible to a finite form, a survey of the general theory for the isotropic strain tensor is presented. Application is specifically made to the first- and second-order strain theory. To motivate the directions in three-dimensional space of the vector invariants of the asymmetric tensors, the absolute notation for vectors and tensors is preferred.

NOTATION

$\mathbf{A}$	asymmetric tensor	$\mathbf{T}$	tensor
$\mathbf{a}$	vector invariant	$\mathbf{u}$	displacement vector
a_i	scalar invariants	x^i	rectilinear coordinates
$\mathbf{E}$	strain tensor	x_i	projection magnitudes
E	$[\mathbf{e}_i\mathbf{e}_j\mathbf{e}_k]$	$\mathbf{\Gamma}/m$	displacement gradient
$\mathbf{e}_i$	basis	δ	Kronecker delta
$\mathbf{e}^i$	reciprocal vectors	$\mathbf{\theta}/m$	rotation
f, g	polynomial functions	$\mathbf{\Lambda}/m$	coordinate gradient
$\mathbf{G}$	symmetric tensor	π	product
h	unit tensor magnitude	ℵ, ב, ג	scalar functions
$\mathbf{I}$	unit tensor	מ	dimensionless number
i, j, k	indices	∇	tensor derivative
O	origin, order	*	transpose
P, Q, R	polynomials	^	adjoint
$\mathbf{r}$	position vector	—	final state
$\mathbf{S}$	skew-symmetric tensor	′	rotated state

1. INTRODUCTION

In 1945 Reiner [1] discussed the most general relation between two symmetric, second-order tensors of the type

$$\mathbf{T} = f(\mathbf{A}) \tag{1.1}$$

where f is an isotropic polynomial tensor function in **A**. By use of the Cayley–Hamilton theorem, (1.1) was reduced to the form

$$\mathbf{T} = \aleph_0\mathbf{I}+\aleph_1\mathbf{A}+\aleph_2\mathbf{A}^2 \tag{1.2}$$

in which the $\aleph$'s are scalar functions expressible as polynomials in the three principal invariants of **A**. An alternative form for (1.2) is

$$\mathbf{T} = \beth_0\mathbf{I}+\beth_1\mathbf{A}+\beth_2\mathbf{A}^{-1}.$$

Equation (1.2) was utilized to establish the general stress–strain relation for the isotropic Hooke solid—the perfectly elastic body. Thus, (1.2) formed a constitutive equation in which **T** and **A**, standing for the stress tensor and strain tensor respectively, were tacitly assumed symmetric.

The theory of the perfectly elastic body involves another fundamental relation, that between strain and the coordinate gradient. Here, as discussed in the following Section, the latter is an asymmetric tensor of second order.

The most general, analytic isotropic function of a three-dimensional tensor **T** in an asymmetric tensor **A** appears as (cf. Truesdell [2, eq. (6.2)])

$$\begin{aligned}\mathbf{T} = f(\mathbf{A}, \mathbf{A}^*) &= \aleph_0\mathbf{I}+\beth_1\mathbf{A}+\beth_2\mathbf{A}^* \\ &+\gimel_1\mathbf{A}^2+\gimel_2\mathbf{A}\mathbf{A}^*+\gimel_3\mathbf{A}^*\mathbf{A}+\gimel_4\mathbf{A}^{*2}+\dots\end{aligned} \tag{1.3}$$

in which **I** is the unit tensor, $\mathbf{A}^*$ denotes the transpose of **A** and the $\aleph$, $\beth$, $\gimel$'s are scalar functions of the six invariants of **A**. It is shown below that the infinite polynomial (1.3) is reducible to a finite form. The result which appears in Theorem 3 below is the key to all further considerations.

We proceed in the following section to discuss the coordinate gradient, the displacement gradient and the strain tensors. Reduction of the infinite polynomial (1.3) is effected in section 3. The rest of the discussion is devoted to the various strain tensors, in use and new.

2. THE COORDINATE GRADIENT, THE DISPLACEMENT GRADIENT AND THE STRAIN TENSORS

In transferring an elastic body from one state to another, the change of coordinates of each point of the body is the only directly observable quantity. The configuration of all points of the body may remain unaltered and the body is rigidly transported through translation and rotation. When the configuration is altered, we speak of *deformation* of the body. The *analysis of deformation* describes the actual changes which take place from one state to the other. As such, it is unique.

Let x^i be the rectilinear coordinates of a point P of the body in the initial undeformed state, with respect to a basis $\mathbf{e}_i$ ($i = 1, 2, 3$) of three non-coplanar unit vectors located at an origin O. By definition, the x^i are the *component* magnitudes of the position vector $\mathbf{r} = \vec{OP}$ with respect to $\mathbf{e}_i$. Introducing a set $\mathbf{e}^i$ reciprocal to $\mathbf{e}_i$ by means of the relations

$$\mathbf{e}^i = E^{-1}\mathbf{e}_j \times \mathbf{e}_k \quad (i, j, k = 1, 2, 3 \text{ cycl})$$
$$E \equiv \mathbf{e}_i \cdot \mathbf{e}_j \times \mathbf{e}_k; \quad E^{-1} = \mathbf{e}^i \cdot \mathbf{e}^j \times \mathbf{e}^k$$

so that

$$\mathbf{e}^i \cdot \mathbf{e}_j = \delta^i_j; \quad [\mathbf{e}_i\mathbf{e}_j\mathbf{e}_k][\mathbf{e}^i\mathbf{e}^j\mathbf{e}^k] = 1.$$

Also, on defining the projection magnitudes x_i of $\mathbf{r}$ as

$$x_i = \mathbf{r} \cdot \mathbf{e}_i,$$

the resolution of the position vector $\mathbf{r}$ with respect to the sets $\mathbf{e}_i$, $\mathbf{e}^i$ is now expressed as

$$\mathbf{r} = x^\alpha \mathbf{e}_\alpha = x_\alpha \mathbf{e}^\alpha\dagger \quad (\alpha = 1 \ldots 3). \tag{2.1}$$

When the body is transferred from the initial to the final deformed state, each point P passes to the new position $\bar{P}$ located by means of the position vector $\bar{\mathbf{r}} = \vec{O\bar{P}}$ or by each set of magnitudes $\bar{x}^i$, $\bar{x}_i$, in conformity with the relation

$$\bar{\mathbf{r}} = \bar{x}^\alpha \mathbf{e}_\alpha = \bar{x}_\alpha \mathbf{e}^\alpha. \tag{2.2}$$

In general, the final rectilinear coordinates $\bar{x}^i$ are functions of the initial coordinates x^i and conversely, namely

$$\bar{x}^i = \bar{x}^i(x^j); \quad x^i = x^i(\bar{x}^j) \quad (i, j = 1, 2, 3).$$

The gradient ∇ of a vector or of a physical tensor of any order expressible as a multi-linear vector form, is a directional derivative defined at the field point P as

$$\nabla(\,) \equiv \frac{\partial(\,)}{\partial x^\alpha}\mathbf{e}^\alpha = \frac{\partial(\,)}{\partial x_\alpha}\mathbf{e}_\alpha. \tag{2.3}$$

Similarly, with respect to the final state we have

$$\bar{\nabla}(\,) \equiv \frac{\partial(\,)}{\partial \bar{x}^\alpha}\mathbf{e}^\alpha = \frac{\partial(\,)}{\partial \bar{x}_\alpha}\mathbf{e}_\alpha. \tag{2.3$_1$}$$

Application of (2.3) and $(2.3)_1$ to the position vectors $\bar{\mathbf{r}}$ and $\mathbf{r}$ respectively leads to the second-order tensors expressible as bi-linear vector

† Repeated Greek indices only should be summed.

forms

$$\nabla\bar{\mathbf{r}} = \frac{\partial\bar{x}^\alpha}{\partial x^\beta}\,\mathbf{e}_\alpha\mathbf{e}^\beta = \frac{\partial\bar{x}_\alpha}{\partial x^\beta}\,\mathbf{e}^\alpha\mathbf{e}^\beta = \frac{\partial\bar{x}^\alpha}{\partial x_\beta}\,\mathbf{e}_\alpha\mathbf{e}_\beta = \frac{\partial\bar{x}_\alpha}{\partial x_\beta}\,\mathbf{e}^\alpha\mathbf{e}_\beta,$$
$$\overline{\nabla}\mathbf{r} = \frac{\partial x^\alpha}{\partial\bar{x}^\beta}\,\mathbf{e}_\alpha\mathbf{e}^\beta = \frac{\partial x_\alpha}{\partial\bar{x}^\beta}\,\mathbf{e}^\alpha\mathbf{e}^\beta = \frac{\partial x^\alpha}{\partial\bar{x}_\beta}\,\mathbf{e}_\alpha\mathbf{e}_\beta = \frac{\partial x_\alpha}{\partial\bar{x}_\beta}\,\mathbf{e}^\alpha\mathbf{e}_\beta. \tag{2.4}$$

Setting

$$\boldsymbol{\Lambda} \equiv \nabla \;;\quad \bar{\boldsymbol{\Lambda}} \equiv \overline{\nabla}\mathbf{r} \tag{2.5}$$

with

$$\lambda^i_j = \frac{\partial\bar{x}^i}{\partial x^j},\quad \lambda_{ij} = \frac{\partial\bar{x}_i}{\partial x^j},\ \text{etc.}$$
$$\bar{\lambda}^i_j = \frac{\partial x^i}{\partial\bar{x}^j},\quad \bar{\lambda}_{ij} = \frac{\partial x_i}{\partial\bar{x}^j},\ \text{etc.} \tag{2.5$_1$}$$

we refer to $\boldsymbol{\Lambda}$ as the *initial* coordinate gradient and to $\bar{\boldsymbol{\Lambda}}$ as the *final* coordinate gradient.

The *displacement* vector $\mathbf{u}$ is defined by

$$\mathbf{u} = \bar{\mathbf{r}} - \mathbf{r}, \tag{2.6}$$

hence

$$u^i = \bar{x}^i - x^i;\quad u_i = \bar{x}_i - x_i. \tag{2.6$_1$}$$

From this, we obtain the initial displacement gradient and the final displacement gradient respectively as

$$\boldsymbol{\Gamma} \equiv \nabla\mathbf{u} = \boldsymbol{\Lambda} - \mathbf{I},$$
$$\bar{\boldsymbol{\Gamma}} \equiv \overline{\nabla}\mathbf{u} = -\bar{\boldsymbol{\Lambda}} + \bar{\mathbf{I}}, \tag{2.7}$$

$$\gamma^i{}_j = \frac{\partial u^i}{\partial x^j} = \lambda^i{}_j - \delta^i{}_j,\quad \gamma_{ij} = \frac{\partial u_i}{\partial x^j} = \lambda_{ij} - \delta_{ij},\ \text{etc.,}$$
$$\bar{\gamma}^i{}_j = \frac{\partial u^i}{\partial\bar{x}^j} = \bar{\delta}^i{}_j - \bar{\lambda}^i{}_j,\quad \bar{\gamma}_{ij} = \frac{\partial u_i}{\partial\bar{x}^j} = \bar{\delta}_{ij} - \bar{\lambda}_{ij},\ \text{etc.} \tag{2.7$_1$}$$

The tensors $\boldsymbol{\Lambda}$, $\bar{\boldsymbol{\Lambda}}$, $\boldsymbol{\Gamma}$, $\bar{\boldsymbol{\Gamma}}$ are all asymmetric.

The analysis of strain postulates an isotropic tensor function between the strain tensor $\mathbf{E}$ and either the coordinate gradient or the displacement gradient, subject to the following provisions:

(i) The strain tensor must be symmetric in order to be related to the stress.

(ii) When expressed as a function of the displacement gradient, the first-order term must be identical with the Cauchy strain tensor, i.e. with the conventional strain of the classical, infinitesimal elasticity.

(iii) The strain tensor must vanish identically when the coordinate gradient is a unit tensor.
(iv) An extension, namely positive components of type e'^{i}_{i}, should be expressed by a gradient component which is positive for an elongation and vice versa.

It immediately follows from the principal postulate of the analysis of strain that the relation strain-gradient is of type (1.3) with **E** replacing **T** and **A** standing for either Λ or Γ, subject to provisions (i)–(iv). It should be pointed out here that in view of the relations

$$\nabla \bar{\mathbf{r}} \cdot \overline{\nabla} \mathbf{r} = \overline{\nabla} \mathbf{r} \cdot \nabla \bar{\mathbf{r}} = \Lambda \cdot \bar{\Lambda} = \bar{\Lambda} \cdot \Lambda = \mathbf{I} \tag{2.8}$$

neither Λ and $\bar{\Lambda}$ nor Γ and $\bar{\Gamma}$ are independent. Thus, expressing **E** in terms of $\bar{\Lambda}$ or $\bar{\Gamma}$ is justifiable only by reasons of simplicity.

Our main concern now reverts to (1.3). The following section treats the expansion of (1.3) and its reduction to a finite closed form.

3. REDUCTION OF THE ISOTROPIC POLYNOMIAL OF AN ASYMMETRIC TENSOR

In discussing polynomial functions of an asymmetric tensor in a three-dimensional space, it becomes preferable to employ the absolute notation for tensors rather than the notation for the tensor components.

By a tensor we shall refer from now on to a three-dimensional, absolute, second-order tensor expressible as a bi-linear vector form with respect to a point in vector space. Given a basis $\mathbf{e}_i$ ($i = 1, 2, 3$) and its reciprocal set $\mathbf{e}^i$, resolution of the tensor **A** into components along the sets $\mathbf{e}_i$ and $\mathbf{e}^i$ yields

$$\mathbf{A} = a^{\alpha\beta}\mathbf{e}_\alpha\mathbf{e}_\beta = a_{\alpha\beta}\mathbf{e}^\alpha\mathbf{e}^\beta = a^\alpha{}_\beta\mathbf{e}_\alpha\mathbf{e}^\beta = a_\alpha{}^\beta\mathbf{e}^\alpha\mathbf{e}_\beta. \tag{3.1}$$

The all-component magnitudes a^{ij} of **A**, the all-projection magnitudes a_{ij}, the mixed component-projection magnitudes a^i_j or a_i^j, and the vector sets $\mathbf{e}_i$, $\mathbf{e}^i$ are all termed *elements* of **A**.

The point tensor **A** is said to be *isotropically invariant under resolution* if, given an arbitrary basis $\mathbf{e}'_i$ and its reciprocal set $\mathbf{e}'^i$ at that point, the following relations hold

$$\mathbf{A} = a^{\alpha\beta}\mathbf{e}_\alpha\mathbf{e}_\beta = a_{\alpha\beta}\mathbf{e}^\alpha\mathbf{e}^\beta = \ldots = a'^{\lambda\mu}\mathbf{e}'_\lambda\mathbf{e}'_\mu = a'_{\lambda\mu}\mathbf{e}'^\lambda\mathbf{e}'^\mu. \tag{3.2}$$

The transformation laws between the various types of magnitudes are directly obtainable from (2.2). They, are, however, of little, if any, concern to the present discussion.

From the elements of **A**, other tensors can be constructed which, as much as **A**, satisfy the invariancy relations (3.2). They are the *tensor invariants* of **A**. Tensor invariants obtained by differentiation operations and known as differential invariants, are presently excluded.

Invariants of **A**, themselves second-order tensors, are:

(i) The *adjoint* $\hat{\mathbf{A}}$(†) of **A** defined as

$$\begin{aligned}\hat{\mathbf{A}} &= \tfrac{1}{2}a^{\alpha\beta}a^{\lambda\mu}(\mathbf{e}_\beta\times\mathbf{e}_\mu)(\mathbf{e}_\alpha\times\mathbf{e}_\lambda) = \tfrac{1}{2}a_{\alpha\beta}a_{\lambda\mu}(\mathbf{e}^\beta\times\mathbf{e}^\mu)(\mathbf{e}^\alpha\times\mathbf{e}^\lambda) = \\ &= \tfrac{1}{2}a^\alpha{}_\beta a^\lambda{}_\mu(\mathbf{e}^\beta\times\mathbf{e}^\mu)(\mathbf{e}_\alpha\times\mathbf{e}_\lambda) = \tfrac{1}{2}a_\alpha{}^\beta a_\lambda{}^\mu(\mathbf{e}_\beta\times\mathbf{e}_\mu)(\mathbf{e}^\alpha\times\mathbf{e}^\lambda).\end{aligned} \tag{3.3}$$

(ii) The transposed tensors

$$\mathbf{A}^* = a_{\beta\alpha}\mathbf{e}^\alpha\mathbf{e}^\beta = a^{\beta\alpha}\mathbf{e}_\alpha\mathbf{e}_\beta = a^\beta{}_\alpha\mathbf{e}^\alpha\mathbf{e}_\beta = a_\beta{}^\alpha\mathbf{e}_\alpha\mathbf{e}^\beta,$$
$$\hat{\mathbf{A}}^* = \tfrac{1}{2}a^{\alpha\beta}a^{\lambda\mu}(\mathbf{e}_\alpha\times\mathbf{e}_\lambda)(\mathbf{e}_\beta\times\mathbf{e}_\mu) = \tfrac{1}{2}a_{\alpha\beta}a_{\lambda\mu}(\mathbf{e}^\alpha\times\mathbf{e}^\lambda)(\mathbf{e}^\beta\times\mathbf{e}^\mu) = \ldots$$

Setting

$$A^{ij} = E^{-2}\text{ cof. of } a_{ji} \text{ in } \|a_{ij}\|, \quad A_{ij} = E^2\text{ cof. of } a^{ji} \text{ in } \|a^{ij}\|$$
$$A^i{}_j = \text{cof. of } a^j{}_i \text{ in } \|a^i{}_j\|, \quad A_i{}^j = \text{cof. of } a_j{}^i \text{ in } \|a_i{}^j\|.$$

Equation (3.3) is rewritten as

$$\hat{\mathbf{A}} = A^{\alpha\beta}\mathbf{e}_\alpha\mathbf{e}_\beta = A_{\alpha\beta}\mathbf{e}^\alpha\mathbf{e}^\beta = A^\alpha{}_\beta\mathbf{e}_\alpha\mathbf{e}^\beta = A_\alpha{}^\beta\mathbf{e}^\alpha\mathbf{e}_\beta. \tag{3.4}$$

The tensors **A**, $\hat{\mathbf{A}}$ admit the vector invariants

$$\begin{aligned}\mathbf{a} &= \tfrac{1}{2}a_{\alpha\beta}\mathbf{e}^\alpha\times\mathbf{e}^\beta = \tfrac{1}{2}a^{\alpha\beta}\mathbf{e}_\alpha\times\mathbf{e}_\beta = \ldots = -\mathbf{a}^* = -a_3^{-1}\hat{\mathbf{A}}\cdot\hat{\mathbf{a}},\\ \hat{\mathbf{a}} &= \tfrac{1}{2}A_{\alpha\beta}\mathbf{e}^\alpha\times\mathbf{e}^\beta = \tfrac{1}{2}A^{\alpha\beta}\mathbf{e}_\alpha\times\mathbf{e}_\beta = \ldots = -\hat{\mathbf{a}}^* = -\mathbf{A}\cdot\mathbf{a}.\end{aligned} \tag{3.5}$$

As a third vector invariant of **A** we choose $\mathbf{a}\times\hat{\mathbf{a}}$. The set, generally oblique,

$$\mathbf{a} \quad \hat{\mathbf{a}} \quad \mathbf{a}\times\hat{\mathbf{a}} \tag{3.6}$$

forms a triplet of vector invariants which may serve as an intrinsic reference system of the second-order tensor **A**, unless **A** is singular.

The six scalar invariants of **A** are

$$\begin{aligned} a_1 &= a^\alpha{}_\beta\mathbf{e}_\alpha\cdot\mathbf{e}^\beta = a^\alpha{}_\alpha = a_\alpha{}^\alpha = \text{tr }\|a^i{}_j\| = \text{tr }\|a_i{}^j\|,\\ a_2 &= A^\alpha{}_\beta\mathbf{e}_\alpha\cdot\mathbf{e}^\beta = A^\alpha{}_\alpha = A_\alpha{}^\alpha = \text{tr }\|A^i{}_j\| = \text{tr }\|A_i{}^j\| = \hat{a}_1,\\ a_3 &= \tfrac{1}{6}a^\alpha{}_\lambda a^\beta{}_\mu a^\gamma{}_\nu(\mathbf{e}_\alpha\times\mathbf{e}_\beta\cdot\mathbf{e}_\gamma)(\mathbf{e}^\lambda\times\mathbf{e}^\mu\cdot\mathbf{e}^\nu) = |a^i{}_j| = |a_i{}^j|,\\ a_4 &= \mathbf{a}\cdot\mathbf{a} = a^2, \qquad a_5 = \mathbf{a}\cdot\hat{\mathbf{a}}, \qquad a_6 = \hat{\mathbf{a}}\cdot\hat{\mathbf{a}} = \hat{a}_4.\end{aligned} \tag{3.7}$$

All other scalar invariants obtained from the elements of **A** are expressible in terms of the a_i $(i = 1\ldots 6)$.

† The matrix of $\hat{\mathbf{A}}$ is known as the *adjugate* matrix [3, IV. 7].

The definition of $\hat{\mathbf{A}}$ readily admits the following relation:

$$\mathbf{A}\cdot(\mathbf{A}-a_1\mathbf{I}) = \hat{\mathbf{A}}-\hat{a}_1\mathbf{I}.$$

On expansion

$$\begin{aligned} \mathbf{A}^2 &= \hat{\mathbf{A}}+a_1\mathbf{A}-\hat{a}_1\mathbf{I}, \\ \hat{\mathbf{A}} &= \mathbf{A}^2-a_1\mathbf{A}+\hat{a}_1\mathbf{I}. \end{aligned} \tag{3.8}$$

The Cayley–Hamilton theorem, in its conventional form for second-order tensors, is obtained on multiplying $(3.8)_1$ further by $\mathbf{A}$.

Sufficient and necessary conditions that $\mathbf{A}$ be:

$$\left.\begin{array}{l} \text{Non-singular} \\ \text{Singular-planar} \\ \text{Singular-linear} \end{array}\right\} \text{ are } \left\{\begin{array}{lll} a_3 \neq 0 & & \\ a_3 = 0 & \hat{\mathbf{A}} \neq 0 & \\ a_3 = 0 & \hat{\mathbf{A}} = 0, & \mathbf{A} \neq 0. \end{array}\right.$$

Non-singular tensors only possess an inverse defined by

$$\mathbf{A}^{-1} = a_3^{-1}\hat{\mathbf{A}}, \qquad \mathbf{A}^{-1}\cdot\mathbf{A} = \mathbf{I}.$$

Since the theory of the asymmetric tensor heavily leans on its skew-symmetric part which is singular, the adjoint $\hat{\mathbf{A}}$ rather than the inverse $\mathbf{A}^{-1}$ is employed throughout.

For a singular tensor we prove:

THEOREM 1. *The adjoint of a singular-planar tensor is a singular-linear tensor.*

Proof. Since every singular-planar tensor can be reduced to the form

$$\mathbf{A} = \mathbf{al}+\mathbf{bm}$$

it follows from the definition of the adjoint that

$$\hat{\mathbf{A}} = (\mathbf{l}\times\mathbf{m})(\mathbf{a}\times\mathbf{b}),$$

which is singular-linear.

Let hereafter $\mathbf{T}$ denote the asymmetric tensor, $\mathbf{G}$ and $\mathbf{S}$ its symmetric and skew-symmetric parts respectively. We further cite:

THEOREM 2 (*Gibbs*).† *Every skew-symmetric tensor is re-expressible in terms of its vector invariant as*

$$\mathbf{S} = -\mathbf{I}\times\mathbf{s}. \tag{3.9}$$

The following Corollaries result immediately:

COROLLARY 2.1. The vector invariant $\mathbf{s} \neq 0$ unless $\mathbf{S} = 0$ and conversely.

† [4, sect. 137].

COROLLARY 2.2. For every skew-symmetric tensor $\mathbf{S}$,

$$\mathbf{S}\cdot\mathbf{s} = 0. \tag{3.10}$$

COROLLARY 2.3. The adjoint of $\mathbf{S}$ is singular-linear and symmetric

$$\hat{\mathbf{S}} = \mathbf{ss}. \tag{3.11}$$

Hence

COROLLARY 2.4. The vector and scalar invariants of the skew-symmetric tensor $\mathbf{S} \neq 0$ are

$$\mathbf{s} \neq 0, \quad \hat{\mathbf{s}} = 0, \quad \mathbf{s}\times\hat{\mathbf{s}} = 0$$

$$s_1 = 0, \quad s_2 = \hat{s}_1 = s^2, \quad s_3 = 0, \quad s_4 = s_2 = s^2, \quad s_5 = s_6 = 0.$$

COROLLARY 2.5. Vanishing of either of the vector invariants is a *sufficient* condition that a tensor be symmetric.

Next, the following relations, of which use is made in the sequel, are established:

$$\begin{gathered}\hat{\mathbf{T}} = \hat{\mathbf{G}}+\hat{\mathbf{S}}+\mathbf{G}\cdot\mathbf{S}+\mathbf{S}\cdot\mathbf{G}-g_1\mathbf{S}\\ (\hat{\mathbf{G}}+\hat{\mathbf{S}})^* = \hat{\mathbf{G}}+\hat{\mathbf{S}} \qquad (\mathbf{G}\cdot\mathbf{S}+\mathbf{S}\cdot\mathbf{G})^* = -(\mathbf{G}\cdot\mathbf{S}+\mathbf{S}\cdot\mathbf{G})\\ \begin{aligned} \mathbf{t} &= \mathbf{s} & \hat{\mathbf{t}} &= -\mathbf{G}\cdot\mathbf{s}\\ t_1 &= g_1 & t_4 &= s_2 = s^2\\ t_2 &= g_2+s_2 = \hat{t}_1 & t_5 &= -\mathbf{s}\cdot\mathbf{G}\cdot\mathbf{s}\\ t_3 &= g_3+t_5 & t_6 &= \mathbf{s}\cdot\mathbf{G}^2\cdot\mathbf{s}.\end{aligned}\end{gathered} \tag{3.12}$$

Let P be an analytic isotropic function of a single asymmetric tensor $\mathbf{T}$, expressible as a polynomial in $\mathbf{T}$ and its transpose $\mathbf{T}^*$. The most general product π in P, which is of extension $2r$,† reads

$$\pi = \mathbf{T}^{i_1}\mathbf{T}^{*j_1}\mathbf{T}^{i_2}\mathbf{T}^{*j_2}, \ \ldots, \ \mathbf{T}^{i_r}\mathbf{T}^{*j_r}.\ddagger$$

The degrees i_m, j_m $(1 \leqslant m \leqslant r)$, in view of the Cayley–Hamilton theorem may only take the values 1, 2 or, alternatively, the values 1, -1 (the value 0 merely reduces the extension of π). Expressing π in terms of the symmetric part $\mathbf{G}$ and the skew-symmetric part $\mathbf{S}$ of $\mathbf{T}$, the most general product π in P assumes the form

$$\pi = \mathbf{G}^{i_1}\mathbf{S}^{j_1}\mathbf{G}^{i_2}\mathbf{S}^{j_2}, \ \ldots, \ \mathbf{G}^{i_r}\mathbf{S}^{j_r}$$

which is of extension $2r$ or less. It was shown by Rivlin [5] that any polynomial in products of two tensors can be expressed as

$$P = Q+\aleph_1\mathbf{GSG}^2\mathbf{S}^2+\aleph_2\mathbf{S}^2\mathbf{G}^2\mathbf{SG}, \tag{3.13}$$

† For the terminology, cf. Rivlin [5].

‡ Henceforth, we omit the dot product if no ambiguity is caused.

where Q is a polynomial of extension 3 or less in **G** and **S**, and $\aleph_1$, $\aleph_2$ are scalar coefficients expressible as polynomials in the six scalar invariants of **T**. Furthermore, the polynomial Q is reducible to

$$\begin{aligned} Q = {} & R+\beth_1\mathbf{GSG}^2+\beth_2\mathbf{G}^2\mathbf{SG}+\beth_3\mathbf{SGS}^2+\beth_4\mathbf{S}^2\mathbf{GS} \\ & +\beth_5\mathbf{GS}^2\mathbf{G}^2+\beth_6\mathbf{G}^2\mathbf{S}^2\mathbf{G}+\beth_7\mathbf{SG}^2\mathbf{S}^2+\beth_8\mathbf{S}^2\mathbf{G}^2\mathbf{S} \end{aligned} \tag{3.14}$$

where R is a polynomial of extension 2 and less in **G**, **S**, and the $\beth$'s are scalar coefficients expressible as polynomials in the scalar invariants of **T**.

All the terms in (3.13), (3.14) are further expressible in polynomials of extension 2 or less. For, it follows from (3.8)–(3.12) that

$$\begin{aligned}
\mathbf{G}\cdot\mathbf{S}\cdot\mathbf{G}^2 &= -(\mathbf{I}\times\hat{\mathbf{t}})\cdot\mathbf{G}^2+g_2\mathbf{S}\cdot\mathbf{G}-g_3\mathbf{S} \\
\mathbf{G}^2\cdot\mathbf{S}\cdot\mathbf{G} &= -\mathbf{G}^2\cdot(\mathbf{I}\times\hat{\mathbf{t}})+g_2\mathbf{G}\cdot\mathbf{S}-g_3\mathbf{S} \\
\mathbf{S}\cdot\mathbf{G}\cdot\mathbf{S}^2 &= (\mathbf{t}\times\hat{\mathbf{t}})\mathbf{t} \quad -t_4\mathbf{S}\cdot\mathbf{G} \\
\mathbf{S}^2\cdot\mathbf{G}\cdot\mathbf{S} &= -\mathbf{t}(\mathbf{t}\times\hat{\mathbf{t}}) \quad -t_4\mathbf{G}\cdot\mathbf{S} \\
\mathbf{G}\cdot\mathbf{S}^2\cdot\mathbf{G}^2 &= -\hat{\mathbf{t}}\hat{\mathbf{t}}\cdot\mathbf{G} \quad -t_4\mathbf{G}^3 \\
\mathbf{G}^2\cdot\mathbf{S}^2\cdot\mathbf{G} &= -\mathbf{G}\cdot\hat{\mathbf{t}}\hat{\mathbf{t}} \quad -t_4\mathbf{G}^3 \\
\mathbf{S}\cdot\mathbf{G}^2\cdot\mathbf{S}^2 &= -\mathbf{G}\cdot(\mathbf{t}\times\hat{\mathbf{t}})\mathbf{t}+g_1(\mathbf{t}\times\hat{\mathbf{t}})\mathbf{t}-t_4\mathbf{S}\cdot\mathbf{G}^2 \\
\mathbf{S}^2\cdot\mathbf{G}^2\cdot\mathbf{S} &= \mathbf{t}(\mathbf{t}\times\hat{\mathbf{t}})\cdot\mathbf{G}-g_1\mathbf{t}(\mathbf{t}\times\hat{\mathbf{t}})-t_4\mathbf{G}^2\cdot\mathbf{S} \\
\mathbf{G}\cdot\mathbf{S}\cdot\mathbf{G}^2\cdot\mathbf{S}^2 &= -\hat{\mathbf{G}}\cdot(\mathbf{t}\times\hat{\mathbf{t}})\mathbf{t}+g_2(\mathbf{t}\times\hat{\mathbf{t}})\mathbf{t}+t_4\mathbf{G}\cdot\mathbf{S}\cdot\mathbf{G}^2 \\
\mathbf{S}^2\cdot\mathbf{G}^2\cdot\mathbf{S}\cdot\mathbf{G} &= \mathbf{t}(\mathbf{t}\times\hat{\mathbf{t}})\cdot\hat{\mathbf{G}}-g_2\mathbf{t}(\mathbf{t}\times\hat{\mathbf{t}})-t_4\mathbf{G}^2\cdot\mathbf{S}\cdot\mathbf{G}.
\end{aligned} \tag{3.15}$$

The polynomial R is composed of the terms

$$\begin{aligned} R = {} & \gimel_0\mathbf{I}+\gimel_1\mathbf{G}+\gimel_2\mathbf{S}+\gimel_3\mathbf{G}^2+\gimel_4\mathbf{S}^2+\gimel_5\mathbf{GS}+\gimel_6\mathbf{SG} \\ & +\gimel_7\mathbf{G}^2\mathbf{S}+\gimel_8\mathbf{SG}^2+\gimel_9\mathbf{GS}^2+\gimel_{10}\mathbf{S}^2\mathbf{G}+\gimel_{11}\mathbf{G}^2\mathbf{S}^2+\gimel_{12}\mathbf{S}^2\mathbf{G}^2 \end{aligned} \tag{3.16}$$

where, again, the $\gimel$'s are scalar coefficients expressible as polynomials in the scalar invariants of **T**.

Equations (3.13)–(3.16) show that there exists a finite set of tensors

$$\mathbf{I},\ \mathbf{G},\ \hat{\mathbf{G}},\ \mathbf{I}\times\mathbf{t},\ \mathbf{I}\times\hat{\mathbf{t}},\ \mathbf{t}\mathbf{t},\ \hat{\mathbf{t}}\hat{\mathbf{t}},\ (\mathbf{t}\times\hat{\mathbf{t}})\mathbf{t},\ \mathbf{t}(\mathbf{t}\times\hat{\mathbf{t}}) \tag{3.17}$$

constituting the *polynomial basis* of P. The tensors **G**, $\hat{\mathbf{G}}$ are non-singular, $\mathbf{I}\times\mathbf{t}$, $\mathbf{I}\times\hat{\mathbf{t}}$ are singular-planar, $\mathbf{tt}$, $\hat{\mathbf{t}}\hat{\mathbf{t}}$, $\mathbf{t}(\mathbf{t}\times\hat{\mathbf{t}})$ and its transpose $(\mathbf{t}\times\hat{\mathbf{t}})\mathbf{t}$ are singular-linear, all of extension 1. Thus, the following has been proved:

THEOREM 3. *The isotropic polynomial of arbitrary extension of an asymmetric tensor and its transpose is expressible as a polynomial of extension 2 or less in the symmetric part of the tensor and its adjoint, in the unit tensor, and in a finite number of singular tensors formed from the vector invariants of the asymmetric tensor.*

COROLLARY 3.1. The products of extension zero (**I**) and extension 1 in P constitute the polynomial basis (3.17).

COROLLARY 3.2. The products of extension 2 in P are composed of either the symmetric part **G** or its adjoint $\hat{\mathbf{G}}$ and one of the singular tensors in the polynomial basis.

It should be remarked that although the last corollary secures finiteness, not all the combinations of **G**, $\hat{\mathbf{G}}$ and the singular tensors (3.17) need be considered, since some of them are expressible in terms of the others. An independent set of products may be chosen as

$$\begin{gathered}\mathbf{G}\cdot(\mathbf{I}\times\mathbf{t}),\ \mathbf{G}\cdot(\mathbf{I}\times\hat{\mathbf{t}}),\ \hat{\mathbf{G}}\cdot(\mathbf{I}\times\mathbf{t}),\ \hat{\mathbf{G}}\cdot(\mathbf{I}\times\hat{\mathbf{t}})\\ \mathbf{G}\cdot\mathbf{tt},\ \mathbf{G}\cdot(\mathbf{t}\times\hat{\mathbf{t}})\mathbf{t},\ \mathbf{G}\cdot\hat{\mathbf{t}}\hat{\mathbf{t}},\ \hat{\mathbf{G}}\cdot\mathbf{tt},\ \hat{\mathbf{G}}\cdot(\mathbf{t}\times\hat{\mathbf{t}})\mathbf{t}\end{gathered} \tag{3.18}$$

and the transpose of $(3.18)_2$. The twenty-three terms in (3.17), (3.18) correspond to the same number of terms in the polynomial P.

A particular case of interest is when one of the principal directions of **G** is parallel to the vector invariant **t**. Such is the case if, in the matrix of **T**, $t_{ij} = 0$ for $i \neq j$ (i is either 1, 2 or 3). All the vector invariants of **T** are then parallel, the adjoints of **G**, **S** possess the properties $\mathbf{G}\hat{\mathbf{S}} = \hat{\mathbf{S}}\mathbf{G}$, $\hat{\mathbf{G}}\hat{\mathbf{S}} = \hat{\mathbf{S}}\hat{\mathbf{G}}$, and the polynomial basis reduces to

$$\mathbf{I},\ \mathbf{G},\ \hat{\mathbf{G}},\ \mathbf{I}\times\mathbf{t},\ \mathbf{tt}. \tag{3.19}$$

The thirteen combinations of products between each two of the set (3.19) are precisely those appearing in (3.16). So we have the following:

LEMMA. If one of the principal directions of the symmetric part of a tensor **T** is parallel to the vector invariant **t**, the isotropic polynomial of arbitrary extension in **T** and **T*** is expressible as a polynomial of extension 2 in the symmetric and skew-symmetric parts of **T**.

4. GENERAL STRAIN TENSOR AND EXAMPLES

The results incorporated in (3.13)–(3.16) and in Theorem 3 furnish the first step towards the construction of the general strain tensor. To recapitulate, they are the manifestation, in a closed form, of the principal postulate that strain is expressible as an isotropic polynomial function of the gradients.

Our next step is to impose symmetry on **E** (provision (i)). To obtain this, it suffices to eliminate the skew-symmetric terms in (3.17), (3.18), leaving the symmetric terms only. Thus, with **G** and **S** now standing for

$$\mathbf{G} = \frac{1}{2}(\Lambda+\Lambda^*) = \frac{1}{2}(\nabla\bar{\mathbf{r}}+\nabla\bar{\mathbf{r}}^*),$$

$$\mathbf{S} = \frac{1}{2}(\Lambda-\Lambda^*) = \frac{1}{2}(\nabla\bar{\mathbf{r}}-\nabla\bar{\mathbf{r}}^*) = -\mathbf{I}\times\boldsymbol{\lambda}, \tag{4.1}$$

$$\boldsymbol{\lambda} = \frac{1}{2}\nabla\times\bar{\mathbf{r}} = -\frac{1}{2}\,\text{curl}\,\bar{\mathbf{r}};\ \hat{\boldsymbol{\lambda}} = -\mathbf{G}\cdot\boldsymbol{\lambda} = \frac{1}{2}\mathbf{G}\cdot\text{curl}\,\bar{\mathbf{r}}$$

we have, on rearranging the coefficients,

$$\begin{aligned}\mathbf{E} = {} & \aleph_0\mathbf{I}+\aleph_1\mathbf{G}+\aleph_2\hat{\mathbf{G}}+\aleph_3\boldsymbol{\lambda}\boldsymbol{\lambda}+\aleph_4\hat{\boldsymbol{\lambda}}\hat{\boldsymbol{\lambda}}+\aleph_5[(\boldsymbol{\lambda}\times\hat{\boldsymbol{\lambda}})\boldsymbol{\lambda}+\boldsymbol{\lambda}(\hat{\boldsymbol{\lambda}}\times\boldsymbol{\lambda})] \\ & +\aleph_6[\mathbf{G}\cdot(\mathbf{I}\times\boldsymbol{\lambda})-(\mathbf{I}\times\boldsymbol{\lambda})\cdot\mathbf{G}]+\aleph_7[\hat{\mathbf{G}}\cdot(\mathbf{I}\times\boldsymbol{\lambda})-(\mathbf{I}\times\boldsymbol{\lambda})\cdot\hat{\mathbf{G}}] \\ & +\aleph_8[\mathbf{G}\cdot\boldsymbol{\lambda}\boldsymbol{\lambda}+\boldsymbol{\lambda}\boldsymbol{\lambda}\cdot\mathbf{G}]+\aleph_9[\hat{\mathbf{G}}\cdot\boldsymbol{\lambda}\boldsymbol{\lambda}+\boldsymbol{\lambda}\boldsymbol{\lambda}\cdot\hat{\mathbf{G}}] \\ & +\aleph_{10}[\mathbf{G}\cdot\hat{\boldsymbol{\lambda}}\hat{\boldsymbol{\lambda}}+\hat{\boldsymbol{\lambda}}\hat{\boldsymbol{\lambda}}\cdot\mathbf{G}]+\aleph_{11}[\mathbf{G}\cdot(\boldsymbol{\lambda}\times\hat{\boldsymbol{\lambda}})\boldsymbol{\lambda}+\boldsymbol{\lambda}(\boldsymbol{\lambda}\times\hat{\boldsymbol{\lambda}})\cdot\mathbf{G}] \\ & +\aleph_{12}[\hat{\mathbf{G}}\cdot(\boldsymbol{\lambda}\times\hat{\boldsymbol{\lambda}})\boldsymbol{\lambda}+\boldsymbol{\lambda}(\boldsymbol{\lambda}\times\hat{\boldsymbol{\lambda}})\cdot\hat{\mathbf{G}}].\end{aligned} \tag{4.2}$$

Along with provisions (ii)–(iv), to be satisfied by a suitable choice of the coefficients $\aleph$, (4.2) may thus be regarded as the *general strain tensor*.

Let some examples of (4.2) be considered. First, setting $\aleph_0 = -1$, $\aleph_1 = 1$ and all other coefficients equal to zero, we obtain

$$\mathbf{E}^C = \mathbf{G}-\mathbf{I} = \frac{1}{2}(\Lambda+\Lambda^*)-\mathbf{I} = \frac{1}{2}(\Gamma+\Gamma^*) \tag{4.3}$$

which is the Cauchy strain tensor—the classical strain tensor of the infinitesimal elasticity. Likewise, and replacing the initial gradients Λ, Γ by the final ones $\bar{\Lambda}, \bar{\Gamma}$, also reversing the signs of $\aleph_0$, $\aleph_1$, (4.3) stands for the Swainger strain tensor

$$\mathbf{E}^S = \bar{\mathbf{I}}-\bar{\mathbf{G}} = \bar{\mathbf{I}}-\tfrac{1}{2}(\bar{\Lambda}+\bar{\Lambda}^*) = \tfrac{1}{2}(\bar{\Gamma}+\bar{\Gamma}^*). \tag{4.4}$$

Murnaghan [6] first derived in tensor form the two finite strain tensors, termed by him the Lagrangian and Eulerian tensors and now commonly known as the Green and Almansi tensors respectively [2, Sec. 15], namely

$$\mathbf{E}^G = \tfrac{1}{2}(\Lambda^*\cdot\Lambda-\mathbf{I}) = \tfrac{1}{2}(\Gamma+\Gamma^*)+\tfrac{1}{2}\Gamma^*\cdot\Gamma \tag{4.5}$$

$$\mathbf{E}^{\bar{A}} = \tfrac{1}{2}(\bar{\mathbf{I}}-\bar{\Lambda}^*\cdot\bar{\Lambda}) = \tfrac{1}{2}(\bar{\Gamma}+\bar{\Gamma}^*)-\tfrac{1}{2}\bar{\Gamma}^*\cdot\bar{\Gamma}. \tag{4.6}$$

Equation (4.5) is obtained on substituting the values $\aleph_0 = -\frac{1}{2}(1+\lambda_2-2\lambda_4)$, $\aleph_1 = \dfrac{\lambda_1}{2}$, $\aleph_2 = \frac{1}{2}$, $\aleph_3 = -\frac{1}{2}$, $\aleph_6 = -\frac{1}{2}$ in (4.2), setting all other coefficients equal to zero. Here, as in (3.7), the λ_i ($i = 1,\ldots,6$) denote the six scalar invariants of $\Lambda = \nabla \bar{\mathbf{r}}$. Similarly, substituting $\aleph_0 = \frac{1}{2}(1-\bar{\lambda}_2-2\bar{\lambda}_4)$, $\aleph_1 = -\frac{\bar{\lambda}_1}{2}$, $\aleph_2 = -\frac{1}{2}$, $\aleph_3 = \frac{1}{2}$, $\aleph_6 = \frac{1}{2}$ in (4.2) with $\bar{\mathbf{G}}$ replacing $\mathbf{G}$, etc., Eq. (4.6) results. It should again be emphasized that, in view of the reciprocity relation between Λ and $\bar{\Lambda}$ or between Γ and $\bar{\Gamma}$, (4.6) may be re-expressed in terms of Λ and Γ, yielding, however, a more complicated expression of the type (4.2).

It is known from the analysis of deformation that the $\mathbf{E}^G$ tensor relates to the initial state of the body, whereas the $\mathbf{E}^{\bar{A}}$ tensor relates to the final state. For that reason a bar, referring all quantities to the final state, was placed over the symbol A, thus $\bar{A}$. Karni and Reiner [7] derived the two complementary tensors, namely for the Green measure of deformation in the final and for the Almansi measure in the initial state, as follows:

$$\mathbf{E}^{\bar{G}} = \tfrac{1}{2}(\Lambda\cdot\Lambda^* - \mathbf{I}) = \tfrac{1}{2}(\Gamma+\Gamma^*)+\tfrac{1}{2}\,\Gamma\cdot\Gamma^*, \tag{4.7}$$

$$\mathbf{E}^{A} = \tfrac{1}{2}(\bar{\mathbf{I}}-\bar{\Lambda}\cdot\bar{\Lambda}^*) = \tfrac{1}{2}(\bar{\Gamma}+\bar{\Gamma}^*)-\tfrac{1}{2}\,\bar{\Gamma}\cdot\bar{\Gamma}^*. \tag{4.8}$$

These again are cases of (4.2) with $\aleph_0 = -\frac{1}{2}(1+\lambda_2-2\lambda_4)$, $\aleph_1 = \frac{\lambda_1}{2}$, $\aleph_2 = \frac{1}{2}$, $\aleph_3 = -\frac{1}{2}$ (as for (4.5)) but $\aleph_6 = \frac{1}{2}$ for (4.7), and the reversed values of the coefficients with the final gradients substituted for the initial ones yielding (4.8).

Later, Karni and Reiner [8] presented two additional strain tensors satisfying all the above-listed provisions, defined as

$$\mathbf{E}_{\mathrm{I}} = \tfrac{1}{2}\,[\tfrac{1}{2}(\Lambda^2+\Lambda^{*2})-\mathbf{I}] = \tfrac{1}{2}(\Gamma+\Gamma^*)+\tfrac{1}{4}(\Gamma^2+\Gamma^{*2}) \tag{4.9}$$

$$\mathbf{E}^{\mathrm{II}} = \tfrac{1}{2}\,[\bar{\mathbf{I}}-\tfrac{1}{2}(\bar{\Lambda}^2+\bar{\Lambda}^{*2})] = \tfrac{1}{2}(\bar{\Gamma}+\bar{\Gamma}^*)-\tfrac{1}{4}(\bar{\Gamma}^2+\bar{\Gamma}^{*2}). \tag{4.10}$$

It is readily observed that (4.9) is a particular case of (4.2) with $\aleph_0 = -\frac{1}{2}(1+\lambda_2)$, $\aleph_1 = \frac{\lambda_1}{2}$, $\aleph_2 = \frac{1}{2}$, $\aleph_3 = \frac{1}{2}$ and all other coefficients vanish; similarly for (4.10) with the signs of the coefficients reversed and final gradients replacing the initial ones.

The application of (4.3)–(4.10) to the two basic types of deformation, namely (i) simple extension and (ii) simple shear, is presented in ref. [8] where the results are shown in a comparative table.

It would be of interest to confine attention from now on to the second-order strain tensors, namely those which include no more than products of *two displacement gradients*. Bearing in mind that $\Lambda = \Gamma+\mathbf{I}$, we readily find by use of (3.5) and (3.7),

$$\begin{aligned} \boldsymbol{\lambda} = \boldsymbol{\gamma}; \quad \hat{\boldsymbol{\lambda}} = -\boldsymbol{\gamma}+\hat{\boldsymbol{\gamma}}; \quad \boldsymbol{\lambda}\times\hat{\boldsymbol{\lambda}} = \boldsymbol{\gamma}\times\hat{\boldsymbol{\gamma}} \\ \lambda_1 = 3+\gamma_1; \quad \lambda_2 = 3+2\gamma_1+\gamma_2; \quad \lambda_3 = 1+\gamma_1+\gamma_2+\gamma_3 \\ \lambda_4 = \gamma_4 = \lambda^2+\gamma^2; \quad \lambda_5 = -\gamma^2+O(\gamma^3); \quad \lambda_6 = \gamma^2+O(\gamma^4) \end{aligned} \tag{4.11}$$

where, as before, γ_i $(i = 1\ldots6)$ stand for the scalar invariants of $\boldsymbol{\Gamma}$; $\boldsymbol{\gamma}$, $\hat{\boldsymbol{\gamma}}$ and $\boldsymbol{\gamma}\times\hat{\boldsymbol{\gamma}}$ are the vector invariants of $\boldsymbol{\Gamma}$ and γ denotes the absolute value of $\boldsymbol{\gamma}$. A glance at (4.2) shows that, since $\boldsymbol{\gamma}$ is of $O(\gamma)$ and $\hat{\boldsymbol{\gamma}}$ of $O(\gamma^2)$, the strain tensor $\mathbf{E}$ of $O(\gamma^2)$ may only consist of

$$\mathbf{E}_{O(\gamma^2)} = \beth_0\mathbf{I}+\beth_1\mathbf{G}+\beth_2\hat{\mathbf{G}}+\beth_3\boldsymbol{\gamma}\boldsymbol{\gamma}+\beth_4[\mathbf{G}\cdot(\mathbf{I}\times\boldsymbol{\gamma})-(\mathbf{I}\times\boldsymbol{\gamma})\cdot\mathbf{G}] \tag{4.12}$$

in which $\mathbf{G}$ now stands for the symmetric part of $\boldsymbol{\Gamma}$, namely $\mathbf{G} = \frac{1}{2}(\boldsymbol{\Gamma}+\boldsymbol{\Gamma}^*)$, and the $\beth$ are scalar functions expressible as polynomials in the invariants γ_1, γ_2, γ_4. More specifically,

$$\begin{aligned} \beth_0 = \beth_0(\gamma_1, \gamma_2, \gamma_4) \quad \beth_1 = \beth_1(\gamma_1) \\ \beth_2 = \beth_2(\text{מ}) \quad \beth_3 = \beth_3(\text{מ}) \quad \beth_4 = \beth_4(\text{מ}). \end{aligned} \tag{4.13}$$

Here, מ denotes the dimensionless number. Thus, (4.12) together with (4.13) present the most general isotropic second-order strain tensor, still subject to provisions (ii)–(iv) to be satisfied by proper choice of the coefficients $\beth$.

The choice of the coefficients $\beth$ now offers a variety of possibilities. As an illustration, let us consider the case of simple shear. If the shear takes place in the x–y-plane, it is known that employment of any of the tensors (4.3)–(4.10) does not entail strain components in the z-direction (ref. [8]). However, setting $\beth_0 = \pm\gamma_4 = \pm\gamma^2$ in addition, yields a lateral second-order strain component e_{zz} on top of the components, of first and second order, e_{xx}, e_{xy}, e_{yy}.

5. DEFORMATION TENSOR AND STRAIN TENSOR

Our final section deals with the question whether or not an additional provision should be imposed on the isotropic strain tensor.

The description of the actual changes in the configuration of the elastic body between the initial and the final, deformed state, forms the subject of the *analysis of deformation*. A comprehensive account of the analysis of deformation is given in Truesdell and Toupin's treatise [9, sects. 26–39].

It is shown and proved that a sphere inscribed in the initial state deforms into an ellipsoid—the *deformation ellipsoid*—in the final state, whatever the amount of the gradients. Likewise, a sphere inscribed in the final state deforms, upon reversing the deformation, into a *reciprocal deformation ellipsoid*. The deformation and reciprocal deformation ellipsoids are expressed, respectively, by the (symmetric) Cauchy and the Green *deformation* tensors

$$\begin{aligned} \overline{\mathbf{D}} &= \overline{\mathbf{I}} - 2\mathbf{E}^{\bar{A}} = \bar{\Lambda}^* \cdot \bar{\Lambda} \\ \mathbf{D} &= \mathbf{I} + 2\mathbf{E}^{G} = \Lambda^* \cdot \Lambda \end{aligned} \tag{5.1}$$

in which the $\mathbf{E}^{\bar{A}}$ and $\mathbf{E}^{G}$ are identified as the Almansi-final and the Green-initial strain tensors discussed above. In general, the orientation of the principal axes of the two ellipsoids differs in space. However, as shown in ref. [10], the principal axes of the *Green-final* strain tensor $\mathbf{E}^{\bar{G}}$, similarly those of the deformation tensor defined by

$$\mathbf{D}' = \mathbf{I} + 2\mathbf{E}^{\bar{G}} = \Lambda \cdot \Lambda^* \tag{5.2}$$

are coaxial with the Cauchy tensor $\overline{\mathbf{D}}$.

If strain is regarded only as a parameter facilitating linking of the deformation and the (symmetric) stress by an isotropic tensor relation, an additional provision should then be imposed on the symmetric strain tensor. Since the stressed state is the final, equilibrium state of the deformed body, we demand

(v) The principal directions of the strain tensor and of the deformation tensors $\overline{\mathbf{D}}$ or $\mathbf{D}'$ should coincide.

Provision (v) limits by far the variety of the strain tensors. In fact, for $\mathbf{E}$ to satisfy all the provisions (i)–(v) it suffices that

$$\mathbf{E} = f(\mathbf{E}^{\bar{G}}) \tag{5.3}$$

or, alternatively,

$$\mathbf{E} = g(\mathbf{E}^{\bar{A}}), \tag{5.4}$$

where f, g are isotropic polynomial functions of the arguments (again, $\mathbf{E}^{\bar{G}}$ and $\mathbf{E}^{\bar{A}}$ are not independent!). What remains is to specify the functional relations f, g.

Evidently, the simplest relations are $\mathbf{E} \equiv \mathbf{E}^{\bar{G}}$ or $\mathbf{E} \equiv \mathbf{E}^{\bar{A}}$. A general procedure, yielding the exact form of the functions f, g so that the principal components of $\mathbf{E}$ correspond to prescribed functions of the principal stretches of the deformation, was established by Hanin and Reiner [11]. As an illustration, the explicit form for the Hencky (logarithmic) strain tensor $\mathbf{E}^{\bar{H}} = g(\mathbf{E}^{\bar{A}})$ was derived.

Rivlin, in a number of papers,† approached the problem from a different angle. Starting from the relation

$$\mathbf{E} = \mathbf{E}(\Lambda) \tag{5.5}$$

where **E** is not necessarily a polynomial function in the argument Λ; next, imposing the conditions that (5.5) is: firstly, invariant under a rigid body rotation; secondly, isotropic—the result turned out to be

$$\mathbf{E} = f(\Lambda \cdot \Lambda^*, \mathbf{I}) \tag{5.6}$$

in conformity with (5.3). By following Rivlin's method it can be shown that, by starting from

$$\mathbf{E} = \bar{\mathbf{E}}(\Lambda) \tag{5.7}$$

instead of (5.5), the final result reads—in conformity with (5.4)—

$$\mathbf{E} = g(\bar{\Lambda}^* \cdot \bar{\Lambda}, \bar{\mathbf{I}}). \tag{5.8}$$

The order of the conditions imposed on (5.5) and leading to (5.6) is commutative, namely, isotropy can be imposed first and the relation then subjected to the condition of invariancy under a rigid-body rotation. To show this, we start from the general isotropy (1.3) rewritten for the symmetric tensor $\mathbf{E}(\Lambda)$ as

$$\mathbf{E} = \mathbf{E}(\Lambda, \Lambda^*) = \aleph_0 \mathbf{I} + \beth_1(\Lambda + \Lambda^*) + \gimel_1(\Lambda^2 + \Lambda^{*2}) + \gimel_2 \Lambda\Lambda^* + \gimel_3 \Lambda^*\Lambda + \ldots \tag{5.9}$$

Let $\mathbf{e}_i$ and $\mathbf{e}_i'$ ($i = 1, 2, 3$) be two bases at the same field point P. The conditions that the two sets differ only by a rigid body rotation (no deflection) are expressed by

$$\begin{aligned} h_{ij} &\equiv \mathbf{e}_i \cdot \mathbf{e}_j = \mathbf{e}'_i \cdot \mathbf{e}'_j \equiv h'_{ij} \\ E \equiv \mathbf{e}_i \cdot \mathbf{e}_j \times \mathbf{e}_k &= \mathbf{e}'_i \cdot \mathbf{e}'_j \times \mathbf{e}'_k \equiv E' \; (i, j, k = 1, 2, 3 \text{ cycl}). \end{aligned} \tag{5.10}$$

The *rotation tensor* $\boldsymbol{\theta}$ is the bi-linear vector form defined, under conditions (5.10), by

$$\boldsymbol{\theta} \equiv \mathbf{e}'^{\alpha}\mathbf{e}_{\alpha} = \mathbf{e}'_{\alpha}\mathbf{e}^{\alpha}. \tag{5.11}$$

Evidently,

$$\boldsymbol{\theta} \cdot \boldsymbol{\theta}^* = \boldsymbol{\theta}^* \cdot \boldsymbol{\theta} = \mathbf{I}.$$

The projection of a vector $\boldsymbol{v}$, similarly of a second-order tensor **T**, on θ/m rotates the vector or the tensor in space, leaving the magnitudes unchanged. Thus, the condition of invariancy under a rigid-body rotation effected upon **E** implies that the rotated tensor **E′** should be the same

† [12, sect. 7], [13], [14.]

original function, now of its rotated arguments. In formula,

$$\mathbf{E}' \equiv \boldsymbol{\theta}\cdot\mathbf{E}\cdot\boldsymbol{\theta}^* = \mathbf{E}(\Lambda', \Lambda'^*) = \mathbf{E}[\nabla(\overline{\mathbf{r}'}), \nabla(\overline{\mathbf{r}'})^*] =$$
$$= \mathbf{E}[\nabla(\boldsymbol{\theta}\cdot\overline{\mathbf{r}}), \nabla(\boldsymbol{\theta}\cdot\overline{\mathbf{r}})^*] = \mathbf{E}[(\boldsymbol{\theta}\cdot\nabla\overline{\mathbf{r}}), (\nabla\overline{\mathbf{r}}^*\cdot\boldsymbol{\theta}^*)] =$$
$$= \mathbf{E}[(\boldsymbol{\theta}\cdot\Lambda), (\Lambda^*\cdot\boldsymbol{\theta}^*)]. \tag{5.12}$$

Setting (5.9) in (5.12), it is readily observed that (5.9) must assume the form

$$\mathbf{E} = f(\Lambda\cdot\Lambda^*, \mathbf{I})$$

where f is a polynomial function of the arguments stated—a result coincident with (5.6). Likewise, subjecting the strain tensor, expressible in terms of the final gradients $\overline{\Lambda}$, $\overline{\Lambda}^*$, to the invariancy condition (5.12), we obtain (5.8).

6. SUMMARY AND CONCLUSIONS

The general isotropic strain tensor, representing the relative change in the geometrical configuration of the deformed body, is expressible as a second-order polynomial tensor function of the gradients, either as coordinitate gradients—initial or final—or as displacement gradients, subject to certain provisions (listed in Section 2).

It is shown that the most general, isotropic polynomial function between two asymmetric tensors is reducible to a finite form consisting of a set of twenty-three independent terms. When symmetrized, it is further reduced to a set of thirteen independent terms. All strain tensors known in the literature are merely particular cases of the general expression.

If, in addition, the condition of invariancy under a rigid-body rotation is imposed upon the expression for the strain tensor, the polynomial basis is narrowed down to a single independent term, the rest being powers (including the zero-th power) of this term. The equivalent statement is that any isotropic strain tensor is expressible as a polynomial function of either the Green-final or the Almansi-final strain tensors.

The absolute notation for tensors, as aggregates of magnitudes and directions pertaining to a physical quantity, has been employed throughout the text. To obtain expressions for all the above formulae in terms of physical components, with respect to orthogonal or oblique bases, or even in terms of generalized coordinates and analytical components, the procedure presented in ref. [15] may be pursued.

ACKNOWLEDGEMENT

The results reported here were partly obtained under an appointment of research associateship at Brown University during 1963, and in the course of research sponsored by the Air Force Office of Scientific Research under Grant AF-EOAR 66–19, through the European Office of Aerospace Research (OAR), United States Air Force.

REFERENCES

1. M. Reiner, "A mathematical theory of dilatancy", *Amer. J. Math.* **67**, 350–62 (1945).
2. C. Truesdell, "The mechanical foundations of elasticity and fluid dynamics", *J. Rational Mech. Anal.* **1**, 125–300 (1952).
3. H. W. Turnbull, *The Theory of Determinants, Matrices and Invariants*, Dover (1960).
4. J. W. Gibbs, *Scientific Papers*, vol. 2, Dover (1961).
5. R. S. Rivlin, "Further remarks on the stress–deformation relations for isotropic materials", *J. Rational Mech. Anal.* **4**, 681–702 (1955).
6. F. O. Murnaghan, "Finite deformations of an elastic solid", *Amer. J. Math.* **59**, 235–60 (1937).
7. Z. Karni and M. Reiner, "Measures of deformation in the strained and in the unstrained state", *Bull. Res. Counc. of Israel* **8c**, 89–92 (1960).
8. Z. Karni and M. Reiner, "The general measure of deformation", Proc. IUTAM symposium on *Second-order Effects in Elasticity, Plasticity and Fluid Dynamics* (edited by Reiner and Abir), Haifa, April 1962, Pergamon and Jerusalem Academic Press, pp. 217–27 (1964).
9. C. Truesdell and R. Toupin, "The classical field theories", in S. Fluegge (Editor), *Encyclopedia of Physics* 3/1, Springer (1960).
10. Z. Karni, "On the analysis deformation and the strain tensors", in Abir, Ollendorff and Reiner (Editors), *Topics in Applied Mechanics*, Elsevier, pp. 287–98 (1965).
11. M. Hanin and M. Reiner, "On isotropic tensor functions and the measure of deformation", *Z.A.M.P.* **7**, 377–93 (1956).
12. R. S. Rivlin and J. L. Ericksen, "Stress–deformation relations for isotropic materials", *J. Rational Mech. Anal.* **4**, 323–425 (1955).
13. A. E. Green and R. S. Rivlin, "The mechanics of non-linear materials with memory", Part 1, *Arch. Rational Mech. Anal.* **1**, 1–21 (1957).
14. R. S. Rivlin, "Constitutive equations for classes of deformations", in *Viscoelasticity: Phenomenological Aspects*, Academic Press, pp. 93–108 (1960).
15. Z. Karni, "A unified equation of motion in the non-relativistic mechanics", *Israel J. Tech.* **2**, 287–94 (1964).

CATEGORIES OF INFORMATION†

PAUL LIEBER

University of California, Berkeley, U.S.A.

ABSTRACT

This paper introduces the concept "Categories of Information", and describes how this concept emerged and how these categories were identified in the course of a comparative study of some of the various known formulations of the principles of classical mechanics. It attempts to show that the existence of these categories of information is an aspect of nature, and that in general, information considered in its usual broad sense consists of manifold and complex natural phenomena which apparently can be structured in terms of distinguishable yet related categories.

A comparative study of the principles of classical mechanics has revealed that they are only conditionally equivalent, and that questions concerning their equivalence and completeness cannot be put with meaning, without naturally evoking a new concept pertaining to the existence in nature of categories of information. The emergence of this concept in the study, shows that the equivalence or non-equivalence of the principles of mechanics, considered as propositions about the world of mechanical experience, should be decided according to the nature of the information which they do and can render explicit about it.

The observations and conclusions noted above were developed by focusing attention on the principles of Newton, of Gauss and of Hertz. In so doing, it has been demonstrated that general and fundamental global information on the distribution of internal forces in many-body systems, which is rendered explicit and without integration by the principles of Gauss and Hertz, at present appears inaccessible in terms of the principles of Newton [10]. This information is obtained within the edifice of the principles of Gauss [3] and Hertz [4], by reintroducing and underlining therein the concept of force, which they sought to eliminate as a primitive notion by its geometrization in terms of geometrical constraints. This has been done for a non-trivial class of mechanical systems which include the gas model used by Maxwell, by establishing and using what appears to be a fundamental connection between non-holonomic unilateral geometrical constraints and the impenetrability of matter. In so doing it is found that the primitive role ascribed by Newton to the concept force is linked with the primitive concept of impenetrability of matter which is the physical basis for its geometrization in terms of the geometrical constraints used by Gauss and Hertz.

This in effect establishes within the edifice of classical mechanics, the impenetrability of matter as the physical foundation of force; and the geometrical constraints as an (geometrical) intermediary between force and impenetrability; and accordingly as a geometrical manifestation of their ontological connection.

† Presented as invited lecture to the Donner Laboratory Seminar of University of California, Berkeley, on October 18, 1965.

The paper is summarized by drawing attention to an eleventh category of information which includes some of the most fundamental propositions of Science and Mathematics.

PRELIMINARY REMARKS AND INTRODUCTION

Information as an aspect of nature appears to be manifold and complex in its manifestations in natural phenomena, particularly in biological phenomena where it is stored, transferred and used with remarkable precision. Biological phenomena are distinguished by the remarkable stability, precision and organization characteristics they display in the storage, in the transfer and especially in the putting to use of information, that is, in performance. I am going to write about categories of information, and describe how this concept emerged and how these categories were identified in the course of a comparative study of some of the various known formulations of the principles of classical mechanics. In so doing I will attempt to show that the existence of these categories of information is an aspect of nature, and that in general, information considered in its usual broad sense consists of manifold and complex natural phenomena which apparently can be structured in terms of distinguishable yet related categories.

The concept "Categories of Information" first emerged in a comparative study of the principles of classical mechanics, concerned with questions pertaining to their equivalence and completeness. This study led inexorably to the concept of categories of information and to the realization that they are an aspect of nature; because it was found that questions concerning the equivalence and completeness of the various known formulations of the propositions of classical mechanics, cannot be posed and hence resolved with meaning, unless this concept is necessarily evoked. More specifically, this concept appears essential in trying to make sense of and reconcile some definite results obtained in this comparative study in classical mechanics, some of which will be reported in this paper. The fact that categories of information appear to be inextricably linked with a most fundamental and well-tested science; namely, classical mechanics, leads to the inference that these categories exist in nature and that they are a fundamental aspect of all natural phenomena.

Once the idea of "Categories of Information" came to mind, we found by producing specific examples entrenched in mechanics and in other realms of experience, that there exist in nature at least ten distinguishable categories of information, and that information considered in its most general aspect, evolves and is structured in these categories. These examples once cited will, I hope, appear familiar in terms of our every day experience as well as our more sophisticated and profound experiences. This I believe attests significantly to their verity since an ultimate criterion for the acceptance of an idea is consensus.

An attempt will be made here to consider in some depth each category which will first be pointed out by an example. For this purpose I will first appeal to specific results obtained in the comparative study noted previously, which focus primarily around the propositions of Newton and those formulated by C. Gauss [3] and H. Hertz [4]. We will also refer to certain aspects of pure and applied mathematics as well as the mechanics of fluids in this regard. In so doing we will demonstrate that the idea of categories of information and an awareness of their existence, can be used to obtain valuable analytical information concerning the behavior of fluids which has been inaccessible until quite recently.

It is also relevant to report here by way of introduction on what appears to be a basic connection between some general information obtained from the comparative study in mechanics noted and the thermodynamics of non-equilibrium processes, as this connection seems to have implications in biophysics and may furnish a basis for finding a general thermodynamical principle conditioning non-equilibrium processes. Some special attention will be given to the concept force, as force, particularly in some global scalar measure, appears from our work to be the basic instrument and vehicle in the production and transfer of information. In this connection we shall briefly consider some questions concerning its ontological basis and how these relate to attempts made by Gauss, Hertz and Einstein to establish force on a purely geometrical foundation. These considerations of course bring us into the realm of physical as distinguished from strictly formal, i.e. abstract geometry. It is by examining such questions that we reach the conclusion that the profound and powerful edifice in which Hertz formulated his variational principle, is nevertheless incomplete, because the geometrical constraints to which he appeals for geometrizing force, is devoid of an ontological basis; that is, a basis for their existence in nature—that is a basis in being. It is hoped that these considerations and results which bear on the nature of force, may contribute toward producing an adequate and increasingly precise concept of force; which according to C. Truesdell always has been and still remains, the central and only essentially non-trivial problem at the foundations of classical mechanics. Whereas I agree that it is indeed a central problem, it is certainly not the only one.

A concept, which from our work appears to be fundamentally related to the nature of force, and which may therefore have to be accommodated by any serious attempt to resolve this central problem which rests at the foundations of classical mechanics, is the concept of impenetrability of matter. An attempt was made by Hamel to axiomatize this idea which can be traced back at least to Leibnitz, but this proved unsuccessful in so far as it did not produce any derivable consequences in the form of theorems; which, of course, is the essential justification for the axiomatic method. Indeed Hamel's formulation of an axiom of impenetrability is unclear and this may account for its impotency in his work [1]. Another of his axioms, which was criticized by McKinsey, Sugar and Suppes in their work on the axiomatization of classical mechanics [2], is one which attempts to relate force to geometry, and in this respect bears some relation to the central idea underlined in Gauss' and Hertz's formulation of the propositions of classical mechanics. McKinsey, Sugar and Suppes, in commenting on these two axioms of Hamel, remark "that an axiom of impenetrability does nothing but complicate proofs" and is therefore omitted in their system, and concerning the second axiom: "One does not see how this axiom could intervene in the proofs of theorems, or in the solution of problems." In other words what we may refer to as Hamel's impenetrability axiom and his geometrical axiom are virtually considered as non-essential and therefore for reasons of convenience are dismissed by McKinsey, Sugar and Suppes. From what we have learned by our work this makes it abundantly clear that they did not realize that both of these axioms are related to the concept force in an essential way, and that what Hamel considered to be separate axioms are, in fact, fundamentally interconnected and thus interdependent. I will try to clarify this statement in what follows.

Also by way of introduction, I should draw attention to the fact that the known propositions of classical mechanics express either directly or indirectly, that classical mechanical systems are everywhere and at all times in equilibrium with respect to all forces acting upon and within them. According to D'Alembert the accelerations produced in dynamical systems are but manifestations of this condition; namely the accelerations are accompanied by forces produced to maintain force equilibrium in such cases where it cannot be maintained statically. Although the equilibrium of forces is a general characteristic of the behavior of classical mechanical systems, and the propositions of mechanics express this fact, stability phenomena, which are equally fundamental and characteristic in the performance of such systems, do not have associated with them general propositions. This is the case even though equilibrium and stability are in fact distinct and mutually independent phenomena. In other words, the known propositions of classical mechanics are general propositions which in effect refer exclusively to the equilibrium of forces and say nothing about stability. Accordingly stability phenomena have so far been examined analytically within the edifice of classical mechanics, by arbitrarily prescribing stability criteria whose selection is guided by examining experiments pertinent to each particular case. These considerations naturally lead to asking whether a general stability principle operates in classical mechanical systems and if it does, what restrictions does it impose upon their performance. It is evident that these restrictions must be compatible with the known equilibrium principles of mechanics which imply micro-temporal reversibility and consequently the absence of a driving force that imparts to them an evolutionary process and thereby historical commitment.

Although the edifice of Hertzian mechanics is incomplete as a scientific theory it is nevertheless conceptually more comprehensive than that of Newton, and can therefore accommodate a new and general proposition which is distinct from and compatible with Hertz's variational principle, and which has the nature of a general stability principle of classical mechanics. Of course, the question of its truth remains to be tested against experience and is therefore open. At this time there does not seem to be a way of formulating its counterpart within the framework of Newtonian mechanics. This situation is but one aspect of the non-equivalence between the edifices of Newton and Hertz as in one we can formulate a general proposition with information content which the former apparently does not accommodate. As far as we can see, the counterpart of the stability principle in Newtonian mechanics would be tantamount to a statement about the time evolution of a scalar measure of the forces acting within the system which in Hertz's edifice are linked with geometrical constraints. Some work we have done in hydrodynamics, by applying the concept of categories of information, points to a connection between the function of the dynamical non-linear terms of the Navier–Stokes equations in the evolution of viscous flow fields, and the stability principle cited above.

So far I have attempted to sketch in outline the materials and considerations that have gone into the study which led to the concept of categories of information and to the identification by examples of some of its members. I will now present some of the specific results which led to this concept and the identification of some of the categories of information.

THEORETICAL CONSIDERATIONS

The various known formulations of the principles of classical mechanics are generally considered as different ways of saying the same thing, and in this rather loose meaning they have been tacitly assumed to be equivalent. Indeed, as far as I know, questions concerning the possibility of their not being equivalent have not been explicitly raised in the literature, even though Lancoz did allude to this possibility in his beautiful book, *The Variational Principles of Mechanics.*

Newtonian mechanics renders without integration, and thus directly, general information on co-operative phenomena produced in many-body systems. As a consequence of Newton's third law, this information is necessarily independent of the internal forces, and conversely; it reports nothing about the internal forces. The reason for this is that Newton's fundamental proposition $\bar{F} = m\bar{a}$ refers strictly to a single spatially isolated particle, and characterizes its mechanical environment which is ascribed to the universe, in which it exists, by a symbol $\bar{F}$. This symbol is understood to represent the sum of all forces that the universe impresses upon the particle. If with D'Alembert we add to the resultant force $\bar{F}$, the inertia force ascribed to the acceleration of the particle, and interpret Newton's proposition as a principle involving the equilibrium of forces; then according to Mach's principle the inertia force is also an expression of the existence of the particle in the universe, since according to Mach its inertia is a global manifestation of its existence in nature. These considerations reflect on the fact that Newton's fundamental proposition which underlines the atomistic rather than the global attributes of mechanical systems, is endowed with fundamental difficulties which Einstein to a large extent resolved. They also remind us that then even the fundamental aspects of what we presently call classical mechanics are apparently incomprehensible in terms of an atomistic conception of nature, which of course is crucial in the formulation of Newton's fundamental proposition. Newton's interpretation of inertia differed fundamentally from Mach's, as Newton regarded inertia as a strictly local rather than global property of a particle. With this in mind, it is incisively clear that Newton's fundamental proposition refers explicitly and solely to a single spatially or isolated particle, and not in any sense to a collection of particles, and to their macro-mechanical processes.

When we consider a collection of material particles from the standpoint of Newton's fundamental proposition, one obtains a set of conditions which express by repetition, the same proposition for each and every particle, rather than one statement (condition) which refers to the mechanical system as a whole, that is, collectively. Newton's fundamental proposition, applied to a mechanical system, consisting of a collection of particles, does not therefore directly express or communicate any information about the global mechanical properties of such a system; and in fact cannot do so directly because the macro-mechanical aspects of the system do not enter at all in the formulation of this proposition.

Newton's third proposition, which expresses a property of symmetry in the *magnitude* of forces of action and reaction between bodies, is the only proposition in his system of mechanics which explicitly refers to the existence of more than one particle. In so doing it underlines that all interactions between such particles are equivalent to sets of binary interactions between them, where the forces produced in each binary interaction comply with the third proposition. Since Newton's third proposition expresses a condition of anti-symmetry (mirror symmetry) with respect to the direction of each pair of forces produced in each binary interaction, any and all global information sought and obtainable by summing over the set of equations which individually express Newton's fundamental proposition for each particle of a system, will necessarily be devoid of the internal forces and therefore cannot in principle render any information about them.

Gauss formulated a general and fundamental principle of mechanics by ascribing to force, a purely geometrical and thus formal basis. This was motivated by an attempt to make precise the concept force, by entrenching it in the primitives of geometry. Gauss' motivation apparently stemmed from the observation that Newton's fundamental proposition may be considered to be a statement which defines force in terms of motion, rather than a principle in which the concept force is regarded as primitive, and which I believe corresponds to Newton's own interpretation. This conclusion is based on the fact that Newton ascribed *a priori*, to the symbol $\bar{F}$, and therefore to what it represents in nature, an unrestricted covariance, in the sense that in the general case the force $\bar{F}$ so to speak belongs to the body on which it is impressed, and is therefore the same in any and all frames of reference to which the motion may be referred. In particular, therefore, it is the same force for non-inertial as well as for inertial frames. The force that the symbol $\bar{F}$ represents in Newton's fundamental proposition is the sum of all the mechanical interactions a material particle experiences and thus senses in the universe, and as such, represents an intrinsic connection between the particle and the universe; as this connection is independent of any particular frame in which its consequences, say in motion, may be examined.

The above way of thinking about force and interpreting the condition $\bar{F} = \bar{F}'$ prescribed *a priori* by Newton, and supported by experiment, is apparently connected with Mach's principle, because the same interpretation was given by Mach to the inertial mass of a body. From this it follows that the force $\bar{F}$ to which Newton assigned *a priori* unlimited covariance is in general more fundamental than is the acceleration of a particle, which limits the covariance of Newton's fundamental proposition to inertial frames. It is interesting to note that the inertial mass in $\bar{F} = m\bar{a}$, which according to Mach is also a manifestation of a connection between the particle and all matter in the universe, enjoys the same unrestricted covariance as does $\bar{F}$, and is met naturally in a formal way because it is a scalar. From these considerations based on classical mechanics we are led to the conclusion that the concept force is indeed more fundamental than is the idea of a particle in motion, which may be thought of as a very special and restricted manifestation of force within realms of classical mechanics. If this is correct then it would be naïve to regard Newton's fundamental proposition as a statement of definition of force.

Even though what motivated Gauss's search for a new formulation of the principles of classical mechanics appears to be invalid, his variational principle of mechanics has imparted to classical mechanics new dimensions and areas of inquiry; particularly as the procedures he uses to geometrize force do not necessarily conflict with the unrestricted covariance which Newton ascribed *a priori* to force.

Gauss' principle is a differential variational principle that consists of one statement which refers to and conditions a global scalar measure of a mechanical system, consisting of an arbitrary number of discrete bodies subject to impressed forces. To my knowledge, it and Hertz's variational principle are the only strictly minimum principles of classical mechanics. Gauss' principle does indeed therefore *directly* express and communicate information which pertains to the macro-mechanical processes that are produced in many-body systems.

As noted, Gauss formulated his new and fundamental variational principle of mechanics by ascribing to force a purely geometrical and thus formal basis; but he did retain in his formulation forces impressed on particles from sources outside of the configuration. Hertz extended and completed Gauss' work by ascribing to all forces in nature a geometrical basis. The reduction of all forces to geometry, as effected by Hertz, is strictly a formal process whose scientific justification now appears to be contingent upon extraformal, i.e. ontological, considerations. This means that the very basis in nature for the geometrization of force by Gauss and Hertz may be extra-mathematical; that is, in principle not amenable to formal representation. This question which concerns the basis in nature and thus the justification in experience for the geometrical constraints used by Hertz to geometrize all forces is most fundamental, because it must be considered, and at least conceptually resolved, in order to relate the formally stated propositions of Hertz and their formally obtained consequences, to experience. Otherwise Hertz's mechanics remains, scientifically speaking, a sterile, i.e. a dead, document. The fact that this question had apparently not been recognized and dealt with sooner may account for the fact that the potential of Hertz's profound conceptual edifice has remained dormant for so long.

A crucial idea, which in my opinion is the key to the resolution of this basic question, is the fundamental connection that exists between the impenetrability of matter and the production in nature of stringent geometrical constraints. This idea which was introduced and used in refs. [5, 6, 7, 8, 9, 10] merges geometry with substance, and underlines the existence in nature of a substantial basis for its geometry which is here attributed to the impenetrability of matter. In other words, impenetrability is here considered as the ontological and thus physical basis and support for the geometrical constraints by which Hertz formally geometrized force. In this view, the geometry of nature is embedded, that is sculptured in matter by virtue of its impenetrability which is also the basis for the production of templates in nature, which are, of course, important in current biological thought. It appears therefore, according to this thinking, that ultimate processes involved in natural phenomena do not only concern the geometrization of matter but equally, if not even more fundamentally, the materialization of geometry.

The notion of impenetrability is essentially ontological and it therefore does not have a formal representation. From the above considerations, it appears to be the crucial concept by which the geometrical ideas formally developed by Hertz can be related to experience and in this way bring the mechanics of Hertz, at least conceptually, to *completion* as a scientific theory. By his important work, Hertz has forced us to think in a way which has revealed what may be a universal connection between force and impenetrability; the geometry of nature being but a manifestation of this universal connection.

A necessary condition, that a scientific theory be complete, is that it give consideration to its ontological basis, that is, to the basis in fact, in experience and thus in nature (which support its fundamental propositions), and that it specify, at least conceptually, the rules and operations by which to relate its formal propositions and their formal consequences to experience. By a scientific theory I mean here, a theory consisting of propositions that are presumed to pertain to the real world, and which can lead to the conception of experimental arrangements and experiments, in which and by which they can be tested.

From these considerations, which are amplified in the section entitled "Concluding Remarks and Observations", it follows that the mechanics of Hertz, which affords greater range and depth in formal expression, and which is more general in its formal structure than is the mechanics of Newton, is nevertheless an incomplete scientific theory. The reason is that it lacks an ontological foundation, since the formal statements of geometrical constraint which it uses to formally geometrize forces in nature are not in any way related to and therefore justified in actual experience. As a consequence, in its original construction, it is strictly a formal theory, consisting of symbols and specified operations upon them which are devoid of a map that imparts to them experimental content. Hertz was, of course, deeply concerned with experimentation, and therefore questions concerning the verity of his mechanical edifice, and the propositions he formulated within it, were of paramount importance to him. In this regard, however, he appealed to the fact that Newton's fundamental proposition, $\bar{F} = m\bar{a}$, can be derived from his proposition, by associating certain combinations of symbols obtained from his theory with the symbol $\bar{F}$, representing the concept force in Newton's mechanics.

In this way we see that Hertz's mechanics which was motivated by a search for a rational basis for force in geometry, and thereby for its elimination in the propositions of mechanics, must nevertheless, though indirectly, refer to force by way of Newtonian mechanics, in order to relate its own propositions and their consequences to mechanical experience. According to this procedure, Newton's mechanics furnishes a bridge and thus a map which relates Hertz's formally stated propositions, to scientific experimentation. Since Hertz's mechanics does not provide such a map within its own edifice, it is therefore incomplete as a scientific theory. However, the fact that Hertz's mechanics yields formally the proposition of Newton attests to the fundamental significance of the formal representation of the geometrical constraints it uses to geometrize force as an aspect of nature; thereby giving indirect but strong support to the concept that the impenetrability of substance is the physical-ontological basis of forces in nature.

The mechanics of Gauss and Hertz as strictly formal theories give fundamental and useful information in abstract form, when they are related to the world of scientific experience by reintroducing into their framework the concept force, which they formally eliminated by abstract geometry. In this way, general and useful information on the distribution of internal forces and global uniformity is made explicit without quadrature for a non-trivial class of mechanical systems to which the model used by Maxwell in his kinetic theory of gases belongs. This information, which bears on uniformity in cooperative phenomena and which is valid for equilibrium as well as non-equilibrium processes, is not and apparently cannot be rendered explicit by Newtonian mechanics as general information, for reasons presented earlier.

In order to make sense of these results and reconcile them with the question concerning the equivalence of the various formulations of the propositions of mechanics, we are compulsively led to the concept of "Categories of Information"; that is, to the idea that there exists in nature various kinds of information and that different formulations of the propositions of mechanics may have the same as well as different information content which belong to these categories.

Thus, for example, whereas Newton's propositions may be equivalent to the propositions of Gauss–Hertz modified by the reintroduction of the concept force, with respect to the category of implicit information, they are not equivalent with respect to the category of explicit information. In other words, Newton's and Gaussian–Hertzian mechanics are only conditionally equivalent, where the conditions refer to different categories of information.

The information content of a theory may be defined as the information it renders explicit which can be related to experience. According to this definition the propositions of Newton and those of Gauss and Hertz are not equivalent in their information content. Newton's propositions are formally derivable by deductive methods from Gauss' and Hertz, but the converse does not seem to be the case.

These considerations which led to the notion of categories of information naturally led to the search for examples of various categories and some examination of the examples found. I shall mention ten such categories of information and briefly discuss examples of each category:

1. The Category of Implicit Information.
2. The Category of Explicit Information.
3. The Category of General Information.

4. The Category of Specific Information.
5. The Category of Qualitative Information.
6. The Category of Quantitative Information.
7. The Category of Exact Information.
8. The Category of Approximate Information.
9. The Category of Abstract Information.
10. The Category of Concrete Information.

An example of information contained in the category of implicit information is the information which is locked in the mathematical formulation of a natural law, which purports to govern and thus furnish a basis for predicting natural phenomena examined in a laboratory. The task of making this information explicit is usually regarded as a strictly mathematical endeavor, and a scientific theory is considered closed when it prescribes all information relevant to a class of natural phenomena, in the category of implicit information. Thus classical mechanics is regarded by most to be a closed science, since all its propositions are presumed known, and in such cases when more then one formulation of its propositions are considered, they have been tacitly assumed to be equivalent without qualification; which according to the thesis of this work is incorrect.

The conditional equivalence of the various propositions of classical mechanics, which stems from the fact that they may be equivalent in one category of information but not in another, can and has been used to transform information from one category to another. In a sense the various formulations of the principles of mechanics complete and thus complement each other from the standpoint of furnishing information in its various categories. With this in mind it becomes clear that the transformation of mechanical information from the implicit to the explicit category is not strictly a mathematical enterprise, as it can be significantly enhanced by feeding information obtained from one formulation of the propositions of mechanics into another formulation, thus rendering explicit the implicit information contained in the latter.

We know of very few examples in theoretical hydrodynamics based on the Navier–Stokes equations where the implicit information which they contain is rendered explicit by strictly mathematical procedures. Indeed it is these equations that gave Von Neumann reason to design a sophisticated computer, because hardly any information sought was made accessible, even to him, on strictly mathematical grounds. Conversely, we know of a number of important examples in which information implicit in the Navier-Stokes equations was rendered explicit by extra-mathematical procedures.

In these cases information is usually drawn by perceptive examination of natural phenomena and then using this information to prescribe simplifying assumptions that are appropriate to the boundary conditions under which such phenomena are produced. In so doing the simplified equations are made mathematically tractable and their solutions are found to agree with experiment with satisfactory approximation, thereby verifying *a posteriori* the simplifying assumptions by which the solutions are obtained. This is in effect tantamount to prescribing explicit information to the Navier–Stokes equations for a given set of boundary conditions, which the solutions to these complete equations would contain were they accessible in the category of explicit information. In other words, it is as though we are telling the Navier–Stokes equations in the category of explicit information what they already know in the category of implicit information. This is in effect the gist of the art of applied mathematics which can gain significant power within the realm of classical mechanics, by benefiting from the fact that the various formulations of its propositions are only conditionally equivalent. Indeed, we have in this way constructed an analytical representation for a viscous-incompressible flow field around a solid obstacle based on the Navier–Stokes equations, which accounts for the production of eddies behind the cylinder and the force imparted by the fluid to the obstacle for a significant range of Reynolds number.

Examples of the categories of general and specific information in classical mechanics can be given by referring to a general theorem on the distribution of internal forces pertaining to a class of mechanical systems [10]. The information rendered by this theorem is appropriate for every member of the class, and in this sense is general. Analytical information which may be obtained concerning the dynamics of a specific body subjected to a specific mechanical environment is in the category of specific information. The study of general dynamics and the theory of dynamical systems is concerned principally with obtaining *explicit* general information which, in many cases, is topological and thus *qualitative*; so that general and qualitative information appear to be very much interconnected. Correspondingly specific and quantitative information appear interconnected and are produced concurrently. Specific information, whether derived by measurement and/or by numerical computation, is necessarily approximate, and this shows a connection between the categories of specific, quantitative and approximate information. The category of exact information is closely merged with processes, i.e. natural phenomena in the mind, which accompany conceptualization in general and abstract thought in particular. These processes are associated with internal experience and as such fall out of the realm of *objective* experience to which science has committed itself as a matter of convention, in order to help with the book-keeping of events. However, these are the processes which lead to mathematical invention, and by which precisely and thus exactly formulated propositions, i.e. axioms, are discovered and conceived in the human mind. The fundamental connection which apparently exists between the category of abstract information whose information content (meaning) is contingent upon and derived from the processes it evokes in the mind, and the category of exact information, both of which are associated with strictly mental experiences, deserves special mention.

Abstract and exact information are produced by the interplay of symbols endowed with prescribed operations, and mental processes; in which case the mind is the site of the accompanying experience, which is strictly internal. A piece of mathematics is completely dead unless it engages a mind in which it invokes processes that respond to its formal language. A vital aspect of processes of mathematical comprehension and invention is constantly maintaining an interface between formal language, i.e. symbols and mental processes, and thus experiences attendant to them which are produced in the mind.

The category of concrete information in a restricted sense consists of information that can be put to use by manipulation with a hand, and is thus related to the kind of objective experience on which science and technology focus attention. However, in a larger and a more profound sense it encompasses information which concerns the nature of being and existence and thus the basis of all experience and natural phenomena. From what we attempted to describe of the interplay between the use of symbols and mental experience which they invoke, we recognize an interface between the categories of abstract, exact and concrete information accompanying internal experience in the mind. Concerning the category of concrete information, I should point out that it must necessarily be represented in every theoretical discourse which purports to relate to human experience, be it internal or external. The most fundamental and unresolved conceptual and philosophical problems, namely questions concerning ontology, i.e. existence in nature, belong to this category. In particular the concept of impenetrability of matter as the ontological basis of all forces and the geometry of the (world) universe belongs to this category. The category of qualitative information is further exemplified by the work of Poincaré on the strictly qualitative aspects of mechanical systems, which led to his important discoveries in topology. This work was undertaken after realizing that explicit quantitative information, which is necessarily approximate, is hard to obtain in analytical representation and is virtually *inaccessible* for systems even with moderate complexity – the three-body problem. Instead, considerable qualitative and general information about dynamical systems has come forth and much of it also belongs to the category of exact information. In the context of these considerations it appears relevant to point out that Jean Piaget has noted connections between group theoretic concepts and the functions of the mind which accompany learning and development. This, I believe, is not fortuitous as it points to a fundamental connection between the function of the mind in the development of a child, that is, in accommodating and manipulating reality with the hand; and the development of abstract ideas, as an aspect of internal human experience, and as an instrument for perceiving truth in its most precise and thus purest aspects. This question which deeply concerned W. K. Clifford, a most brilliant and inventive man of the nineteenth century, is very much entrenched in Kant's thesis; that the mind can perceive truth *a priori*; which I think was without justification shattered by the development of non-euclidean geometries.

I would like finally to remark on an eleventh category of information which is distinguished by circumstance that it contains propositions, all of which are cannot statements. This category includes the most fundamental propositions in science such as the natural laws, as well as the most fundamental theorems quite recently established at the foundations of mathematics. It is becoming increasingly clear that the so-called atomic propositions, that is the irreducible ones which Wittgenstein struggled to identify, will fall into this very special category. These so-called atomic propositions now appear to be those which assert the existence of a universal constant in nature and which therefore are the epitome of a cannot proposition, that is, the most elementary kind of cannot statement; since they assert in the most primitive manner the immutability of a condition in nature and thereby that this condition *cannot be changed* – cannot be manipulated. The propositions that belong to this fundamental category are distinguished by *informing* what cannot be done, and in particular therefore what man cannot do.

CONCLUDING OBSERVATIONS AND REMARKS

If we pose the problem, whether actual mechanical systems satisfy formal propositions, such as, for example, those formulated by G. Hamel [1] and by McKinsey, Sugar and Suppes [2], to axiomatize classical mechanics, we must first give physical definitions, that is, consider the ontological basis of the terms appearing therein as well as of the axioms themselves. In other words, we must indicate the circumstances in nature that correspond to such terms and to the axioms that refer to them. This concerns essentially the production of a map which relates the formally constructed axioms, to the domain of experience to which science, by its own convention, is committed.

The same requirement holds equally for any acceptable formulation of a scientific theory, axiomatized or ortherwise, which presents propositions in formal language, that are understood to have bearing on the domain of scientific experience. With this in mind, we will further demonstrate in some detail that Hertz's mechanics is incomplete as a scientific theory, even though its formal edifice appears more comprehensive and complete than that of Newton's, in so far as it affords a larger and, I believe, a deeper range of conception and expression than does the formal structure of Newtonian mechanics. We will also demonstrate that Newtonian mechanics does indeed satisfy the above conditions for the completeness of a scientific theory, even though its formal structure appears to be more restricted than that of Hertz.

If we consider impenetrability to be the physico-ontological basis for the production of geometrical constraints in nature, then it follows almost directly from Hertz's mechanics, that is, from his general and strictly formal connections between geometrical constraint and force, that the impenetrability of substance is also the ontological basis for force. To amplify, this follows, because Hertz's mechanics has established a mathematical, that is a strictly formal connection between abstract geometrical restrictions and the symbol, $\bar{F}$, as it appears formally in Newton's fundamental proposition $\bar{F} = m\bar{a}$. This connection, in turn, is in itself formal, as it is strictly speaking a statement relating symbols, and therefore does not either explicitly or tacitly refer to what they mean in terms of experience, that is, to their ontological basis.

Now the ontological basis of the symbol $\bar{F}$, that is, what it represents in actual experience, is derived from its operational definition, which, of course, refers to experimentation directly. The combination of symbols obtained from Hertz's mechanics, and which are strictly speaking only formally related to Newton's symbol $\bar{F}$, would by the above reasoning have their ontological basis in the impenetrability of matter, which accounts for the production in nature of the geometrical constraints from which these symbols formally emerge. We see, therefore, that Hertz's mechanics establishes in this way a fundamental connection through geometry, between force and impenetrability. Moreover, in this fundamental connection we see that impenetrability is the ontological basis of the geometry of mechanical space as well as of the forces produced in this space; the geometry is therefore an intermediary, i.e. a manifestation of the connection between force and impenetrability. By this thinking physical space does not have its ontological basis in geometry but rather in impenetrability and its geometry is, so to speak, sculptored in the substance of which it is made, and whose ontological basis is impenetrability.

Leibnitz wrote seriously about impenetrability, about the geometry of the space of the universe, and about substance as the ontological basis of this space. What apparently went unnoticed by Leibnitz, and since then as well, is that the ontological basis of substance is its impenetrability and consequently that impenetrability is also the ontological basis of space itself and of events produced therein.

There also appears to be a connection between impenetrability considered here as the ontological basis of physical space and, more generally, of action space, and the notion of an aether.

The ontological basis of the concept aether was never resolved, and this may account for its adoption and subsequent rejection in the history of physics. Its adoption had a strong intuitive basis and motivation for, without it, certain facts in nature appeared ambiguous and to defy comprehension. Attempts were, of course, made to describe the aether by ascribing to it properties suggested by our physical experience, such as elasticity and fluidity; but these proved unsuccessful, as models for the aether endowed with such properties were unable to accommodate experimental facts. Even though Lorentz was indeed able to reconcile the existence of an aether, without ascribing to it specific physical attributes, with the constancy of the velocity of light in all Galilean frames, the aether concept lost favor in theoretical physics when Einstein discovered the special theory of relativity whose formulation is independent of an aether concept.

It is important to emphasize, however, that strictly speaking, neither the Michelson–Morely experiment or the special theory of relativity deny existence of a physical space, consisting of a substance which is the site and support of all events, actions and phenomena in nature. What this fundamental experiment does, however, make evident is *that this space*, if it does exist, cannot be identified by appealing to the idea that the fundamental processes in nature can be ascribed to irreducible bodies which move relative to it, along trajectories. If the Michelson–Morely experiment denies incisively anything at all, it does deny the validity of this conception of process as an elementary aspect of nature which quantum mechanics denies as well. Furthermore, the special theory of relativity does not challenge the existence of such a space, and in fact cannot do so, because its propositions are completely independent of such a concept, and therefore can neither support nor deny it.

Even though the success of the special theory may suggest that an aether concept is extraneous to the development of physics, it now seems that the physical basis for its fundamental propositions, which have by some authors been treated as axioms, may in fact emerge from the existence of such a space. This space is indeed tantamount to an aether, whose ontological basis we ascribe here to impenetrability of its substance; which is understood here to be a property of position rather than of extension. This inference is drawn by interpreting the dimensional universal constants of physics as the manifestations of the fundamental, i.e. irreducible processes and facts of nature; and by searching for an ontological basis of physical space which can accommodate these facts which are in themselves the most fundamental propositions of physics.

The method of scientific experimentation, and consequently information derived by such experiments, depends crucially on the process of identification, that is, on the identification of an object examined by the experiment. This process appears to be in principle extra-mathematical, in the sense that it cannot be given a mathematical, i.e. a strictly formal, representation. The reason is, that mathematics by its own convention, and from which it derives its power and freedom, concerns only classes of abstract objects endowed with properties that are indirectly defined by the method of axiomatization. None of its axioms pertain to a particular member of a class of objects which they condition, nor are rules prescribed by which a particular member can be chosen, i.e. selected, from a set to which it belongs. Such rules are, of course, essential in the process of identification.

It is obvious that if the process of identification is extra-formal, then in principle, it cannot be proven or disproven by a strictly formal procedure. As a concluding remark, it may be of interest to point out that if we consider the impenetrability of physical space to be its ontological basis then, it seems to follow that events produced in this space are necessarily connected by mirror-symmetry relations. The reason for this is that the impenetrability of physical space is an ontological basis for the production of mirror-symmetries in this space.

In summing up, it is important to emphasize that quantitative information is not necessarily precise; it is in fact necessarily approximate when obtained by measurement and/or by numerical computation. It is correspondingly and equally important to emphasize that qualitative informa tion is not necessarily approximate and that indeed fundamental and precise information seem necessarily to belong to this category. The category of precise information which is most fundamental in the mental processes accompanying mathematical thought and invention, i.e. mathematical experience, and which belong to the realm of internal experience, appears equally fundamental with respect to the stability, control, organization and precise performance of all biological processes. Réné Thom is attempting to establish a correspondence between the processes in the mind attendant to mathematical thought and some fundamental manifestations of the life process, and thus conceptualize them in the category of precise information, which apparently is crucial in their performance.

A central and most challenging problem which has faced man for years is to find a way of reconciling two notions which appear equally primitive in the function of the mind, and which it necessarily evokes when it contemplates and/or reasons. These are constancy, that is, elementarity, and change, that is, process. The problem is to produce a conceptual framework in which these two primitive notions are reconciled. The problem still remains and its resolution seems essential for comprehending the life process and something of the nature of the mind which is the most advanced development of the evolutionary process. The universal constants of physics are the most significant example produced in scientific fact, of the primitive notion of constancy and thus of elementarity. These constants, I believe, are the universal constants of nature and therefore in particular of the life process. It is in these constants and in what their existence manifests that the processes which we arbitrarily ascribe to the inanimate and animate worlds merge, and where we must search for their synthesis and thus their comprehension.

It seems appropriate to bring this paper to a head by pointing out that a study based on the universal constants of physics shows that the dimensional constants are more fundamental than are the dimensionless constants which are considered, for example, by Eddington. This implies that the fundamental propositions which emerge from the existence of the dimensional constants in nature cannot in principle be given a numerical and thus a quantitative representation, as is the case for the dimensionless constants. The fundamental propositions determined by the dimensional universal constants consequently belong to the category of qualitative information. This is another example which shows that the most fundamental propositions of physics evidently belong to the category of qualitative information.

I shall close by citing a profound remark by Poincaré, made after pondering for years the question of mathematical invention. He said: "the process of mathematical invention does not have a mathematical representation."

CONCLUDING REMARK

This paper is but a small token of my esteem for Professor Markus Reiner—a noble man and a distinguished scientist.

REFERENCES

1. G. HAMEL, *Handbuch der Physik*, Vol. 5 (1927).
2. J. C. C. McKINSEY, A. C. SUGAR and PATRICK SUPPES, *Journal Rational Mechanics and Analysis* (1953).
3. CARL F. GAUSS, "On a new general fundamental principle of mechanics", *Creele's Journal f. Math.* **4,** 232 (1829). Also appears in *Werke*, **5,** 23.
4. HEINRICH HERTZ, *Collected Works*, Col. III, *Principles of Mechanics*, Macmillan, New York (1896).
5. PAUL LIEBER and K. WAN, "A minimum dissipation principle for real fluids", *Proc. IX Int. Congress of Mechanics*, 1957.
6. PAUL LIEBER and K. S. WAN, *A Proof and Generalization of the Principle of Minimum Dissipation*, Air Force Office of Scientific Research, Report TN 57–479, AD 136 471 (Sept. 1957).
7. PAUL LIEBER and ARTHUR FARMER,"Studies of wave propagation in granular media", *Trans. American Geophysical Union* **39,** No. 2 (Apr. 1958).
8. PAUL LIEBER and K. S. WAN, "An integral invariant of steady-state viscous-incompressible flows", *Trans. American Geophysical Union* **43,** No. 4 (Dec. 1962).
9. PAUL LIEBER, "The mechanical evolution of clusters by binary elastic collisions and conception of a crucial experiment on turbulence", IUTAM Symposium on *Second-order Effects in Elasticity, Plasticity and Fluid Dynamics*, Haifa, April 1962 (Ed. M. REINER and D. ABIR), Pergamon Press, Oxford (1964).
10. PAUL LIEBER, *A Principle of Maximum Uniformity Obtained as a Theorem on the Distribution of Internal Forces*, Institute of Engineering Research, University of California, Berkeley, Nonr-222(87), No. MD-63-8, April 1963, and *Israel Journal of Technology* **5,** No. 3 (1967).

THE POSTULATES IN THE THEORY OF SUPERELASTICITY

MELVIN MOONEY †

Mountain Lakes, New Jersey, U.S.A.

ABSTRACT

The Mooney and the Rivlin theories on W, the strain energy function of an ideal superelastic material, are compared. The two theories agree on the form of W, particularly that W must be even in the principle stretches, the λ_i or $1+e_i$. However, there is a basic difference in the postulated mechanical properties of the superelastic material. Both authors postulate incompressibility, isotropy, and zero hysteresis; but Mooney postulates also Hooke's law in simple shear over a finite range of shear. The main question examined here is whether or not the Rivlin theory is correct and the postulate of Hooke's law is unnecessary.

It is concluded that the Rivlin theory contains an error and that the extra postulate in the Mooney theory is both necessary and sufficient to establish the form of W. The error in Rivlin lies in treating a λ_1, λ_2, λ_3 strain, followed by a 180° rotation about the 3-axis, as fully equivalent to the double negative strain, $-\lambda_1$, $-\lambda_2$, λ_3. Ignoring the physical impossibility of any negative stretch, the point is made here that in the second operation, that of the double negative stretch, the atoms in the sample are not rotated, while of course in the first operation they are.

A new method is outlined for testing the self-consistency of a basic set of experimental data, consisting of measurements in simple extension, compression and shear.

1. INTRODUCTION

There have been published two independent analyses of superelasticity, one by M. Mooney [1] and the other by R. S. Rivlin [2a, b], in which there was developed, as the first approximation, the following expression for W, the isothermal strain energy, the Helmholtz free energy, for an incompressible superelastic material:

$$W = C_1 J_1 + C_2 J_2, \tag{1}$$

wherein C_1 and C_2 are constants, and

$$\left\{\begin{aligned} I_1 &= \lambda_1^2+\lambda_2^2+\lambda_3^2 = J_1+3, \\ I_2 &= \lambda_2^2\lambda_3^2+\lambda_3^2\lambda_1^2+\lambda_1^2\lambda_2^2 \\ &= 1/\lambda_1^2+1/\lambda_2^2+1/\lambda_3^2 = J_2+3. \end{aligned}\right. \tag{2}$$

† The Editor regrets the passing away of Dr. Melvin Mooney shortly before the publication of this book.

The λ_i are the principal stretches, or $1+e_i$. The second form of I_2 results from the postulate of incompressibility, or

$$I_3 = \lambda_1^2\lambda_2^2\lambda_3^2 = 1. \tag{3}$$

The I_i are the standard strain invariants expressed, as by Rivlin [2b], in terms of the principal stretches.

In both *M*, or Mooney [1], and *R*, or Rivlin [2b], the analysis is carried beyond the first approximation to yield expressions for the general strain energy function. *M* is limited to the incompressibility case. When *R* is also so limited, the two analyses yield general expressions which, though different in form, are equivalent.

While agreeing in the final results, the two analyses employ quite different mathematical procedures, because of a difference in the postulates adopted concerning the idealized superelastic material. The primary purpose of the present article is to decide, by mathematical analysis, whether the extra postulate in *M* is or is not necessary for the derivation of the strain energy function of a superelastic material. For this purpose it will be sufficient first to deal primarily with the linear approximation, Eq. (1).

From a purely practical point of view the difference between *M* and *R* seems to be of no importance, since the ultimate results are the same; but the difference is of theoretical importance, because the questions that are raised relate to the fundamentals of the science of elasticity.

2. THE POSTULATES; THE LINEAR CASE

Common to both *M* and *R* are the postulates:

1. The superelastic material is isotropic.
2. The elasticity is perfect; that is, the hysteresis is zero.
3. The material is incompressible.

The additional postulate, made in *M* but not in *R*, is:

4. In a simple shear the stress–strain relationship is Hookean, that is, linear, not merely for a small shear, but for shears up to 2 or more. Mathematically,

$$\tau = G\nu \tag{4}$$

where τ is the shear stress, G is the modulus of rigidity, and ν is the shear.

The wide range of validity included in the last postulate is a statement of experimental results [3]. The significance, in the theory, of this wide range is perhaps obscure, for the range is not explicitly used anywhere in the derivation of the strain energy. However, the wide range implies that the strain energy function derived from (4) will also have a wide range of validity; and this is experimentally found to be the case.

To test whether Eq. (1) can be correctly derived from the first three postulates without any use of Eq. (4), it is necessary to examine carefully the pertinent section in the mathematical argument in R. This is the section which deals with the direct consequences of the isotropy of the superelastic material, deduced in R by using the usual strain invariants in the modern version of the classical theory of elasticity. The argument proceeds as follows:

3. THE STRAIN INVARIANTS AND SYMMETRY

In this section incompressibility is not assumed except where indicated.

The strain energy per unit volume in the uniformly strained material must be a function exclusively of the principal stretches λ_1, λ_2, λ_3; for these three variables are necessary and sufficient to describe completely the state of strain. Furthermore, to satisfy the condition of isotropy, the strain energy must be symmetrical in the λ_i, thus being invariant under an interchange of any pair, λ_i and λ_j. There being three independent λ_i, the strain energy must be a function of three independent symmetric functions of the λ_i. Such a set of three functions would constitute a complete set of strain invariants.

There is an infinite number of sets of invariants, some of which are discussed later; but from the argument in R it is concluded that the strain energy must be even in each of the λ_i; and hence the invariants in terms of which the strain energy is expressed must likewise be even in each of the λ_i. The I_i of Eqs. (2) and (3) constitute the simplest set of invariants that satisfies both conditions of evenness and symmetry. Therefore, it is argued, the strain energy in the general, non-linear but incompressible case must be expressible as a power series in the closely related J_i. Thus,

$$W = \sum_{m, n=0} C_{mn} J_1^m J_2^n, \tag{5}$$

with $C_{00} = 0$. Clearly $W = 0$ at $\lambda_i = 1$, as should be the case. As a consequence of Eq. (1), the invariant J_3, or $I_3 - 1$, vanishes and does not appear in Eq. (5).

4. NEGATIVE STRETCH

In the argument in R that W must be even in the λ_i, a novel concept is introduced, that of the negative stretch, which is a generalization of the usual physical concept. In the past it has always been held that the λ_i are restricted to positive, non-zero values; for if any stretch is negative, it can become so only by passing through the value 0. Surely, an elastic body compressed to zero thickness without destruction is not possible in the realm of physical reality. Nevertheless, the generalization in R consists in admitting negative values of the λ_i under certain conditions.

To present here the R concept of the negative stretch, Fig. 1 (A, B, and C) are taken from R, with some minor alterations; and Fig. 1 (D) is added.

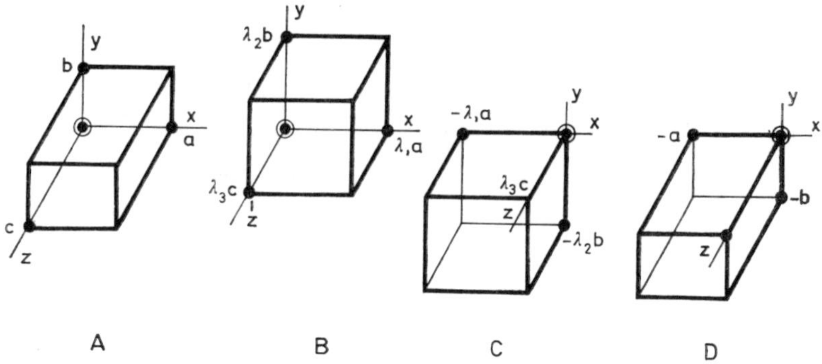

FIG. 1. Strained cuboids. A. Original cuboid. B. A subjected to $\lambda_1, \lambda_h, \lambda_3$ strain. C. B rotated 180° around z-axis, an *sr* operation; or B subjected to strain -1, -1, 1, an *ns* operation; or A subjected to strain $-\lambda_1$, $-\lambda_2$, $-\lambda_3$ an *ns* operation. D. C after stress release.

Now let the cuboid A, of dimensions a, b, c, be subjected to the stretches λ_1, λ_2, and λ_3, respectively, with $\lambda_1 \lambda_2 \lambda_3 = 1$. Cuboid A then becomes a cuboid of different dimensions, as shown by B. If further the cuboid B is now rotated 180° about the z-axis, the cuboid assumes the position and orientation shown as C.

But this rotation is equivalent to changing λ_1, λ_2 to $-\lambda_1$, $-\lambda_2$ or equivalent to applying to B the stretches -1, -1, 1. Cuboid C can be produced also by applying to A the stretches $-\lambda_1$, $-\lambda_2$, λ_3, according to the definition of a stretch.† The cuboids C obtained by these different operations are equivalent not only as to the shape, orientation and position of C, but also as to the final location in C of any initial chosen point in A.

$W(\lambda_1, \lambda_2, \lambda_3)$, the strain energy of cuboid B, is not altered by the rigid rotation to the cuboid C. But this strain energy must equal the strain energy of the equivalent C produced by the double negative stretch $-\lambda_1$, $-\lambda_2$, λ_3 applied to A. Hence we must have

$$W(\lambda_1, \lambda_2, \lambda_3) = W(-\lambda_1, -\lambda_2, \lambda_3). \tag{6}$$

† The explanation in R [2b], p. 355, is not clear on some of these details; but [2b], p. 359, and [2a], p. 384, justify the interpretation given here.

From this it follows that W must be even in λ_1 and in λ_2. By the same argument it must be even also in λ_2 and in λ_3; hence in all three λ_i. It must therefore have the form $W(\lambda_1^2, \lambda_2^2, \lambda_3^2)$.

In further discussion cuboid C will henceforth be referred to as of type *sr* or *ns*, depending on whether it was produced from A by a positive stretch plus a rigid rotation, *sr*, or by a double negative stretch, *ns*, the latter being mathematically identical with the reversal of the stretches λ_1, λ_2 existing in B.

5. ERRORS IN THE ARGUMENT

In the preceding argument there are unfortunately several things that are not correct. In the first place, there is a confusion between the use of λ as a vector, which is a length and direction from the origin, and its use as a stretch, which is the ratio of two lengths, after and before a deformation, or before and after a reverse deformation to the original length. Thus, when the cuboid C of type *sr* is released, it will take the form D; and hence the stretches in C are

$$\left\{\begin{aligned} \frac{-\lambda_1 a}{-a} &= \lambda_1, \\ \frac{-\lambda_2 b}{-b} &= \lambda_2, \\ \frac{+\lambda_3 c}{+c} &= \lambda_3. \end{aligned}\right. \tag{7}$$

Hence, in this case there is no justification for the two negative λ_i in Eq. (6).

In the case of the cuboid C of type *ns*, the same argument and conclusion apply. If, as stated in *R*, cuboids C of the two different types are identical, they will behave alike on release; and C of type *ns* will also take the form D, not the initial form A. Again we find no justification for the negative λ_i in Eq. (6).

There remains only one possible way to justify these negative stretches; and that is to assume that an ideal superelastic material can be forced to pass reversibly from a positive to a negative double stretch, and vice versa.

It may appear meaningless and wasteful of time to analyze an assumption that is in reality impossible; but in the present instance we shall at least cast a revealing light on a false idea that has so far been accepted without comment in the published literature. This idea is that the two different cuboids of form C are identical in all respects. Certain similarities have been noted; but in one respect, an important one, they are dissimilar. To demonstrate the dissimilarity, we consider first a stretch reversal, from λ_1 to $-\lambda_1$; and we consider at present only the ultimate effects, not the intermediate stages of the continuous transition.

The ultimate effect can be described as that of slicing the cuboid B into infinitely thin slices normal to the x-axis; then reassembling the slices in reverse order on the other side of the yz-plane.

If, following R, we consider only a double reverse stretch, from λ_1, λ_2 to $-\lambda_1$, $-\lambda_2$, the slicing operation will be two-dimensional, yielding infinitely thin sticks, to be reassembled in reverse order in both the x- and the

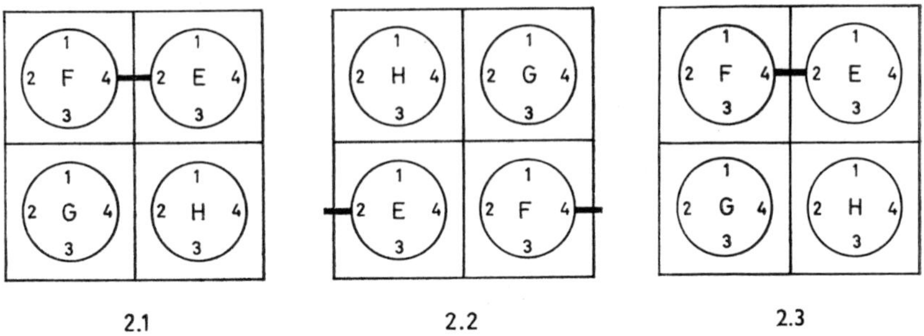

FIG. 2. Atomic cubes in strain reversal.

FIG. 2.1. Group of four atomic cubes, or carbon atoms, lying approximately in an xy-plane. State of strain λ_1, λ_2, λ_3 with λ_1 and λ_2 very small but positive. Atoms E and F connected by C—C bond.

FIG. 2.2. Group of Fig. 2.1 after double change of stretch sign, λ_1, λ_2 going to $-\lambda_1$, $-\lambda_2$. The C—C bond is broken.

FIG. 2.3. Group of Fig. 2.2 after 180° rotation of each carbon atom; or group of Fig. 2.1 after rotation of the group about the group center.

y-directions. It is convenient to consider the slicing to be three-dimensional, yielding infinitely small cubes; but the original sequence is preserved in the z-direction in the reassembling operation.

In applying these operations to a real superelastic material, consisting of tangled long-chain molecules with occasional cross-links, we must place a lower limit on the size of the cubes; for we cannot accept the idea of chopping up the atoms into pieces. The ultimate cubes are not mere geometric points. They have finite dimensions, and hence have the property of orientation.† It is in this property, not hitherto mentioned in this analysis, that the two types of cuboid C are dissimilar.

† The argument applies also to continuous media; for the sliced thickness only approaches zero, never reaches zero.

Figure 2 shows the essential details of the orientation relations. In Fig. 2.1 EFGH constitute a group of four carbon atoms lying approximately in an xy-plane in cuboid B of Fig. 1. Each of these four atoms is chemically bonded to two side groups and to two other carbon atoms in a chain molecule. E and F are bonded to each other. Double bonds would make no essential difference.

In Fig. 2.2 the four carbon atoms are shown as they would lie in cuboid C of type *ns*. In Fig. 2.3 they are shown as they would lie in cuboid C of type *sr*. In Fig. 2.2 the E—F bond is broken; and all other chemical bonds would likewise be broken except those that happen to lie parallel to the z-axis. Preservation of the bonds requires rotation of each of the atomic cubes in Fig. 2.2 through 180° about the axis through its center and parallel to the z-axis. The result would be the configuration of Fig. 2.3, or that in cuboid C, type *sr*.

It is not surprising that a rotation must be added to the operation *ns* to produce a result fully equivalent to the operation *sr*. The conclusion on this point is in agreement with the classical division of a general strain into a pure strain and a rigid rotation. There is no possibility of producing any rotation by a pure strain, or by any sequence of pure strains all having the same orientation of the strain axes. The equivalence of *sr* and *ns*, claimed in *R*, is not only disproved here, but is also contrary to the classical analysis of strain.

The atomic configuration and orientation in Fig. 2.2 show that a negative double stretch would necessitate the total disruption of the molecular chain structure and the cross-links. This structural break-up would occur in the assumed transition, or jump, by some inconceivable process, from a very small positive $a\lambda_1$ and $b\lambda_2$, of atomic dimensions, to the equally small negative values, $-a\lambda_1$ and $-b\lambda_2$. The resulting carbon ions or free radicals would be highly reactive and would recombine quickly, but in a completely different chain configuration, with zero strain energy at the small negative values of $-a\lambda_1$ and $-b\lambda_2$. Thus we see the importance of the difference between C *sr* and C *sn* in the atomic orientations. The assumption of a reversible double negative stretch or its reversible reversal to the stretch 1, 1, 1 is untenable. "Reversible" is used here in the thermodynamic sense, signifying in this case zero hysteresis.

This picture of the double negative stretch is the most favorable that is conceivable; and it is possible only if the stretches are produced by a balanced x and y pressure. If they are produced by a z tension, rupture of the sample would be inevitable.

6. PRESENT KNOWLEDGE AND IGNORANCE

We summarize here what we know and what we do not know, from pure theory, about the elastic properties of an isotropic superelastic material.

Known Properties

1. The strain energy, W, is a function of the three strain invariants of a set of three, each one of which is a symmetric function of the λ_i, the principal stretches.

2. In all cases, the $\lambda_i > 0$.

3. If the postulate of incompressibility is added, Eq. (3) is obeyed, or $\lambda_1\lambda_2\lambda_3 = 1$.

4. As a consequence of isotropy, the shear stress in simple shear is an odd function of the shear; and it is the same function for all orientations of the shear axes in the specimen.

Unknown Properties

1. What set of strain invariants may or must be used in the function W?

2. Is W analytic in these invariants at $J_i = 0$, and hence expansible as a power series in the J_i invariants? Each invariant is assumed to include an additive constant, if necessary, to reduce its value to zero at $\lambda_i = 1$.

3. If W is expansible as a power series in the invariants, does the series begin with terms of the first degree, or is it zero in these terms, beginning with non-zero terms of higher degree?

4. Is the shear stress in simple shear analytic in the shear and hence expansible as a power series in the shear?

5. If the shear stress is expansible as a power series in the shear, does the series begin with a term of the first degree, or is it zero in this term, beginning with a non-zero term of higher degree?

Questions 4 and 5 are not independent of 2 and 3; but their precise relationships do not concern us.

It is obvious that if any mathematical equation, without derivation or proof, goes beyond the four listed Known Properties, the equation constitutes a physical assumption or postulate. This is true, for example, of Eq. (4), which goes beyond the Known Property 4 and postulates that τ the shearing stress, has no singularity at $\gamma = 0$, and is non-zero in the first power of ν. It is true also of Eq. (1), which assumes answers to all of the listed Unknown Properties. Eq. (1), in fact, assumes much more than does Eq. (4). It assumes not only Eq. (4), which is derivable from (1) by an easy substitution and differentiation, it assumes also the stress–strain relationships in all other types of deformation, all of them being likewise directly derivable from Eq. (1). The normal stresses are determinable in an incompressible material only as to their differences, as is well known.

It might be said, concerning Eqs. (1) and (4), that neither assumes more than the other, because either can be derived from the other. This is true, but it is not the whole truth. The statement ignores the vast difference between deriving Eq. (4) from (1) and deriving (1) from (4). The former derivation is direct and easy, making use only of the known preperties common to all potential functions. The derivation of Eq. (1) from (4), on the other hand is not, by comparison, either short or easy. It requires repeated application of the symmetry requirement, and involves the solution of a functional equation and a method of solution which, comparatively speaking, is surely not obvious.

The theory of the strain energy function, either as derived in *M* or postulated in *R*, is a geometric and mathematical theory. It can predict the form of the function, but not the absolute or relative values of the constants in the function. However, the derived form becomes a requirement which must be satisfied by any molecular theory of superelasticity. The kinetic theory of the cross-linked chain molecules leads to the term C_1J_1 in the strain energy; and in this respect the kinetic theory is correct. However, this molecular theory, in its present form, remains incomplete until it can derive the term C_2J_2; for experimental tests have shown that C_2 is not zero.

7. DISCUSSION: THE LINEAR CASE

On the basis of the foregoing analysis it can be stated that the property of isotropy in a superelastic material does not in itself give any direct information about the stress–strain relationships or the strain energy function, except for the symmetry requirements. Isotropy does not even rule out singularities in these functions; and if we are inclined to think that singularities cannot exist in the properties of real materials, and we so assume, we would be making an assumption that is merely suggested by the empirical laws known to us. The assumption is not in fact a matter of logical or mathematical necessity.

Since any two-dimensional strain of zero dilatation is reducible to a simple shear relative to properly oriented shear axes, Eq. (4) is a postulate covering all two-dimensional incompressible strains. Any alternative, different postulate leading to Eq. (1) must therefore deal with three-dimensional deformations. One-dimensional deformations are inconsistent with incompressibility. Any three-dimensional postulate must contain C_1 and C_2 of Eq. (1) as separate constants, whereas in the two-dimensional postulate they combine as a single constant, C_1+C_2. This is because in the two-dimensional deformation J_1 and J_2 are equal.

The conclusion is that, isotropy being already postulated, Hooke's law in simple shear is the minimum additional postulate that is sufficient for the derivation of Eq. (1). We can now state the theorem:

THEOREM I. *For an incompressible, isotropic superelastic material, Hooke's law in simple shear, or $\tau = G\gamma$, is the necessary and sufficient condition for the strain energy to have the form*

$$W = C_1J_1+C_2J_2 \tag{8}$$

where $$C_1J_1+C_2J_2 \equiv C_1(\lambda_1^2+\lambda_2^2+\lambda_3^2-3)+C_2\left(\frac{1}{\lambda_1^2}+\frac{1}{\lambda_2^2}+\frac{1}{\lambda_3^2}-3\right), \tag{9}$$

with the λ_i subject to Eq. (3), $\lambda_1\lambda_2\lambda_3 = 1$.

It has been stated by a number of writers on superelasticity that Eq. (1) was suggested or proposed or postulated in *M*. On the contrary, as is quite clear in *M* and as is emphasized here, this equation was deduced or proved, on the basis of a much simpler postulate, Hooke's law in simple shear. The statements referred to are therefore not quite correct, and they are misleading as to what was postulated and what was deduced.

8. THE NON-LINEAR, GENERAL CASE

Going beyond the linear case stated by Eq. (4), we now consider the non-linear case, stated by

$$\tau = \sum_{n=2} G_n\gamma^{n-1}. \tag{10}$$

The n are even integers. If n goes to infinity, it will be assumed that the range of convergence of the infinite series is at least as large as the greatest γ attainable in experiment. We now define a group of sets of three independent, symmetric strain invariants by the equations

$$\begin{cases} I_{n1} = \lambda_1^n+\lambda_2^n+\lambda_3^n = J_{n1}+3, \\ I_{n2} = \lambda_2^n\lambda_3^n+\lambda_3^n\lambda_1^n+\lambda_1^n\lambda_2^n = J_{n2}+3, \\ I_{n3} = \lambda_1^n\lambda_2^n\lambda_3^n = J_{n3}+1, \end{cases} \tag{11}$$

with all $\lambda_i > 0$. Each n, a positive integer, designates a particular set of invariants hereafter referred to as the set of order n. In the incompressibility case, $J_3 = 0$, and the above set of equations reduces to the pair

$$\left\{\begin{aligned} J_{n1} &= \lambda_1^n+\lambda_2^n+\lambda_3^n-3, \\ J_{n2} &= \frac{1}{\lambda_1^n}+\frac{1}{\lambda_2^n}+\frac{1}{\lambda_3^n}-3. \end{aligned}\right. \tag{12}$$

In connection with the theory in R that W must be even in the λ_i, two questions of immediate interest are: first, can W in the general case, Eq. (10), be of the form $W(J_{n1}, J_{n2})$ and, second, if so, must n be limited to even values?

The answer to both questions is positive, according to M. Eq. (46) of this reference is, in present notation,

$$W_n = C_{n1}J_{n1}+C_{n2}J_{n2}, \tag{13}$$

with n even. From this, the total strain energy is given by the sum

$$W = \sum_{n=2}(C_{n1}J_{n1}+C_{n2}J_{n2}). \tag{14}$$

If the upper limit of n in Eq. (10) is finite, say p, then p is also the upper limit of n in Eq. (14), n being here the order of the invariant set J_{ni}.

It is now obvious that the group of invariant sets defined by Eq. (11) is of fundamental importance in the theory of superelasticity; for these sets arise as necessities in the M theory and they become the initial functions in terms of which the strain energy is expressed. The invariance property of these functions seems not to have been previously noted.

Since only even values of n are included in Eq. (14), the conclusion in the R theory on the form of W is justified if Eq. (10) is given and is convergent; but without Eq. (10) no conclusion is possible.

This might appear to be the end of the story, but there are still some other questions left hanging. It will soon be shown that if n is odd, a term of the form (13) leads only to odd powers of γ in τ, Eq. (10). Why, then, must odd n be ruled out?

By means of Eqs. (3), (8) and some equations from the classical theory of strains, the strain energy term W_n, Eq. (13), can be expressed in the case of simple shear, in terms of γ. The result is

$$W_n = (C_{n1}+C_{n2})\left[\frac{(\sqrt{(4+\gamma^2)}+\gamma)^n+(\sqrt{(4+\gamma^2)}-\gamma)^n}{2^n}-2\right]. \tag{15}$$

Now if the two binomials to the power n are expanded and combined, all odd powers of γ cancel. Hence W_n is even in γ; and τ, which is

$$\tau = \frac{dW_n}{d\tau}, \tag{16}$$

is odd, as was stated above. But here an important difference arises between n even and n odd. For n even, W_n reduces to a polynomial of nth degree in γ. For n odd, W_n, expressed in powers of γ, includes an infinite series, arising from $\sqrt{(4+\gamma^2)}$. The series becomes divergent at $\gamma = 2$. This is true likewise of τ. It follows, therefore, that n odd is ruled out by the assumption that τ is convergent at least for values of γ somewhat greater than 2. This shows that convergent Eq. (10) is not only sufficient, but is also necessary to ensure that n in Eq. (14) must be even.

An exception to this conclusion may seem to arise if W_n, Eq. (13), is non-linear in the J_{ni}, n odd; for in certain combinations of the J_{ni}, the radical $\sqrt{(4+\gamma^2)}$ can be eliminated. But what this signifies is that such a combination is equal to a J_{mi}, with m even and greater than n. We thus have completed a circle and return to the original conclusion that n odd is ruled out.

As an example of the situation described here, the following relationships between the I_{1i} and the I_{2i} are given:

$$\left\{\begin{aligned} I_{21} &= I_{11}^2 - 2I_{12}, \\ I_{22} &= I_{12}^2 - 2I_{13}^2(I_{11}^2 - 2I_{12}), \\ I_{23} &= I_{13}^2. \end{aligned}\right. \tag{17}$$

$$\left\{\begin{aligned} I_{11}^2 &= I_{21} + 2\sqrt{(I_{22} + 2I_{21}I_{23})}, \\ I_{12}^2 &= I_{22} + 2I_{21}I_{23}, \\ I_{13}^2 &= I_{23}. \end{aligned}\right. \tag{18}$$

These equations can be derived by eliminating the λ_i from I_{1i} and I_2 defined by Eq. (11). The set I_{1i}, like I_{2i}, can be derived directly by an appropriate analysis of the geometry of a general strain. Such an analysis was initiated by Love [4], p. 67, Eqs. (18) and (19); though the analysis was not carried through to the equations for the invariants.

The preceding analysis establishes the theorem:

THEOREM II. *For an incompressible, isotropic superelastic material, the shear stress equation for simple shear,*

$$\tau = \sum_{n=2}^{p} K_n \gamma^{n-1}, \tag{19}$$

with n even and τ convergent in the range of interest, is the necessary and sufficient condition for the strain energy to have the form

$$W = \sum_{n=2}^{p} (C_{n1}J_{n1} + C_{n2}J_{n2}), \tag{20}$$

p being either finite or infinite.

The necessity for the stated condition on τ in this theorem will perhaps be more clearly understood if an example is given which does not satisfy the condition. If $n = 1$, Eq. (15) becomes

$$W_1 = (C_{11}+C_{12})[\sqrt{(4+\gamma^2)}-2]. \tag{21}$$

If Eq. (10) is assumed, this form of W cannot exist; but if (10) is not assumed, W_1 of Eq. (19), for simple shear, is possible; and also W_1 of Eq. (13), for any type of incompressible strain. The latter form is

$$W_1 = C_{11}(\lambda_1+\lambda_2+\lambda_3-3)+C_{12}\left(\frac{1}{\lambda_1}+\frac{1}{\lambda_2}+\frac{1}{\lambda_3}-3\right). \tag{22}$$

These two expressions for W can be proved invalid only by comparison with experimental data.

GENERAL DISCUSSION

To a physicist's way of thinking, the theory that isotropy requires W to be even in the λ_i is immediately suspect; because there seems to be no physical, causal mechanism connecting the two properties. It is no surprise that a careful analysis fully justifies this scepticism.

It has been stated that the analysis in R includes the compressible case. This case is included also in an article by Mooney [5] on the thermodynamics of elastomers, which develops the equation of state as a function of the temperature, mean pressure, degree of crystallinity, and the stretch variables. The stretches, by a modified definition, are made independent of the volume and are determined exclusively by the shape of the deformed specimen. With the λ_i so defined, the analysis and theorems in the present article all still apply.

It has been stated above that the general form of W developed in M and R are equivalent. Proof of this is not given here; but it can be pointed out that the investigator of superelasticity has a choice among three equivalent forms of W. These are: Eq. (14) above, which uses the invariant sets J_{n1} and J_{n2} of different orders, or values of n; the same form slightly altered, in M; and Eq. (5) above, given in R, which uses powers of J_{21} and J_{22}.

In M the final, general expression for the strain energy is, with changes to agree with the present notation,

$$W = \sum_{n=2}^{p}\frac{Gn}{2n}(J_{n1}+J_{n2})+\sum_{n=2}^{p}\frac{H_n}{2n}(J_{n1}-J_{n2}), \tag{23}$$

wherein n is even and

$$\left.\begin{aligned} Gn/n &= C_{n1}+C_{n2},\\ H_n/n &= C_{n1}-C_{n2}. \end{aligned}\right\} \tag{24}$$

With W expressed in this form certain interesting facts become obvious or can be easily established. As is shown by Eq. (49) in M, W for a simple shear is a power series in which the coefficients, the K_n of Eq. (19), are complicated sums of the G_n only, the H_n not appearing. Consequently, while the convergence of τ, Eq. (19), with p infinite, may guarantee the convergence of W, Eq. (23), so far as the G_n terms are concerned, this guaranteed convergence of W does not apply to the H_n terms; for the H_n are independent of the K_n. However, the point is of little practical significance, because experimental results are normally represented mathematically by polynomials, not infinite power series.

In the case of a deformation produced by a uni-axial tension or pressure, W can be expressed as a function of the axial stretch exclusively. After the change of variable, $\lambda = e^x$, it is found that the $J_{n1}+J_{n2}$ are even in x, while the $J_{n1}-J_{n2}$ are odd. The opposite is true of the derivative functions of $J_{n1}+J_{n2}$ and $J_{n1}-J_{n2}$ that occur in the expression for σ, the tension, Eq. (52) in M. Thus, if the observed σ is resolved into an even and an odd function of x, each of these two functions can be represented, theoretically, as a sum of known functions of x, with the coefficients in terms of the G_n in one case and the H_n in the other case. By using enough terms, probably not many, the two theoretical curves can be made to fit the experimental data within experimental error.

In this curve fitting process the numerical values of the G_n and H_n are determined. Then the K_n of Eq. (19) can be computed from the G_n by using Eq. (49) in M. Now the big question is, Does the stress–strain curve for simple shear thus obtained agree with the experimental stress–strain curve?

The procedure here outlined constitutes a test of the self-consistency of the experimental data for simple tension, compression and shear. Only a few complete sets of data of this kind have been published, and in no case has this self-consistency test been used. Perhaps investigators will begin using it, now that it has been described.

REFERENCES

1. M. Mooney, *J. Appl. Physics*, **11**, 582 (1940).
2. R. S. Rivlin, (a) *Phil. Trans. Roy. Soc.* (London) A **241**, 379 (1948); (b) *Rheology*, F. R. Eirich (Ed)., vol. I, p. 351, Academic Press, New York (1956).
3. M. Mooney, *Rub. Chem. Tech.* **35**, p. xxvii (1962).
4. A. E. H. Love, *Mathematical Theory of Elasticity*, 4th ed., Dover Publications, New York (1956).
5. M. Mooney, *J. Appl. Physics*, **19**, 434 (1948).

ORTHOGONAL INTEGRITY BASIS FOR N SYMMETRIC MATRICES

R. S. RIVLIN AND G. F. SMITH

Center for the Application of Mathematics, Lehigh University, Bethlehem, Pennsylvania, U.S.A.

1. INTRODUCTION

In a paper published in 1945, Reiner [1] considered a constitutive equation for a non-Newtonian fluid in which the Cauchy stress matrix $\boldsymbol{\sigma}$ is expressed in terms of the strain-velocity matrix $\mathbf{d}$ by

$$\boldsymbol{\sigma} = \sum_{n=0}^{N} \alpha_n \mathbf{d}^n,$$

where the α's are functions of the invariants of $\mathbf{d}$ under the orthogonal group, i.e. of tr $\mathbf{d}$, tr $\mathbf{d}^2$ and tr $\mathbf{d}^3$. He used the Hamilton–Cayley theorem to express σ in the canonical form

$$\boldsymbol{\sigma} = \aleph_0 \mathbf{I} + \aleph_1 \mathbf{d} + \aleph_2 \mathbf{d}^2,$$

where the $\aleph$'s are functions of tr $\mathbf{d}$, tr $\mathbf{d}^2$, tr $\mathbf{d}^3$. Since then an extensive theory of canonical forms for non-linear constitutive equations in mechanics and other branches of physics has grown up.

It was shown by Smith and Rivlin [2] and by Pipkin and Rivlin [3] that if we make an initial constitutive assumption in the form of a polynomial relationship between one tensor and a number of other tensors, the problem of determining explicitly the restrictions imposed on the constitutive equation by material symmetry may be reduced to one of finding an integrity basis, under the group describing the material symmetry, for an appropriate set of tensors. The determination of an irreducible integrity basis (i.e. an integrity basis with no redundant elements which may be expressed as polynomials in the remaining ones) for an arbitrary number of second-order, symmetric, three-dimensional tensors under the full (or proper) orthogonal group was carried out by Rivlin [4], Spencer and Rivlin [5, 6, 7] and by Spencer [8]. The irreducibility of the integrity basis obtained by Spencer and Rivlin was proven by Smith [9]. Since the algebraic reasoning leading to the results of Spencer and Rivlin was presented in a rather elaborate manner, complicated by a good deal of extraneous material and alternative procedures, it appears worth while to recapitulate in a single paper the essential features of the argument.

It is first shown that any polynomial invariant, under the full or proper orthogonal group, of N (say) symmetric, second-order, three-dimensional tensors (i.e. 3×3 symmetric matrices) may be expressed as a polynomial in traces of products formed from them. Thus, the set of traces of all possible products, which can be formed from the N matrices, forms an integrity basis, albeit not a finite one, for N second-order symmetric tensors under the orthogonal group.

It is shown in Sections 6 to 8 how traces of products of certain types can be expressed as polynomials in traces of products of lower degree. One of these results (Corollary to Lemma 12) tells us that the trace of any product of degree greater than or equal to seven may be expressed as a polynomial in traces of products of degree less than seven. The problem of determining an irreducible integrity basis for the N tensors is thus reduced to that of determining an irreducible integrity basis for six tensors. This is done by means of the results developed in Sections 6 to 9. These results are combined in a theorem (Corollary to Theorem 1), which enables us to write down (in Table 1) a finite set of products in terms of whose traces, the trace of any product of one, two, . . ., six matrices may be expressed as a polynomial. This table then provides us with a finite integrity basis for the N matrices by substituting for the matrices occurring in the table all selections from the N matrices taken one, two, . . ., six at a time and taking the trace of each of the products so obtained. The integrity basis given in this manner by Table 1, though finite, is still reducible. It is shown in Section 10 that certain of its elements are expressible as polynomials in the remaining ones. Eliminating from Table 1 the products which give rise to these elements, we obtain a table of products (Table 2) from which an irreducible integrity basis can be read off. It is shown in Section 12 how a canonical form for a symmetric isotropic matrix polynomial in the N matrices can be obtained from this.

2. SOME DEFINITIONS

We consider N, $r \times r$ matrices $\mathbf{a}$, $\mathbf{b}$, $\mathbf{c}$, . . ., none of which is the unit matrix. Let $\mathbf{\Pi}$ be any product of these. It can be expressed as

$$\mathbf{\Pi} = \mathbf{u}^{\alpha_1}\mathbf{v}^{\alpha_2}\ldots\mathbf{z}^{\alpha_n}, \tag{2.1}$$

where the α's are positive integers and the matrices **u**, **v**, ..., **z** are a selection from the matrices **a**, **b**, **c**, ..., including repetitions, but no two adjacent matrices in the sequence **u**, **v**, ..., **z** are the same. The *factors* of the product are $\mathbf{u}^{\alpha_1}$, $\mathbf{v}^{\alpha_2}$, ..., $\mathbf{z}^{\alpha_n}$. The *powers* of the factors in $\mathbf{\Pi}$ are α_1, α_2, ... The *total degree* (or *degree*) α of the product is defined by

$$\alpha = \alpha_1+\alpha_2+\ldots+\alpha_n.$$

Let $\beta_1, \beta_2, \beta_3, \ldots$ be the sum of the exponents occurring in the factors which are powers of **a**, **b**, **c**, ..., respectively. Then $\beta_1, \beta_2, \beta_3, \ldots$ are said to be the *partial degrees* of the product in **a**, **b**, **c**, ..., respectively. For brevity we can say that the *partial degree* of $\mathbf{\Pi}$ in **a**, **b**, **c**, ... is $\{\beta_1\beta_2\beta_3\ldots\}$.

For example, $\mathbf{a}^2\mathbf{b}^3\mathbf{c}$ and $\mathbf{b}\mathbf{a}^2\mathbf{b}^2\mathbf{c}$ are each of partial degrees 2, 3 and 1 in **a**, **b**, **c**, respectively, and have partial degree {231}.

It is evident that

$$\alpha = \beta_1+\beta_2+\beta_3+\ldots.$$

The number of factors occurring in the product $\mathbf{\Pi}$ is called its *extension.* The extension of $\mathbf{\Pi}$ defined by Eq. (2.1) is then n. The extensions of $\mathbf{a}^2\mathbf{b}^3\mathbf{c}$ and $\mathbf{b}\mathbf{a}^2\mathbf{b}^2\mathbf{c}$ are 3 and 4, respectively.

3. ELEMENTARY PROPERTIES OF THE TRACE OF A PRODUCT

From the definition of the trace of a matrix it follows that for any two $r\times r$ matrices **a**, **b**,

$$\text{tr}\,\mathbf{ab} = \text{tr}\,\mathbf{ba}.$$

Therefore,

$$\text{tr}\,\mathbf{\Pi} = \text{tr}\,\mathbf{u}^{\alpha_1}(\mathbf{v}^{\alpha_1}\ldots\mathbf{z}^{\alpha_n}) = \text{tr}\,\mathbf{v}^{\alpha_2}\ldots\mathbf{z}^{\alpha_n}\mathbf{u}^{\alpha_1}.$$

We thus have

LEMMA 1. The trace of a product of any number of matrices is unaltered by cyclic permutation of the factors in the product.

Now consider that the matrices **a**, **b**, **c**, ..., from which the product $\mathbf{\Pi}$ is formed, are symmetric. Let $\mathbf{\Pi}^t$ denote the transpose of $\mathbf{\Pi}$. It follows directly from the definition of matrix multiplication that

$$\mathbf{\Pi}^t = \mathbf{z}^{\alpha_n}\ldots\mathbf{v}^{\alpha_2}\mathbf{u}^{\alpha_1}.$$

We obtain immediately

LEMMA 2. The trace of a product of any number of symmetric matrices is unaltered by reversal of the order of the factors in the product.

4. POLYNOMIAL INVARIANTS AND ISOTROPIC MATRIX POLYNOMIALS

A scalar polynomial (or polynomial) P in the N, $r \times r$ matrices **a**, **b**, **c**, ... is an expression of the form

$$P = \sum_{\alpha=0}^{\beta} A_{i_1 j_1 i_2 j_2 \ldots i_\alpha j_\alpha} u_{i_1 j_1} v_{i_2 j_2} \ldots z_{i_\alpha j_\alpha}, \tag{4.1}$$

where $\mathbf{u} (= ||u_{ij}||)$, $\mathbf{v} (= ||v_{ij}||)$, etc., is a selection from the matrices $\mathbf{a} (= ||a_{ij}||)$, $\mathbf{b} (= ||b_{ij}||)$, etc., including repetitions, and $A_{i_1 j_1 \ldots i_\alpha j_\alpha}$ are constants. The subscripts i_1, j_1, i_2, j_2, etc., take the values $1, 2, \ldots, r$ and the Einstein summation convention is understood. The term in Eq. (4.1) obtained by taking $\alpha = 0$ is interpreted as a scalar constant A (say).

A matrix polynomial $\mathbf{P} (= ||P_{ij}||)$ in the N, $r \times r$ matrices **a**, **b**, **c**, ... is an expression of the form

$$P_{ij} = \sum_{\alpha=0}^{\beta} A_{iji_1 j_1 \ldots i_\alpha j_\alpha} u_{i_1 j_1} v_{i_2 j_2} \ldots z_{i_\alpha j_\alpha}, \tag{4.2}$$

where $A_{iji_1 \ldots j_\alpha}$ are constants. The term in Eq. (4.2) obtained by taking $\alpha = 0$ is interpreted as a constant matrix A_{ij}. In other respects the notation in Eq. (4.2) is analogous to that in (4.1).

Let $\mathbf{G} (= ||G_{ij}||)$ be a generic transformation of a group g, which is the orthogonal group or a subgroup of it. We define $\bar{\mathbf{u}} (= ||\bar{u}_{ij}||)$, $\bar{\mathbf{v}} (= ||\bar{v}_{ij}||)$, ..., $\bar{P}$ and $\bar{\mathbf{P}} (= ||\bar{P}_{ij}||)$ by

$$\bar{\mathbf{u}} = \mathbf{G}\mathbf{u}\mathbf{G}^t, \quad \bar{\mathbf{v}} = \mathbf{G}\mathbf{v}\mathbf{G}^t, \quad \text{etc.},$$

$$\bar{P} = \sum_{\alpha=0}^{\beta} A_{i_1 j_1 \ldots i_\alpha j_\alpha} \bar{u}_{i_1 j_1} \bar{v}_{i_2 j_2} \ldots \bar{z}_{i_\alpha j_\alpha} \tag{4.3}$$

and

$$\bar{P}_{ij} = \sum_{\alpha=0}^{\beta} A_{iji_1 j_1 \ldots i_\alpha j_\alpha} \bar{u}_{i_1 j_1} \bar{v}_{i_2 j_2} \ldots \bar{z}_{i_\alpha j_\alpha}.$$

$\mathbf{G}^t$ denotes the transpose of **G**. **u** and $\bar{\mathbf{u}}$ are then the matrices formed by the components of a second-order tensor in rectangular Cartesian coordinate systems related by the transformation **G**.

P is said to be an *isotropic scalar polynomial*, or *polynomial invariant*, with respect to g, if and only if

$$P = \bar{P}, \tag{4.4}$$

for all transformations **G** of the group g.

Similarly, **P** is said to be an *isotropic matrix polynomial* with respect to g, if and only if

$$\bar{\mathbf{P}} = \mathbf{G}\mathbf{P}\mathbf{G}^t, \tag{4.5}$$

for all transformations **G** of the group g.

It follows from Eqs. (4.1) and (4.3) that the condition (4.4) is satisfied if and only if†

$$A_{i_1 j_1 \ldots i_\alpha j_\alpha} = G_{i_1 p_1} G_{j_1 q_1} \ldots G_{i_\alpha p_\alpha} G_{j_\alpha q_\alpha} A_{p_1 q_1 \ldots p_\alpha q_\alpha}. \tag{4.6}$$

Similarly, it follows from Eqs. (4.2) and (4.3) that the condition (4.5) is satisfied if and only if

$$A_{ij i_1 j_1 \ldots i_\alpha j_\alpha} = G_{ip} G_{jq} G_{i_1 p_1} \ldots G_{j_\alpha q_\alpha} A_{pq p_1 q_1 \ldots p_\alpha q_\alpha}. \tag{4.7}$$

Equation (4.6) states that $A_{i_1 j_1 \cdots i_\alpha j_\alpha}$ is a tensor, the components of which are unaltered by transformations of the group g. Such a tensor is called an *invariant tensor*‡ with respect to g. For each group g there exists a finite set of invariant tensors in terms of which any invariant tensor may be expressed as the sum of outer products with scalar coefficients [9]. This set is called an *invariant tensor basis* for the group g.

We shall now consider that the matrices **a**, **b**, **c**, ... are 3×3 matrices (i.e. $r = 3$) and that the group g is the full orthogonal group. The invariant tensor basis then consists of the three-dimensional Kronecker delta δ_{ij}. Any invariant tensor with respect to the full orthogonal group is expressible as a sum of outer products of Kronecker deltas with scalar coefficients. Thus,

$$A_{i_1 j_1 \cdots i_\alpha j_\alpha} = \sum C^{(p_1 \cdots q_\alpha)} \delta_{p_1 q_1} \delta_{p_2 q_2} \ldots \delta_{p_\alpha q_\alpha}, \tag{4.8}$$

where $p_1 q_1 \ldots p_\alpha q_\alpha$ is a permutation of $i_1 j_1 \ldots i_\alpha j_\alpha$ and the summation is carried out over all such permutations. For example, if $\alpha = 2$

$$A_{i_1 j_1 i_2 j_2} = B \delta_{i_1 j_1} \delta_{i_2 j_2} + C \delta_{i_1 i_2} \delta_{j_1 j_2} + D \delta_{i_1 j_2} \delta_{j_1 i_2}, \tag{4.9}$$

where B, C, D are constants.

Introducing Eq. (4.8) into (4.1) it is readily seen that P is now expressible as a polynomial in traces of products formed from **a**, **b**, **c**, ... and their transposes.

If the group g is the proper orthogonal group, then any even-order invariant tensor (but not one of odd order) is still expressible as a sum of outer products of Kronecker deltas with scalar coefficients as indicated in Eq. (4.8). Consequently P is still expressible as a polynomial in traces of products formed from **a**, **b**, **c**, ... and their transposes. We thus have

LEMMA 3. Any polynomial invariant with respect to the full or proper orthogonal group of any number of 3×3 matrices may be expressed as a

† We assume here, and in deriving Eq. (4.7), that identical terms under the summation sign in Eq. (4.1), or (4.2), are added to give a single coefficient.

‡ Elsewhere the expression *isotropic tensor* has been used when g is the full or proper orthogonal group and *anisotropic tensor* when it is not.

polynomial in traces of products formed from the matrices and their transposes.

If the matrices are symmetric, we obtain immediately the

COROLLARY. Any polynomial invariant with respect to the full or proper orthogonal group of any number of symmetric 3×3 matrices may be expressed as a polynomial in traces of products formed from the matrices.

We note that the trace of any product formed from matrices and their transposes is a polynomial invariant, under the full or proper orthogonal group.

Starting with Eqs. (4.2) and (4.8) and proceeding in a manner analogous to that leading to Lemma 3, we obtain

LEMMA 4. Any isotropic matrix polynomial with respect to the full or proper orthogonal group of any number of 3×3 matrices may be expressed as the sum of a number of products (including the unit matrix) formed from the matrices and their transposes, with coefficients which are polynomials in traces of products formed from the matrices and their transposes.

If the matrices are symmetric, we obtain the

COROLLARY. Any isotropic matrix polynomial with respect to the full or proper orthogonal group of any number of symmetric 3×3 matrices may be expressed as the sum of a number of products (including the unit matrix) formed from the matrices, with scalar coefficients which are polynomials in traces of products formed from the matrices.

We note that any product formed from the matrices and their transposes is an isotropic matrix polynomial, under the full or proper orthogonal group.

5. REDUCIBILITY AND EQUIVALENCE

We say that the trace of a product of 3×3 matrices, or the trace of the sum of a number of such matrix products of equal partial degrees in each of the matrices, is *reducible* if and only if it can be expressed as a polynomial in traces of products of lower total degree.† If $\mathbf{\Pi}$ is a product of ma-

† We note that there is some conflict between this definition of reducible and that which is usually used in discussing integrity bases. According to the usual usage, an integrity basis is said to be reducible if one element is expressible as a polynomial in the remaining ones. We shall use the word reducible with the usual meaning in connection with integrity bases.

trices, we shall denote the fact that tr $\mathbf{\Pi}$ is reducible by writing

$$\operatorname{tr} \mathbf{\Pi} \approx 0,$$

with a similar notation indicating reducibility of the trace of a sum of matrix products. For example, from the Hamilton–Cayley theorem for a 3×3 matrix $\mathbf{a}$, expressed by Eq. (6.1) below, we obtain, by multiplying throughout by $\mathbf{a}$ and taking the trace, and expression for tr $\mathbf{a}^4$ as a polynomial in tr $\mathbf{a}$, tr $\mathbf{a}^2$, and tr $\mathbf{a}^3$. Thus,

$$\operatorname{tr} \mathbf{a}^4 \approx 0.$$

We say that a product of any number of 3×3 matrices, or the sum of a number of such matrix products of equal partial degrees in each of the matrices, is *reducible* if and only if it is expressible as the sum of a number of products, each of lower total degree in the matrices, with coefficients which are polynomial invariants. We shall denote the fact that a product $\mathbf{\Pi}$ of matrices is reducible by writing

$$\mathbf{\Pi} \approx 0,$$

with a similar notation indicating reducibility of a sum of products. Thus, the Hamilton–Cayley theorem for a 3×3 matrix $\mathbf{a}$ expressed by Eq. (6.1) below, tells us that $\mathbf{a}^3$ is reducible, so that we may write

$$\mathbf{a}^3 \approx 0.$$

The result deduced from it, and given in Eq. (6.2) below, tells us that if $\mathbf{a}$ and $\mathbf{b}$ are two 3×3 matrices, $\mathbf{aba}+\mathbf{a}^2\mathbf{b}+\mathbf{ba}^2$ is reducible and we may write

$$\mathbf{aba}+\mathbf{a}^2\mathbf{b}+\mathbf{ba}^2 \approx 0.$$

The following lemma follows immediately from the definition of reducibility:

LEMMA 5. If $\mathbf{\Pi}$ is a product of 3×3 matrices $\mathbf{a}, \mathbf{b}, \mathbf{c}, \ldots$ which is reducible and $\mathbf{x}$ is another 3×3 matrix, then tr $\mathbf{x\Pi}$ and tr $\mathbf{\Pi x}$ are reducible.

Let $\mathbf{\Pi}$ and $\hat{\mathbf{\Pi}}$ be two matrix products formed from any number of matrices $\mathbf{a}, \mathbf{b}, \mathbf{c}, \ldots$ of the same partial degrees in each of them. If and only if $\mathbf{\Pi} \pm \hat{\mathbf{\Pi}}$ is reducible, we say that $\mathbf{\Pi}$ is *equivalent* to $\mp \hat{\mathbf{\Pi}}$ and denote this

$$\mathbf{\Pi} \approx \mp \hat{\mathbf{\Pi}}.$$

If and only if tr $\mathbf{\Pi} \pm$ tr $\hat{\mathbf{\Pi}}$ is reducible, we say that tr $\mathbf{\Pi}$ is *equivalent* to $\mp$ tr $\hat{\mathbf{\Pi}}$ and denote this

$$\operatorname{tr} \mathbf{\Pi} \approx \mp \operatorname{tr} \hat{\mathbf{\Pi}}.$$

It is understood that in these definitions of equivalence either the upper signs are to be taken or the lower signs are to be taken. The definitions may be extended in an obvious manner to the equivalence of linear combinations of products of matrices of equal partial degree and to that of linear combinations of traces of such products.

The remainder of this paper is concerned entirely with polynomial invariants and isotropic matrix polynomials of 3×3 matrices with respect to the full or proper orthogonal groups. We shall use the terms polynomial invariant and isotropic matrix polynomial, without qualification, with the understanding that they are used in this sense. Also, we shall understand that the type of representation for them given by Lemmas 3 and 4 or their corollaries is implied.

6. SOME IMPLICATIONS OF THE HAMILTON–CAYLEY THEOREM

Let $\mathbf{a}$ be any 3×3 matrix. The Hamilton–Cayley theorem states that

$$\mathbf{a}^3-\mathbf{a}^2 \operatorname{tr} \mathbf{a}+\tfrac{1}{2}\mathbf{a}[(\operatorname{tr}\mathbf{a})^2-\operatorname{tr}\mathbf{a}^2] -\mathbf{I}\tfrac{1}{6}[(\operatorname{tr}\mathbf{a})^3-3\operatorname{tr}\mathbf{a}\operatorname{tr}\mathbf{a}^2+2\operatorname{tr}\mathbf{a}^3] = 0, \qquad (6.1)$$

where $\mathbf{I}$ denotes the unit matrix. Now, let $\mathbf{a}$ and $\mathbf{b}$ be two 3×3 matrices. Replacing $\mathbf{a}$ in Eq. (6.1) by $\mathbf{a}+\mathbf{b}$ and by $\mathbf{a}-\mathbf{b}$ and subtracting the two resulting relations, we obtain

$$\mathbf{aba}+\mathbf{a}^2\mathbf{b}+\mathbf{ba}^2 = \mathbf{G}(\mathbf{a},\mathbf{b}), \qquad (6.2)$$

where

$$\begin{aligned}\mathbf{G}(\mathbf{a},\mathbf{b}) = {} & \mathbf{a}(\operatorname{tr}\mathbf{ab}-\operatorname{tr}\mathbf{a}\operatorname{tr}\mathbf{b})+\tfrac{1}{2}\mathbf{b}[\operatorname{tr}\mathbf{a}^2-(\operatorname{tr}\mathbf{a})^2]\\ & +(\mathbf{ab}+\mathbf{ba})\operatorname{tr}\mathbf{a}+\mathbf{a}^2\operatorname{tr}\mathbf{b}\\ & +\mathbf{I}[\operatorname{tr}\mathbf{a}^2\mathbf{b}-\operatorname{tr}\mathbf{a}\operatorname{tr}\mathbf{ab}+\tfrac{1}{2}\operatorname{tr}\mathbf{b}\{(\operatorname{tr}\mathbf{a})^2-\operatorname{tr}\mathbf{a}^2\}]. \qquad (6.3)\end{aligned}$$

Replacing $\mathbf{a}$ by $\mathbf{a}+\mathbf{c}$ and $\mathbf{a}-\mathbf{c}$, we obtain the relations

$$(\mathbf{a}+\mathbf{c})\mathbf{b}(\mathbf{a}+\mathbf{c})+(\mathbf{a}+\mathbf{c})^2\mathbf{b}+\mathbf{b}(\mathbf{a}+\mathbf{c})^2 = \mathbf{G}(\mathbf{a}+\mathbf{c},\mathbf{b})$$

and

$$(\mathbf{a}-\mathbf{c})\mathbf{b}(\mathbf{a}-\mathbf{c})+(\mathbf{a}-\mathbf{c})^2\mathbf{b}+\mathbf{b}(\mathbf{a}-\mathbf{c})^2) = \mathbf{G}(\mathbf{a}-\mathbf{c},\mathbf{b}). \qquad (6.4)$$

Subtracting these, we have

$$\mathbf{abc}+\mathbf{bca}+\mathbf{cab}+\mathbf{cba}+\mathbf{acb}+\mathbf{bac} = \mathbf{H}(\mathbf{a},\mathbf{b},\mathbf{c}), \qquad (6.5)$$

where

$$\begin{aligned}\mathbf{H}(\mathbf{a},\mathbf{b},\mathbf{c}) &= \tfrac{1}{2}[\mathbf{G}(\mathbf{a}+\mathbf{c},\mathbf{b})-\mathbf{G}(\mathbf{a}-\mathbf{c},\mathbf{b})]\\ &= \mathbf{a}(\operatorname{tr}\mathbf{bc}-\operatorname{tr}\mathbf{b}\operatorname{tr}\mathbf{c})+\mathbf{b}(\operatorname{tr}\mathbf{ca}-\operatorname{tr}\mathbf{c}\operatorname{tr}\mathbf{a})\\ &\quad +\mathbf{c}(\operatorname{tr}\mathbf{ab}-\operatorname{tr}\mathbf{a}\operatorname{tr}\mathbf{b})+(\mathbf{bc}+\mathbf{cb})\operatorname{tr}\mathbf{a}\\ &\quad +(\mathbf{ca}+\mathbf{ac})\operatorname{tr}\mathbf{b}+(\mathbf{ab}+\mathbf{ba})\operatorname{tr}\mathbf{c}\\ &\quad +\mathbf{I}(\operatorname{tr}\mathbf{a}\operatorname{tr}\mathbf{b}\operatorname{tr}\mathbf{c}-\operatorname{tr}\mathbf{a}\operatorname{tr}\mathbf{bc}-\operatorname{tr}\mathbf{b}\operatorname{tr}\mathbf{ca}\\ &\qquad -\operatorname{tr}\mathbf{c}\operatorname{tr}\mathbf{ab}+\operatorname{tr}\mathbf{abc}+\operatorname{tr}\mathbf{cba}). \qquad (6.6\end{aligned}$$

We note that

$$2\mathbf{G}(\mathbf{a}, \mathbf{b}) = \mathbf{H}(\mathbf{a}, \mathbf{b}, \mathbf{a}) \tag{6.7}$$

and that Eq. (6.2) results from Eq. (6.5) by taking $\mathbf{c} = \mathbf{a}$, while Eq. (6.1) results from Eq. (6.2) by taking $\mathbf{b} = \mathbf{a}$.

From the definition of $\mathbf{H}(\mathbf{a}, \mathbf{b}, \mathbf{c})$, it is apparent that Eq. (6.5) implies that the expression on its left hand side is reducible. We thus have

LEMMA 6. If $\mathbf{a}$, $\mathbf{b}$, $\mathbf{c}$ are 3×3 matrices, then $\mathbf{abc}+\mathbf{bca}+\mathbf{cab}+\mathbf{cba}+\mathbf{acb}+\mathbf{bac}$ is reducible.

From this we have, using Lemma 5, the

COROLLARY. If $\mathbf{a}$, $\mathbf{b}$, $\mathbf{c}$, $\mathbf{y}$ are 3×3 matrices, then tr $\mathbf{y}(\mathbf{abc}+\mathbf{bac}+\mathbf{cab}+\mathbf{cba}+\mathbf{acb}+\mathbf{bac})$is reducible.

Now, in (6.5) take $\mathbf{c} = \mathbf{a}^2$. We obtain

$$\mathbf{aba}^2+\mathbf{a}^2\mathbf{ba}+2(\mathbf{a}^3\mathbf{b}+\mathbf{ba}^3) = \mathbf{H}(\mathbf{a}, \mathbf{b}, \mathbf{a}^2). \tag{6.8}$$

We employ the Hamilton–Cayley theorem (6.1) to substitute for $\mathbf{a}^3$ on the left-hand side of Eq. (6.8) and obtain

LEMMA 7. If $\mathbf{a}$ and $\mathbf{b}$ are two 3×3 matrices, then $\mathbf{aba}^2\times\mathbf{a}^2\mathbf{ba}$ is reducible.

Using Lemma 5 we obtain the

COROLLARY 1. If $\mathbf{a}$, $\mathbf{b}$ and $\mathbf{x}$ are three 3×3 matrices, then tr $(\mathbf{aba}^2+\mathbf{a}^2\mathbf{ba})\mathbf{x}$ is reducible. We now suppose that $\mathbf{a}$, $\mathbf{b}$ and $\mathbf{x}$ are symmetric 3×3 matrices. From Lemmas 1 and 2, we have

$$\operatorname{tr} \mathbf{a}^2\mathbf{bax} = \operatorname{tr} \mathbf{xa}^2\mathbf{ba} = \operatorname{tr} \mathbf{aba}^2\mathbf{x}.$$

Thus,

$$\operatorname{tr} (\mathbf{aba}^2+\mathbf{a}^2\mathbf{ba})\mathbf{x} = 2 \operatorname{tr} \mathbf{aba}^2\mathbf{x} = 2 \operatorname{tr} \mathbf{a}^2\mathbf{bax}.$$

Then, with Corollary 1, we obtain

COROLLARY 2. If $\mathbf{a}$, $\mathbf{b}$ and $\mathbf{x}$ are three 3×3 symmetric matrices, tr $\mathbf{aba}^2\mathbf{x}$ and tr $\mathbf{a}^2\mathbf{bax}$ are reducible.

7. REDUCIBILITY OF $\mathbf{a}^2\mathbf{xb}^2$

Let $\mathbf{a}$, $\mathbf{b}$, $\mathbf{x}$ be three 3×3 matrices. Then, replacing $\mathbf{a}$ by $\mathbf{b}$ and $\mathbf{b}$ by $\mathbf{ax}$ in Eq. (6.2) and rearranging terms, we obtain

$$\mathbf{axb}^2 = -[\mathbf{baxb}+\mathbf{b}^2\mathbf{ax}-\mathbf{G}(\mathbf{b}, \mathbf{ax})]. \tag{7.1}$$

Multiplying throughout by $\mathbf{a}$, we have

$$\begin{aligned}\mathbf{a}^2\mathbf{xb}^2 &= -(\mathbf{aba})\mathbf{xb}-(\mathbf{ab}^2\mathbf{a})\mathbf{x}+\mathbf{aG}(\mathbf{b}, \mathbf{ax})\\ &= (\mathbf{a}^2\mathbf{b}+\mathbf{ba}^2)\mathbf{xb}+(\mathbf{a}^2\mathbf{b}^2+\mathbf{b}^2\mathbf{a}^2)\mathbf{x}-\mathbf{G}(\mathbf{a}, \mathbf{b})\mathbf{xb}-\mathbf{G}(\mathbf{a}, \mathbf{b}^2)\mathbf{x}+\mathbf{aG}(\mathbf{b}, \mathbf{ax})\\ &= \mathbf{a}^2(\mathbf{bxb})+\mathbf{b}(\mathbf{a}^2\mathbf{x})\mathbf{b}+(\mathbf{a}^2\mathbf{b}^2+\mathbf{b}^2\mathbf{a}^2)\mathbf{x}-\mathbf{G}(\mathbf{a}, \mathbf{b})\mathbf{xb}-\mathbf{G}(\mathbf{a}, \mathbf{b}^2)\mathbf{x}+\mathbf{aG}(\mathbf{b}, \mathbf{ax})\\ &= -2\mathbf{a}^2\mathbf{xb}^2+3\mathbf{K}(\mathbf{a}, \mathbf{b},\mathbf{x}),\end{aligned} \tag{7.2}$$

where

$$3\mathbf{K}(\mathbf{a}, \mathbf{b}, \mathbf{x}) = \mathbf{a}^2\mathbf{G}(\mathbf{b}, \mathbf{x}) + \mathbf{G}(\mathbf{b}, \mathbf{a}^2\mathbf{x}) - \mathbf{G}(\mathbf{a}, \mathbf{b})\mathbf{x}\mathbf{b} - \mathbf{G}(\mathbf{a}, \mathbf{b}^2)\mathbf{x} + \mathbf{a}\mathbf{G}(\mathbf{b}, \mathbf{a}\mathbf{x}). \quad (7.3)$$

In the steps of Eq. (7.2), relations of the type (6.2) are used. From Eq. (7.2), we obtain

$$\mathbf{a}^2\mathbf{x}\mathbf{b}^2 = \mathbf{K}(\mathbf{a}, \mathbf{b}, \mathbf{x}). \quad (7.4)$$

It is evident from the definition of $\mathbf{K}(\mathbf{a}, \mathbf{b}, \mathbf{x})$ that Eq. (3.4) implies the reducibility of $\mathbf{a}^2\mathbf{x}\mathbf{b}^2$. We thus have

LEMMA 8. If $\mathbf{a}$, $\mathbf{b}$, $\mathbf{x}$ are 3×3 matrices, the product $\mathbf{a}^2\mathbf{x}\mathbf{b}^2$ is reducible. With Lemma 5 and Lemma 1, we obtain the

COROLLARY. If $\mathbf{a}$, $\mathbf{b}$, $\mathbf{x}$ and $\mathbf{y}$ are 3×3 matrices, then tr $\mathbf{y}\mathbf{a}^2\mathbf{x}\mathbf{b}^2$ and tr $\mathbf{a}^2\mathbf{x}\mathbf{b}^2\mathbf{y}$ are reducible.

Now let $\mathbf{a}$, $\mathbf{b}$, $\mathbf{c}$ be three symmetric 3×3 matrices. In the Corollary to Lemma 6 replace $\mathbf{y}$, $\mathbf{a}$, $\mathbf{b}$, $\mathbf{c}$ by $\mathbf{a}$, $\mathbf{a}$, $\mathbf{b}^2$, $\mathbf{c}^2$ respectively. We obtain

$$\mathrm{tr}\ \mathbf{a}(\mathbf{a}\mathbf{b}^2\mathbf{c}^2 + \mathbf{b}^2\mathbf{c}^2\mathbf{a} + \mathbf{c}^2\mathbf{a}\mathbf{b}^2 + \mathbf{c}^2\mathbf{b}^2\mathbf{a} + \mathbf{a}\mathbf{c}^2\mathbf{b}^2 + \mathbf{b}^2\mathbf{a}\mathbf{c}^2) \approx 0. \quad (7.5)$$

With Lemmas 1 and 2, this may be rewritten as

$$4\ \mathrm{tr}\ \mathbf{a}^2\mathbf{b}^2\mathbf{c}^2 + \mathrm{tr}\ \mathbf{a}\mathbf{c}^2\mathbf{a}\mathbf{b}^2 + \mathrm{tr}\ \mathbf{a}\mathbf{b}^2\mathbf{a}\mathbf{c}^2 \approx 0. \quad (7.6)$$

Using the Corollary to Lemma 8, we obtain

LEMMA 9. If $\mathbf{a}$, $\mathbf{b}$, $\mathbf{c}$ are three symmetric 3×3 matrices then tr $\mathbf{a}^2\mathbf{b}^2\mathbf{c}^2$ is reducible.

8. REDUCIBILITY OF THE TRACE OF A PRODUCT OF SEVEN MATRICES

Let $\mathbf{a}$, $\mathbf{b}$, $\mathbf{c}$, $\mathbf{d}$, $\mathbf{e}$, $\mathbf{f}$, $\mathbf{g}$ be seven 3×3 matrices. From the Corollary to LEMMA 8, we have

$$\mathrm{tr}\ \mathbf{a}^2\mathbf{c}\mathbf{e}^2\mathbf{g} \approx 0. \quad (8.1)$$

Replacing $\mathbf{e}$ by $\mathbf{e}+\mathbf{f}$, we obtain, using results of the type (8.1),

$$\mathrm{tr}\ \mathbf{a}^2\mathbf{c}(\mathbf{e}\mathbf{f} + \mathbf{f}\mathbf{e})\mathbf{g} \approx 0. \quad (8.2)$$

We thus have

LEMMA 10. If $\mathbf{a}$, $\mathbf{b}$, $\mathbf{c}$, $\mathbf{d}$, $\mathbf{e}$ are five 3×3 matrices then tr $\mathbf{a}^2\mathbf{b}\mathbf{c}\mathbf{d}\mathbf{e}$ is equivalent to $-$tr $\mathbf{a}^2\mathbf{b}\mathbf{d}\mathbf{c}\mathbf{e}$.

Now, in (8.2) we replace $\mathbf{c}$ by $\mathbf{c}\mathbf{d}$ and obtain

$$\mathrm{tr}\ \mathbf{a}^2\mathbf{c}\mathbf{d}(\mathbf{e}\mathbf{f} + \mathbf{f}\mathbf{e})\mathbf{g} \approx 0. \quad (8.3)$$

Again, replacing $\mathbf{e}$ by $\mathbf{d}\mathbf{e}$ in Eq. (8.2), we obtain

$$\mathrm{tr}\ \mathbf{a}^2\mathbf{c}(\mathbf{d}\mathbf{e}\mathbf{f} + \mathbf{f}\mathbf{d}\mathbf{e})\mathbf{g} \approx 0. \quad (8.4)$$

Yet again, replacing **e** by **d** and **g** by **eg** in Eq. (8.2), we obtain

$$\operatorname{tr} \mathbf{a}^2\mathbf{c}(\mathbf{df}+\mathbf{fd})\mathbf{eg} \approx 0. \tag{8.5}$$

Add Eqs. (8.3) and (8.4) and subtract (8.5). We obtain

$$\operatorname{tr} \mathbf{a}^2\mathbf{cdefg} \approx 0. \tag{8.6}$$

In Eq. (8.6) we replace **a** by **a**+**b** and use results of the type (8.6) to obtain

$$\operatorname{tr} (\mathbf{ab}+\mathbf{ba})\mathbf{cdefg} \approx 0. \tag{8.7}$$

Using Lemma 1, we obtain

LEMMA 11. If **a**, **b**, . . ., **g** are seven 3×3 matrices, then tr **abcdefg** is equivalent to the negative of the trace of any product formed from **abcdefg** by an odd permutation of the factors and to the trace of any product formed by an even permutation of the factors.

In the Corollary to Lemma 6, we replace **a**, **b**, **c**, **y** by **ab**, **cd**, **ef**, **g**, respectively. We obtain, with Lemma 1,

$$\operatorname{tr} (\mathbf{abcdef}+\mathbf{cdefab}+\mathbf{efabcd}+\mathbf{efcdab}+\mathbf{abefcd}+\mathbf{cdabef})\,\mathbf{g} \approx 0. \tag{8.8}$$

Each of the matrix products on the left-hand side of Eq. (8.8) is an even permutation of **abcdefg**. Hence by Lemma 11, we obtain

LEMMA 12. If **a**, **b**, . . ., **g** are seven 3×3 matrices, tr **abcdefg** is reducible.

By taking two or more of the matrices **a**, . . ., **g** equal, we obtain the

COROLLARY. The trace of a product of total degree ≥ 7, formed from any number of 3×3 matrices is reducible.

9. REDUCIBILITY OF THE TRACE OF AN ARBITRARY PRODUCT

We now consider tr **Π**, where **Π** is a product of *N*, 3×3 matrices **a**, **b**, **c**, . . . defined as in (2.1). We see immediately by using the Hamilton–Cayley theorem, together with Lemma 5, that the following lemma is valid:

LEMMA 13. The trace of a product of 3×3 matrices is reducible if the power of any factor is greater than 2 or the product is the cube of one of the matrices.

From Eq. (6.2) we see that

$$\mathbf{aba}+\mathbf{a}^2\mathbf{b}+\mathbf{ba}^2 \approx 0. \tag{9.1}$$

From Lemma 5, it follows that if **x** is a 3×3 matrix

$$\operatorname{tr} \mathbf{x}(\mathbf{aba}+\mathbf{a}^2\mathbf{b}+\mathbf{ba}^2) \approx 0. \tag{9.2}$$

With Lemma 1, we obtain

LEMMA 14. The trace of a product of any number of 3×3 matrices in which two factors are identical is equivalent to a linear combination of traces of products in which no two factors are identical.

We now combine Lemmas 1, 13 and 14 and the Corollaries to Lemmas 8 and 12 to obtain

THEOREM 1. If a product of any number of 3×3 matrices has any of the following properties:

(i) it has a repeated factor (Lemma 14);
(ii) any factor has power greater than 2, but the product is not the cube of a single matrix (Lemma 13);
(iii) it contains two non-adjacent factors of power 2 (Corollary to Lemma 8 and Lemma 1);
(iv) it has total degree greater than 6 (Corollary to Lemma 12),

then its trace is either reducible or it is equivalent to a linear combination of products which do not have the properties (i) to (iv).

Combining Theorem 1 with Corollary 2 to Lemma 7, we obtain the

COROLLARY. If a product of any number of symmetric 3×3 matrices has any of the following properties:

(i) it has a repeated factor;
(ii) any factor has power greater than 2, but the product is not the cube of a single matrix;
(iii) it contains two non-adjacent factors of power 2;
(iv) it has total degree greater than 6;
(v) two of its factors are of powers 1 and 2 respectively in the same matrix and are separated by a symmetric factor,

then its trace is either reducible or it is equivalent to a linear combination of traces of products which do not have the properties (i) to (v).

We now write down in Table 1 all those products involving one, two, ..., six symmetric 3×3 matrices which do not have any of the properties (i)–(v) listed in the Corollary to Theorem 1. In doing so we omit any products whose traces can be seen to be equivalent to any of those listed by using Lemmas 1, 2, 9 and 10. In Table 1 certain of the products are underlined. It will be shown in Section 10 that the traces of these are equivalent to linear combinations of traces of those which are not underlined.

10. EQUIVALENCE OF THE TRACES OF CERTAIN PRODUCTS IN TABLE 1

(i) *Line 6 of Table 1*

We employ the notation

$$\sum \mathbf{abc} = \mathbf{abc}+\mathbf{bca}+\mathbf{cab}+\mathbf{cba}+\mathbf{acb}+\mathbf{bac}. \tag{10.1}$$

From the Corollary to Lemma 6, we obtain by appropriate substitutions the relation

$$\text{tr }\mathbf{a}\sum \mathbf{bcd} \approx 0. \tag{10.2}$$

Using Lemmas 1 and 2, we obtain

$$\text{tr }(\mathbf{abcd}+\mathbf{abdc}+\mathbf{acbd}) \approx 0. \tag{10.3}$$

Thus tr **acbd** is equivalent to a linear combination of tr **abcd** and tr **abdc**.

(ii) *Lines 7 to 10 of Table 1*

By substituting **a**, **b**, **c**, $\mathbf{d}^2$ for **b**, **c**, **d**, **a** respectively in Eqs. (10.3) we obtain with Lemmas 1 and 2,

$$\text{tr }(\mathbf{abcd}^2+\mathbf{acbd}^2+\mathbf{bacd}^2) \approx 0. \tag{10.4}$$

TABLE 1

One matrix **a**:

(1) $\mathbf{a}, \mathbf{a}^2, \mathbf{a}^3$.

Two matrices **a**, **b**:

(2) $\mathbf{ab}, \mathbf{a}^2\mathbf{b}, \mathbf{b}^2\mathbf{a}, \mathbf{a}^2\mathbf{b}^2$.

Three matrices **a**, **b**, **c**:

(3) **abc**;
(4) $\mathbf{bca}^2, \mathbf{cab}^2, \mathbf{abc}^2$;
(5) $\mathbf{ab}^2\mathbf{c}^2, \mathbf{bc}^2\mathbf{a}^2; \mathbf{ca}^2\mathbf{b}^2$.

Four matrices **a**, **b**, **c**, **d**:

(6) $\mathbf{abcd}$, $\mathbf{abdc}$, $\mathbf{acbd}$;
(7) $\mathbf{abcd}^2$, $\mathbf{acbd}^2$, $\overline{\mathbf{bacd}^2}$;
(8) $\mathbf{bcda}^2$, $\mathbf{bdca}^2$, $\overline{\mathbf{cbda}^2}$;
(9) $\mathbf{cdab}^2$, $\mathbf{cadb}^2$, $\overline{\mathbf{dcab}^2}$;
(10) $\mathbf{dabc}^2$, $\mathbf{dbac}^2$, $\overline{\mathbf{adbc}^2}$;
(11) $\mathbf{abc}^2\mathbf{d}^2$, $\mathbf{acb}^2\mathbf{d}^2$, $\mathbf{bca}^2\mathbf{d}^2$, $\mathbf{adb}^2\mathbf{c}^2$, $\mathbf{bda}^2\mathbf{c}^2$, $\mathbf{cda}^2\mathbf{b}^2$;
(12) $\underline{\mathbf{bac}^2\mathbf{d}^2}$, $\underline{\mathbf{cab}^2\mathbf{d}^2}$, $\underline{\mathbf{cba}^2\mathbf{d}^2}$, $\underline{\mathbf{dab}^2\mathbf{c}^2}$, $\underline{\mathbf{dba}^2\mathbf{c}^2}$, $\underline{\mathbf{dca}^2\mathbf{b}^2}$;
(13) $\mathbf{bacda}^2$, $\mathbf{cbdab}^2$, $\mathbf{dcabc}^2$, $\mathbf{adbcd}^2$;

and *products obtained from these by permuting* **b**, **c**, **d** *in the first product*, **c**, **d**, **a** *in the second and so on.*

TABLE 1 (*cont.*)

Five matrices **a, b, c, d, e**:

(14) **abcde,**	**abdec,**	**abecd;**	
(15) **adbce,**	**acbed,**	**acdbe;**	
(16) **acebd,**	**acbde,**	**adcbe;**	
(17) **abedc,**	**abced,**	**abdce;**	
(18) $\mathbf{abcde^2}$,	$\mathbf{dacbe^2}$,	$\mathbf{acdbe^2}$,	$\mathbf{adbce^2}$;
(19) $\mathbf{cabde^2}$,	$\mathbf{cdabe^2}$;		

and *products obtained from (18) and (19) by cyclically permuting* **a, b, c, d, e.**

Six matrices **a, b, c, d, e, f**:

(20) **acfebd,**	**adcbfe,**	**adcfbe,**	**adfbce,**	**adfcbe;**
(21) **aebdcf,**	**aecbdf,**	**aecdbf,**	**aedbcf,**	**aedcbf;**
(22) **abcdef,**	**abcdfe,**	**abcedf,**	**abcefd,**	**abcfde;**
(23) **abcfed,**	**abdcef,**	**abdcfe,**	**abdecf,**	**abdefc;**
(24) **abdfce,**	**abdfec,**	**abecdf,**	**abecfd,**	**abedcf;**
(25) **abedfc,**	**abefcd,**	**abefdc,**	**abfcde,**	**abfced;**
(26) **abfdce,**	**abfdec,**	**abfecd,**	**abfedc,**	**acbdef;**
(27) **acbdfe,**	**acbedf,**	**acbefd,**	**acbfde,**	**acbfed;**
(28) **acdbef,**	**acdbfe,**	**acdebf,**	**acdfbe,**	**acebdf;**
(29) **acebfd,**	**acedbf,**	**acefbd,**	**acfbde,**	**acfbed;**
(30) **acfdbe,**	**adbcef,**	**adbcfe,**	**adbecf,**	**adbfce;**
(31) **adcbef,**	**adcebf,**	**adebcf,**	**adecbf,**	**aebcdf.**

Thus, the trace of $\mathbf{bacd^2}$, the underlined matrix in line 7 of Table 1, is equivalent to a linear combination of the traces of the remaining two matrices in line 7. Similar results evidently apply to lines 8, 9 and 10.

(iii) *Lines 11 and 12 of Table 1*

Now in Eq. (10.3) we replace **c, d** by $\mathbf{c^2}$, $\mathbf{d^2}$, respectively. We obtain

$$\text{tr}\,(\mathbf{abc^2d^2}+\mathbf{abd^2c^2}+\mathbf{ac^2bd^2}) \approx 0. \tag{10.5}$$

From the Corollary to Lemma 8, we have

$$\text{tr}\,\mathbf{ac^2bd^2} \approx 0. \tag{10.6}$$

Then, with Eq. (10.5) and Lemmas 1 and 2, we obtain

$$\text{tr}\,\mathbf{bac^2d^2} \approx -\text{tr}\,\mathbf{abc^2d^2}. \tag{10.7}$$

Thus, the trace of each of the products in line 12 of Table 1 is equivalent to the negative of the corresponding matrix in line 11.

(iv) *Line 13 of Table 1*

Let **a**, **b**, **c**, **d**, be symmetric 3×3 matrices.

It follows from Corollary 1 to Lemma 7 that

$$\text{tr } \mathbf{b}(\mathbf{acda}^2+\mathbf{a}^2\mathbf{cda}) \approx 0. \tag{10.8}$$

Using Lemmas 1 and 2, we obtain

$$\text{tr } \mathbf{bacda}^2+\text{tr } \mathbf{badca}^2 \approx 0. \tag{10.9}$$

Again, from Lemmas 10, 1 and 2, we see that

$$\text{tr } \mathbf{bacda}^2 \approx -\text{tr } \mathbf{bcada}^2 = -\text{tr } \mathbf{dacba}^2. \tag{10.10}$$

Combining the results expressed by Eqs. (10.9) and (10.10), we obtain

LEMMA 15. If **a**, **b**, **c**, **d** are symmetric 3×3 matrices, then the trace of a product formed from $\mathbf{bacda}^2$ by an odd permutation of **b**, **c**, **d** is equivalent to the negative of tr $\mathbf{bacda}^2$ and the trace of a product formed by an even permutation is equivalent to tr $\mathbf{bacda}^2$.

(v) *Lines 14 to 17 of Table 1*

From the Corollary to Lemma 6, we obtain by appropriate substitutions the six relations

$$\begin{aligned} &\text{tr } \mathbf{de} \textstyle\sum \mathbf{abc} \approx 0, \ \text{tr } \mathbf{ce} \textstyle\sum \mathbf{abd} \approx 0, \ \text{tr } \mathbf{cd} \textstyle\sum \mathbf{abe} \approx 0, \\ &\text{tr } \mathbf{be} \textstyle\sum \mathbf{acd} \approx 0, \ \text{tr } \mathbf{bd} \textstyle\sum \mathbf{ace} \approx 0, \ \text{tr } \mathbf{bc} \textstyle\sum \mathbf{ade} \approx 0. \end{aligned} \tag{10.11}$$

Using Lemmas 1 and 2 these relations may be rewritten as

$$\begin{aligned} &\text{tr } (\mathbf{abcde}+\mathbf{abced}+\mathbf{abdec}+\mathbf{abedc}+\mathbf{acbde}+\mathbf{acbed}) \approx 0, \\ &\text{tr } (\mathbf{abced}+\mathbf{abdce}+\mathbf{abdec}+\mathbf{abecd}+\mathbf{acebd}+\mathbf{adbce}) \approx 0, \\ &\text{tr } (\mathbf{abcde}+\mathbf{abdce}+\mathbf{abecd}+\mathbf{abedc}+\mathbf{acdbe}+\mathbf{adcbe}) \approx 0, \\ &\text{tr } (\mathbf{abecd}+\mathbf{abedc}+\mathbf{acbed}+\mathbf{acdbe}+\mathbf{acebd}+\mathbf{adcbe}) \approx 0, \\ &\text{tr } (\mathbf{abdce}+\mathbf{abdec}+\mathbf{acbde}+\mathbf{acdbe}+\mathbf{acebd}+\mathbf{adbce}) \approx 0, \\ &\text{tr } (\mathbf{abcde}+\mathbf{abced}+\mathbf{acbde}+\mathbf{acbed}+\mathbf{adbce}+\mathbf{adcbe}) \approx 0. \end{aligned} \tag{10.12}$$

In Eqs. (10.12) there are twelve matrix products. These are the same as the twelve matrix products listed in lines 14, 15, 16 and 17 of Table 1. The traces of the six of these listed in lines 16 and 17 can be shown from Eqs. (10.12) to be equivalent to linear combinations of the traces of the six listed in lines 14 and 15. This is evident from a consideration of the rank of the determinant of the coefficients of the former.

(v) *Lines 18 and 19 of Table 1*

From Lemma 9 we obtain by replacing **a**, **b**, **c** by **a**+**b**, **c**+**d** and **e**, respectively,

$$\operatorname{tr}(\mathbf{a}+\mathbf{b})^2(\mathbf{c}+\mathbf{d})^2\mathbf{e}^2 \approx 0. \tag{10.13}$$

From this we have with Lemma 9, and using relations of the type (10.7) together with Lemma 1,

$$\operatorname{tr}(\mathbf{ab}+\mathbf{ba})(\mathbf{cd}+\mathbf{dc})\mathbf{e}^2 \approx 0. \tag{10.14}$$

With Lemmas 1, 2 and 10, we may rewrite this as

$$\operatorname{tr}(\mathbf{abcd}-\mathbf{adbc}-\mathbf{dacb}+\mathbf{cdab})\mathbf{e}^2 \approx 0. \tag{10.15}$$

Thus the trace of the product $\mathbf{cdabe}^2$ in line 19 of Table 1 is equivalent to a linear combination of traces of the products in line 18. In a similar manner an analogous result can be obtained for the trace of the product $\mathbf{cabde}^2$ in line 19.

(vi) *Lines 20 to 31 of Table 1*

From the Corollary to Lemma 6, we have

$$\operatorname{tr}\mathbf{ab}\sum(\mathbf{cd})\,\mathbf{ef} \approx 0. \tag{10.16}$$

From the Corollary to Lemma 8, we have

$$\operatorname{tr}(\mathbf{a}+\mathbf{b})^2\mathbf{e}(\mathbf{c}+\mathbf{d})^2\mathbf{f} \approx 0. \tag{10.17}$$

With the Corollary to Lemma 8, Lemma 1 and Lemma 10, we obtain from Eq. (10.17)

$$\operatorname{tr}(\mathbf{ab}+\mathbf{ba})\mathbf{e}(\mathbf{cd}+\mathbf{dc})\mathbf{f} \approx 0. \tag{10.18}$$

Again, from Lemma 9, we have

$$\operatorname{tr}(\mathbf{a}+\mathbf{b})^2(\mathbf{c}+\mathbf{d})^2(\mathbf{e}+\mathbf{f})^2 \approx 0. \tag{10.19}$$

Using Lemma 9 and relations of the type (10.7) and (10.14), we obtain from Eq. (10.19)

$$\operatorname{tr}(\mathbf{ab}+\mathbf{ba})(\mathbf{cd}+\mathbf{dc})(\mathbf{ef}+\mathbf{fe}) \approx 0. \tag{10.20}$$

For brevity, we shall employ the notation

$$\begin{aligned} B(\mathbf{a},\mathbf{b};\mathbf{c},\mathbf{d}:\mathbf{e},\mathbf{f}) &= \operatorname{tr}\mathbf{ab}\sum(\mathbf{cd})\mathbf{ef},\\ C(\mathbf{a},\mathbf{b};\mathbf{c},\mathbf{d}:\mathbf{e},\mathbf{f}) &= \operatorname{tr}(\mathbf{ab}+\mathbf{ba})\mathbf{e}(\mathbf{cd}+\mathbf{dc})\mathbf{f},\\ D(\mathbf{a},\mathbf{b};\mathbf{c},\mathbf{d};\mathbf{e},\mathbf{f}) &= \operatorname{tr}(\mathbf{ab}+\mathbf{ba})(\mathbf{cd}+\mathbf{dc})(\mathbf{ef}+\mathbf{fe}). \end{aligned} \tag{10.21}$$

We take the fifty reducibility relations

$$
\begin{aligned}
&B(\mathbf{b}, \mathbf{c}; \mathbf{d}, \mathbf{e}: \mathbf{a}, \quad \approx 0, \quad B(\mathbf{b}, \mathbf{c}; \mathbf{d}, \mathbf{f}: \mathbf{a}, \mathbf{e}) \approx 0, \quad B(\mathbf{b}, \mathbf{c}; \mathbf{e}, \mathbf{f}: \mathbf{a}, \mathbf{d}) \approx 0,\\
&B(\mathbf{b}, \mathbf{d}; \mathbf{e}, \mathbf{f}: \mathbf{a}, \mathbf{c}) \approx 0, \quad B(\mathbf{c}, \mathbf{d}; \mathbf{e}, \mathbf{f}: \mathbf{a}, \mathbf{b}) \approx 0;\\
&C(\mathbf{c}, \mathbf{d}; \mathbf{e}, \mathbf{f}: \mathbf{a}, \mathbf{b}) \approx 0, \quad C(\mathbf{c}, \mathbf{f}; \mathbf{d}, \mathbf{e}: \mathbf{a}, \mathbf{b}) \approx 0, \quad C(\mathbf{b}, \mathbf{e}; \mathbf{d}, \mathbf{f}: \mathbf{a}, \mathbf{c}) \approx 0,\\
&C(\mathbf{b}, \mathbf{f}; \mathbf{d}, \mathbf{e}: \mathbf{a}, \mathbf{c}) \approx 0, \quad C(\mathbf{b}, \mathbf{c}; \mathbf{e}, \mathbf{f}: \mathbf{a}, \mathbf{d}) \approx 0, \quad C(\mathbf{b}, \mathbf{f}; \mathbf{c}, \mathbf{e}: \mathbf{a}, \mathbf{d}) \approx 0,\\
&C(\mathbf{b}, \mathbf{c}; \mathbf{d}, \mathbf{f}: \mathbf{a}, \mathbf{e}) \approx 0, \quad C(\mathbf{b}, \mathbf{d}; \mathbf{c}, \mathbf{f}: \mathbf{a}, \mathbf{e}) \approx 0, \quad C(\mathbf{b}, \mathbf{d}; \mathbf{c}, \mathbf{e}: \mathbf{a}, \mathbf{f}) \approx 0,\\
&C(\mathbf{b}, \mathbf{e}; \mathbf{c}, \mathbf{d}: \mathbf{a}, \mathbf{f}) \approx 0, \quad C(\mathbf{a}, \mathbf{d}; \mathbf{e}, \mathbf{f}: \mathbf{b}, \mathbf{c}) \approx 0, \quad C(\mathbf{a}, \mathbf{e}; \mathbf{d}, \mathbf{f}: \mathbf{b}, \mathbf{c}) \approx 0,\\
&C(\mathbf{a}, \mathbf{f}; \mathbf{d}, \mathbf{e}: \mathbf{b}, \mathbf{c}) \approx 0, \quad C(\mathbf{a}, \mathbf{c}; \mathbf{e}, \mathbf{f}: \mathbf{b}, \mathbf{d}) \approx 0 \quad C(\mathbf{a}, \mathbf{e}; \mathbf{c}, \mathbf{f}: \mathbf{b}, \mathbf{d}) \approx 0,\\
&C(\mathbf{a}, \mathbf{f}; \mathbf{c}, \mathbf{e}: \mathbf{b}, \mathbf{d}) \approx 0, \quad C(\mathbf{a}, \mathbf{c}; \mathbf{d}, \mathbf{f}: \mathbf{b}, \mathbf{e}) \approx 0, \quad C(\mathbf{a}, \mathbf{d}; \mathbf{c}, \mathbf{f}: \mathbf{b}, \mathbf{e}) \approx 0,\\
&C(\mathbf{a}, \mathbf{f}; \mathbf{c}, \mathbf{d}: \mathbf{b}, \mathbf{e}) \approx 0, \quad C(\mathbf{a}, \mathbf{c}; \mathbf{d}, \mathbf{e}: \mathbf{b}, \mathbf{f}) \approx 0, \quad C(\mathbf{a}, \mathbf{d}; \mathbf{c}, \mathbf{e}: \mathbf{b}, \mathbf{f}) \approx 0,\\
&C(\mathbf{a}, \mathbf{e}; \mathbf{c}, \mathbf{d}: \mathbf{b}, \mathbf{f}) \approx 0, \quad C(\mathbf{a}, \mathbf{b}; \mathbf{e}, \mathbf{f}: \mathbf{c}, \mathbf{d}) \approx 0, \quad C(\mathbf{a}, \mathbf{e}; \mathbf{b}, \mathbf{f}: \mathbf{c}, \mathbf{d}) \approx 0,\\
&C(\mathbf{a}, \mathbf{f}; \mathbf{b}, \mathbf{e}: \mathbf{c}, \mathbf{d}) \approx 0, \quad C(\mathbf{a}, \mathbf{b}; \mathbf{d}, \mathbf{f}: \mathbf{c}, \mathbf{e}) \approx 0, \quad C(\mathbf{a}, \mathbf{d}; \mathbf{b}, \mathbf{f}: \mathbf{c}, \mathbf{e}) \approx 0,\\
&C(\mathbf{a}, \mathbf{f}; \mathbf{b}, \mathbf{d}: \mathbf{c}, \mathbf{e}) \approx 0, \quad C(\mathbf{a}, \mathbf{b}; \mathbf{d}, \mathbf{e}: \mathbf{c}, \mathbf{f}) \approx 0, \quad C(\mathbf{a}, \mathbf{d}; \mathbf{b}, \mathbf{e}: \mathbf{c}, \mathbf{f}) \approx 0,\\
&C(\mathbf{a}, \mathbf{e}; \mathbf{b}, \mathbf{d}: \mathbf{c}, \mathbf{f}) \approx 0, \quad C(\mathbf{a}, \mathbf{b}; \mathbf{c}, \mathbf{f}: \mathbf{d}, \mathbf{e}) \approx 0, \quad C(\mathbf{a}, \mathbf{c}; \mathbf{b}, \mathbf{f}: \mathbf{d}, \mathbf{e}) \approx 0,\\
&C(\mathbf{a}, \mathbf{f}; \mathbf{b}, \mathbf{c}: \mathbf{d}, \mathbf{e}) \approx 0, \quad C(\mathbf{a}, \mathbf{b}; \mathbf{c}, \mathbf{e}: \mathbf{d}, \mathbf{f}) \approx 0, \quad C(\mathbf{a}, \mathbf{c}; \mathbf{b}, \mathbf{e}: \mathbf{d}, \mathbf{f}) \approx 0,\\
&C(\mathbf{a}, \mathbf{e}; \mathbf{b}, \mathbf{c}: \mathbf{d}, \mathbf{f}) \approx 0, \quad C(\mathbf{a}, \mathbf{b}; \mathbf{c}, \mathbf{d}: \mathbf{e}, \mathbf{f}) \approx 0, \quad C(\mathbf{a}, \mathbf{c}; \mathbf{b}, \mathbf{d}: \mathbf{e}, \mathbf{f}) \approx 0,\\
&C(\mathbf{a}, \mathbf{d}; \mathbf{b}, \mathbf{c}: \mathbf{e}, \mathbf{f}) \approx 0;\\
&D(\mathbf{a}, \mathbf{b}; \mathbf{c}, \mathbf{e}; \mathbf{d}, \mathbf{f}) \approx 0, \quad D(\mathbf{a}, \mathbf{c}; \mathbf{b}, \mathbf{d}; \mathbf{e}, \mathbf{f}) \approx 0, \quad D(\mathbf{a}, \mathbf{d}; \mathbf{b}, \mathbf{e}; \mathbf{c}, \mathbf{f}) \approx 0.\\
&D(\mathbf{a}, \mathbf{e}; \mathbf{b}, \mathbf{f}; \mathbf{c}, \mathbf{d}) \approx 0, \quad D(\mathbf{a}, \mathbf{f}; \mathbf{b}, \mathbf{c}; \mathbf{d}, \mathbf{e}) \approx 0.
\end{aligned}
\tag{10.22}
$$

With Lemmas 1 and 2, it can be seen that these relations may be rearranged to express the equivalence of fifty linear combinations of the traces of the fifty products listed in lines 22 to 31 of Table 1 to linear combinations of the traces of the ten products listed in lines 20 and 21. It follows from the fact that the 50×50 determinant of coefficients on the left-hand side of these equivalence relations does not vanish, that the trace of each of the products in lines 22 to 31 of Table 1 is equivalent to a linear combination of traces of products listed in lines 20 and 21.

11. INTEGRITY BASIS FOR N SYMMETRIC MATRICES

Omitting the underlined products in Table 1, we obtain Table 2.

According to the Corollary to Lemma 3, any polynomial invariant, with respect to the full, or proper, orthogonal group, of N symmetric 3×3 matrices **a**, **b**, **c**, . . . may be expressed as a polynomial in traces of products formed from them. Now, it follows from the Corollary to Theorem 1—and this fact is reflected in Table 2—that the trace of any product of degree greater than 6 is expressible as a polynomial in traces of products of degree 6 or less. Such a product can involve at most six matrices.

The trace of any power of **a** may be expressed as a polynomial in traces of the products in Table 2 which involve **a** only, i.e. those listed in line 1. The trace of any power of **b** may be expressed as a polynomial in traces of the products obtained from line 1 of Table 2 by replacing **a** by **b**. Similar results apply for traces of powers of any other symmetric 3×3 matrix. The trace of any product formed from two symmetric 3×3 matrices **a** and **b** is expressible as a polynomial in the traces of the products given in line 2 of Table 2, of those given in line 1 and of those obtained from line 1 by substituting **b** for **a**. Analogous results apply to any other pair of symmetric 3×3 matrices.

TABLE 2

One matrix **a**:

(1) $\mathbf{a}$, $\mathbf{a}^2$, $\mathbf{a}^3$.					

Two matrices **a**, **b**:

(2) $\mathbf{ab}$, $\mathbf{a}^2\mathbf{b}$, $\mathbf{b}^2\mathbf{a}$, $\mathbf{a}^2\mathbf{b}^2$.					

Three matrices **a**, **b**, **c**:

(3) $\mathbf{abc}$;					
(4) $\mathbf{bca}^2$,	$\mathbf{cab}^2$,	$\mathbf{abc}^2$;			
(5) $\mathbf{ab}^2\mathbf{c}^2$,	$\mathbf{bc}^2\mathbf{a}^2$,	$\mathbf{ca}^2\mathbf{b}^2$.			

Four matrices **a**, **b**, **c**, **d**:

(6) $\mathbf{abcd}$,	$\mathbf{abdc}$;				
(7) $\mathbf{abcd}^2$,	$\mathbf{acbd}^2$;				
(8) $\mathbf{bcda}^2$,	$\mathbf{bdca}^2$;				
(9) $\mathbf{cdab}^2$,	$\mathbf{cadb}^2$;				
(10) $\mathbf{dabc}^2$,	$\mathbf{dbac}^2$;				
(11) $\mathbf{abc}^2\mathbf{d}^2$,	$\mathbf{acb}^2\mathbf{d}^2$,	$\mathbf{bca}^2\mathbf{d}^2$,	$\mathbf{adb}^2\mathbf{c}^2$,	$\mathbf{bda}^2\mathbf{c}^2$,	$\mathbf{cda}^2\mathbf{b}^2$;
(12) $\mathbf{bacda}^2$,	$\mathbf{cbdab}^2$,	$\mathbf{dcabc}^2$,	$\mathbf{adbcd}^2$.		

Five matrices **a**, **b**, **c**, **d**, **e**:

(13) $\mathbf{abcde}$,	$\mathbf{abdec}$,	$\mathbf{abecd}$;			
(14) $\mathbf{adbce}$,	$\mathbf{acbed}$,	$\mathbf{acdbe}$;			
(15) $\mathbf{abcde}^2$,	$\mathbf{dacbe}^2$,	$\mathbf{acdbe}^2$,	$\mathbf{adbce}^2$;		

and *products obtained from these by cyclically permuting* **a**, **b**, **c**, **d**, **e**.

Six matrices **a**, **b**, **c**, **d**, **e**, **f**:

(16) $\mathbf{acfebd}$,	$\mathbf{adcbfe}$,	$\mathbf{adcfbe}$,	$\mathbf{adfbce}$,	$\mathbf{adfcbe}$;	
(17) $\mathbf{aebdcf}$,	$\mathbf{aecbdf}$,	$\mathbf{aecdbf}$,	$\mathbf{aedbcf}$,	$\mathbf{aedcbf}$.	

In this way, it is apparent that the trace of any product formed from N symmetric 3×3 matrices **a**, **b**, **c**, ..., and therefore any polynomial invariant of them, under the full or proper orthogonal group, may be expressed as a polynomial in traces of the products obtained from Table 2 in the following way. We replace **a** in line 1 by each of the N matrices **a**, **b**, **c**, ... in turn. We replace **a** and **b** in line **2** in turn by $\binom{N}{2}$ pairs of different matrices which can be selected from the N matrices, and so on. Finally, we replace **a**, **b**, ..., **f** in lines 16 and 17 and by the $\binom{N}{6}$ selections of six matrices which can be made from the set of N matrices.

The set of traces of products obtained in this way constitutes an integrity basis for the N symmetric 3×3 matrices under the full, or proper, orthogonal group. It has been shown by Smith [9] that this integrity basis is *irreducible*, i.e. none of its elements is expressible as a polynomial in the remaining ones. The traces of the products listed in Table 2 form a *table of typical invariants* for the irreducible integrity basis.

12. CANONICAL FORM FOR SYMMETRIC ISOTROPIC MATRIX POLYNOMIALS

Let **P** be a symmetric isotropic matrix polynomial under the orthogonal group of the N symmetric 3×3 matrices **a**, **b**, **c**, ... According to the Corollary to Lemma 4, it can be expressed in the form

$$\mathbf{P} = \sum \alpha_R \, \mathbf{\Pi}_R, \tag{12.1}$$

where $\mathbf{\Pi}_R$ is a product formed from the N symmetric matrices ($\mathbf{\Pi}_0 = \mathbf{I}$) and the α's are scalar polynomials in traces of products formed from these matrices. Since **P** is symmetric, we may write

$$\mathbf{P} = \tfrac{1}{2} \sum_R \alpha_R(\mathbf{\Pi}_R + \mathbf{\Pi}'_R), \tag{12.2}$$

where $\mathbf{\Pi}'_R$ denotes the transpose of $\mathbf{\Pi}_R$. We recall that since **a**, **b**, **c**, ... are symmetric, $\mathbf{\Pi}'_R$ is the product formed from $\mathbf{\Pi}_R$ by reversing the order of the factors.

Now, let **x** be a symmetric 3×3 matrix, not contained in the set **a**, **b**, **c**, ... Then, from Eq. (12.1)

$$\operatorname{tr} \mathbf{P}\,\mathbf{x} = \sum_R \alpha_R \operatorname{tr} \mathbf{\Pi}_R \, \mathbf{x}. \tag{12.3}$$

Since **P x** is an isotropic matrix polynomial in the $N+1$ symmetric matrices **x**, **a**, **b**, **c**, ..., tr **P x** is an orthogonal polynomial invariant of these. It may therefore be expressed as a polynomial in the elements of an irreducible integrity basis for the $N+1$ symmetric 3×3 matrices. These elements may be obtained from Table 2 in the manner described in the previous section. Some of these elements are independent of **x** and we shall denote these by $I_1, I_2, \ldots, I_\mu$; others are linear in **x** and we shall denote them by

$J_1, J_2, \ldots, J_\nu$; yet others are of higher degree in $\mathbf{x}$ but we shall have no occasion to use them.

Since tr $\mathbf{Px}$ is linear in $\mathbf{x}$, it must be expressible in the form

$$\text{tr } \mathbf{Px} = \sum_{\gamma=1}^{\nu} \beta_\gamma J_\gamma, \tag{12.3}$$

where the β's are polynomials in $I_1, I_2, \ldots, I_\mu$. From (12.3) we obtain, bearing in mind that $\mathbf{x}$ and $\mathbf{P}$ are symmetric,†

$$\mathbf{P} = \frac{1}{2} \sum_{\gamma=1}^{\nu} \beta_\gamma \left[\frac{\partial J_\gamma}{\partial \mathbf{x}} + \left(\frac{\partial J_\gamma}{\partial \mathbf{x}} \right)^t \right]. \tag{12.4}$$

Now, $\partial J_\gamma / \partial \mathbf{x}$ $(\gamma = 1, 2, \ldots, \nu)$ are products formed from $\mathbf{a}, \mathbf{b}, \mathbf{c}, \ldots$ and they can easily be read off from the elements of an irreducible integrity basis for $\mathbf{x}, \mathbf{a}, \mathbf{b}, \mathbf{c}, \ldots$ which are linear in $\mathbf{x}$. These can in turn be read off from Table 2. For example, one element is tr $\mathbf{abcx}$. Differentiating with respect to $\mathbf{x}$, we obtain the corresponding expression for $\partial J_\gamma/\partial \mathbf{x}$ as $\mathbf{cba}$. Since $\mathbf{a}, \mathbf{b}, \mathbf{c}$ are symmetric, we have $(\partial J_\gamma/\partial \mathbf{x})^t = \mathbf{abc}$.

In Table 3 we give the typical matrix products corresponding to $\partial J_\gamma/\partial \mathbf{x}$ which are obtained from Table 2 in this way. The products $\partial J_\gamma/\partial \mathbf{x}$ in the canonical form (12.4) for a symmetric isotropic matrix polynomial in N symmetric 3×3 matrices $\mathbf{a}, \mathbf{b}, \mathbf{c}, \ldots$ can then be obtained from Table 3 in the following manner. We first have the unit matrix. Next, we substitute for $\mathbf{a}$ in the products involving $\mathbf{a}$ only, each of the N matrices $\mathbf{a}, \mathbf{b}, \mathbf{c}, \ldots$ in turn. Then we substitute for $\mathbf{a}$ and $\mathbf{b}$, in the products involving only these, each of the $\binom{N}{2}$ pairs of different matrices which can be selected from $\mathbf{a}$, $\mathbf{b}, \mathbf{c}, \ldots$, and so on. Finally, we replace $\mathbf{a}, \mathbf{b}, \ldots, \mathbf{e}$, in the products involving these five matrices, by the $\binom{N}{5}$ selections of five different matrices which can be made from the N matrices $\mathbf{a}, \mathbf{b}, \mathbf{c}, \ldots$.

TABLE 3

(1) $\mathbf{I}$,				
(2) $\mathbf{a}, \mathbf{a}^2$,				
(3) $\mathbf{ab}$,				
(4) $\mathbf{ab}^2, \mathbf{a}^2\mathbf{b}$,				
(5) $\mathbf{a}^2\mathbf{b}^2$,				
(6) $\mathbf{abc}, \mathbf{cab}$,				
(7) $\mathbf{a}^2\mathbf{bc}$,	$\mathbf{ca}^2\mathbf{b}$,			
(8) $\mathbf{ab}^2\mathbf{c}$,	$\mathbf{b}^2\mathbf{ca}$,			
(9) $\mathbf{abc}^2$,	$\mathbf{bac}^2$,			
(10) $\mathbf{b}^2\mathbf{c}^2\mathbf{a}$,	$\mathbf{a}^2\mathbf{c}^2\mathbf{b}$,	$\mathbf{a}^2\mathbf{b}^2\mathbf{c}$,		
(11) $\mathbf{a}^2\mathbf{bac}$,	$\mathbf{ab}^2\mathbf{cb}$,	$\mathbf{cabc}^2$,		
(12) $\mathbf{abcd}$,	$\mathbf{cabd}$,	$\mathbf{cdab}$,		
(13) $\mathbf{adbc}$,	$\mathbf{dacb}$,	$\mathbf{acdb}$,		
(14) $\mathbf{a}^2\mathbf{bcd}$,	$\mathbf{bdca}^2$,	$\mathbf{ca}^2\mathbf{bd}$,	$\mathbf{cda}^2\mathbf{b}$,	
$\mathbf{ab}^2\mathbf{cd}$,	$\mathbf{db}^2\mathbf{ac}$,	$\mathbf{adb}^2\mathbf{c}$,	$\mathbf{b}^2\mathbf{cad}$,	
$\mathbf{abc}^2\mathbf{d}$,	$\mathbf{c}^2\mathbf{bda}$,	$\mathbf{c}^2\mathbf{dab}$,	$\mathbf{ac}^2\mathbf{db}$,	
$\mathbf{abcd}^2$,	$\mathbf{bad}^2\mathbf{c}$,	$\mathbf{bcad}^2$,	$\mathbf{cabd}^2$,	
(15) $\mathbf{ebdac}$,	$\mathbf{eadcb}$,	$\mathbf{beadc}$,	$\mathbf{bcead}$,	$\mathbf{cbead}$,
$\mathbf{aebdc}$,	$\mathbf{aecbd}$,	$\mathbf{aecdb}$,	$\mathbf{aedbc}$,	$\mathbf{aedcb}$.

† We use the notation $\partial J_\gamma/\partial \mathbf{x}$ to denote the matrix $\|\partial J_\gamma/\partial x_{ij}\|$, where $\mathbf{x} = \|x_{ij}\|$. $(\partial J_\gamma/\partial \mathbf{x})^t$ denotes the transpose of the matrix $\partial J_\gamma/\partial \mathbf{x}$.

The coefficients β_γ in the canonical form (12.4) are, of course, polynomials in the elements of an irreducible integrity basis for **a**, **b**, **c**, ..., under the orthogonal group.

It can easily be shown that the canonical form obtained in the above manner is irreducible; i.e. none of the terms $\partial J_\gamma/\partial \mathbf{x}+(\partial J_\gamma/\partial \mathbf{x})^t$ can be expressed as a linear combination of the remaining ones with polynomial invariant coefficients. Let us suppose, for the moment, that the term corresponding to $\gamma = \lambda$ can be so expressed. It then follows, since

$$\operatorname{tr} \mathbf{x} \frac{\partial J_\gamma}{\partial \mathbf{x}} = J_\gamma,$$

that J_λ can be expressed as a polynomial in the I's and the remaining J's. But this cannot be the case since the I's and J's are elements of an irreducible integrity basis. Thus, the canonical form obtained from Table 3 provides an irreducible representation for an arbitrary symmetric isotropic matrix polynomial in N symmetric 3×3 matrices.

ACKNOWLEDGEMENT

This paper was written with the support of a grant from the National Science Foundation.

REFERENCES

1. M. Reiner, *Amer. J. Math.* **67**, 350 (1945).
2. G. F. Smith and R. S. Rivlin, *Arch. Rational Mech. Anal.* **1**, 107 (1957).
3. A. C. Pipkin and R. S. Rivlin, *Arch. Rational Mech. Anal.* **4**, 129 (1959).
4. R. S. Rivlin, *J. Rational Mech. Anal.* **4**, 681 (1955).
5. A. J. M. Spencer and R. S. Rivlin, *Arch. Rational Mech. Anal.* **2**, 309 (1959).
6. A. J. M. Spencer and R. S. Rivlin, *Arch. Rational Mech. Anal.* **2**, 435 (1959).
7. A. J. M. Spencer and R. S. Rivlin, *Arch. Rational Mech. Anal.* **4**, 214 (1960).
8. A. J. M. Spencer, *Arch. Rational Mech. Anal.* **7**, 64 (1961).
9. G. F. Smith, *Arch. Rational Mech. Anal.* **5**, 382 (1960).

ELASTICITY AND PLASTICITY

ENERGY THEOREMS FOR FINITE DEFORMATION OF ORIGINALLY PLANE, PLASTIC, STRAIN-HARDENING MEMBRANES

F. K. G. ODQVIST

The Royal Institute of Technology, Stockholm, Sweden

CONSIDER a plane membrane of unit thickness occupying a finite domain D in the xy-plane, bounded by a single closed curve C of continuous curvature and acted upon by forces p_x, p_y, p_z per unit area. Neglecting moments, we have forces $N_x, N_y, N_{xy} = N_{yx}$ per unit length of sections perpendicular to the x- and y-axis. If u, v, w are the components of displacement of a point x, y in the unstrained plane, and—following the well-known von Kármán hypothesis—retaining only those second-order terms which contain derivatives of the normal deflection w, we obtain for the finite strains

$$\left.\begin{aligned} \varepsilon_x &= \frac{\partial u}{\partial x}+\frac{1}{2}\left(\frac{\partial w}{\partial x}\right)^2, \quad \varepsilon_y = \frac{\partial v}{\partial y}+\frac{1}{2}\left(\frac{\partial w}{\partial y}\right)^2 \\ \varepsilon_{xy} &= \frac{1}{2}\left(\frac{\partial u}{\partial y}+\frac{\partial v}{\partial x}+\frac{\partial w}{\partial x}\frac{\partial w}{\partial y}\right) \end{aligned}\right\}, \tag{1}$$

the full strain tensor being

$$\varepsilon_{ij} = \begin{pmatrix} \varepsilon_x, & \varepsilon_{xy}, & 0 \\ \varepsilon_{xy}, & \varepsilon_y, & 0 \\ 0, & 0, & -(\varepsilon_x+\varepsilon_y) \end{pmatrix} \tag{2}$$

corresponding to incompressible deformation of the material [1]. The stress deviation tensor will be

$$S_{ij} = \begin{pmatrix} \dfrac{2N_x-N_y}{3}, & N_{xy}, & 0 \\ N_{xy}, & \dfrac{2N_y-N_x}{3}, & 0 \\ 0, & 0, & -\dfrac{N_x+N_y}{3} \end{pmatrix}. \tag{3}$$

We may now introduce strain and stress deviation invariants ε_e = "effective strain" and N_e = "effective stress" by (summation convention for subsoripts i and j)

$$\varepsilon_e^2 = \tfrac{2}{3}\,\varepsilon_{ij}\varepsilon_{ij} = \tfrac{4}{3}\,(\varepsilon_x^2+\varepsilon_y^2+\varepsilon_x\varepsilon_y+\varepsilon_{xy}^2), \tag{4}$$

$$N_e^2 = \tfrac{3}{2}\,S_{ij}S_{ij} = N_x^2+N_y^2-N_xN_y+3N_{xy}^2. \tag{5}$$

Further, we may define energy W and complementary energy $\overline{W}$ as non-negative functions for $\varepsilon_e > 0$ and $N_e > 0$, respectively,

$$W = W(\varepsilon_e), \tag{6}$$

$$\overline{W} = \overline{W}(N_e), \tag{7}$$

such that

$$W+\overline{W} = N_e\varepsilon_e; \tag{8}$$

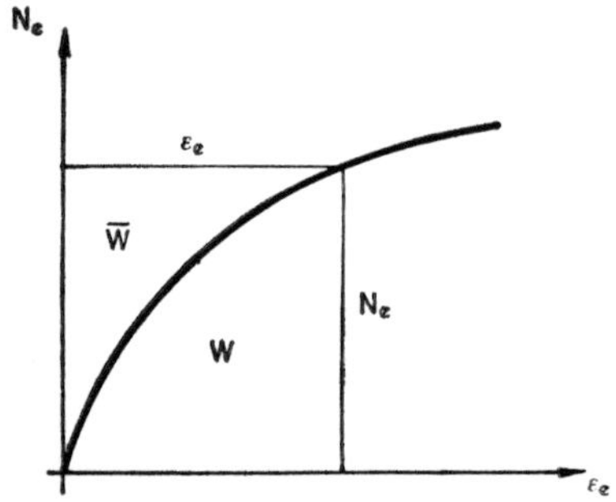

FIG. 1.

by definition, we have (cf. Fig. 1)

$$dW/d\varepsilon_e = N_e \quad \text{and} \quad d\overline{W}/dN_e = \varepsilon_e. \tag{9}$$

The functions $W(\varepsilon_e)$ and $\overline{W}(N_e)$ shall be continuous and monotonically increasing (for physical reasons, unloading must be excluded at every point of the membrane). The functions shall possess derivatives up to and including second order. Finally, we assume $W(0) = 0$, $\overline{W}(0) = 0$ and (cf. Fig. 1)

$$\frac{d^2W}{d\varepsilon_e^2} > 0 \quad \text{and hence also} \quad \frac{d^2\overline{W}}{dN_e^2} > 0. \tag{10}$$

Figure 1 illustrates the strain-hardening property of the material $N_e = N_e(\varepsilon_e)$, which is supposed to be isotropic.

As body relations we will assume

$$\varepsilon_{ij} = \frac{d\overline{W}}{\partial S_{ii}} = \frac{d\overline{W}}{dN_e}\cdot\frac{\partial N_e}{\partial S_{ij}} = \varepsilon_e\frac{\partial N_e}{\partial S_{ij}}. \tag{11}$$

From Eq. (5) we obtain

$$N_e \frac{\partial N_e}{\partial S_{ij}} = \frac{3}{2} S_{ij} \tag{12}$$

and thus, using Eq. (11)

$$\varepsilon_{ij} = \frac{3}{2} \frac{d\overline{W}}{dN_e} \frac{S_{ij}}{N_e} = \frac{3}{2} \varepsilon_e \frac{S_{ij}}{N_e} \tag{13}$$

so that

$$\varepsilon_{ij} S_{ij} = \frac{d\overline{W}}{dN_e} \cdot \frac{1}{N_e} \cdot \frac{3}{2} S_{ij} S_{ij} = N_e \frac{d\overline{W}}{dN_e} = N_e \varepsilon_e. \tag{14}$$

The equations of equilibrium yield

$$\left.\begin{aligned}
&\frac{\partial N_x}{\partial x} + \frac{\partial N_{xy}}{\partial y} + p_x = 0 \\
&\frac{\partial N_{xy}}{\partial x} + \frac{\partial N_y}{\partial y} + p_y = 0 \\
&N_x \frac{\partial^2 w}{\partial x^2} + 2N_{xy} \frac{\partial^2 w}{\partial x \partial y} + N_y \frac{\partial^2 w}{\partial y^2} + p_z - p_x \frac{\partial w}{\partial x} - p_y \frac{\partial w}{\partial y} = 0
\end{aligned}\right\} . \tag{15}$$

Equations (3) may be solved for N_x, . . ., in the form

$$\begin{pmatrix} N_x, & N_{xy} \\ N_{xy}, & N_y \end{pmatrix} = \begin{pmatrix} 2S_{11} + S_{22}, & S_{12} \\ S_{12}, & 2S_{22} + S_{11} \end{pmatrix} \tag{16}$$

and expressed in terms of the strain components by means of Eqs. (13), (9) and (4).

The boundary conditions may refer to the displacement vector u, v, w over part C_u of the curve C and to the boundary tractions T_x, T_y, T_z along the rest $C_T = C - C_u$ of the curve (see Fig. 2) where α, β are the direction cosines of the outer normal to the curve, and we have

$$\left.\begin{aligned}
T_x &= N_x \alpha + N_{xy} \beta \\
T_y &= N_{xy} \alpha + N_y \beta \\
T_z &= \left(N_x \frac{\partial w}{\partial x} + N_{xy} \frac{\partial w}{\partial y}\right) \alpha \\
&\quad + \left(N_{xy} \frac{\partial w}{\partial x} + N_y \frac{\partial w}{\partial y}\right) \beta
\end{aligned}\right\} . \tag{17}$$

The system of differential equations (15), together with the boundary conditions just mentioned, determines—after elimination of all other quantities and utilizing Eqs. (1) through (16) a complete, mixed non-linear *boundary value problem E* for the functions u, v, w.

The system is non-linear for three reasons. First, Eqs. (1) introduce *geometrical* non-linearity; further, the last equation (15) involves *static* non-linearity; finally, (9) and (13) define the *physical* non-linearity inherent in the material in connection with Fig. 1.

We shall assume that a solution of the boundary value problem E exists.

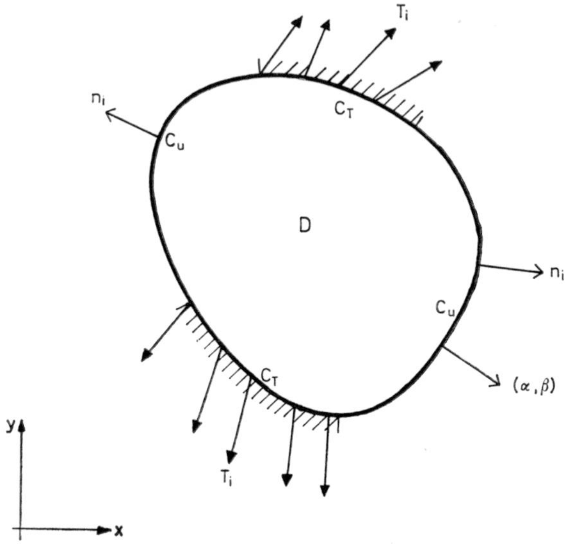

FIG. 2.

We can now construct the potential U of internal and external forces, as well as a modified complementary potential $\hat{U}$, in the following way, different from the conventional form of this function. Let us write

$$U = \iint_D [W(\varepsilon_e) - p_x u - p_y v - p_z w] dx dy - \int_{C_T} (T_x u + T_y v + T_z w) ds, \quad (18)$$

$$\hat{U} = \iint_D [\overline{W}(N_e) + p_z w/2] dx dy + \int_C T_z w/2 \, ds - \int_{C_u} (T_x u + T_y v + T_z w) ds, \quad (19)$$

ds being the line element of the curve C having an outer normal with direc, tion cosines α, β. If we add the two Eqs. (18) and (19), we obtain

$$U + \hat{U} = \iint_D [W + \overline{W} - p_x u - p_y v - p_z w/2] \, dx dy - \int_C T_x u + T_y v + T_z w/2) ds. \quad (20)$$

From Eqs. (8) and (14) we have, substituting Eq. (1),

$$W+\overline{W} = S_{ij}\varepsilon_{ij} = N_x\varepsilon_x+2N_{xy}\varepsilon_{xy}+N_y\varepsilon_y$$

$$= \frac{\partial}{\partial x}(uN_x+vN_{xy})+\frac{\partial}{\partial y}(uN_{xy}+vN_y)+\frac{\partial}{\partial x}\frac{w}{2}\left(N_x\frac{\partial w}{\partial x}+N_{xy}\frac{dw}{\partial y}\right)$$

$$+\frac{\partial}{\partial y}\frac{w}{2}\left(N_{xy}\frac{\partial w}{\partial x}+N_y\frac{\partial w}{\partial y}\right)-u\left(\frac{\partial N_x}{\partial x}+\frac{\partial N_{xy}}{\partial y}\right)-v\left(\frac{\partial N_{xy}}{\partial x}+\frac{\partial N_y}{\partial y}\right)$$

$$-\frac{w}{2}\left[\frac{\partial}{\partial x}\left(N_x\frac{\partial w}{\partial x}\right)+\frac{\partial}{\partial y}\left(N_{xy}\frac{\partial w}{\partial x}\right)+\frac{\partial}{\partial x}\left(N_{xy}\frac{\partial w}{\partial y}\right)+\frac{\partial}{\partial y}\left(N_y\frac{\partial w}{\partial y}\right)\right]. \tag{21}$$

Substituting this in Eqs. (20) we obtain, using Green's theorem,

$$U+\hat{U} = -\iint_D\left\{u\left[\frac{\partial N_x}{\partial x}+\frac{\partial N_{xy}}{\partial y}+p_x\right]+v\left[\frac{\partial N_{xy}}{\partial x}+\frac{\partial N_y}{\partial y}+p_y\right]\right.$$

$$+\frac{w}{2}\left[\frac{\partial}{\partial x}\left(N_x\frac{\partial w}{\partial x}\right)+\frac{\partial}{\partial y}\left(N_{xy}\frac{\partial w}{\partial x}\right)+\frac{\partial}{\partial x}\left(N_{xy}\frac{\partial w}{\partial y}\right)\right.$$

$$\left.\left.+\frac{\partial}{\partial y}\left(N_y\frac{\partial w}{\partial y}\right)+p_z\right]\right\}dx\,dy$$

$$+\int_C\left\{u\,[N_x\alpha+N_{xy}\beta-T_x]+v[N_{xy}\alpha+N_y\beta-T_y]\right.$$

$$\left.+\frac{w}{2}\left[\left(N_x\frac{\partial w}{\partial x}+N_{xy}\frac{\partial w}{\partial y}\right)\alpha+\left(N_{xy}\frac{\partial w}{\partial x}+N_y\frac{\partial w}{\partial y}\right)\beta-T_z\right]\right\}ds. \tag{22}$$

In view of Eq. (15) and (17), both integrals in Eq. (22) vanish and we obtain

$$U+\hat{U} = 0. \tag{23}$$

On the other hand, if we give an arbitrary variation δu, δv, δw to u, v, w with $u+\delta u, \ldots$ satisfying the boundary conditions on C_u, so that we have on C_u

$$\delta u = 0, \quad \delta v = 0, \quad \delta w = 0, \tag{24}$$

while leaving $p_x, \ldots$, and the stress system $N_x, \ldots$ unchanged in D and $T_x, \ldots$, on C_T, we obtain for the variation δU of U in the neighbourhood of the true solution u, v, w of the boundary value problem E

$$\delta U = \iint_D(\delta W-p_x\,\delta u-p_y\,\delta v-p_z\,\delta w)\,dx\,dy$$

$$-\int_{C_T}(T_x\,\delta u+T_y\,\delta v+T_z\,\delta w)\,ds. \tag{25}$$

Now we have, carrying out the variation as prescribed and integrating by parts,

$$
\begin{aligned}
\delta W &= N_x\,\delta\varepsilon_x + N_y\,\delta\varepsilon_y + 2N_{xy}\,\delta\varepsilon_{xy} \\
&= N_x\left(\frac{\partial \delta u}{\partial x} + \frac{\partial w}{\partial x}\frac{\partial \delta w}{\partial x}\right) + N_y\left(\frac{\partial \delta v}{\partial y} + \frac{\partial w}{\partial y}\frac{\partial \delta w}{\partial y}\right) \\
&\quad + N_{xy}\left(\frac{\partial \delta u}{\partial y} + \frac{\partial \delta v}{\partial x} + \frac{\partial w}{\partial x}\frac{\partial \delta w}{\partial y} + \frac{\partial w}{\partial y}\frac{\partial \delta w}{\partial x}\right) = \\
&= \frac{\partial}{\partial x}(N_x\,\delta u) - \frac{\partial N_x}{\partial x}\delta u + \frac{\partial}{\partial x}\left(N_x\frac{\partial w}{\partial x}\delta w\right) - \delta w\frac{\partial}{\partial x}\left(N_x\frac{\partial w}{\partial x}\right) + \dots \\
&\quad + \frac{\partial}{\partial y}(N_{xy}\,\delta u) - \frac{\partial N_{xy}}{\partial y}\delta u + \frac{\partial}{\partial y}\left(N_{xy}\frac{\partial w}{\partial x}\delta w\right) - \delta w\frac{\partial}{\partial y}\left(N_{xy}\frac{\partial w}{\partial x}\right) + \dots
\end{aligned}
\tag{26}
$$

Substituting Eq. (26) in Eq. (25) and using Green's theorem, we obtain

$$
\begin{aligned}
\delta U = &-\iint_D \Bigg\{\left[\frac{\partial N_x}{\partial x} + \frac{\partial N_{xy}}{\partial y} + p_x\right]\delta u + \left[\frac{\partial N_{xy}}{\partial x} + \frac{\partial N_y}{\partial y} + p_y\right]\delta v \\
&+ \left[\frac{\partial}{\partial x}\left(N_x\frac{\partial w}{\partial x}\right) + \frac{\partial}{\partial y}\left(N_{xy}\frac{\partial w}{\partial x}\right) + \frac{\partial}{\partial x}\left(N_{xy}\frac{\partial w}{\partial y}\right)\right. \\
&\left. + \frac{\partial}{\partial y}\left(N_y\frac{\partial w}{\partial y}\right) + p_z\right]\delta w\Bigg\}\,dx\,dy \\
&+ \int_{C_T}\Bigg\{[N_x\alpha + N_{xy}\beta - T_x]\,\delta u + [N_{xy}\alpha + N_y\beta - T_y]\,\delta v \\
&+ \left[\left(N_x\frac{\partial w}{\partial x} + N_{xy}\frac{\partial w}{\partial y}\right)\alpha + \left(N_{xy}\frac{\partial w}{\partial x} + N_y\frac{\partial w}{\partial y}\right)\beta - T_z\right]\delta w\Bigg\}\,ds \\
&+ \int_{C_u}\Bigg\{[N_x\alpha + N_{xy}\beta - T_x]\,\delta u + [N_{xy}\alpha + N_y\beta - T_y]\,\delta v \\
&+ \left[\left(N_x\frac{\partial w}{\partial x} + N_{xy}\frac{\partial w}{\partial y}\right)\alpha + \left(N_{xy}\frac{\partial w}{\partial x} + N_y\frac{\partial w}{\partial y}\right)\beta - T_z\right]\delta w\Bigg\}\,ds.
\end{aligned}
\tag{27}
$$

In Eq. (27) the integrand of the double integral will vanish due to Eq. (15), that of the curve integral along C_T due to Eq. (17), and that along C_u due to Eq. (24), so that we obtain

$$\delta U = 0. \tag{28}$$

As for the variation $\delta^2 U$, we have

$$\delta U^2 = \iint_D \delta^2 W\,dx\,dy = \frac{1}{2}\iint_D \frac{d^2 W}{d\varepsilon_e^2}(\delta\varepsilon_e)^2\,dx\,dy, \tag{29}$$

which is obviously positive for arbitrary variations δu, δv, δw due to ieqnualtiy (10).

We now can state

THEOREM A. *The potential U will have a true minimum for variations of the state of displacement in the neighbourhood of the solution u, v, w of the boundary value problem E with finite displacements, provided the varied system satisfies the boundary conditions for the displacements on those parts of the boundary where these are prescribed, and provided the given surface forces and boundary tractions remain unchanged during the variation.*

On the other hand, while leaving the displacement components u, v, w unchanged, we may impose on the system of internal forces N_x, N_y, N_{xy}, associated with the solution of boundary value problem E, a variation δN_x, . . ., satisfying the equilibrium conditions

$$\left.\begin{aligned} &\frac{\partial \delta N_x}{\partial x}+\frac{\partial \delta N_{xy}}{\partial y}=0\\ &\frac{\partial \delta N_{xy}}{\partial x}+\frac{\partial \delta N_y}{\partial y}=0\\ &\delta N_x\frac{\partial^2 w}{\partial x^2}+2\delta N_{xy}\frac{\partial^2 w}{\partial x\partial y}+\delta N_y\frac{\partial^2 w}{\partial y^2}=0 \end{aligned}\right\} \tag{30}$$

and such that the boundary tractions formed according to Eq. (17) will be

$$\delta T_x=0,\quad \delta T_y=0,\quad \delta T_z=0 \tag{31}$$

on the part C_T of the boundary curve C.

We may now form the corresponding variation $\delta\hat{U}$ of the modified complementary potential $\hat{U}$

$$\delta\hat{U}=\iint_D \delta\overline{W}\,dx\,dy-\int_{C_u}(\delta T_x\,u+\delta T_y\,v+\delta T_z\cdot w/2)\,ds. \tag{32}$$

Carrying out the variation as prescribed we obtain, according to Eq. (1),

$$\begin{aligned} \delta\overline{W} &= \varepsilon_x\,\delta N_x+2\varepsilon_{xy}\,\delta N_{xy}+\varepsilon_y\,\delta N_y=\\ &=\frac{\partial}{\partial x}(u\delta N_x)-u\frac{\partial\delta N_x}{\partial x}+\frac{\partial}{\partial x}\left(\frac{w}{2}\frac{\partial w}{\partial x}\,\delta N_x\right)-\frac{w}{2}\frac{\partial}{\partial x}\left(\frac{\partial}{\partial x}\,\delta N_x\right)\\ &\quad+\frac{\partial}{\partial y}(u\delta N_{xy})-u\frac{\partial\delta N_{xy}}{\partial y}+\frac{\partial}{\partial x}\left(\frac{w}{2}\frac{\partial w}{\partial y}\,\delta N_{xy}\right)-\frac{w}{2}\frac{\partial}{\partial x}\left(\frac{\partial w}{\partial y}\,\delta N_{xy}\right)\\ &\quad+\frac{\partial}{\partial x}(v\delta N_{xy})-v\frac{\partial\delta N_{xy}}{\partial x}+\frac{\partial}{\partial y}\left(\frac{w}{2}\frac{\partial w}{\partial x}\,\delta N_{xy}\right)-\frac{w}{2}\frac{\partial}{\partial y}\left(\frac{\partial w}{\partial x}\,\delta N_{xy}\right)\\ &\quad+\frac{\partial}{\partial y}(v\delta N_y)-v\frac{\partial\delta N_y}{\partial y}+\frac{\partial}{\partial y}\left(\frac{w}{2}\frac{\partial w}{\partial y}\,\delta N_y\right)-\frac{w}{2}\frac{\partial}{\partial y}\left(\frac{\partial w}{\partial y}\,\delta N_y\right). \end{aligned}$$

This, substituted in Eq. (32), yields the following expression for $\delta\hat{U}$ if due regard is taken for Eq. (30) and for Green's theorem:

$$\delta\hat{U} = \int_{C_T}\Big\{(\delta N_x\,\alpha+\delta N_{xy}\,\beta)\,u+(\delta N_{xy}\,\alpha+\delta N_y\,\beta)v$$
$$+\left[\left(\delta N_x\frac{\partial w}{\partial x}+\delta N_{xy}\frac{\partial w}{\partial y}\right)\alpha+\left(\delta N_{xy}\frac{\partial w}{\partial x}+\delta N_y\frac{\partial w}{\partial y}\right)\beta\right]\frac{w}{2}\Big\}\,ds$$
$$+\int_{C_u}\Big\{(\delta N_x\,\alpha+\delta N_{xy}\,\beta-\delta T_x)\,u+(\delta N_{xy}\,\alpha+\delta N_y\,\beta-\delta T_y)\,v$$
$$+\left[\left(\delta N_x\frac{\partial w}{\partial x}+\delta N_{xy}\frac{\partial w}{\partial y}\right)\alpha+\left(\delta N_{xy}\frac{\partial w}{\partial x}+\delta N_y\frac{\partial w}{\partial y}\right)\beta-\delta T_z\right]\frac{w}{2}\Big\}\,ds. \quad (33)$$

In Eq. (33) the first integral will vanish due to Eq. (31) and the second due to Eq. (17). The second variation $\delta^2\hat{U}$ of $\hat{U}$ will be

$$\delta^2\hat{U} = \iint \delta^2\overline{W}\,dx\,dy = \frac{1}{2}\iint\frac{d^2\overline{W}}{dN_e^2}(\delta N_e)^2\,dx\,dy$$

and this quantity will obviously remain non-negative irrespective of δN_x, ... Thus we obtain

THEOREM B. *The modified complementary potential $\hat{U}$ given by Eq.* (19) *has a true minimum for variations in the state of internal forces N_x, N_{xy}, N_y associated with the solution of the boundary value problem E, provided the variations δN_x, ... satisfy the equilibrium conditions* (30), *and the corresponding boundary tractions $T_x+\delta T_x$, ..., for the varied system satisfy the same boundary conditions for the forces on the boundary C_T as the solution, whereas the displacement vector u, v, w remains unchanged.*

Theorems A and B, together with the relation (23), will permit estimates of upper and lower bounds for the potential U and the modified complementary potential $\hat{U}$ for the true solution of the boundary value problem E. If we denote by U^* and $\hat{U}^*$ the values of the potentials obtained for the varied systems considered in the derivation of Theorems A and B above, we obviously have

$$U^* > U,$$
$$\hat{U}^* > \hat{U}.$$

But then we also have

$$-U^* < -U,$$
$$-\hat{U}^* < -\hat{U},$$

and hence Eq. (23) immediately yields

$$\begin{aligned} -U^* &< \hat{U} < \hat{U}^*, \\ -\hat{U}^* &< U < U^*. \end{aligned} \quad (34)$$

In this way we have obtained, at least from a formal point of view, upper and lower bounds for U and $\hat{U}$. Unfortunately, in this case the upper bound for $\hat{U}$ and the lower bound for U will contain the unknown deflection w.

Practical application of Theorems A and B for solution of specific problems will be restricted by prohibitive computational work. So far, only Theorem A has been utilized by L. Wallin in a study of the deflection of a rectangular membrane under constant lateral pressure, [2], in the special case where N_e is of the form

$$N_e = N_0 \varepsilon_e^{1/n},$$

N_0 and $n > 1$ being material constants. He tested the accuracy of his computations by treating the corresponding problem for a circular membrane, in which case an exact solution is available for comparison. The comparison turns out to be favourable. Wallin's work is reviewed in the author's book [1].

REFERENCES

1. F. K. G. Odqvist, *Mathematical Theory of Creep and Creep Rupture*, Clarendon Press, Oxford (1966).
2. L. Wallin, "Large deflections of non-linear elastic rectangular plates", *Arkiv för Fysik*, K. V. A., Stockholm, Bd. 17, no. 4, pp. 89–95 (1960).

ON CERTAIN CASES OF ELASTIC NON-HOMOGENEITY

W. OLSZAK and J. RYCHLEWSKI

Polish Academy of Sciences, Warsaw

1. Problems of elastic non-homogeneity have, for a number of years, been the object of interest of many scientists (cf., for instance, refs. [1], [2], [3]). This is certainly due to the interesting theoretical aspect of such problems and also to the needs of practice. As a typical example let us quote the problem of propagation of seismic waves in a non-homogeneous medium.

2. In ref. [4], a semi-inverse method for tackling problems of non-homogeneous elasticity was proposed.

Let us express Hooke's law in the form

$$\varepsilon_{ij} = ns_{ij} + m\sigma\delta_{ij} \tag{1}$$

$$s_{ij} \equiv \sigma_{ij} - \sigma\delta_{ij}, \quad \sigma \equiv \tfrac{1}{3}\sigma_{ii}$$

where the elastic coefficients are

$$n = n(x_i) \equiv \frac{1}{2\mu(x_i)}, \quad m = m(x_i) \equiv \frac{1}{2K(x_i)}, \tag{2}$$

μ is the shear modulus and K the bulk modulus.

On substituting Eq. (1) in the compatibility equations

$$e_{ijk}e_{pqr}e_{jq,kr} = 0 \tag{3}$$

we obtain, after some familiar operations, the Beltrami–Michell equations for a non-homogeneous body,

$$(ns_{ij} + m\sigma\delta_{ij})_{,kk} + [(2n+m)\sigma]_{,ij} = n_{i,k}\sigma_{kj}. \tag{4}$$

The idea put forward in ref. [4] is as follows. Let us consider a body G with purely static boundary conditions, and a statically admissible stress field $\sigma^0_{ij}(x_i)$ (such that $\sigma^0_{ij,j} + \varrho F_i = 0$ in G, $\sigma^0_{ij}\nu_j = T_i$ on the edge). On substituting σ^0_{ij} in Eq. (4), we obtain a set of six partial differential equations of the second order for two functions $n(x_i)$, $m(x_i)$. *Every solution of this set of equations, that is not in contradiction with the conditions of physical validity, determines the type of non-homogeneity for which* σ_{ij}^0 *represents a real field of stresses.*

Let us point out two facts: (1) no boundary-value problem is formulated for the set of equations obtained; (2) the solutions of a set of six partial equations in two unknown functions constitute, in general, a relatively narrow class. (For special states of stress the situation is completely different.)

The proposed semi-inverse method has two advantages. On the one hand, it yields, in a relatively simple manner, a number of closed-form solutions, and permits qualitative study of the influence of elastic non-homogeneity on the stress field (analysis). On the other, it enables us to formulate the problem of determining the type of non-homogeneity producing a desired state of stress (synthesis). It is obvious that the above idea can be generalized to a more general approach, in which only certain elements of the stress field are assumed.

Let us consider, according to ref. [5], an instructive example of application of the idea in ref. [4]

Let a body G be loaded by a constant hydrostatic pressure p. Taking for σ_{ij}^0 the stresses

$$\sigma_{ij} = -p\delta_{ij} \tag{5}$$

we obtain from Eq. (4)

$$m_{,ij} = 0. \tag{6}$$

This result may be formulated as follows: a hydrostatic state of stress (5) in G may occur only if the inverse value of the bulk modulus is a linear function of Cartesian coordinates in G,

$$m = a_0 + a_i x_i. \tag{7}$$

The conclusions of this statement on the occurrence of the first plastic strain in a body under hydrostatic pressure are presented in refs. [5] and [6].

Another example is considered in ref. [4]. It is shown that in order to realize, in a thick-walled sphere subjected to internal pressure, a uniform reduced stress (in the sense of Huber–Mises), the rigidity must vary as follows

$$E = E(r) = E_0\left\{1 - \frac{1}{\eta}\ln\frac{b}{r}\right\}^{3\eta} \tag{8}$$

$$\eta \equiv \frac{1-\nu}{1-2\nu}, \quad \nu = \text{const},$$

where E is Young's modulus and ν Poisson's ratio.

3. In the general case of plane problem we have instead of a set of six equations (4), a single compatibility equation expressed in the form (cf. ref. [4]).

$$\begin{aligned}&\nabla^2[\bar{M}\nabla^2\Omega+(2\bar{M}-\bar{N})U+\bar{\varepsilon}\alpha\theta]\\&-(\bar{N}_{,yy}\,\Omega_{,xx}+\bar{N}_{,xx}\,\Omega_{,yy}-2\bar{N}_{,xy}\,\Omega_{,xy})=0,\end{aligned}\tag{9}$$

where $\nabla(\ )=(\ )_{,xx}+(\ )_{,yy}$,

Ω = Airy's stress function,

U = potential of mass forces,

θ = temperature,

α = coefficient of thermal elongation,

and

$$\bar{M}=\frac{1-\nu}{E},\quad \bar{N}=\frac{(1+\nu)(1-\nu)}{E},\quad \bar{\varepsilon}=1+\nu\tag{10}$$

for plane strain, and

$$\bar{M}=\frac{1}{E},\quad \bar{N}=\frac{1+\nu}{E},\quad \bar{\varepsilon}=1\tag{11}$$

for plane stress.

By prescribing a statically admissible state of stress $\Omega(x, y)$ satisfying the boundary conditions, we obtain a single partial differential equation of the second order for two sought-for functions. *Thus the state of stress will be the same for a very large class of non-homogeneities.*

Referring the reader, for other details, to refs. [4], [7] we shall only quote the following example: a prescribed state of stress determined by $\Omega(x, y)$ is realized for any distribution of the temperature and mass forces, if the elastic non-homogeneity belongs to the class determined by the relations

$$\begin{aligned}\bar{M}&=\bar{M}(x,y)=\frac{\Phi(x,y)+\bar{N}U(x,y)-\bar{\varepsilon}\alpha\theta(x,y)}{\nabla^2\Omega(x,y)+2U(x,y)},\\ \bar{N}&=\bar{N}(x,y)=n_0+n_1x+n_2y,\end{aligned}\tag{12}$$

where $\Phi(x, y)$ is a harmonic function and n_0, n_1, n_2 are constants. This is a direct consequence of Eq. (9).

A statically admissible stress field may contain a number of independent parameters t_i ($i = 1,\ldots, s$) determining the intensity of the forces and l_i ($i = 1,\ldots, k$) determining the shape of the body. For the solution to be valid over any range of t_i, we must require that the non-homogeneity types obtained be independent of t_i which leads to further reduction of the class of solutions of Eq. (9). In some cases we can introduce also independence of $\bar{M}, \bar{N}$ of l_i. A solution for the case of complete independence of the non-homogeneity of the parameters is given in [8]. If $U \equiv \theta \equiv 0$, and if a statically admissible stress field contains the factor t_1, the solutions of Eq. (9) are already independent of t_1.

In the case of ν = const, Eq. (9) becomes

$$(\sigma_{yy}-\bar{\nu}\sigma_{xy})\bar{M}_{,xx}-2(1+\bar{\nu})\sigma_{xy}\bar{M}_{,xy}+ \\ +(\sigma_{xx}-\bar{\nu}\sigma_{yy})\bar{M}_{,yy}+L(\bar{M}, U, \theta, \sigma_{ij})=0, \tag{13}$$

the operator L not being in explicit form (it is an operator of first order for $\bar{M}$). The type of Eq. (13) for the function $\bar{M}$ is determined by the sign of the determinant

$$\Delta \equiv \frac{1}{(1-\nu)^2}\left[T^2-3\frac{(1-2\nu)}{1+\nu}\sigma^2\right] \tag{14}$$

for plane strain, and

$$\Delta \equiv (1+\nu)\left[T^2-3\frac{1+\nu^3}{(1+\nu)^3}\sigma^2\right] \tag{15}$$

for plane stress, where σ, T are the first and second stress invariants respectively. The extreme cases are as follows: for pure shear, $\sigma = 0$, Eq. (13) is hyperbolic; for $T = 0$ it is elliptic. It is interesting to observe that for plane strain in an incompressible body Eq. (13) is always hyperbolic for $T \neq 0$, its characteristics being the trajectories of the maximum shear stresses.

An example illustrating Eq. (13) is funished by the homogeneous state of pure shear

$$\sigma_{xy} = \text{const}, \quad \sigma_{xx} = \sigma_{yy} = 0 \tag{16}$$

obtained, as follows from Eq. (13), for any non-homogeneity of the type

$$\bar{M} = f(x)+g(y), \tag{17}$$

where f, g are arbitrary functions of their arguments.

4. Formulating the inverse problem in Section 2, in general terms, we mentioned the physical validity of the resulting non-homogeneity functions. We have taken into account that, in the material considered, the well-known inequalities for elastic moduli should hold.

Thus we have

$$0 \leqslant n < \infty, \quad 0 \leqslant m < \infty. \tag{18}$$

The left-hand inequality in the second formula refers to an incompressible body.

The right-hand limitations should be strengthened and expressed thus

$$n \leqslant n_0 < \infty, \quad m \leqslant m_0 < \infty, \tag{19}$$

where n_0, m_0 must satisfy the following condition: for a prescribed intensity of surface forces, the stress following from (1) should be so small that the linear theory of elasticity, which is the basis of our analysis, would hold.

Let us consider also the left-hand inequality in the first of the formulae (18). If $n = 0$ and $m = 0$ in a certain region, this is perfectly rigid. If $n = 0$, $m = 0$ on a line or surface, the latter are perfectly rigid.

Admission of the possibility of the elastic coefficients n, m vanishing in certain sets of points leads to an interesting idea of describing structures by means of geometric constants. Let us consider, as an example, a few periodic structures.

Taking n, m in the form

$$f = c(A_1 + A_2 + A_3), \tag{20}$$

where

$$A_i = \left[\frac{1}{2}\left(1 + \sin 2\pi \frac{x_i}{a}\right)\right]^p, \quad c = \text{const},$$

it is seen that there exists, inside the body, a regular set of points, of spacing a, at which $n = m = 0$.

If n, m have the form

$$f = c(A_1A_2 + A_1A_3 + A_2A_3) \tag{21}$$

the body contains three orthogonal rectilinear families of rigid equidistant fibres.

If n, m are

$$f = cA_1A_2A_3 \tag{22}$$

the body contains a set of three orthogonal families of rigid equidistant planes.

If we take the limit for $p \to 0$, then in all three cases we obtain ideal images in which perfectly rigid point, linear or surface elements are included in a homogeneous body.

Refraining from detailed discussion of the problems arising in this connection, let us indicate some interesting aspects calling for special attention. An example is to be found in ref. [9]. The approach adopted there independently, and the example considered, coincide with the contents of Section 6 of our ref. [4]. The author also discusses some previous cases not analysed in ref. [4].

In particular, for a wedge as presented in Fig. 1, whose non-homogeneity type is

$$E = E_1 x^{-1} = E_1 (r \cos \theta)^{-1}, \quad \nu = \text{const}, \tag{23}$$

the following interesting radial stress distribution was obtained

$$\sigma_r = -\frac{1}{2r \cos \theta} \left| \frac{P_x}{\beta} + \frac{P_y}{g(\beta)} \theta \right|, \quad \sigma_\theta = \tau_{r\theta} = 0, \tag{24}$$

where

$$g(\beta) = \int_0^\beta \theta \tan \theta \, d\theta. \tag{25}$$

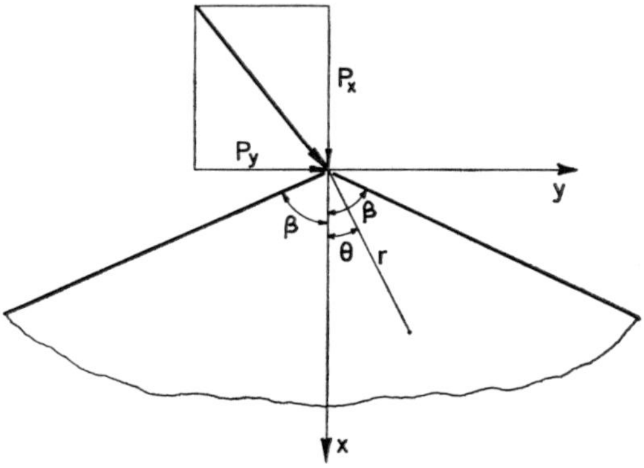

FIG. 1.

It is easily seen that the integral equilibrium conditions

$$P_x = -\int_{-\beta}^{+\beta} \sigma_r r \cos \theta \, d\theta, \qquad P_y = -\int_{-\beta}^{+\beta} \sigma_r r \sin \theta \, d\theta \tag{26}$$

are satisfied for any β. The non-homogeneity of type (23) constitutes a special case of a "hyperbolic" non-homogeneity analysed in ref. [10].

Passing to the case of a half-plane, $\beta \to \pi/2$, the author writes, in view of $g(\beta) \to \infty$,

$$\sigma_r = -\frac{P_x}{\pi x} \tag{27}$$

and infers that the force acting along the boundary ($P_x = 0$) does not produce any stress in the half-plane whose non-homogeneity is represented by (23). He attributes this to inaccuracy of the linear theory of elasticity.

This conclusion calls for comment. Indeed, if $\sigma_r \equiv 0$ for every $-\pi/2 \leqslant \theta \leqslant +\pi/2$ and $P_y \neq 0$, we obtain a contradiction in the second part of

Eq. (26). However, since that equation holds for every β, the statement of $\sigma_r \equiv 0$ is not valid. This paradox is due to inaccuracy of the limit operation. Indeed, if $\beta \to \pi/2$ we have

$$\lim_{\beta\to\pi/2} \sigma_r = -\frac{P_x}{\pi x} - \frac{P_y}{2r} \lim_{\beta\to\pi/2} \theta \left(\cos\theta \int_0^\beta \theta \tan\theta \, d\theta\right)^{-1}. \tag{28}$$

The limit of the second term is a *generalized function*, zero at every point except for the edge, for which we have

$$\pm \lim_{\beta\to\pi/2} \beta \left(\cos\beta \int_0^\beta \theta \tan\theta \, d\,\theta\right)^{-1} = \pm \lim_{\beta\to\pi/2} \frac{\dfrac{d}{d\beta}\left(\dfrac{\beta}{\cos\beta}\right)}{\beta\tan\beta} = \pm\infty. \tag{29}$$

Thus, for the stress distribution in the wedge we have not Eq. (27) from ref. [9], but the following expression

$$\sigma_r = -\frac{P_x}{\pi x} - \frac{P_y}{2r}\left[\delta\left(\theta - \frac{\pi}{2}\right) - \delta\left(\theta + \frac{\pi}{2}\right)\right] \tag{30}$$

where $\delta(\xi)$ is Dirac's function. Equations (26) are satisfied in view of the properties of the function δ, [11], for integration interval

$$[-\pi/2 - \varepsilon, \quad \pi/2 + \varepsilon], \quad \varepsilon > 0.$$

The mechanical significance of the above remarks is obvious. In a wedge of apex angle $\beta < \pi/2$ we have $E = \infty$ at the apex only. On the other hand, there occurs in the half-plane, and in the wedge with $\beta = \pi/2 + \varepsilon$ in agreement with Eq. (23), an infinitely rigid layer over $x = 0$. This layer carries, without deformation (cf. Eqs. (23), (24)), the entire horizontal component of the force; there, in agreement with Eq. (30), an infinite compressive stress occurs in its right-hand part and an infinite tensile stress "equal" to it in its left-hand part.

REFERENCES

1. S. G. Mikhlin, "Plane elastic problem of non-homogeneous media", *Tr. Sejsmol. Inst.* No. 66 (1935).
2. J. Golecki, "On the foundations of the theory of elasticity of plane incompressible non-homogeneous bodies", *Arch. Mech. Stos.* **11**, 4 (1959).
3. P. P. Teodorescu, and M. Predeleanu, "Ueber das ebene Problem nicht-homogener elastischer Kœrper," *Acta Techn. Acad. Hung.* **27**, 3–4 (1959).
4. W. Olszak and J. Rychlewski, "Nichthomogenitaets-Probleme im elastischen und vorplastischen Bereich", *Oest. Ing.-Archiv*, b. XV, h. 1–4 (1961).
5. J. Rychlewski, "Note on the beginning of plastic deformation in a body under uniform pressure", *Arch. Mech. Stos.* **17**, 3 (1965).

6. J. OSTROWSKA, "The yielding of an elastically non-homogeneous sphere under uniform pressure", *Arch. Mech. Stos.* **17,** 3 (1965).
7. W. OLSZAK and J. RYCHLEWSKI, "On plane states of equilibrium in non-homogeneous elastic and plastic media", *Proceedings of Int. Symp. Appl. Theory of Functions, Tbilisi,* 1963, *"Nauka" Publ. House* (Moscow, 1965). vol. 1, pp. 289–308.
8. J. OSTROWSKA, "Propagation of plastic zones in an elastically non-homogeneous wedge", *Bull. Acad. Polon. Sci.*, Serie Sci. Techn. **11,** 7 (1963).
9. S. G. LEKHNITZKY, "Radial stress field in a wedge and a half-plane with variable elastic modulus", *Prikl. Mat. Mekh.* **26,** 1 (1962).
10. W. OLSZAK and W. URBANOWSKI, "Plastic non-homogeneity: A survey of theoretical and experimental research, Non-homogeneity in elasticity and plasticity", *Proc. IUTAM Symp.*, Pergamon Press, Oxford (1959).
11. I. M. GELFAND and G. E. SHILOV, *Generalized Functions and Their Operations* vol. 1, Fizmatgiz (Moscow, 1959).

OPTIMAL PLASTIC DESIGN OF RINGS

WILLIAM PRAGER

IBM Zürich Research Laboratory, Rüschlikon, Zürich, Switzerland, and Department of the Aerospace and Mechanical Engineering Sciences, University of California, San Diego, U.S.A.

ABSTRACT

Optimal plastic design of a circular sandwich ring is discussed under the following assumptions:

(1) The loading is symmetric with respect to two orthogonal diameters of the ring;
(2) a minimum plastic resistance is prescribed for the cross sections;
(3) the cost of any stronger cross section is proportional to the excess of the plastic resistance of the actual cross section over the minimum plastic resistance.

1. INTRODUCTION

Optimal plastic design is usually based on the assumption that the *specific cost* is proportional to the *plastic resistance*. For example, the weight per unit length of a beam with continuously varying cross section is assumed to be proportional to the plastic moment. The resulting optimal design will have vanishing plastic moment at certain cross sections and is therefore regarded as unrealistic. To arrive at a more realistic design, one may prescribe a minimum cross section and only consider the difference between the costs of the actual and the minimum cross sections. When this difference is a convex function of the difference between the actual and the minimum plastic resistances, an optimal design may be obtained by a method developed in an earlier paper [1]. In the following, this method is applied to the optimal design of circular rings.

2. GENERALIZED STRESSES AND STRAIN RATES

The discussion will be restricted to states of loading that are symmetrical with respect to the vertical and horizontal diameters of the ring. The radius of the circular center line will be denoted by R, and a cross section will be specified by its polar angle θ (Fig. 1). The ring is to have a rectangular sandwich section consisting of two thin strips of the variable thickness $T(\theta)$ and the constant breadth B that are separated by a core of the constant height $2H$.

To describe a state of plastic flow of the ring, we give the circumferential velocity $u(\theta)$ and the radial velocity $v(\theta)$ of the points on the center line and adopt Bernoulli's hypothesis, which stipulates that material cross sections of the undeformed ring remain plane and normal to the center line as this undergoes a deformation.

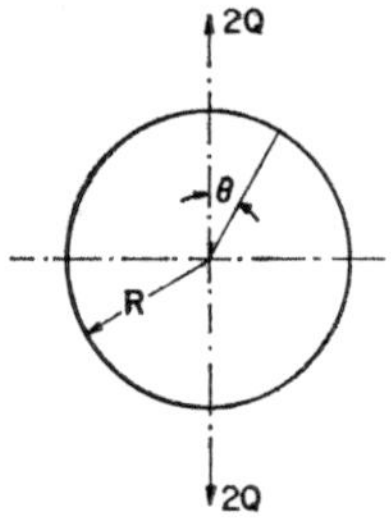

FIG. 1. Ring under doubly symmetric loading.

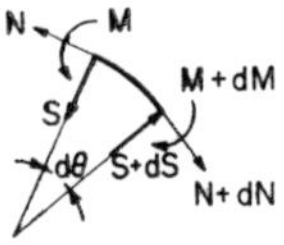

FIG. 2. Equilibrium of ring element.

As generalized stresses at the typical cross section (Fig. 2), we take the axial force $N(\theta)$ and the bending moment $M(\theta)$ but not the shear force $S(\theta)$, which is regarded as a reaction to the kinematic constraint of Bernoulli's hypothesis. If $p(\theta)$ and $q(\theta)$ denote the distributed loads per unit central angle, the equations of equilibrium are

$$\begin{aligned} N'+S &= -Rp, \\ S'-N &= -Rq, \\ M'-RS &= 0, \end{aligned} \tag{1}$$

where the prime indicates differentiation with respect to θ. At a cross section to which the concentrated circumferential and radial loads P and Q are applied, the axial force and the shear force decrease by P and Q, respectively.

To derive the correct expressions for the generalized strain rates ε and $\varkappa$ that correspond to the generalized stresses N and M, we may consider a ring that only carries distributed loads and use the virtual work identity

$$\int (pu+qv)\, d\theta = \int (N\varepsilon + M\varkappa)\, d\theta, \tag{2}$$

in which the integration is extended over the entire ring. Eliminating the reaction S between Eqs. (1), substituting the resulting expressions for p and q into Eq. (2), using integration by parts to suppress the derivatives of M and N, and comparing coefficients, we obtain

$$\varepsilon = \frac{1}{R}(u' + v), \qquad \varkappa = \frac{1}{R^2}(v' - v''). \tag{3}$$

3. PRINCIPLES OF OPTIMAL PLASTIC DESIGN

For the considered sandwich section, plastic flow cannot occur at the cross section θ unless the bending moment $M(\theta)$ and the axial force $N(\theta)$ satisfy the *yield condition*

$$|M(\theta)| + H|N(\theta)| = M_p(\theta), \tag{4}$$

where the *plastic resistance* $M_p(\theta) = 2BHT(\theta)\sigma_0$ is the bending moment for plastic flexure and σ_0 the yield stress in simple tension or compression.

In the *stress plane* with the rectangular coordinates M and HN, the yield condition (4) is represented by a square in diamond position (Fig. 3). In a

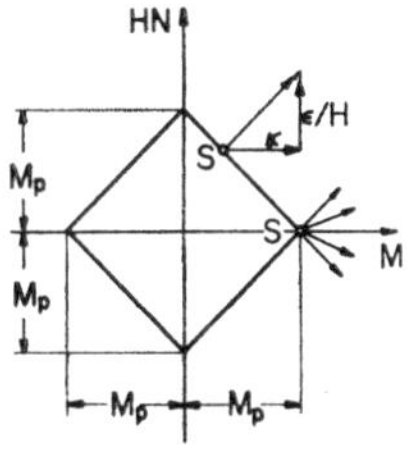

FIG. 3. Stress points S on yield locus and associated strain rate vectors.

state of stress represented by a point S of this *yield locus* that does not coincide with a vertex, the *strain rate vector* with the components $\varkappa$ and ε/H has the direction of the exterior normal to the yield locus at S. If the *stress point* S is at a vertex of the yield locus, the strain rate vector may have the direction of the exterior normal to one or the other side through S or any intermediate direction (see Fig. 3).

Denoting the prescribed minimum plastic resistance by Y, we assume that the excess c of the actual cost per unit central angle over the minimum cost is given by

$$c = \alpha(M_p - Y), \tag{5}$$

where α is a constant and $M_p \geqslant Y$ the actual plastic resistance. Applying the method of Marçal and Prager [1], we may then obtain an optimal design from stress distributions $M(\theta)$ and $N(\theta)$ that are in equilibrium with the given loads and compatible with circumferential and radial velocities satisfying

$$
\begin{aligned}
|\varepsilon| &= \frac{1}{R}|u'+v| = \alpha H, \\
|\varkappa| &= \frac{1}{R^2}|u'-v''| = \alpha,
\end{aligned}
\tag{6}
$$

where the plastic resistance M_p exceeds Y, and $\varepsilon = \varkappa = 0$, where $M_p = Y$. In this context, the term "compatible" indicates that stress point and strain rate vector correspond to each other in the manner discussed in connection with Fig. 3. The design procedure will now be discussed for the ring shown in Fig. 1.

4. STATISTICS OF OPTIMAL DESIGN

If the load $2Q$ at $\theta = 0$ is replaced by two equal loads Q at $\theta - 0$ and $\theta + 0$, no shear force will be transmitted in the symmetry cuts $\theta = 0$ and $\theta = \pi/2$. The upper right-hand quarter of the ring is therefore in equilib-

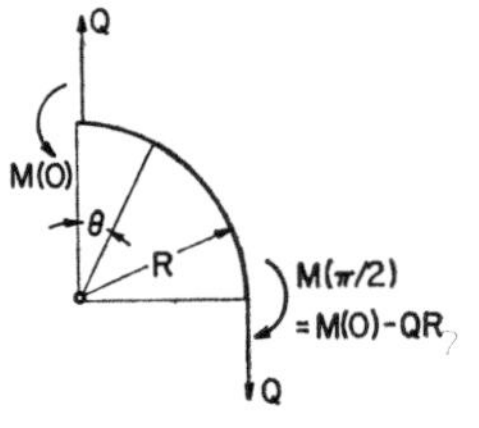

FIG. 4. Equilibrium of ring quadrant.

rium under the forces and couples shown in Fig. 4, and bending moment $M(\theta)$ and axial force $N(\theta)$ vary according to

$$
\begin{aligned}
M(\theta) &= M(0) - QR \sin\theta, \\
N(\theta) &= Q \sin\theta.
\end{aligned}
\tag{7}
$$

This stress distribution involves a single statically indeterminate quantity, the bending moment $M(0)$. Whereas the axial force $N(\theta)$ is non-negative throughout the considered quadrant, the bending moment $M(\theta)$ is likely to be positive at $\theta = 0$ and negative at $\theta = \pi/2$.

For sufficiently small positive values of Q, say for $0 \leqslant Q \leqslant Q_0$, the minimum plastic resistance Y will be adequate for the entire ring. For $Q > Q_0$, however, the minimum plastic resistance will only suffice in some interval $\theta_0 \leqslant \theta \leqslant \theta_1$, whereas greater plastic resistances will be needed in $0 \leqslant \theta < \theta_0$ and $\theta_1 < \theta \leqslant \pi/2$. In view of Eq. (4) and the preceding remarks about the signs of M and N, we have

$$\begin{aligned} M(\theta_0)+HN(\theta_0) &= Y, \\ -M(\theta_1)+HN(\theta_1) &= Y. \end{aligned} \tag{8}$$

The values of θ_0 and θ_1 must be determined from these equations and the following kinematic considerations.

5. KINEMATICS OF OPTIMAL DESIGN

From the conditions stated at the end of Section 3 and the remarks made in Section 4 concerning the signs of $M(\theta)$ and $N(\theta)$, it follows that the displacements $u(\theta)$ and $v(\theta)$ satisfy the following conditions:

$$\text{Interval 1 } (-\theta_0 < \theta < \theta_0)\text{: } \begin{aligned} u'+v &= \alpha RH, \\ u'-v'' &= \alpha R^2; \end{aligned} \tag{9}$$

$$\text{Interval 2 } (\theta_0 < \theta < \theta_1)\text{: } \begin{aligned} u'+v &= 0, \\ u'-v'' &= 0; \end{aligned} \tag{10}$$

$$\text{Interval 3 } (\theta_1 < \theta < \pi-\theta_1)\text{: } \begin{aligned} u'+v &= \alpha RH, \\ u'-v'' &= -\alpha R^2. \end{aligned} \tag{11}$$

Moreover, symmetry imposes the conditions

$$\begin{aligned} u(0) &= 0, & v'(0) &= 0, \\ u(\pi/2) &= 0, & v'(\pi/2) &= 0. \end{aligned} \tag{12}$$

Finally, u, v and v' must be continuous at $\theta = \theta_0$ and $\theta = \theta_1$.

6. OPTIMAL DESIGN

As unknowns, we have the statically indeterminate moment $M(0)$ in (7), the angles θ_0 and θ_1 in (8), and three constants of integration for each of the systems (9) through (11). To determine these twelve unknowns, we have at our disposal the two equations (8), the four symmetry conditions (12), and three conditions of continuity at both $\theta = \theta_0$ and $\theta = \theta_1$, that is a total of twelve equations.

As a rule, the designer is interested in the collapse mechanism only to the extent to which it determines the optimal design. He would therefore like to extract the necessary information from these twelve equations without completely solving them. That this is possible will now be shown.

Integrating the second equations (9), (10) and (11) once with respect to θ, we obtain

$$\begin{aligned} u-v' &= \alpha R^2\theta+c, && \text{in Interval 1,} \\ u-v' &= c_2 && \text{in Interval 2,} \\ u-v' &= -\alpha R^2\theta+c_3 && \text{in Interval 3.} \end{aligned} \tag{13}$$

Now, $u-v'$ must be continuous at $\theta = \theta_0$ and $\theta = \theta_1$ and vanish at $\theta = 0$ and $\theta = \pi/2$ [see Eq. (12)]. Thus,

$$\begin{aligned} \alpha R^2\theta_0+c_1 = c_2 = -\alpha R^2\theta_1+c_3, \\ c_1 = 0, \quad c_3 = \alpha R^2\pi/2, \end{aligned} \tag{14}$$

and hence

$$\theta_1 = \frac{\pi}{2}-\theta_0. \tag{15}$$

Substituting Eqs. (7) and (15) into Eq. (8) and eliminating $M(0)$, we obtain

$$\cos\theta_0-\sin\theta_0+\frac{H}{R}(\cos\theta_0+\sin\theta_0) = \frac{2Y}{QR}\,. \tag{16}$$

Setting

$$\theta_0 = \frac{\pi}{4}-\delta, \tag{17}$$

we write Eq. (16) as

$$\sin\delta+\frac{H}{R}\cos\delta = \frac{Y\sqrt{2}}{QR} \tag{18}$$

or, to within higher order terms in H/R,

$$\delta = \text{arc sin}\frac{Y\sqrt{2}}{QR}-\frac{H}{R}\,. \tag{19}$$

With the values of θ_0 that follow from Eqs. (18) or (19), (17) and (15), we find

$$\begin{aligned} M(0) &= Y+QR\left(1-\frac{H}{R}\right)\sin\theta_0, \\ &= -Y+QR\left(1+\frac{H}{R}\right)\sin\theta_1. \end{aligned} \tag{20}$$

From Eqs. (4) and (7), we finally obtain the optimal design

$$
\begin{aligned}
M_p(\theta) &= Y+Q\left(1-\frac{H}{R}\right)(\sin\theta_0-\sin\theta) \quad \text{in Interval 1,}\\
M_p(\theta) &= Y \quad \text{in Interval 2,}\\
M_p(\theta) &= Y+Q\left(1+\frac{H}{R}\right)(\sin\theta-\sin\theta_1) \quad \text{in Interval 3.}
\end{aligned} \tag{21}
$$

These formulas are only valid when Q exceeds the critical value Q_0. For $Q = Q_0$, the Interval 3 comprises the entire quadrant $0 \leqslant \theta \leqslant \pi/2$, and we have $\theta_0 = 0$, $\theta_1 = \pi/2$, and hence $\delta = \pi/4$. Using this value of δ in (18), solving for Q, and setting the result equal to Q_0, we find

$$
Q_0 = \frac{2Y}{R+H}. \tag{22}
$$

On the other hand, Interval 2 disappears when $\delta = 0$. Equation (18) shows that this is the case when Q has the intensity

$$
Q_1 = \frac{Y\sqrt{2}}{H}. \tag{23}
$$

For $Q > Q_1$, the second equation (13) must be dropped and θ_1 must be set equal to θ_0. Continuity and symmetry arguments then furnish the relation

$$
\theta_0 = \theta_1 = \pi/4. \tag{24}
$$

The corresponding optimal design is specified by

$$
\begin{aligned}
M_p(\theta) &= QR\left[\sin\frac{\pi}{4}-\left(1-\frac{H}{R}\right)\sin\theta\right] \quad \text{for} \quad 0 \leqslant \theta < \frac{\pi}{4},\\
M_p(\theta) &= QR\left[\left(1+\frac{H}{R}\right)\sin\theta-\sin\frac{\pi}{4}\right] \quad \text{for} \quad \pi/4 < \theta \leqslant \pi/2.
\end{aligned} \tag{25}
$$

REFERENCE

1. P. V. MARÇAL and W. PRAGER, "A method of optimal plastic design", *Journal de Mécanique* 3, 509–30 (1964).

ON STABILITY OF VISCOPLASTIC SYSTEMS WITH THERMO-MECHANICAL COUPLING†

DAVID RUBIN and D. C. DRUCKER

Brown University, Providence, R.I., U.S.A.

ABSTRACT

The stability of equilibrium configurations of thermo-mechanically coupled systems in the plastic and the viscoplastic range is examined through simple examples and by dimensional analysis. A uniform fully insulated thin-walled tube twisted into the inelastic range is studied first. The material is idealized as linear work-hardening viscoplastic with a current yield stress which decreases linearly with increase in temperature. Any annealing or age-hardening or similar influence of time at temperature is supposed negligible. A local stability criterion $\beta > 1$ governs, where β is the ratio of the heat stored in the material to the heat generated by inelastic deformation with a given temperature rise. Neither viscosity nor heat conduction enters into the stability or instability of this homogeneous system. They both play a role in successive modifications of the simple model which demonstrate the stabilizing effect of heat transfer to the surroundings and of the heat capacity in regions of material below yield.

INTRODUCTION

Stability of a material system is determined by its overall characteristics, not by the properties of the material alone or of the sub-elements which make up the system. A system composed entirely of stable material can be unstable; a system containing unstable elements may be stable in its response.

A work-hardening bar under end compression provides a familiar example of a purely mechanical system which becomes unstable despite the stability of the material. Above some critical maximum load, the increase in bending moment with increase in deflection exceeds the capacity of the material to resist the added deformation. On the other hand,

† The results presented in this paper were obtained in the course of research sponsored by the Office of Naval Research under Contract Nonr-562(20) with Brown University.

a statically indeterminate truss may be stable despite the incipient instability of one or more of its compression members.

The test for stability or instability of a system in static equilibrium is to perturb the system and determine its subsequent behavior. An equilibrium state is said to be stable in the small if the response to each permissible perturbation is small. More precisely, in a mechanical system, the ratio of the displacement to the perturbing force is to be finite in the limit as the magnitude of the force goes to zero. Stability in this sense guarantees that the system will not depart from its equilibrium state spontaneously. However, as Shanley [1] pointed out in his treatment of the buckling of columns, it does not necessarily guarantee stability of a path of quasi-equilibrium states traversed by the system.

Instability of material is equivalent in its effect to the instability caused by change of geometry. With a few exceptions, such as upper yield point instabilities, significant material instability seems to occur only as a consequence of the decrease in yield or flow strength with an increase in temperature. Therefore, any system instability due to material instability is likely to be associated with the real coupling between thermal and mechanical behavior. The heat generated by inelastic deformation raises the temperature of the material, reduces the yield strength, and so tends to produce additional deformation which in turn generates heat. This instability can be offset locally by work-hardening which accompanies the deformation and can be balanced globally through radiation or convection of heat from the deforming body and through conduction by the transfer of heat to free boundaries and by storage of heat in regions of low stress.

The balance between the stabilizing and the destabilizing effects of the properties of the material and the system as a whole [2] is examined in the subsequent sections. Parameters on which stability depends are exhibited by examples and dimensional analysis. A fully insulated model under homogeneous stress provides a simple starting point. Then the effect of heat loss to the outside world is added, and consideration is given to inhomogeneous states of stress and the heat capacity of regions below yield. The perturbing "forces" to test stability of a thermo-mechanically coupled system include both real forces and changes in temperature.

CONSTITUTIVE RELATIONS

The complete coupling of thermal and mechanical effects involves both the influence of temperature on the stress–strain relations and the influence of stress and strain on the thermal properties. Both the viscous response and the plastic response at a given stress level tend to increase appreciably as the temperature goes up; but the very interesting reciprocal effect of the level of stress on the heat capacity or the thermal conductivity is not nearly as significant here. This is because the energy changes associated with the thermal terms, for a given increase in temperature above an existing equilibrium state, far exceed the energy associated with the mechanical terms [3]. The heat stored for a mere one degree rise in temperature is of the order of 300 in. lb/in^3 for steel or aluminum and so equals the total mechanical energy dissipation in a strain increment of the order of 1%.

It is permissible, therefore, in the study of the thermo-mechanical stability of a loaded metallic structure in equilibrium at a sensibly uniform or moderate range of temperature, to consider the heat capacity c and the thermal conductivity k to be constants independent of stress and of temperature. Furthermore, if stability in the small is to be examined, a linearized viscoplastic form of the appropriate general stress–strain–time–temperature relation should be satisfactory. Any annealing or age-hardening or similar influence of time at temperature is supposed negligible.

For a simple viscoplastic work-hardening material under a one-dimensional state of stress as in pure torsion of a thin-walled tube, the inelastic strain rate $\dot{\gamma}$ at a stress τ may be taken as proportional to the increase in stress over τ_0^*, the current value of the yield stress:

$$\dot{\gamma} = (\tau - \tau_0^*)/\eta, \tag{1}$$

where η is a coefficient of viscosity. Within the range of small *change* in plastic strain γ and small *change* in temperature θ from a previous equilibrium yield state τ_0, γ_0, θ_0, τ_0^* may be written as

$$\tau_0^* = \tau_0 + G_t\gamma - b\theta, \tag{2}$$

a linear combination of the hardening due to change in strain and a softening due to change in temperature. Both G_t, the plastic tangent modulus in the given range of stress, and b are constants evaluated at τ_0, γ_0, θ_0.

Substitution of (2) into (1) gives the one-dimensional form of the stress–strain relation to be employed for the inelastic strain rate $\dot{\gamma}$:

$$\begin{aligned} \eta\dot{\gamma} &= \tau - \tau_0 - G_t\gamma + b\theta \quad \text{for} \quad \tau \geqslant \tau_0^*, \\ \dot{\gamma} &= 0 \quad\quad\quad\quad\quad\quad\quad\;\; \text{for} \quad \tau \leqslant \tau_0^*. \end{aligned} \tag{3}$$

Small departures from equilibrium are considered so that when plastic deformation occurs τ does not differ much from τ_0. Consequently, the rate of energy dissipation $\tau\dot{\gamma}$ is given by $\tau_0\dot{\gamma}$ to first order. Little of the work done is stored [4]. The one-dimensional heat conduction equation [5] then may be written in the form

$$c\frac{\partial\theta}{\partial t} = k\frac{\partial^2\theta}{\partial x^2} + \tau_0\dot{\gamma}, \tag{4}$$

where c is the heat capacity per unit volume. The conversion factor for units of work to units of heat is suppressed; each term is supposed expressed in units of work.

General relations for three dimensions and multiaxial states of stress can take many forms. Perhaps the simplest are the isotropic J_2 type, homogeneous in stress of degree one, which reduce to the formulation by Prager [6] when the material is inviscid.

$$2\eta\dot{\varepsilon}_{ij} = (J_2^{\frac{1}{2}} - f^*)\, s_{ij}/J_2^{\frac{1}{2}} \tag{5}$$

for $J_2^{\frac{1}{2}} \equiv \sqrt{(s_{mn}s_{mn}/2)} \geq f^* \geq f_0$, where $f_0 = \tau_0$ is the value of $J_2^{\frac{1}{2}}$ in the previous equilibrium yield state σ_{ij}^0, ε_{ij}^0, θ_0 and f^* is the current yield value

$$f^* = f_0 + G_t \int \sqrt{(2\dot{\varepsilon}_{mn}\dot{\varepsilon}_{mn})}\, dt - b\theta. \tag{6}$$

The equation of heat conduction in three dimensions for an isotropic material is

$$c\dot{\theta} = k\nabla^2\theta + \dot{W} \tag{7}$$

in which ∇^2 is the Laplacian operator and $\dot{W}$ is the rate of energy dissipation.

$$\begin{aligned} \dot{W} = \sigma_{ij}\dot{\varepsilon}_{ij} = s_{ij}\dot{\varepsilon}_{ij} &= J_2^{\frac{1}{2}}(J_2^{\frac{1}{2}} - f^*)/\eta \\ = J_2^{\frac{1}{2}}\sqrt{(2\dot{\varepsilon}_{ij}\dot{\varepsilon}_{ij})} &= f_0\sqrt{(2\dot{\varepsilon}_{ij}\dot{\varepsilon}_{ij})} \quad \text{(to first order).} \end{aligned} \tag{8}$$

The parameters c, k, η, G_t, $f_0 = \tau_0$, and b describe the material behavior within this first-order isotropic theory for three dimensions as well as one.

CRITERION OF STABILITY FOR FULLY INSULATED THIN-WALLED TUBE IN TORSION

A homogeneous isotropic thin-walled circular tube in pure torsion provides a suitable starting point for a discussion of stability of thermo-mechanically coupled systems. If the wall thickness is constant, the state of stress is closely one of simple shear τ everywhere. If in addition the walls and the ends are fully insulated, the temperature is uniform throughout in an equilibrium configuration.

Suppose the tube to be twisted very slowly into the plastic range and to be in equilibrium at τ_0, γ_0, θ_0 at a given twisting moment. The stability of this equilibrium configuration is tested most severely by a uniform perturbation. The temperature can be imagined as increased abruptly everywhere by θ_u, or equivalently the twisting moment can be increased abruptly to produce a shear stress τ slightly greater than τ_0. In either case, $\tau-\tau_0^*$ of Eqs. (1) and (2) is positive; a shear strain rate $\dot{\gamma}$ is induced; heat is generated as the shear strain is increased by γ; and the temperature rises by θ as the material work-hardens.

The process clearly is unstable if the work-hardening is absent or is insufficient to balance the loss of yield strength with increase in temperature. Positive stability requires that $G_t\gamma$ more than balance $b\theta$ in Eq. (2) where θ is the temperature rise caused by the energy dissipation $\int \tau\dot{\gamma}dt$ or $\tau_0\gamma$. The stability criterion is given by

$$G_t\gamma > b\theta = b(\tau_0\gamma/c)$$

or

$$\frac{G_t c}{b\tau_0} \equiv \beta > 1, \tag{9}$$

where $\beta = 1$ represents neutral or indifferent stability.

The criterion $\beta > 1$ is the same result reported earlier [3] for stability of material. β is a local parameter, independent of heat conduction and of the coefficient of viscosity. $\beta > 1$ is necessary for the stability of this particular system because of the assumed homogeneity of material, stress, and temperature. It is sufficient for the stability of a loaded body with any inhomogeneous state of stress, provided geometric instability is ruled out. The discussion so far, however, does not shed any other light on the system aspect of thermo-mechanical coupling, on the role of k in heat conduction to regions of low stress and of heat transfer to the outside which involves time rate of strain and η as a consequence.

Although $\beta > 1$ is the necessary and sufficient condition for stability in this simplest of all problems, several aspects of system stability do appear when variations of θ with position x along the axis of the tube are assumed. This is because variations of θ with x are equivalent to variations of γ with x and to a variation of the stress difference $\tau-\tau_0^*$. The combination of Eqs. (3) and (4) for constant $\tau \geqslant \tau_0^*$ gives the same linear partial differential equation for θ as for γ

$$\begin{aligned} (D)\theta &= 0 \\ (D)\gamma &= 0 \end{aligned} \tag{10}$$

where

$$(D) \equiv c\eta\frac{\partial^2}{\partial t^2}+(cG_t-b\tau_0)\frac{\partial}{\partial t}-kG_t\frac{\partial^2}{\partial x^2}-k\eta\frac{\partial^3}{\partial x^2\partial t}.$$

The transformation of the differential operator (10) from x and t to non-dimensional form [7]

$$(D_d) \equiv \beta \frac{\partial^2}{\partial T^2} + (\beta - 1)\frac{\partial}{\partial T} - \frac{\partial^2}{\partial X^2} - \frac{\partial^3}{\partial X^2 \partial T} \tag{11}$$

with independent dimensionless variables X and T

$$X = \frac{x}{\sqrt{(k\eta/b\tau_0)}} \qquad T = \frac{t}{\eta/G_t} \tag{12}$$

shows clearly that stability depends only upon β. Conductivity affects the scale of distance, viscosity alters both the time and distance scales.

Nevertheless, particular solutions of $(D_d)\theta = 0$ which are Fourier components of the general solution are informative:

$$\theta_n = A_n e^{\mu_n T} \cos nX, \tag{13}$$

where

$$2\beta\mu_n = -(\beta - 1 + n^2) \pm [(\beta - 1 + n^2)^2 - 4n^2\beta]^{1/2}. \tag{14}$$

In Eq. (14), $n = 0$ corresponds to the uniform perturbation and confirms, as it must, that $\beta > 1$ is needed for stability. Were $\beta < 1$, the real part of μ_n would be positive and θ would be unbounded with time. An individual θ_n term, for $n > 0$, represents more than just the response to an initial temperature distribution $A_n \cos nX$ for which the stability requirement $\beta > 1 - n^2$ is met more easily than when $n = 0$ and is satisfied automatically for $n > 1$. Although perturbation in all modes, including $n = 0$, must be taken into account, and $\beta > 1$ remains the requirement for stability, the analogy between variation of θ and variation of $\tau - \tau_0^*$ or γ shows the strong stabilizing effect of neighboring regions of stress below current yield when a uniform temperature rise is imposed. Neighboring distances are distances of the order of $\sqrt{(k\eta/b\tau_0)}$; $n = 1$ in Eq. (13) corresponds to a cycle over a distance $2\pi\sqrt{(k\eta/b\tau_0)}$.

EFFECT OF HEAT TRANSFER TO THE SURROUNDINGS (CONVECTION)

If the insulation on the lateral surface of the tube is removed, and the tube is immersed in a reservoir at temperature θ_0, heat will be transferred out as plastic deformation proceeds. One-dimensionality of the problem is retained with the usual assumption for thin-walled tubes that the temperature remains uniform through the thickness of the tube wall. Linearity of the differential equation is retained with the usual assumption that the heat loss to the reservoir per unit of surface area is proportional to the temperature rise θ in the tube. This heat loss per unit area $h\theta$ is converted to a heat loss per unit volume $2h\theta/w$ by multiplying by the area per unit of length $2(2\pi r)$ and dividing by volume per unit of length $2\pi rw$, where r is the radius of the tube and w is the wall thickness.

The net effect is to replace $\tau_0\dot{\gamma}$ in Eq. (4) by $\tau_0\dot{\gamma}-2h\theta/w$. A modified form of the differential operator (11) is found and the differential equation for temperature takes the form

$$\beta\frac{\partial^2\theta}{\partial T^2}+(\beta-1+H)\frac{\partial\theta}{\partial T}-\frac{\partial^2\theta}{\partial X^2}-\frac{\partial^3\theta}{\partial X^2\partial T}+H\theta = 0, \tag{15}$$

where $H = 2\eta h/wb\tau_0$ is a non-dimensional parameter.

The stability criterion now is

$$\beta > 1-H. \tag{16}$$

The range $1-H < \beta < 1$, corresponding to an unstable material and a stable system, demonstrates the stabilizing effect of the loss of heat to the surroundings. If heat can be convected away fast enough, $H > 1$, the system is stable for all values of β. Although the tube is under homogeneous conditions of stress and temperature in the most critical test of stability, time enters in an essential manner and the viscosity coefficient η has a strong influence on the stability of the system.

SEVERAL DIMENSIONLESS VARIABLES OF THE GENERAL PROBLEM

The stability of a general thermo-mechanical system is governed by its geometry and the applied loads (including temperatures) as well as by the properties of its constituent materials. Thermal parameters of material behavior c, k, and h appear along with mechanical parameters η, G_t, b, τ_0 or f_0 for the simplest forms of isotropic behavior. Elastic coefficients and other thermal and mechanical properties also may have influence. All should be combined with geometric quantities such as volume V, area of surface A, and a number of characteristic lengths L. The present value of the loads and their past history are of importance; the equilibrium state of temperature, stress, strain and displacement whose stability in the small is studied is independent of history only for fully elastic response.

A large number of dimensionless parameters or variables can be written down in a formal way but their individual usefulness in a class of problems stated so broadly is uncertain. Dimensionless variables of position such as

$x/\sqrt{(k\eta/b\tau_0)}$ are useful in solving particular problems but they are not stability criteria. The same comment applies to the dimensionless time $t/(\eta/G_t)$. On the other hand, a dimensionless local parameter of material properties, such as $\beta = G_t c/b\tau_0$, ignores any system aspect. It provides an overriding sufficient condition of stability of equilibrium in the small in the absence of geometric instability. When the state of stress is inhomogeneous, the entire volume V of the body can store the heat which is generated in that portion V^P which deforms plastically. Perhaps, then, $\beta > V^P/V$ is enough of a safeguard for stability of equilibrium; but V^P is not determined easily.

A dimensionless system parameter $H = \eta hA/Vb\tau_0$, independent of the solution whose stability is studied, seems a reasonable measure of one stabilizing effect of convection to the outside world; V/A is a characteristic dimension which specializes to $w/2$ in Eqs. (15) and (16). Here too, replacement of V by V^P would seem appropriate when feasible for a particular solution.

Conductivity should enter in two ways. One is through the transfer of heat to a free surface for transfer out of the body; the other is transfer to storage in a region of low stress. Lengths of travel L should be scaled, just as position x is scaled in Eq. (12), to give dimensionless parameters of the solution $K = [\sqrt{(k\eta/b\tau_0)}]/L$ for the transfer of heat from regions of generation to low stress regions of storage. Storage is represented also by the ratio of V to V^P. The ratio of conduction to a free surface to heat transfer from the surface is measured by k/hL or kA/hV or kV/hAL^2. The last of these forms has the appearance of K^2/H and serves as a reminder that independent dimensionless parameters should be sought. However, each characteristic length L gives an additional dimensionless parameter. The L associated with K is a characteristic length of the solution to a particular boundary value problem and so may be very different from the geometric parameter of the body V/A or the L in k/hL.

CONCLUDING REMARKS

A thin-walled circular cylinder at yield under any combination of tension, torsion, and interior pressure has stability characteristics identical with a tube in torsion alone. The question which then arises is whether the local parameter $\beta \equiv G_t c/b\tau_0$ and the system parameter $H \equiv \eta hA/Vb\tau_0$ need be considered in the design of such structures as pressure vessels. A calculation of β for the common structural metals indicates strong thermo-mechanical stability. Values of the order of 100 are found, far above the sufficient value of unity. However, it is possible that some of the recently developed high-strength steels with rather low tangent moduli may become thermally unstable even at moderate strain rates and temperatures. It would be of interest to determine their heat capacity in conjunction with tension or torsion tests at various temperatures and strain rates.

Inhomogeneity of stress and loss of heat to the surroundings increases the probability of thermo-mechanical stability. Once again, in the absence of geometric instability, the stability of an equilibrium configuration against small disturbances is assured if $\beta > 1$ at each point. Should this sufficient local condition not be met, consideration must be given to the dimensionless parameters of the system and of the solution asdiscussed in the paper.

REFERENCES

1. F. R. Shanley, "Inelastic column theory", *Journal of the Aeronautical Sciences* **14,** 261 (1947).
2. I. J. Gruntfest, "A note on thermal feedback and the fracture of solids", in *Fracture of Solids*, Edited by D. C. Drucker and J. J. Gilman, Metallurgical Society Conferences, vol. 20, Interscience–Wiley, pp. 189–93 (1963).
3. D. C. Drucker, "Extension of the stability postulate with emphasis on temperature changes", in *Plasticity*, edited by E. H. Lee and P. S. Symonds, Pergamon Press, pp. 170–84 (1960).
4. G. I. Taylor and W. S. Farren, "The heat developed during plastic extension of metals", *Proc. Roy. Soc.* A **107**, 422–51 (1925).
5. H. S. Carslaw and J. C. Jaeger, *Conduction of Heat in Solids*, Clarendon Press, Oxford (1947).
6. W. Prager, "Non-isothermal plastic deformation", *Akad. van Wetenschappen,* Series B, **61** (3), 176–82 (1958).
7. D. Rubin, "Stability of a thermo-mechanically coupled system", Sc.M. Thesis, Division of Engineering, Brown University, 1965.

NON-LINEAR THEORY FOR AXISYMMETRIC DEFORMATIONS OF HETEROGENEOUS SHELLS OF REVOLUTION

YEHUDA STAVSKY

Department of Mechanics, Technion—Israel Institute of Technology, Haifa

ABSTRACT

A general nonlinear theory, based on the Euler–Bernoulli hypothesis, is established for heterogeneous orthotropic shells of revolution that are subject to rotationally symmetric mechanical and thermal loads. Two simultaneous second-order ordinary differential equations are developed, in terms of a deformation variable and a stress-resultant variable, which exhibit a stronger coupling of these two basic variables, than occurs in the corresponding equations for homogeneous shells of revolution. The general theory is specialized for small finite deformations, and specific equations for large finite deflections of circular heterogeneous plates are shown. The possibility of using a modified system of shell equations, resulting from a simplified compatibility relation, is indicated.

NOTATION

A_{ij}	elastic area
A_{ij}^*	modified extensional rigidity
B_{ij}	elastic statical moment
B_{ij}^*, C_{ij}^*	extensional–flexural coupled rigidities
D_{ij}	elastic moment of inertia
D_{ij}^*	modified flexural rigidity
E_{ij}	elastic stiffness modulus
h	shell thickness
h_1, h_2	distances to bounding surfaces of shell
H	radial stress resultant
L_{ij}	functional operator
m	meridional body moment
M_ξ, M_θ	meridional and circumferential bending moments, respectively
N_ξ, N_θ	meridional and circumferential stress resultants, respectively
N_{iT}, M_{iT}	thermal quantities defined by Eq. (4.13)

p_H, p_V load intensity components in radial and axial directions, respectively
Q transverse shear resultant
r radial distance to undeformed shell reference surface
R_ξ, R_θ principal radii of curvature of undeformed shell reference surface
s_m arc length
T temperature change from initial stress-free state
V axial stress resultant
u, w displacement components in radial and axial directions, respectively
x, y Cartesian coordinates in horizontal plane
z axial distance to undeformed shell reference surface
α length parameter
A_ξ, A_θ coefficients of thermal expansion in stress–strain relations
β change of slope angle of meridian curve, at a point, due to deformation
$\varepsilon_\xi, \varepsilon_\theta$ strain tensor components in meridional and circumferential directions, respectively
ε_{i0} strain tensor component at the reference surface
ζ axis normal to reference surface
θ polar angle
$\varkappa_\xi, \varkappa_\theta$ curvature changes
ξ meridional coordinate
τ_ξ, τ_θ stress tensor components in meridional and circumferential directions, respectively
ϕ, Φ slope angle of undeformed and deformed meridian curve, respectively
Ψ a stress resultant function defined by Eq. (7.1)

1. INTRODUCTION

This study is concerned with the theory of rotationally orthotropic shells of revolution, heterogeneous in the thickness and meridional directions and subject to symmetrical deformations. Such heterogeneity may be achieved, for example, by composing a multi-layer shell [1], [2], which permits adaptation of material properties to the design rather than of the design to the material. Hence, the analysis of layered shells gained much interest in recent years, as indicated, for example, by Ambartsumyan's survey [3]. Most of the research in heterogeneous shells was towards the development of linear theories [4], [5]. A theory of finite symmetric bending of homogeneous orthotropic shallow shells of revolution, remaining shallow after deformation, was given by Reissner [6]. As the reference surface was not taken to coincide with the middle surface of the shell, the obtained equations hold, almost directly, for shallow shells of revolution that are non-homogeneous in the thickness and meridional directions.

The present paper is concerned with the general non-linear thermoelastic theory for axisymmetric deformations of non-shallow heterogeneous orthotropic shells of revolution, extending previous work by Reissner [7], [8] for corresponding homogeneous shells. The theory is formulated in terms of two simultaneous non-linear ordinary differential equations for a basic deformation variable Φ and a basic stress resultant variable Ψ. The equations for small finite deflections are obtained from the general theory by specialization, and reduction to the von Kármán theory is indicated. The general system of non-linear equations for large finite deflections of circular layered plates is shown by a continuous transition from the shell equations. The analysis is restricted to thin elastic shells, and it is assumed that no coupling occurs between elastic and heat-conduction phenomena. The theory developed is based on the classical Euler–Bernoulli deformation hypothesis.

2. GEOMETRY OF SHELL

Consider a multi-layer elastic shell bounded by two coaxial surfaces of revolution, and loaded in such a manner that it remains a shell of revolution. The geometry of the shell is described by a "reference surface of revolution" and by a coordinate ζ measuring the distance of any material point from this reference surface ($\zeta = 0$). Note that the reference surface does not necessarily coincide with the middle surface of the shell. A similar situation exists in heterogeneous plates and shells [9], [10] and plates with variable thickness [11], where the reference plane was conveniently located, at the bottom face of the plate or shell, not at its middle surface.

Using cylindrical coordinates r, θ, z, the parametric equations of the undeformed reference surface of the shell (see Fig.1) are given by

$$r = r(\xi), \quad z = z(\xi). \tag{2.1}$$

The parameter ξ and the polar angle θ in the x–y-plane are thus the coordinates on the reference surface. The bounding surfaces of the shell are

$$\zeta = -h_1(\xi), \quad \zeta = +h_2(\xi) \tag{2.2}$$

and h, the total thickness of the shell, is the sum of h_1 and h_2. Points of the undeformed shell are designated by ξ, θ and ζ, which constitute a system of orthogonal curvilinear coordinates in space. The range of applicability of the following shell theory is given by the order-of-magnitude relation.

$$|h/R| \ll 1, \tag{2.3}$$

where R is the smaller of the two radii of curvature of the reference surface.

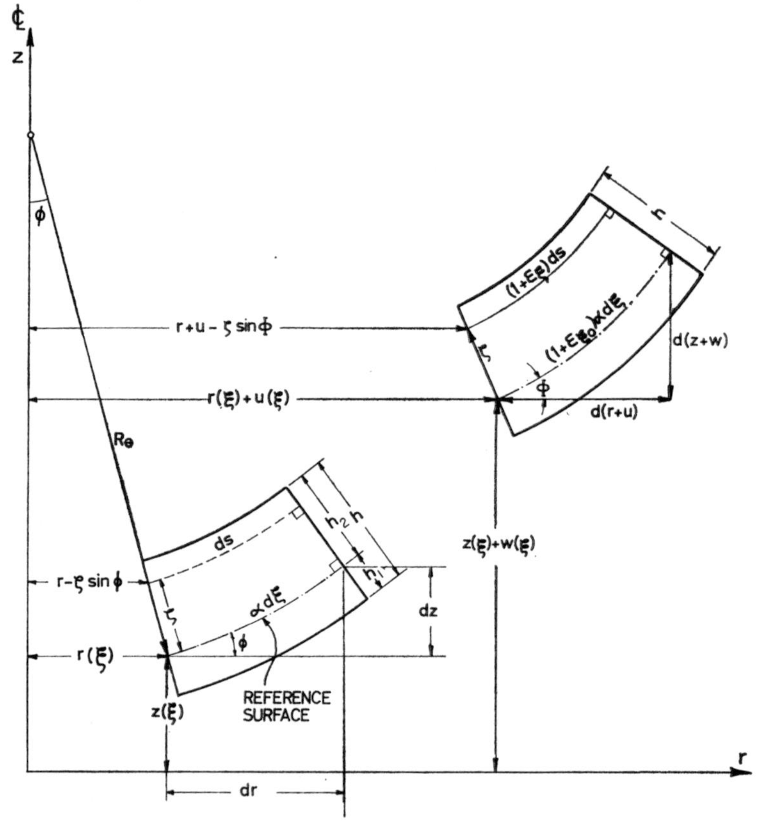

FIG. 1. Shell element before and after deformation.

The meridional differential arc length ds_m on the undeformed reference surface is given, in terms of r and z, by

$$(ds_m)^2 = (dr)^2+(dz)^2 = (\alpha d\xi)^2. \tag{2.4}$$

Then

$$\alpha^2 = (r')^2+(z')^2, \tag{2.5}$$

where primes indicate differentiation with respect to the parameter ξ. Other useful geometrical relations are the following:

$$r' = \alpha \cos \phi, \quad z' = \alpha \sin \phi, \tag{2.6}$$

$$\tan \phi = \frac{z'}{r'}, \quad \tan \Phi = \frac{(z+w)'}{(r+u)'} \tag{2.7}$$

where ϕ and Φ denote the slope angle of the tangent to a meridian curve before and after deformation of shell, whereas u and w are the radial and axial displacement components respectively.

The radius of curvature of the curve generating the undeformed reference surface is given by

$$R_\xi = \alpha/\phi' \tag{2.8}$$

and the other principal radius of curvature is given by the length of the normal between the generating curve and the axis of rotation,

$$R_\theta = r/\sin \phi. \tag{2.9}$$

3. STRAIN–DISPLACEMENT RELATIONS

Let the shell deform in a rotationally symmetric pattern, so that its points move radially and axially but *not* in the circumferential direction. Assuming that the Euler–Bernoulli hypothesis holds for the heterogeneous shells under consideration, we follow Reissner [12] in writing the following expressions for the meridional strain ε_ξ and the circumferential strain ε_θ:

$$\varepsilon_\xi = \frac{\varepsilon_{\xi 0}+\zeta\varkappa_\xi}{1-\zeta/R_\xi}, \quad \varepsilon_\theta = \frac{\varepsilon_{\theta 0}+\zeta\varkappa_\theta}{1-\zeta/R_\theta}, \tag{3.1}$$

where $\varepsilon_{\xi 0}$ and $\varepsilon_{\theta 0}$ are the reference surface strains given by

$$\varepsilon_{\xi 0} = \left(1+\frac{u'}{r'}\right)\frac{\cos \phi}{\cos \Phi} - 1 \tag{3.2}$$

or alternatively,

$$\varepsilon_{\xi 0} = \cos(\Phi-\phi) - 1 + \frac{u'}{\alpha}\cos \Phi + \frac{w'}{\alpha}\sin \Phi \tag{3.2'}$$

and

$$\varepsilon_{\theta 0} = \frac{u}{r}. \tag{3.3}$$

The reference-surface curvature changes $\varkappa_\xi$, $\varkappa_\theta$ are of the form

$$\varkappa_\xi = -\frac{1}{\alpha}(\Phi'-\phi'), \quad \varkappa_\theta = -\frac{1}{r}(\sin \Phi - \sin \phi), \tag{3.4}$$

whereas the radial and axial displacement components are given, respectively, by the following expressions:

$$u = r\varepsilon_{\theta 0}, \quad w = \int \alpha[(1+\varepsilon_{\xi 0}) \sin \Phi - \sin \phi] \, d\xi. \tag{3.5}$$

4. STRESS–STRAIN RELATIONS

The shell material is taken to be thermoelastic, rotationally orthotropic and heterogeneous. Hooke–Neumann's law takes the form

$$\tau_\xi = E_{\xi\xi}\varepsilon_\xi + E_{\xi\theta}\varepsilon_\theta + A_\xi T, \tag{4.1}$$

$$\tau_\theta = E_{\theta\xi}\varepsilon_\xi + E_{\theta\theta}\varepsilon_\theta + A_\theta T, \tag{4.2}$$

$$\tau_{\xi\theta} = 0. \tag{4.3}$$

In view of the shell heterogeneity, we note that

$$E_{ij} = E_{ij}(\xi, \zeta), \quad A_i = A_i(\xi, \zeta), \quad (i, j = \xi, \theta), \tag{4.4}$$

where the E_{ij} are the components of the symmetric elastic stiffness tensor, the A_i are the appropriate thermal coefficients in the meridional and circumferential directions, and T is a given function of ξ and ζ.

In order to establish a two-dimensional shell theory, stress resultants and couples are defined as follows:

$$N_\xi(1+\varepsilon_{\theta 0})\, r d\theta = \int_{-h_1}^{h_2} \tau_\xi(1+\varepsilon_\theta)\, r(1-\zeta/R_\theta)\, d\theta \, d\zeta. \tag{4.5}$$

Neglecting $\varepsilon_{\theta 0}$, ε_θ and ζ/R_θ compared with unity, one finds

$$N_\xi = \int_{-h_1}^{h_2} \tau_\xi d\zeta = N_\xi^* + N_{\xi T}, \quad N_{\xi T} = \int_{-h_1}^{h_2} A_\xi T d\zeta. \tag{4.6}$$

Similarly

$$N_\theta = \int_{-h_1}^{h_2} \tau_\theta d\zeta, \quad (M_\xi, M_\theta) = \int_{-h_1}^{h_2} (\tau_\xi, \tau_\theta)\, \zeta d\zeta. \tag{4.7}$$

The only transverse shear stress resultant that occurs in the symmetric shell problem is defined by

$$Q = \int_{-h_1}^{h_2} \tau_{\xi\zeta} d\zeta. \tag{4.8}$$

It is convenient to introduce axial and radial components of N_ξ and Q, denoted by V and H respectively. The following relations hold between these resultants, in the deformed state of the shell,

$$N_\xi = H \cos \Phi + V \sin \Phi, \tag{4.9}$$

$$Q = -H \sin \Phi + V \cos \Phi. \tag{4.10}$$

Substitution of Eqs. (4.1), (4.2) in Eqs. (4.6)–(4.7) yields, by Eqs. (3.1)–(3.4), the following relations:

$$\begin{bmatrix} N_\xi^* \\ N_\theta^* \\ \\ M_\xi^* \\ M_\xi^* \end{bmatrix} = \begin{bmatrix} A_{\xi\xi} & A_{\xi\theta} & B_{\xi\xi} & B_{\xi\theta} \\ A_{\theta\xi} & A_{\theta\theta} & B_{\theta\xi} & B_{\theta\theta} \\ \cdot & \cdot & \cdot & \cdot \\ B_{\xi\xi} & B_{\xi\theta} & D_{\xi\xi} & D_{\xi\theta} \\ B_{\theta\xi} & B_{\theta\theta} & D_{\theta\xi} & D_{\theta\theta} \end{bmatrix} \begin{bmatrix} \varepsilon_{\xi 0} \\ \varepsilon_{\theta 0} \\ \\ \varkappa_\xi \\ \varkappa_\theta \end{bmatrix} \tag{4.11}$$

which can be written more compactly in the form

$$\begin{bmatrix} N^* \\ \\ M^* \end{bmatrix} = \begin{bmatrix} A & B \\ \cdots & \\ B & D \end{bmatrix} \begin{bmatrix} \varepsilon_0 \\ \\ \varkappa \end{bmatrix}, \tag{4.12}$$

where

$$N_i^* = N_i - N_{iT}, \quad M_i^* = M_i - M_{iT}, \quad (i = \xi, \theta) \tag{4.13}$$

and

$$(N_{iT}, M_{iT}) = \int_{-h_1}^{h_2} (1, \zeta) A_i \, T d\zeta \quad (\mathrm{i}=\zeta, \theta). \tag{4.14}$$

The elastic areas, elastic statical moments and elastic moments of inertia are given, respectively, by

$$(A_{ij}, B_{ij}, D_{ij}) = \int_{-h_1}^{h_2} (1, \zeta, \zeta^2) E_{ij} \, d\zeta \quad (i, j = \xi, \theta). \tag{4.15}$$

5. STATICS OF SHELL ELEMENT

Force- and moment-equilibrium conditions of the deformed shell element (Fig. 2), which hold for arbitrarily large deformations, take the form

$$[r(1+\varepsilon_{\theta 0})H]' + r\alpha p_H(1+\varepsilon_{\xi 0})(1+\varepsilon_{\theta 0}) - \alpha N_\theta(1+\varepsilon_{\xi 0}) = 0 \tag{5.1}$$

$$[r(1+\varepsilon_{\theta 0})V]' + r\alpha p_V(1+\varepsilon_{\xi 0})(1+\varepsilon_{\theta 0}) = 0 \tag{5.2}$$

$$[r(1+\varepsilon_{\theta 0})M_\xi]' - \alpha M_\theta(1+\varepsilon_{\xi 0}) \cos \Phi - r\alpha Q(1+\varepsilon_{\theta 0}) + r\alpha m(1+\varepsilon_{\xi 0})(1+\varepsilon_{\theta 0}) = 0, \tag{5.3}$$

where p_H, p_V components of surface load intensity, and m is the surface or body couple per unit area.

Assuming that the strains $\varepsilon_{\xi 0}$, $\varepsilon_{\theta 0}$ are small compared whit unity, the following equilibrium equations hold:

$$(rH)' + r\alpha p_H - \alpha N_\theta = 0, \tag{5.4}$$

$$(rV)' + r\alpha p_V = 0, \tag{5.5}$$

$$(rM_\xi)' - \alpha M_\theta \cos \Phi - r\alpha Q + r\alpha m = 0. \tag{5.6}$$

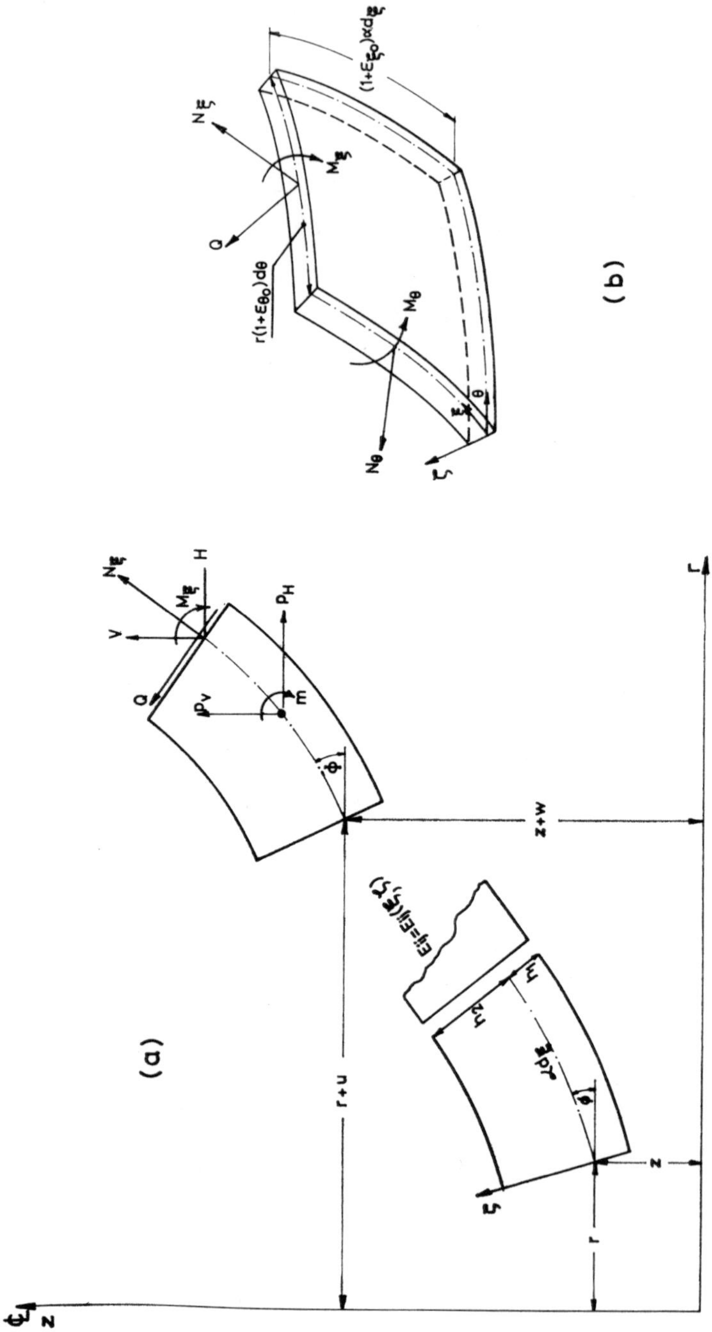

FIG. 2. Shell element with applied loads, stress resultants and couples.

Equations (5.4) and (5.6), together with Eqs. (4.11) and in view of Eqs. (4.9), (4.10), and (3.2)–(3.4), constitute a system of seven equations for three stress resultants N_ξ, N_θ, Q (or H, V, N_θ), two stress couples M_ξ, M_θ and two deformation variables u, Φ.

6. COMPATIBILITY EQUATION FOR STRAINS

Noting that

$$u' = (r\varepsilon_{\theta 0})', \tag{6.1}$$

and substituting this relation in Eq. (3.2), one finds the following compatibility condition,

$$(r\varepsilon_{\theta 0})' - r'\varepsilon_{\xi 0} = \alpha(1+\varepsilon_{\xi 0})(\cos\Phi - \cos\phi), \tag{6.2}$$

which is valid for large strains.

As long as $\varepsilon_{\xi 0}$ is assumed to be small compared with unity, a simplified compatibility relation holds,

$$(r\varepsilon_{\theta 0})' = \alpha[(\cos\Phi - \cos\phi) + \varepsilon_{\xi 0}\cos\phi], \tag{6.3}$$

where the circumferential strain $\varepsilon_{\theta 0}$ can still be large.

This logical modification of the compatibility equation (6.2) was suggested by Reissner in ref. [8], arguing that the possibility of this simplification manifested itself in previous applications of the homogeneous shell theory in the course of analysis of the differential equation obtained from (6.2). Whether or not Eq. (6.3) should be used instead of Eq. (6.2) remains an open question pending availability of solutions of the present heterogeneous shell theory for some representative boundary value problems.

7. REDUCTION TO TWO SIMULTANEOUS EQUATIONS

Equations (5.4)–(5.6) and (4.11), in view of Eqs. (3.2)–(3.4) and (4.9), (4.10), constitute a system of seven equations for seven unknowns H, V, N_θ, M_ξ, M_θ, Φ and u. The reduction of this system to two simultaneous equations in terms of the meridian angle Φ, in the deformed shell, and a stress resultant function Ψ defined by

$$\Psi = rH, \tag{7.1}$$

follows along similar lines to ref. [7].

In terms of Φ and Ψ, the listed expressions are obtained for the stress resultants

$$rV = -\int r\alpha p_V \, d\xi, \tag{7.2}$$

$$rN_\xi = \Psi\cos\Phi + rV\sin\Phi \tag{7.3}$$

$$rQ = -\Psi\sin\Phi + rV\cos\Phi \tag{7.4}$$

$$\alpha N_\theta = \Psi' + r\alpha p_H. \tag{7.5}$$

The stress–strain relations (4.11) are transformed into the form

$$\begin{bmatrix} \varepsilon_{\xi 0} \\ \varepsilon_{\theta 0} \\ \cdot \\ M_\xi^* \\ M_\theta^* \end{bmatrix} = \begin{bmatrix} A_{\xi\xi}^* & A_{\xi\theta}^* & B_{\xi\xi}^* & B_{\xi\theta}^* \\ A_{\theta\xi}^* & A_{\theta\theta}^* & B_{\theta\xi}^* & B_{\theta\theta}^* \\ \cdot & \cdot & \cdot & \cdot \\ C_{\xi\xi}^* & C_{\xi\theta}^* & D_{\xi\xi}^* & D_{\xi\theta}^* \\ C_{\theta\xi}^* & C_{\theta\theta}^* & D_{\theta\xi}^* & D_{\theta\theta}^* \end{bmatrix} \begin{bmatrix} N_\xi^* \\ N_\theta^* \\ \cdot \\ \varkappa_\xi \\ \varkappa_\theta \end{bmatrix} \tag{7.6}$$

or more compactly

$$\begin{bmatrix} \varepsilon_0 \\ \\ M^* \end{bmatrix} = \begin{bmatrix} A^* | B^* \\ \cdot\ \cdot\ \cdot \\ C^* | D^* \end{bmatrix} \begin{bmatrix} N^* \\ \\ \varkappa \end{bmatrix}. \tag{7.7}$$

The asterisked matrices are given in terms of the unasterisked matrices and their inverses, as follows:

$$A^* = A^{-1}, \quad B^* = -A^{-1}B, \quad C^* = BA^{-1} = -B^{*T}, \quad D^* = D - BA^{-1}B. \tag{7.8}$$

Note that whereas the A^* and D^* are symmetric matrices, B^* and C^* are not necessarily so.

The third equilibrium equation (5.6) and the compatibility relation (6.2) are expressed, correspondingly, by the following two simultaneous equations in terms of Φ and Ψ:

$$L_{11}\Phi + L_{12}\Psi + L_{13}(\Phi, \Psi) = L_{14}(\Phi, p_H, p_V, m) + L_{15}(\Phi, T), \tag{7.9}$$

$$L_{21}\Phi + L_{22}\Psi + L_{23}(\Phi, \Psi) = L_{24}(\Phi, p_H, p_V) + L_{25}(\Phi, T). \tag{7.10}$$

The functional operators L_{ij} are so written that each term is non-dimensional, and are in the form

$$L_{11}\Phi = r^2(\Phi'' - \phi'') + \frac{r\alpha}{D_{\xi\xi}^*}(rD_{\xi\xi}^*/\alpha)'(\Phi' - \phi') + \frac{r\alpha}{D_{\xi\xi}^*}(D_{\xi\theta}^*)'(\sin\Phi - \sin\phi)$$
$$+ (r\alpha D_{\xi\theta}^*/D_{\xi\xi}^*)(\cos\Phi - \cos\phi)\phi' - (\alpha^2 D_{\theta\theta}^*/D_{\xi\xi}^*)\cos\Phi\,(\sin\Phi - \sin\phi), \tag{7.11}$$

$$L_{12}\Psi = -(r^2 C_{\xi\theta}^*/D_{\xi\xi}^*)\Psi'' - \frac{r\alpha}{D_{\xi\xi}^*}(rC_{\xi\theta}^*/\alpha)'\Psi', \tag{7.12}$$

$$L_{13}(\Phi, \Psi) = -\frac{r\alpha}{D_{\xi\xi}^*}(C_{\xi\xi}^* - C_{\theta\theta}^*)\Psi'\cos\Phi - \frac{r\alpha}{D_{\xi\xi}^*}\Psi[(C_{\xi\xi}^*\cos\Phi)'$$
$$-(\alpha C_{\theta\xi}^*/r)\cos^2\Phi + \alpha\sin\Phi], \tag{7.13}$$

$$L_{14}(\Phi, p_H, p_V, m) = [(r\alpha C_{\xi\xi}^*/D_{\xi\xi}^*)\sin\Phi](rV)' + (r\alpha/D_{\xi\xi}^*)[(C_{\xi\xi}^*\sin\Phi)'$$
$$-(\alpha C_{\theta\xi}^*/r)\sin\Phi\cos\Phi - \alpha\cos\Phi](rV) + (r\alpha C_{\xi\theta}^*/D_{\xi\xi}^*)(r^2 p_H)'$$
$$+(r\alpha/D_{\xi\xi}^*)[(C_{\xi\theta}^*)' - \frac{\alpha}{r}C_{\theta\theta}^*\cos\Phi](r^2 p_H) + (r^2\alpha^2/D_{\xi\xi}^*)\,m, \tag{7.14}$$

$$L_{15}(\Phi, T) = (r\alpha/D_{\xi\xi}^*)[-(C_{\xi\xi}^* rN_{\xi T})' - (C_{\xi\theta}^* rN_{\theta T})' + (rM_{\xi T})'$$
$$+\alpha\cos\Phi\,(C_{\theta\xi}^* N_{\xi T} + C_{\theta\theta}^* N_{\theta T} - M_{\theta T})], \tag{7.15}$$

$$L_{21}\Phi = -\frac{r}{\alpha} B^*_{\theta\xi}(\Phi''-\phi'') + \left[B^*_{\xi\xi} \cos\Phi - \left(\frac{r}{\alpha} B^*_{\theta\xi}\right)' \right] (\Phi'-\phi')$$

$$-[B^*_{\theta\theta}(\sin\Phi - \sin\phi)]' + \frac{\alpha}{r} B^*_{\xi\theta} \cos\Phi(\sin\Phi - \sin\phi) - \alpha(\cos\Phi - \cos\phi), \quad (7.16)$$

$$L_{22}\Psi = \left(\frac{r}{\alpha} A^*_{\theta\theta}\right)\Psi'' + \left(\frac{r}{\alpha} A^*_{\theta\theta}\right)'\Psi', \quad (7.17)$$

$$L_{23}(\Phi, \Psi) = \left[(A^*_{\theta\xi} \cos\Phi)' - \frac{\alpha}{r} A^*_{\xi\xi} \cos^2\Phi \right] \Psi, \quad (7.18)$$

$$L_{24}(\Phi, p_H, p_V) = -A^*_{\theta\theta}(r^2 p_H)' + \left[\frac{\alpha}{r} A^*_{\xi\theta} \cos\Phi - (A^*_{\theta\theta})' \right] (r^2 p_H)$$

$$-(A^*_{\theta\xi} \sin\Phi)(rV)' + \left[\frac{\alpha}{r} A^*_{\xi\xi} \sin\Phi \cos\Phi - (A^*_{\theta\xi} \sin\Phi)' \right] (rV), \quad (7.19)$$

$$L_{25}(\Phi, T) = (rA^*_{\theta\xi} N_{\xi T})' + (rA^*_{\theta\theta} N_{\theta T})' - \alpha \cos\Phi \, [A^*_{\xi\xi} N_{\xi T} + A^*_{\xi\theta} N_{\theta T}]. \quad (7.20)$$

Equation (7.10) may be considerably simplified if the modified compatibility relation (6.3) is used instead of (6.2) and Φ is replaced by ϕ in Eq. (7.3). In the simplified equation (7.10) all terms with Φ, in the coefficients of Ψ, p_H, p_V, become terms with ϕ. Since such simplification occurs also in some of the B terms of Eq. (7.10), it is felt that the possibility of the indicated modification should be further examined in the light of the analysis of the differential equations (7.9), (7.10).

The two simultaneous heterogeneous shell equations (7.9), (7.10) for Φ and Ψ are valid for arbitrarily large changes of the tangent angles of the shell. The homogeneous shell equations are obtained as a special case of the present shell theory by omitting all B^* and C^* terms in the system (7.9), (7.10), whereas the reduction to the isothermal case is obtained by setting $L_{15} = L_{25} = 0$.

8. SMALL FINITE DEFLECTION THEORY

In many applications, it suffices to consider rather small finite changes of the tangent angle ϕ. Denoting by

$$\beta = \phi - \Phi \quad (8.1)$$

and retaining terms up to the second degree in β and Ψ in Eqs. (7.9), (7.10), we write

$$\cos\Phi = \cos(\phi-\beta) = \cos\phi + \beta\sin\phi - \tfrac{1}{2}\beta^2\cos\phi, \quad (8.2)$$

$$\sin\Phi = \sin(\phi-\beta) = \sin\phi - \beta\cos\phi - \tfrac{1}{2}\beta^2\sin\phi. \quad (8.3)$$

After some computations, the system of two differential equations (7.9), (7.10) assumes the following form:

$$L_{11}\beta+L_{12}\Psi+L_{13}(\beta,\Psi) = L_{14}(\beta, p_H, p_V, m)+L_{15}(\beta, T), \tag{8.4}$$

$$L_{21}\beta+L_{22}\Psi+L_{23}(\beta,\Psi) = L_{24}(\beta, p_H, p_V)+L_{25}(\beta, T). \tag{8.5}$$

The functional operators in the equilibrium equation (8.4) are now defined as follows:

$$L_{11}\beta = r^2\beta''+\frac{r\alpha}{D^*_{\xi\xi}}\left(\frac{r}{\alpha}D^*_{\xi\xi}\right)'\beta'+\left[\frac{r\alpha}{D^*_{\xi\xi}}\left(\frac{r'}{\alpha}D^*_{\xi\theta}\right)'-(r')^2(D^*_{\theta\theta}/D^*_{\xi\xi})\right]\beta$$
$$+\frac{1}{2}\left[\frac{r\alpha}{D^*_{\xi\xi}}\left(\frac{z'}{\alpha}D^*_{\xi\theta}\right)'-3r'z'(D^*_{\theta\theta}/D^*_{\xi\xi})\right]\beta^2, \tag{8.6}$$

$$L_{12}\Psi = (r^2C^*_{\xi\theta}/D^*_{\xi\xi})\Psi''+\left[(rr'/D^*_{\xi\xi})(C^*_{\xi\xi}-C^*_{\theta\theta})+\frac{r\alpha}{D^*_{\xi\xi}}\left(\frac{r}{\alpha}C^*_{\xi\theta}\right)'\right]\Psi'$$
$$+\left[\frac{r\alpha}{D^*_{\xi\xi}}\left(\frac{r'}{\alpha}C^*_{\xi\xi}\right)'+\frac{r\alpha^2}{D^*_{\xi\xi}}\sin\phi-(r')^2(C^*_{\theta\xi}/D^*_{\xi\xi})\right]\Psi, \tag{8.7}$$

$$L_{13}(\beta,\Psi) = \frac{rz'}{D^*_{\xi\xi}}(C^*_{\xi\xi}-C^*_{\theta\theta})\beta\Psi'+\left[\frac{r\alpha}{D^*_{\xi\xi}}\left(\frac{z'}{\alpha}C^*_{\xi\xi}\right)'-\frac{r\alpha^2}{D^*_{\xi\xi}}\cos\phi\right.$$
$$\left.-2r'z'(C^*_{\theta\xi}/D^*_{\xi\xi})\right]\beta\Psi+\frac{r\alpha}{D^*_{\xi\xi}}\left(\frac{z'}{\alpha}C^*_{\xi\xi}\right)\beta'\Psi, \tag{8.8}$$

$$L_{14}(\beta, p_H, p_V, m) = [r\alpha(C^*_{\xi\xi}/D^*_{\xi\xi})(\beta\cos\phi-\sin\phi)](rV)'+\left[\frac{r\alpha^2}{D^*_{\xi\xi}}(\cos\phi\right.$$
$$+\beta\sin\phi)+(\alpha^2C^*_{\theta\xi}/D^*_{\xi\xi})\left(\tfrac{1}{2}\sin 2\phi-\beta\cos 2\phi\right)+\frac{r\alpha}{D^*_{\xi\xi}}\{C^*_{\xi\xi}(\beta\cos\phi$$
$$\left.-\sin\phi)\}'\right](rV)+\left[(\alpha^2C^*_{\theta\theta}/D^*_{\xi\xi})(\cos\phi+\beta\sin\phi)-\left(\frac{r\alpha}{D^*_{\xi\xi}}\right)C^{*\prime}_{\xi\theta}\right]r^2p_H$$
$$-(r\alpha C^*_{\xi\theta}/D^*_{\xi\xi})(r^2p_H)'-\frac{r^2\alpha^2}{D^*_{\xi\xi}}m, \tag{8.9}$$

$$L_{15}(\beta, T) = \frac{r\alpha}{D^*_{\xi\xi}}[(rC^*_{\xi\xi}N_{\xi T})'+(rC^*_{\xi\theta}N_{\theta T})'-\alpha(\cos\phi+\beta\sin\phi)(C^*_{\theta\xi}N_{\xi T}$$
$$+C^*_{\theta\theta}N_{\theta T}-M_{\theta T})-(rM_{\xi T})']. \tag{8.10}$$

The functional operators in the compatibility equation (8.5) are of the form

$$L_{21}\beta = \left(\frac{r}{\alpha}B^*_{\theta\xi}\right)\beta''+\left[\frac{r'}{\alpha}(B^*_{\theta\theta}-B^*_{\xi\xi})+\left(\frac{r}{\alpha}B^*_{\theta\xi}\right)'\right]\beta'+\left[-\alpha\sin\phi\right.$$
$$\left.+\left(\frac{r'}{\alpha}B^*_{\theta\theta}\right)'-\frac{(r')^2}{r\alpha}B^*_{\xi\theta}\right]\beta+\left[\frac{z'}{\alpha}(B^*_{\theta\theta}-B^*_{\xi\xi})\right]\beta\beta'+\tfrac{1}{2}\left[\alpha\cos\phi\right.$$
$$\left.+\left(\frac{z'}{\alpha}B^*_{\theta\theta}\right)'-3\frac{r'z'}{r\alpha}B^*_{\xi\theta}\right]\beta^2, \tag{81.1}$$

$$L_{22}\,\Psi = \left(\frac{r}{\alpha}A^{*}_{\theta\theta}\right)\Psi'' + \left(\frac{r}{\alpha}A^{*}_{\theta\theta}\right)'\Psi' + \left[\left(\frac{r'}{\alpha}A^{*}_{\theta\xi}\right)' - \frac{(r')^{2}}{r\alpha}A^{*}_{\xi\xi}\right]\Psi, \tag{8.12}$$

$$L_{23}(\beta, \Psi) = \left[\left(\frac{z'}{\alpha}A^{*}_{\theta\xi}\right)' - 2\frac{r'z'}{r\alpha}A^{*}_{\xi\xi}\right]\beta\Psi + \left(\frac{z'}{\alpha}A^{*}_{\theta\xi}\right)\beta'\Psi, \tag{8.13}$$

$$L_{24}(\beta, p_{H}, p_{V}) = \left(-\frac{z'}{\alpha}A^{*}_{\theta\xi}\right)(rV)' + \left[\frac{r'z'}{r\alpha}A^{*}_{\xi\xi} - \left(\frac{z'}{\alpha}A^{*}_{\theta\xi}\right)'\right](rV)$$

$$+\left[\left(\frac{r'}{\alpha}A^{*}_{\theta\xi}\right)' + \frac{(z')^{2}-(r')^{2}}{r\alpha}A^{*}_{\xi\xi}\right]\beta(rV) + \left(\frac{r'}{\alpha}A^{*}_{\theta\xi}\right)(\beta rV)' - A^{*}_{\theta\theta}(r^{2}p_{H})'$$

$$+\left[\frac{r'+z'\beta}{r}A^{*}_{\xi\theta} - (A^{*}_{\theta\theta})'\right](r^{2}p_{H}), \tag{8.14}$$

$$L_{25}(\beta T) = (rA^{*}_{\theta\xi}N_{\xi T})' + (rA^{*}_{\theta\theta}N_{\theta T})' - \alpha(\cos\phi + \beta\sin\phi)(A^{*}_{\xi\xi}N_{\xi T} + A^{*}{}_{\xi\theta}N_{\theta T}). \tag{8.15}$$

The small finite deflection theory of heterogeneous orthotropic shells of revolution is formulated in terms of two coupled non-linear differential equations (8.4) and (8.5) in β and ψ. These equations may be considered as an extension of the classical von Kármán theory for small finite deflections of homogeneous plates. The system (8.4), (8.5) can be shown to include, as special cases, Reissner's homogeneous shell equations of ref. [7] and the shallow shell equations of ref. [6]. There is a continuous transition from Eqs. (7.9), (7.10) or (8.4), (8.5) to the case of shallow heterogeneous shells, which remain shallow after deformation. The details require, however, some discussion which is postponed to another occasion.

9. LARGE FINITE DEFLECTION THEORY OF CIRCULAR HETEROGENEOUS PLATES

The system of circular plate equations, for the case of large deflections, are obtained from Eqs. (7.9), (7.10) by setting

$$r = \xi, \quad z = 0, \quad \alpha = 1, \quad \sin\phi = 0, \quad \cos\phi = 1. \tag{9.1}$$

In particular, for the case of a composite circular plate of uniform thickness, with applied edge moments and horizontal edge forces, the following equations hold:

$$\xi^{2}\Phi'' + \xi\Phi' - \frac{1}{2}(D^{*}_{\theta\theta}/D^{*}_{\xi\xi})\sin 2\Phi - (\xi^{2}C^{*}_{\xi\theta}/D^{*}_{\xi\xi})\Psi'' - (\xi/D^{*}_{\xi\xi})[(C^{*}_{\xi\xi}$$

$$-C^{*}_{\theta\theta})\cos\Phi + C^{*}_{\xi\theta}]\Psi' - (\xi/D^{*}_{\xi\xi})(\sin\Phi - C^{*}_{\xi\xi}\Phi'\sin\Phi - \frac{1}{\xi}C^{*}_{\theta\xi}\cos^{2}\Phi)\Psi = 0, \tag{9.2}$$

$$\xi A^*_{\theta\theta}\Psi'' + A^*_{\theta\theta}\Psi' - (A^*_{\theta\xi}\Phi' \sin\Phi + \frac{1}{\xi} A^*_{\xi\xi} \cos^2\Phi)\,\Psi - \xi B^*_{\theta\xi}\Phi''$$

$$+ [(B^*_{\xi\xi} - B^*_{\theta\theta}) \cos\Phi - B^*_{\theta\xi}]\,\Phi' + \frac{1}{2\xi} B^*_{\xi\theta} \sin 2\Phi + 1 - \cos\Phi = 0. \quad (9.3)$$

These coupled non-linear equations in Φ and Ψ are reduced to Eqs. (V′), (VI′) of ref. [7] by assuming isotropy and omitting all B^* and C^* terms. Extension of von Kármán's classical plate equations for small finite deflections of heterogeneous circular plates is obtained from the system (9.2), (9.3) on replacing $\sin\Phi$ with Φ and $\cos\Phi$ with $(1-\frac{1}{2}\Phi^2)$.

REFERENCES

1. A. G. H. Dietz *et al.*, *Micromechanics of Fibrous Composites*, Report MAB-207-M, Materials Advisory Board, Ad Hoc Committee on Micromechanics and Fibrous Composites, National Academy of Sciences, National Research Council, Washington, D.C. (May 1965).
2. S. W. Tsai, *Structural Behavior of Composite Materials*, National Aeronautics and Space Administration, CR-71 (July 1964).
3. S. A. Ambartsumyan, "Contributions to the theory of anisotropic layered shells", *Applied Mechanics Reviews* **15**, 245–9 (April 1962).
4. S. A. Ambartsumyan, *Theory of Anisotropic Shells*, National Aeronautics and Space Administration, Technical Translation F-118, 1964 (of Russian 1961 issue).
5. V. I. Korolev, *Layered Anisotropic Plates and Shells of Reinforced Plastics*, Izdatelstvo "Mashynostroyenie", Moskva, 1965 (in Russian).
6. E. Reissner, "Symmetric bending of shallow shells of revolution", *Journal of Mathematics and Mechanics* **7**, 121–40 (March 1958).
7. E. Reissner, "On axisymmetrical deformations of thin shells of revolution", *Proceedings of Symposia in Applied Mathematics* **3**, 27–52 (1950).
8. E. Reissner, "On the equations for finite symmetrical deflections of thin shells of revolution", *Progress in Applied Mechanics*, Prager Anniversary Volume, Macmillan, pp. 171–8 (1963).
9. Y. Stavsky, "Thermoelasticity of heterogeneous aeolotropic plates", *Journal of the Engineering Mechanics Division*, Proceedings of the American Society of Civil Engineers, **89**, EM2, 89–105 (April 1963).
10. Y. Stavsky, "On finite deformations of heterogeneous shallow shells", *Proceedings of the Fifth International Symposium on Space Technology and Science*, Tokyo, pp. 543–60 (1964).
11. Y. Stavsky, "On the general theory of heterogeneous aeolotropic plates", *The Aeronautical Quarterly* **15**, 29–38 (Feb. 1964).
12. E. Reissner, "On the theory of thin elastic shells," H. Reissner Anniversary Volume, *Contributions to Applied Mechanics*, J. W. Edwards, Ann Arbor, Michigan, pp. 231–47 (1949).

DYNAMICS

PRESSURE DISTRIBUTION ON CONICAL DELTA WINGS IN SUPERSONIC-MODERATE HYPERSONIC FLOW

ELIE CARAFOLI and CORNEL BERBENTE

Institute of Fluid Mechanics "Traian Unia", Academy of the Socialist Republic of Romania, Bucharest

1. EFFECT OF YAW ON THE PRESSURES BEHIND THE SHOCK WAVE

In an earlier paper [1] we have established, on the basis of the general theory of oblique shock waves, the following formula for calculating the pressures on aerodynamic profiles in supersonic-moderate hypersonic flow:

$$\frac{C_p}{m^2 \sin^2 \tau} = \frac{\gamma+1}{2} + \frac{2}{K}\sqrt{\left[1+\left(\frac{\gamma+1}{4}\right)^2 K^2\right]}, \tag{1}$$

also valid both for compression and expansion, where C_p is the pressure coefficient, τ the free stream deflection (positive in the case of compression and negative in the case of expansion), γ the ratio of specific heats, while

$$K = mM_\infty \sin \tau \quad \left(m = \frac{M_\infty}{\sqrt{(M_\infty^2-1)}}\right) \tag{2}$$

represents the similarity parameter which is valid over the entire supersonic-moderate hypersonic flow range. We have designated by M_∞ the Mach number of the free stream.

If the wing of infinite aspect ratio has also a yaw χ (Fig. 1), we resolve the velocity U_∞ into the components U_l and U_n oriented in the span direction and normal to the span, respectively. In the absence of friction, the component U_l has no effect, while the normal component U_n is deflected by the angle τ_n, connected through the relation

$$\tau_n \cos \chi = \tau, \tag{3}$$

with the deflection τ defined previously.

The Mach number of the normal component is

$$M_n = M_\infty \cos \chi. \tag{4}$$

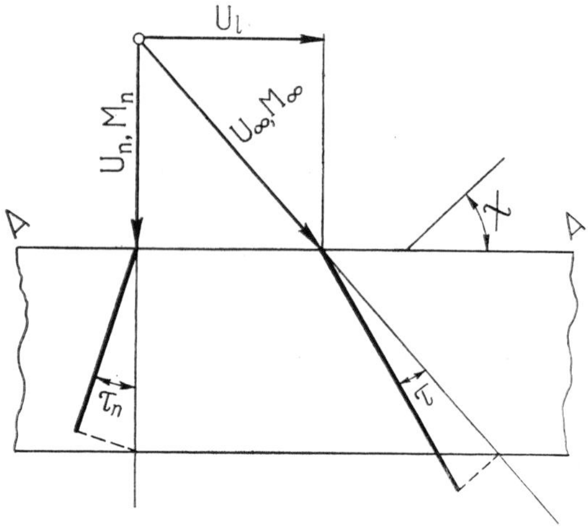

FIG. 1

Further, using the notation

$$K_n = m_n M_n \sin \tau_n \left(m_n = \frac{M_n}{\sqrt{(M_n^2 - 1)}} \right), \tag{5}$$

the pressure coefficient C_{pn} referred to the velocity U_n will be yielded by Eq. (1):

$$\frac{C_{pn}}{m_n^2 \sin^2 \tau_n} = \frac{\gamma+1}{2} + \frac{2}{K_n} \sqrt{\left[1 + \left(\frac{\gamma+1}{4}\right)^2 K_n^2\right]}. \tag{6}$$

Reverting to the conditions of the free stream U_∞, we obtain, after several simple computations:

$$\frac{\lambda^2 C_p}{m^2 \sin^2 \tau} = \frac{\gamma+1}{2} + \frac{2\lambda}{K} \sqrt{\left[1 + \left(\frac{\gamma+1}{4}\right)^2 \frac{K^2}{\lambda^2}\right]}, \left(\lambda = \sqrt{1 - \frac{\tan^2 \chi}{M_\infty^2 - 1}}\right). \tag{7}$$

On comparing Eqs. (1) and (7), one observes that the yaw effect is taken into consideration when the deflection τ is replaced by an *equivalent deflection* θ given by the relation

$$\theta = \frac{\sin \tau}{\lambda} \approx \frac{\tau}{\lambda}. \tag{8}$$

Equation (7) can be also written in terms of the axial disturbance velocity u, deduced within the framework of the linearized theory. Indeed,

u and θ are connected through the relation

$$\frac{u}{U_\infty} = -\frac{m}{M_\infty}\theta, \tag{9}$$

so that Eq. (7) becomes

$$M_\infty^2 C_p = \frac{\gamma+1}{2}\left(M_\infty^2 \frac{u}{U_\infty}\right)^2 - 2\left(M_\infty^2 \frac{u}{U_\infty}\right)\sqrt{\left[1+\left(\frac{\gamma+1}{4}\right)^2\left(M_\infty^2 \frac{u}{U_\infty}\right)^2\right]}. \tag{10}$$

This formula has a general character, as shown in our previous paper [2].

2. APPLICATION TO A CONICALLY SHAPED WING WITH SUPERSONIC EDGES

We now intend to extend Eq. (10) to a delta wing with supersonic leading edges, whose incidence varies according to a zero-order homogeneous function (Fig. 2)

$$\tau^* = \tau^*(y), \quad \left(y = \frac{x_2}{x_1}\right). \tag{11}$$

At point P on the generatrix OP, representing the ridge of an elementary dihedron, the stream undergoes the deflection $d\tau^*$ and passes to P' on the infinitely close generatrix OP', while the *elementary equivalent deflection*

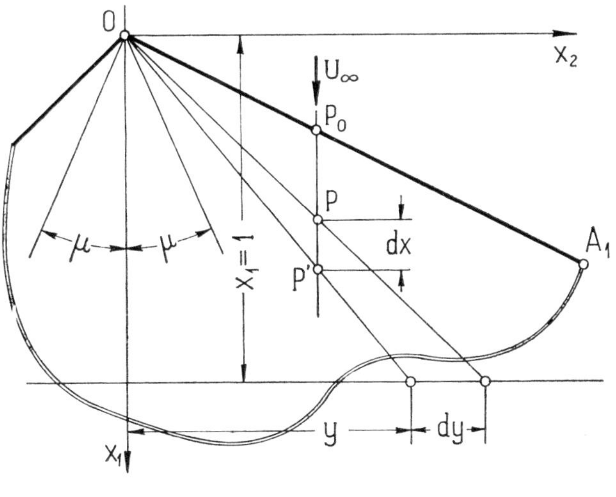

FIG. 2

will be given by the same Eq. (8):

$$d\theta^* = \frac{d\tau^*}{\lambda^*}; \quad \left[\lambda^* = \sqrt{\left(1 - \frac{1}{y^2(M_\infty^2 - 1)}\right)}\right]. \tag{12}$$

The *equivalent total deflection* results easily by integrating from P_0 to P:

$$\theta^* = \theta_0^* + \int_{\tau_0^*}^{\tau^*} \frac{d\tau^*}{\lambda^*}. \tag{13}$$

On the other hand, by the theory of conical motions [3] we have:

$$\frac{du}{U_\infty} = -\frac{m}{M_\infty} d\theta^*, \quad \frac{u}{U_\infty} = -\frac{m}{M_\infty} \theta^*, \tag{14}$$

hence u is formed in the same manner as θ^*, while Eq. (10) is also perfectly valid for conical delta wings with supersonic leading edges.

3. EXPERIMENTAL VERIFICATION

Let us consider a conical delta wing with parabolic directrix, lying in a supersonic stream having $M_\infty = 4$ (Fig. 3).

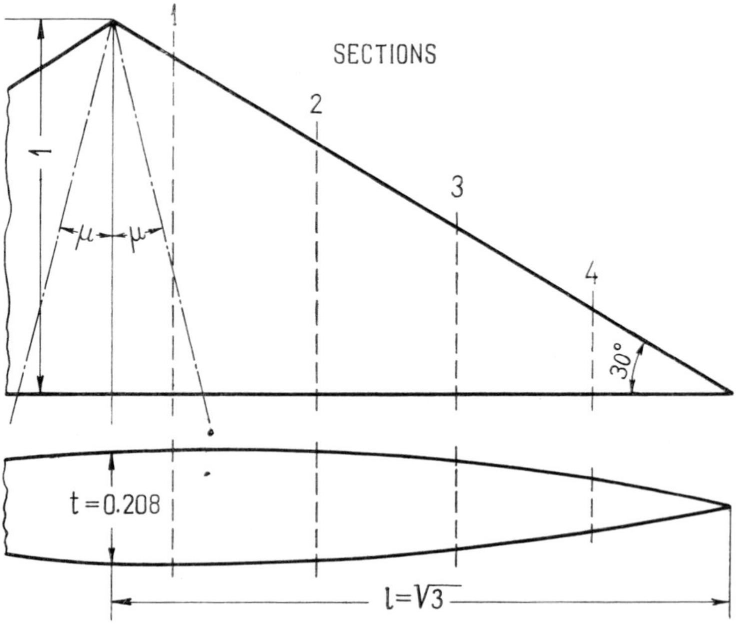

FIG. 3

In such a case, the axial disturbance velocity u is given by [4]:

$$\frac{u}{U_\infty} = -Re\frac{2}{\pi}\frac{0{\cdot}95\tau_0 l}{\sqrt{B^2l^2-1}}\left[1+\frac{a^2}{B^2l^2}+\frac{b(2-B^2l^2)}{B^4l^4}\right]\cos^{-1}\sqrt{\frac{1-B^2y^2}{B^2(l^2-y^2)}}$$

$$+\text{Re}\frac{2}{\pi}\frac{0{\cdot}95\tau_0 l}{B^2l^2}\left\{2\left(a+\frac{2b}{B^2l^2}\right)Ch^{-1}\sqrt{\frac{1+By}{2By}}\right.$$

$$\left.-\sqrt{1-B^2y^2}\left[\left(a+(2+B^2y^2)\frac{b}{B^2l^2}\right)\left(\ln\sqrt{\frac{l^2-y^2}{y^2}}-\frac{i\pi}{2}\right)+\frac{1}{2}b\right]\right\}, \quad (15\text{-a})$$

$$l = \text{cotan}\,\chi,\quad B^2 = M_\infty^2-1,\quad a = 0{\cdot}525;\quad b = 0{\cdot}575. \qquad (15\text{-b})$$

Substituting u in Eq. (10), and including the effect of the boundary layer,

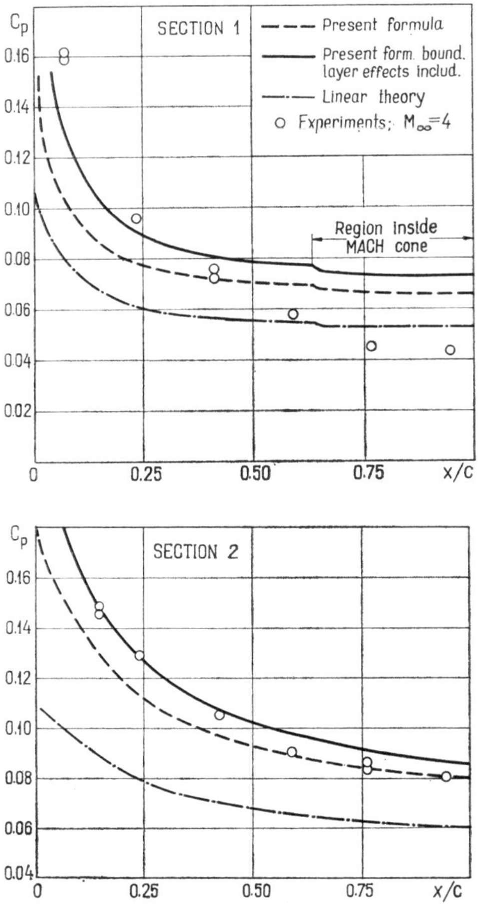

FIG. 4

one finds the actual value of C_p which is plotted in Figs. 4 and 5 for four sections of the wing.

One can observe fair agreement between theory and experiment [5] except for the points close to the trailing edge of the wing, where the stream

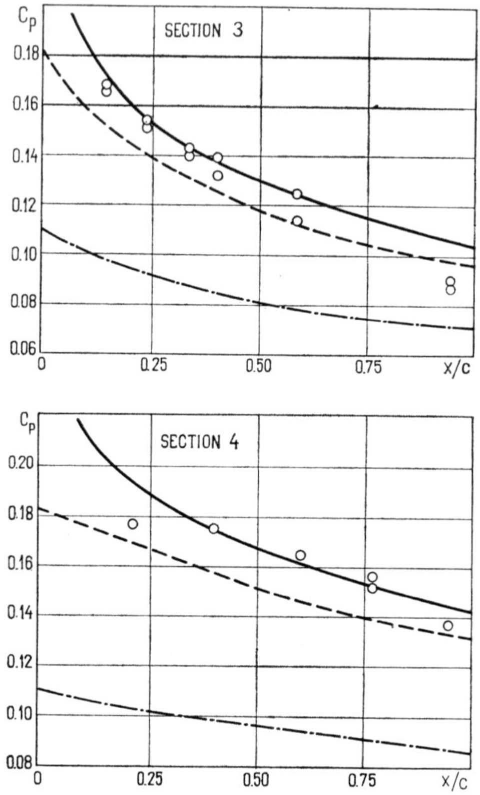

FIG. 5

expands suddenly, producing a pressure decrease transmitted upstream through the boundary layer.

This effect is more important in Section 1, which lies partially in the Mach cone of the wing.

REFERENCES

1. E. CARAFOLI, "On a unitary formula for compression–expansion in supersonic–hypersonic flow", *Revue de Mécanique Appliquée*, t. vii, no. 5, Bucharest (1962).
2. E. CARAFOLI and C. BERBENTE, "Determination of pressures and aerodynamic

characteristics of delta wings in supersonic–moderate hypersonic flow", *Revue de Mécanique Appliquée*, t. xi, no. 3, Bucharest (1966).

3. E. Carafoli, *High-speed Aerodynamics*, Editura Tehnică, Bucharest (1956).
4. E. Carafoli and D. Mateescu, "Sur une classe d'ailes delta à incidence et pente variables d'après des fonctions homogènes, en régime supersonique", *Revue de Mécanique Appliquée*, t. xviii, no. 2, Bucharest (1965).
5. M. Landhal, G. Drougge and B. Beane, "Theoretical and experimental investigation of second order supersonic wing–body interference", *J.A.S.* **27**, no. 9 (1960).

SIMILARITY RULE IN THE HYDRODYNAMICALLY TRANSCRITICAL RÉGIME IN AN IONIZED MULTI-FLUID GAS†

M. Z. v. KRZYWOBLOCKI‡

Michigan State University, U.S.A.

ABSTRACT

The purpose of the present work is to derive a similarity rule in the hydrodynamically transcritical régime ($\sim M = 1$) in an ionized multi-fluid gas. Reduction of the results to conditions existing in classical gas dynamics furnishes the von Kármán transonic similarity rule. The derived similarity rule refers to an entire family of flow patterns around thin (small-perturbation) bodies satisfying certain similarity requirements. A brief illustrative example closes the work.

INTRODUCTION

In classical gas dynamics, the transonic similarity law was originated by v. Kármán and developed by Oswatitsch [8, 9]. The similarity law was further modified by Spreiter [15] to accommodate the regions of higher subsonic and lower supersonic motions. The transonic similarity states: If a series of bodies with the same dimensionless thickness distribution is placed in flows of different free-stream Mach numbers and different values of γ, in such a way that a certain similarity parameter remains constant, then the flow patterns are similar in the sense that the same dimensionless potential function describes several flows. For affinely related profiles, the camber and the angle of attack are adjusted in proportion to the thickness ratio.

Some attempts have been made to investigate whether the similarity in magnetogasdynamics, analogous to the gas-dynamic similarity in the transonic region, does exist. We briefly mention the following case: Tamada [16] investigates two-dimensional transonic flow of a perfectly conducting,

† Dedicated by the author to Professor Dr. Markus Reiner on his eightieth anniversary, with deep respect for his invaluable contributions to science and higher education.

‡ With the collaboration of H. R. Kim.

inviscid compressible fluid past a thin body with aligned magnetic field, based on the small-perturbation theory in the hodograph plane. He shows that the equations of motion can be reduced to those of ordinary flow by a suitable affine transformation. He concludes that the v. Kármán similarity law is extensible to the class of magnetogasdynamic flows in question. The Tamada investigation is based on the fact that the ionized gas is assumed to be a single fluid. In an ionized gas, in which currents and charges are of importance, the single-fluid technique may give a poor description of the phenomenon. To improve the description governing the electrical current density [11], the fundamental assumption in the present work is that of a multi-fluid structure. We deal with an inviscid, non-heat-conducting, steady, uniform, electrically conducting (finite conductivity) perfect fluid flow in a uniform electromagnetic field.

In Section 1 we compile the results of the transonic similarity law in gas dynamics and magnetogasdynamics. We define the thermodynamic properties of a component fluid in relation to the gross fluid and derive the governing equations of each component fluid in Section 2. The resulting system of equations is approximated by means of small-perturbation theory in the transonic region. The constant coefficients in this system are invariant under the coordinate transformations. Equations in the system become functions of coordinates and invariant constant coefficients. We find a modified similarity law, a limiting case of which becomes the v. Kármán similarity law. Using the generalized Bernoulli equation, we approximate the pressure coefficient of each component flow and of the pressure coefficient of the gross fluid flow. Additionally, a special case, in which the free-stream velocity and the magnetic field are parallel, is treated. An illustrative example closes the present work.

1. TRANSONIC SIMILARITY LAW

1.1. Transonic Similarity Law in Gasdynamics

The quasi-linear partial differential equation for the velocity potential Φ of a steady, isentropic, two-dimensional flow of a perfect inviscid gas [4, 13] is simplified with the use of the critical speed of sound and becomes:

$$(\gamma+1)(1-M_1)\varphi_{,x}\varphi_{,xx}+2(1-M_1)\varphi_{,y}\varphi_{,xy}-\alpha_*\varphi_{,yy} = 0, \qquad (1.1.1)$$

where φ is the perturbed velocity potential, $M_1 = u_0\alpha_*^{-1}$, $\alpha_* =$ critical velocity, $u_0 =$ free-stream velocity. Introducing:

$$\xi = xl^{-1}, \quad \eta = yl^{-1}[(\gamma+1)\delta]^{1/3}, \quad \varphi = \alpha_* lf(\xi, \eta), \qquad (1.1.2)$$

into Eq. (1.1.1), one obtains a differential equation for the transonic flow with small perturbation in the dimensionless form:

$$-2^{4/3}Kf_{,\xi}f_{,\xi\xi}+f_{,\eta\eta} = 0; \tag{1.1.3}$$

$$K \equiv (1-M_0)[2^{-1}(\gamma+1)\delta]^{-2/3} \quad \text{or} \quad \cong 2^{-1}(1-M_0^2)[2^{-1}(\gamma+1)\delta]^{-2/3}. \tag{1.1.4}$$

Boundary conditions to Eq. (1.1.3) are:

$$f_{,\xi} = -1; \quad f_{,\eta} = 0 \quad \text{at} \quad \xi = \eta = \pm\infty, \tag{1.1.5a}$$

$$2Kf_{,\eta} = 2^{-1/3}h(\xi) \quad \text{at} \quad \eta = 0; \quad 0 \leqq \xi \leqq 1, \tag{1.1.5b}$$

where $h(\xi)$ is the ratio of the slope at the surface of the body to the thickness ratio.

Spreiter [15] presents a differential equation which is the most satisfactory at Mach numbers above and below unity from the viewpoint of agreement with experiments and with the exact theoretical solution. His parameter K is given by Eq. $(1.1.4)_2$.

The pressure coefficient is approximated by:

$$C_p = (p-p_0)(\tfrac{1}{2}\varrho_0 u_0^2)^{-1} \cong -2(u-u_0)u_0^{-1} =$$
$$= -2(1-M_1)f_{,\xi} = -4(\gamma+1)^{-1}(1-M_0)f_{,\xi}, \tag{1.1.6}$$

or

$$C_p = 2(1-M_0)(\gamma+1)P(K, \xi, \eta) = \delta^{\frac{2}{3}}[2^{-1}(\gamma+1)]^{-\frac{2}{3}}KP, \tag{1.1.7}$$

where P is a function of K, ξ, η and the form of P is determined by the dimensionless thickness distribution of the family of profiles in question.

1.2. Similarity Law in Magnetogasdynamics

It is found in ref. [12] that if one deals with a two-dimensional flow of a perfectly-conducting, inviscid, compressible fluid past a thin body in the presence of a static magnetic field parallel to the free-stream flow in the hodograph plane, the potential function, Φ, and stream function, Ψ, have relations:

$$\Phi_{,q} = (M^2-1)(A^2-1)(A^2\varrho q)^{-1}\Psi_{,\theta}, \tag{1.2.1$_1$}$$

$$\Phi_{,\theta} = (A^2-1)[(A^2+M^2-1)A^2-1]^{-1}\varrho^{-1}q\Psi, \tag{1.2.1$_2$}$$

$$M = q\alpha^{-1}, \quad A^2 = q^2(\mu H^2)^{-1}4\pi\varrho. \tag{1.2.1$_3$}$$

Tamada [16] defines a relation between a local velocity q and the critical velocity α_*, and dimensionless coordinates ξ and ϑ by:

$$q\alpha_*^{-1} = 1+\nu\xi; \quad \theta = \delta\vartheta; \quad \nu = (\gamma+1)^{-\frac{1}{3}}\delta^{\frac{2}{3}}(1-A_*^{-2})^{\frac{1}{3}}, \tag{1.2.2}$$

where δ is thickness ratio and A_* is the value of A for $q = \alpha_*$. He also presents the dimensionless potential and stream function as:

$$\Phi = l\alpha_*(1-A_*^{-2})\varphi(\xi, \eta); \quad \Psi = l\varrho_*\alpha_*(\gamma+1)^{-\frac{1}{3}}\delta^{-\frac{1}{3}}(1-A_*^{-2})^{-\frac{2}{3}}\psi, \quad (1.2.3)$$

where l is a characteristic length of the body in question. Using (1.2.2) and (1.2.3) in Eq. (1.2.1) with an assumption of the small-perturbation method, excluding simultaneously $M = 1$ and $A = 1^\dagger$, yields a Tricomi equation:

$$\psi_{,\xi\xi} - \xi\psi_{,\vartheta\vartheta} = 0. \quad (1.2.4)$$

Thus, the fundamental equations have been reduced to those of ordinary gas dynamics [3, 4, 8, 9, 13]. There is introduced a specific transonic parameter:

$$\xi_0 = (\gamma+1)^{-\frac{2}{3}}\delta^{-\frac{2}{3}}(M_0^2-1)(1-A_*^{-2})^{-\frac{1}{3}}. \quad (1.2.5)$$

The pressure coefficient is approximated by:

$$C_p = -2\nu\xi. \quad (1.2.6)$$

Therefore, if we find a solution of Eq. (1.2.4) past a body $\xi = F(\vartheta)$, with a definite value of ξ_0, we may obtain a whole family of conventional and magnetogasdynamic flows past geometrically similar bodies with different free-stream Mach number M_0, different Alfvén number in the undisturbed flow A_0 and specific heat ratios of fluid flows γ, provided that the transonic parameter (1.2.5) remains fixed. Geffen [1] arrives at the same conclusion for the transonic flow in the physical plane.

2. MULTI-FLUID MAGNETOGASDYNAMICS

2.1. Introduction

We consider an inviscid, non-heat-conducting, steady, uniform, electrically conducting, perfect fluid flow which consists of n components such as electrons, ions, neutral particles, etc., in a uniform electromagnetic field. In reality, the ambient flow does not need to be electrically conducting, but we assume that the free stream flow is already partially ionized, i.e. is an electrically conducting gas mixture. It is assumed that in the free-stream there are no electric currents. Thus, it implies that a uniform electric field balances the total induced electromotive force [6].‡ The numbers of ions and electrons are assumed to be equal to each other. Intro-

† See ref. [1] for the investigation of other singularities in Eq. (1.2.1).
‡ In one case $(\mathbf{E}+\mu_e\mathbf{q}\times\mathbf{H}) = 0$, in the other $\mathbf{q}\,||\,\mathbf{H}$.

ducing the disturbance by placing a body in the flow, we assume that the disturbed flow in the vicinity of the body can be explained by means of the small-perturbation method. We consider that, excluding Joule heat, there is no energy (heat) added to or subtracted from a particle along the streamline of the gross fluid. Each component fluid obeys equations of state, continuity, momentum, and the first law of thermodynamics. We assume that the characteristic properties of the quasi-three-dimensional flow depend on the coordinates (x, y), where the uniform flow direction is taken as the x-axis and the y-axis is vertical. The direction transverse to the (x, y)-axes is the z-axis.

We assume that the thermodynamic properties of the gross fluid have the following relations with respect to those of each component fluid:

$$\varrho = \sum_{s=1}^{n} \varrho_s = \sum_{s=1}^{n} m_s \nu_s, \quad s = 1, 2, \ldots, n, \tag{2.1.1}$$

$$p = \sum_{s=1}^{n} p_s, \tag{2.1.2}$$

$$T = \left(\sum_{s=1}^{n} \nu_s\right)^{-1} \left(\sum_{s=1}^{n} \nu_s T_s\right), \tag{2.1.3}$$

where m_s = mass of a particle in sth component fluid, ν_s = the number density, ϱ_s = density of the sth fluid component; p_s is the pressure, T_s is the temperature of the sth fluid. For the gross fluid, the density ϱ, the pressure and the temperature T obey the perfect gas law:

$$p = \varrho R_A m^{-1} T = \varrho R_g T, \quad R_g = \text{const}, \tag{2.1.4}$$

where R_A is the universal gas constant and m is the mass of the gross fluid. Each component fluid may not necessarily obey the perfect gas law, depending on a function W_s which represents a deviation from the state of a perfect gas, i.e.

$$p_s = \varrho_s R_A m_s^{-1}(1+W_s)T_s = \varrho_s R_s T_s; \quad R_s = R_A m_s^{-1}(1+W_s). \tag{2.1.5}$$

We also assume that the mass rate per unit area is

$$\varrho \mathbf{q} = \sum_{s=1}^{n} \varrho_s \mathbf{q}_s, \tag{2.1.6}$$

where $\mathbf{q}$, $\mathbf{q}_s$ are velocities of the gross fluid and the component fluid, respectively. It is assumed that the specific internal energy of the sth component is given by:

$$U_s = c_{vs}(1+A_s)T_s, \tag{2.1.7}$$

where A_s is a function representing the deviation of the specific internal energy of the sth component from that of a perfect gas; c_{vs} is the specific heat capacity at constant volume of sth fluid in a state of a perfect gas (i.e. $A_s = 0$). Then:

$$K_s = c_{ps}c_{vs}^{-1} = 1+R_A m_s^{-1}c_{vs}^{-1} = 1+R_{ps}c_{vs}^{-1}; \quad R_{ps}= R_A m_s^{-1}. \qquad (2.1.8)$$

Let us introduce the definition of the specific enthalphy, $I_s = U_s+R_sT_s$. Then, using Eqs. (2.1.5), (2.1.7), we get:

$$I_s = U_s+R_sT_s = c_{ps}(1+D_s)T_s; \quad D_s = K_s^{-1}[A_s+(K_s-1)W_s]. \qquad (2.1.9)$$

The entropy of the s-fluid S_s is

$$T_s\, dS_s = dQ_s; \quad dQ_s = dI_s-\varrho_s^{-1}dp_s, \qquad (2.1.10)$$

where Q_s involves the energy (heat) due to the electromagnetic origin and to the sources from the outside of the system. Substituting (2.1.7), $(2.1.10)_1$ in $(2.1.10)_2$, dividing the resulting equation by T_s and using (2.1.5), we have:

$$dS_s = c_{vs}(1+A_s)T_s^{-1}dT_s+c_{vs}dA_s-R_{ps}(1+W_s)\varrho_s^{-1}d\varrho_s; \qquad (2.1.11)$$

integrating from a zero-subscript initial state to some end state, and performing elementary operations, we have:

$$TT_{s0}^{-1} = G_s(\varrho_s\varrho_{s0}^{-1})^{K_s-1} \exp\,[(S_s-S_{s0})c_{vs}^{-1}]\,; \qquad (2.1.12)$$

$$G_s = \exp\left[-(A_s-A_{s0})-\int_0^s A_sT_s^{-1}dT_s+(K_s-1)\int_0^s W_s\varrho_s^{-1}d\varrho_s\right]. \qquad (2.1.13)$$

The pressure–density–entropy relation, and the speed of sound, may be expressed by using Eqs. (2.1.4) to (2.1.13):

$$p_s = p_{s0}\varrho_{s0}^{-K_s}(1+W_s)(1+W_{s0})^{-1}G_s \exp\,(-S_{s0}c_{vs}^{-1})\varrho_s^{K_s} \exp\,(S_sc_{vs}^{-1}); \quad (2.1.14)$$

$$(\alpha_s)^2 = \left(\frac{\partial p_s}{\partial \varrho_s}\right)_{S_s} = K_sp_s\varrho_s^{-1}(1+N_s)+p_sM_s; \qquad (2.1.15)$$

$$N_s = (K_s-1)K_s^{-1}(W_s-A_s)(1+A_s)^{-1};$$

$$M_s = \frac{\partial}{\partial \varrho_s}\log\,(1+W_s)(1+A_s)^{-1}\Big|_{S_s=\text{const.}} \qquad (2.1.16)$$

The corresponding expressions for the gross fluid can be calculated by using the above relations for component fluids summed on s. In the disturbed region, we consider that characteristic quantities are perturbed. The perturbed quantities yield those for the gross fluid by means of relations (2.1.1), (2.1.2), (2.1.3) and (2.1.6). We assume that in the free stream each component fluid obeys the perfect gas law and that the number density $\nu_{s0} = \nu_0$, where s refers only to charged component fluids. There is uniform speed $u_{s0} = u_0$ on all streamlines, but T_{s0} may be different for al species. Gross fluid temperature T_0 calculated by (2.1.3) is the same on all streamlines. This means that:

$$\tfrac{1}{2}\Big(\sum_{s=1}^{n}\nu_s\Big)\Big(\sum_{s=1}^{n}\nu_s\Big)^{-1}u_0^2+KR_g(K-1)^{-1}\Big(\sum_{s=1}^{n}\nu_sT_{s0}\Big)\Big(\sum_{s=1}^{n}\nu_s\Big)^{-1}= C(=\text{const.}), \qquad (2.1.17)_1$$

$$\nu_1\Big(\sum_{s=1}^{n}\nu_s\Big)^{-1}\big[\tfrac{1}{2}u_0^2+KR_g(K-1)^{-1}T_{10}\big]+\dots$$
$$+\nu_n\Big(\sum_{s=1}^{n}\nu_s\Big)^{-1}\big[\tfrac{1}{2}u_0^2+KR_g)(K-1)^{-1}T_{n0}\big] = C. \qquad (2.1.17)_2$$

Let C_s be the Bernoulli constant for the sth fluid. Then

$$\nu_1C_1\Big(\sum_{s=1}^{n}\nu_s\Big)^{-1}+\dots+\nu_nC_n\Big(\sum_{s=1}^{n}\nu_s\Big)^{-1}=C;\quad \Big(\sum_{s=1}^{n}\nu_s\Big)^{-1}\Big(\sum_{s=1}^{n}\nu_sC_s\Big)=C. \quad (2.1.18)$$

In the perturbation region, there occur some sort of momentum and energy exchange phenomena between the component fluids. Hence, a species flow is diabatic. Such a flow is called a "multi-diabatic" flow.

2.2. Governing Equations of Fluid Flow

For each component fluid, we have [2]:
continuity equation:

$$(\varrho_su_s),_x+(\varrho_sv_s),_y=\sigma_s,\quad \sum_{s=1}^{n}\sigma_s=0, \qquad (2.2.1)$$

momentum equation:

$$\frac{d\mathbf{q}_s}{dt}+\varrho_s^{-1}\nabla p_s=\varrho_s^{-1}\mathbf{F}_s+\sigma_s\varrho_s^{-1}(\mathbf{Z}_s-q_s)-\varrho_s^{-1}\mathbf{X}_s. \qquad (2.2.2)$$

Equation of state and the first law of thermodynamics are:

$$p_s=\varrho_sR_sT_s,\quad dQ_s=dI_s-\varrho_s^{-1}dp_s,\quad \nabla Q_s=\nabla I_s-\varrho_s^{-1}\nabla p_s, \quad (2.2.3)$$

where the used symbols denote:

$\mathbf{q}_s$ = (u_s, v_s, w_s) = the velocity vector with components u_s, v_s, w_s of the sth fluid flow in the chosen coordinate system;
σ_s = mass source of the sth fluid per unit volume;
$\sigma_s\mathbf{Z}_s$ = momentum source (chemical origin, ionization, etc.) per unit volume associated with σ_s;

$$\mathbf{X}_s=\left(\frac{\partial X_{sxx}}{\partial x}+\frac{\partial X_{sxy}}{\partial y},\quad \frac{\partial X_{syx}}{\partial x}+\frac{\partial X_{syy}}{\partial y}\right);$$

X_{sij} = stress tensor, $i, j = x, y, z$;
$\mathbf{F}_s$ = external forces per unit volume;

Q_s = the energy of the sth fluid due to the electromagnetic origin and from outside sources;

$$\frac{d}{dt} = \mathbf{q}_s \cdot \nabla . \tag{2.2.4}$$

The external forces are assumed to be an electromagnetic force and the forces due to the interactions of other fluid components:

$$\mathbf{F}_s = \varrho_{es}(\mathbf{E}+\mu_e \mathbf{q}_s \times \mathbf{H}) + \sum_{r=1}^{n} \alpha_{rs}(\mathbf{q}_r - \mathbf{q}_s), \tag{2.2.5}$$

$\varrho_{es} = e_s \nu_s$ = charge density; e_s = charge of each species particle; $\mathbf{E}$ = electric field intensity vector; μ_e = magnetic permeability, $\mathbf{H}$ = magnetic field vector, α_{rs} = proportionality constant of the interaction forces with $\alpha_{rs} = \alpha_{sr} = \alpha$ = const, if we choose a particular case of isotropic and homogeneous interactions, $\mathbf{q}_r$ = velocity vector of fluid components other than the sth fluid.

In addition to Eqs. (2.2.1), (2.2.2) and (2.2.3), the Maxwell equations become part of the system of equations:

$$\nabla \times \mathbf{E} = -\mathbf{B}_{,t}, \quad \nabla \cdot \mathbf{D} = \varrho_e, \quad \varrho_e = \sum_{s=1}^{n} \varrho_{es}, \tag{2.2.6}$$

$$\nabla \times \mathbf{H} = \mathbf{J} + \mathbf{D}_{,t}, \quad \nabla \cdot \mathbf{B} = 0, \tag{2.2.7}$$

with $\mathbf{B}_{,t} = \mathbf{D}_{,t} = 0$ in the present case, where $\mathbf{D} = \varepsilon \mathbf{E}$ is the displacement current vector, ε is the inductive capacity, assumed to be constant; $\mathbf{B} = \mu_e \mathbf{H}$ = magnetic flux density vector. It is easily shown that by multiplying Eq. (2.2.1) by $\gamma_s = e_s m_s^{-1}$ = const, and taking summation over s we obtain the conservation of charge density, and that $(2.2.6)_2$, $(2.2.7)_2$ can be derived from $(2.2.6)_1$, $(2.2.7)_1$, respectively. The system of Eqs. (2.2.1), (2.2.2), (2.2.3), $(2.2.6)_1$ and $(2.2.7)_1$ consists of $(6n+6)$ equations in $(6n+6)$ unknowns, ϱ_s, $\mathbf{q}_s$, p_s, T_s, $\mathbf{E}$, and $\mathbf{H}$. In the present case of steady flow, Eqs. $(2.2.6)_2$, $(2.2.7)_2$ should be taken into account. The current density vector is defined by:

$$\mathbf{J} = \sum_{s=1}^{n} \varrho_{es} \mathbf{q}_s = \sigma(\mathbf{E} + \mu_e \mathbf{q} \times \mathbf{H}) + \varrho_e \mathbf{q}. \tag{2.2.8}$$

We use the current density in the first form of the sum of products of the charge density and velocity of each species.

We calculate $\varrho_s^{-1} \nabla p_s$ from (2.2.2) and substitute in $(2.2.3)_3$, and we obtain:

$$\frac{d\mathbf{q}_s}{dt} + \nabla I_s - \nabla Q_s - \varrho_s^{-1}\mathbf{F}_s - \sigma_s \varrho_s^{-1}(\mathbf{Z}_s - \mathbf{q}_s) - \varrho_s^{-1}\mathbf{X}_s = 0. \tag{2.2.9}$$

Multiplying the x-component of (2.2.2) by u_s, the y-component of (2.2.2) by v_s, adding the resulting equations together and rearranging terms, we obtain an expression:

$$(\alpha_s^2-u_s^2)u_{s,x}-u_s v_s(u_{s,y}+v_{s,x})+(\alpha_s^2-v_s^2)v_{s,y}-\sigma_s\varrho_s^{-1}(\alpha_s^2+u_s^2+v_s^2)$$
$$+\varrho_s^{-1}[(F_{sx}-X_{sx}+\sigma_s Z_{sx})-(K_s-1)(1+W_s)(1+A_s)^{-1}Q_{s,x}-L_{sx}]u_s$$
$$+\varrho_s^{-1}[(F_{sy}-X_{sy}+\sigma_s Z_{sy})-(K_s-1)(1+W_s)(1+A_s)^{-1}Q_{s,y}-L_{sy}]v_s=0, \tag{2.2.10}$$

where the definition of speed of sound, (2.1.15), is used and L_{sx}, L_{sy} are given by:

$$L_{sx} = p_s\sum_{j=1}^{k_s}\frac{\partial}{\partial\xi_s^j}\log(1+W_s)(1+A_s)^{-1}\frac{\partial\xi_s^j}{\partial x}, \tag{2.2.11}$$

$$L_{sy} = p_s\sum_{j=1}^{k_s}\frac{\partial}{\partial\xi_s^j}\log(1+W_s)(1+A_s)^{-1}\frac{\partial\xi_s^j}{\partial y}. \tag{2.2.12}$$

2.3. Electromagnetic Field

In Section 2.1 we assumed that in the subdomain of the free stream there exists a uniform electric field to balance the induced electromotive force. We take the uniform magnetic field as (H_{0x}, H_{0y}, H_{0z}). Then we have:

$$\mathbf{E}_0 = -\mu_e\mathbf{q}_0\times\mathbf{H}_0 = (0,\quad \mu_e u_0 H_{0z}, -\mu_e u_0 H_{0y}); \tag{2.3.1}$$

$$\mathbf{E} = \mathbf{E}_0+\mathbf{E}_p;\quad \mathbf{E}_{0x} = 0;\quad \mathbf{H} = \mathbf{H}_0+\mathbf{H}_p, \qquad (2.3.2), (2.3.3)$$

where terms with subscript p denote those of the disturbances. We also assume that:

$$\varrho_{es} = \varrho_{es0}+\varrho_{esp};\quad \varrho_{e0} = \sum_{s=1}^{n}\varrho_{es0} = 0. \tag{2.3.4}$$

The Maxwell system (2.2.6), (2.2.7) is satisfied in the free stream. In general, the quantities in the free stream are much larger than the perturbed quantities in (2.3.2), (2.3.3) and (2.3.4).

From Eqs. $(2.2.7)_{1,2}$ and (2.3.2), we have:

$$E_z = E_{0z};\quad E_{yp,x}-E_{xp,y} = 0,\quad E_{xp,x}-E_{yp,y} = \varepsilon^{-1}\sum_{s=1}^{n}\varrho_{es}, \tag{2.3.5}$$

or

$$\nabla^2 G_1 = -\varepsilon^{-1}\sum_{s=1}^{n}\varrho_{es};\quad \nabla^2 = \frac{\partial^2}{\partial x^2}+\frac{\partial^2}{\partial y^2},\quad \mathbf{E}_p = -\nabla G_1(x, y), \tag{2.3.6}$$

with B.C.:

$$G_1 = \nabla G_1 = 0 \quad\text{at}\quad r = \infty,\quad r^2 = x^2+y^2. \tag{2.3.7}$$

The relation $(2.3.2)_1$ becomes:

$$E_x = -G_{1,x}; \quad E_y = E_{0y}(1-G_{1,y}E_{0y}^{-1}); \quad E_z = E_{0z}. \tag{2.3.8}$$

Expanding (2.2.7), we have:

$$H_{z,y} = J_x, \quad -H_{z,x} = J_y, \quad H_{y,x}-H_{x,y} = J_z; \quad H_{x,x}+H_{y,y} = 0. \tag{2.3.9}$$

Then we have:

$$H_z = H_{0z}+H_{zp} = H_{0z}(1+H_{zp}H_{0z}^{-1}); \quad H_{zp} = \int_0^s (J_x\, dy - J_y\, dx). \tag{2.3.10}$$

Introducing a new function $G_2(x, y)$, such that:

$$H_{xp} = -G_{2,y}, \quad H_{yp} = G_{2,x}, \tag{2.3.11}$$

we obtain:

$$\nabla^2 G_2 = J_z; \quad \nabla^2 = \frac{\partial^2}{\partial x^2}+\frac{\partial^2}{\partial y^2}, \tag{2.3.12}$$

$$G_2 = \nabla G_2 = 0, \quad \text{at} \quad r = \infty, \quad r^2 = x^2+y^2. \tag{2.3.13}$$

From $(2.3.2)_2$ and (2.3.11), we get:

$$H_x = H_{0x}(1-G_{2,y}H_{0x}^{-1}); \quad H_y = H_{0y}(1+G_{2,x}H_{0y}^{-1}). \tag{2.3.14}$$

2.4. Generalized Bernoulli and Crocco Equations and Vorticity Function

From (2.2.5) and (2.2.9) we have:

$$\frac{d\mathbf{q}_s}{dt}+\nabla I_s-\nabla Q_s-\sigma_s\varrho_s^{-1}(\mathbf{Z}_s-\mathbf{q}_s)-\varrho_s^{-1}\mathbf{X}_s$$
$$-\varrho_s^{-1}\sum_{r=1}^{n}\alpha(\mathbf{q}_r-\mathbf{q}_s)-\gamma_s(\mathbf{E}+\mu_e\mathbf{q}_s\times\mathbf{H}) = 0. \tag{2.4.1}$$

Taking a scalar product with $\mathbf{q}_s$ and multiplying by dt, we obtain:

$$d(\tfrac{1}{2}q_s^2)+dI_s-dQ_s-\sigma_s\varrho_s^{-1}(\mathbf{Z}_s-\mathbf{q}_s)\cdot d\mathbf{r}_s-\varrho_s^{-1}\mathbf{X}_s\cdot d\mathbf{r}_s$$
$$-\varrho_s^{-1}\left[\sum_{r=1}^{n}\alpha(\mathbf{q}_r-\mathbf{q}_s)\right]\cdot d\mathbf{r}_s-\gamma_s\mathbf{E}\cdot d\mathbf{r}_s = 0, \tag{2.4.2}$$

where $d\mathbf{r}_s/dt = \mathbf{q}_s$, $\nabla(\;)\cdot d\mathbf{r}_s = d(\;\;)$, $d\mathbf{r}_s$ is an element along the streamline. We denote the variables in the free stream by the subscript "0": $\mathbf{q}_s = \mathbf{u}_0$, $I_s = I_{s0}$, with $W_s \equiv W_{s0} \equiv 0$, $A_s \equiv A_{s0} \equiv 0$, $Q_s \equiv Q_{s0} \equiv 0$, $\sigma_{s0} \equiv 0$, $X_s \equiv X_{s0} \equiv 0$. Integrating (2.4.2) along a streamline from the "0" state to a certain state "s" in the perturbation, we obtain:

$$\tfrac{1}{2}q_s^2+I_s-\int_0^s dQ_s-\int_0^s \varrho_s^{-1}\sigma_s(\mathbf{Z}_s-\mathbf{q}_s)\cdot d\mathbf{r}_s-\int_0^s \varrho_s^{-1}\mathbf{X}_s\cdot d\mathbf{r}_s$$
$$-\int_0^s \varrho_s^{-1}\left[\sum_{r=1}^{n}\alpha(\mathbf{q}_r-\mathbf{q}_s)\right]\cdot d\mathbf{r}_s-\int_0^s \gamma_s\mathbf{E}\cdot d\mathbf{r}_s = \tfrac{1}{2}u_0^2+I_{s0} = \mathcal{H}_{s0}, \tag{2.4.3}$$

where the quantity $\mathscr{H}_{s0}$ is constant for a species flow along the streamline and the free stream is the isentropic flow. We remodel $\mathscr{H}_{s0}$:

$$\mathscr{H}_{s0} = \tfrac{1}{2}u_0^2+I_{s0} = \tfrac{1}{2}u_0^2+\alpha_{s0}^2(K_s-1)^{-1} = \tfrac{1}{2}(K_s+1)(K_s-1)^{-1}\alpha_{s*}^2, \tag{2.4.4}$$

where $u_0 = \alpha_{s*}$ at Mach number $= 1$.

Using (2.1.8), (2.1.9), (2.1.15) and (2.1.16), we obtain:

$$I_s\alpha_s^{-2} = (1+P_s)(K_s-1)^{-1};$$

$$P_s = \{(1+W_s)(1+A_s)^{-1}[1+\varrho_sK_s(1+N_s)^{-1}M_s]\}^{-1}-1. \tag{2.4.5}$$

Then we can write (2.4.3) with (2.4.4) and (2.4.5):

$$\tfrac{1}{2}q_s^2+\alpha_s^2(1+P_s)(K_s-1)^{-1}-\int_0^s dQ_s-\int_0^s \mathbf{Y}_s'\cdot d\mathbf{r}_s-\int_0^s \gamma_s\mathbf{E}\cdot d\mathbf{r}_s = \mathscr{H}_{s0}, \tag{2.4.6}$$

$$\mathbf{Y}_s' = \varrho_s^{-1}\left[\sigma_s(\mathbf{Z}_s-\mathbf{q}_s)+\mathbf{X}_s+\sum_{r=1}^{n}\alpha(\mathbf{q}_r-\mathbf{q}_s)\right]. \tag{2.4.7}$$

Let:

$$\zeta_s = \alpha_s^{-2}\left[\int_0^s dQ_s+\int_0^s \mathbf{Y}'\cdot d\mathbf{r}_s+\int_0^s \gamma_s\mathbf{E}\cdot d\mathbf{r}_s\right], \tag{2.4.8}$$

then (2.4.6) becomes:

$$\tfrac{1}{2}(K_s-1)\,q_s^2+\alpha_s^2[1+P_s-\zeta_s(K_s-1)] = \tfrac{1}{2}(K_s+1)\alpha_{s*}^2. \tag{2.4.9}$$

From (2.4.4) we solve for $\alpha_{s0}\alpha_{s*}^{-1}$:

$$\alpha_{s0}^2 = \tfrac{1}{2}(K_s+1)\,\alpha_{s*}^2-\tfrac{1}{2}(K_s-1)u_0^2;$$

$$\alpha_{s0}\alpha_{s*}^{-1} = [1+2(K_s-1)(K_s+1)^{-1}(M_0-1)+(K_s-1)(K_s+1)^{-1}(M_0-1)^2]^{-\frac{1}{2}}. \tag{2.4.10}$$

Using the binomial theorem expansion in terms of (M_0-1), retaining first-order terms, we have:

$$\alpha_{s0}\alpha_{s*}^{-1} \cong 1+(K_s-1)(K_s+1)^{-1}(1-M_0). \tag{2.4.11}$$

Substituting this in the previous expression, we obtain, if terms of first order in $(1-M_0)$ are retained,

$$1-M_1 \cong 2(K_s+1)^{-1}(1-M_0). \tag{2.4.12}$$

The relation (2.2.4) with its vector identity, and the vorticity vector ω_s:

$$(\mathbf{q}_s\cdot\nabla)\,\mathbf{q}_s = \nabla(\tfrac{1}{2}q_s^2)-\mathbf{q}_s\times\boldsymbol{\omega}_s;\quad \boldsymbol{\omega}_s = \nabla\times\mathbf{q}_s, \tag{2.4.13}$$

substituted in (2.4.1), yield:

$$\nabla H_s-\nabla Q_s-\mathbf{Y}_s''-\mathbf{q}_s\times\boldsymbol{\omega}_s = 0; \tag{2.4.14}$$

$$H_s = I_s+\tfrac{1}{2}q_s^2;\quad \mathbf{Y}''_s = \gamma_s(\mathbf{E}+\mu_e\mathbf{q}_s\times\mathbf{H})+\mathbf{Y}_s'. \tag{2.4.15}$$

Equation (2.4.14) can be written [7]:

$$\nabla \times \mathbf{q}_s \times \mathbf{q}_s = \nabla Q_s - \nabla H_s'. \tag{2.4.16}$$

$$H_s' = I_s + \tfrac{1}{2} q_s^2 + A; \quad A = -\int \mathbf{Y}'' \cdot d\mathbf{r}_s, \tag{2.4.17}$$

which is a generalized form of Crocco's equation. Divergence operator on (2.4.14) yields:

$$\nabla^2 H_s - \nabla^2 Q_s - \nabla \cdot \mathbf{Y}_s'' - \nabla \cdot (\mathbf{q}_s \times \omega_s) = 0. \tag{2.4.18}$$

The system of equations in question consists of Eqs. (2.2.1), $(2.2.2)_3$, (2.2.10), $(2.3.6)_1$, $(2.3.12)_1$, and (2.4.18) in unknowns ϱ_s, u_s, v_s, w_s, G_1, and G_2, since the vorticity vector ω_s and the velocity vector $\mathbf{q}_s$ have the relation given by $(2.4.13)_2$.

3. SIMILARITY LAW OF CRITICAL FLOW IN ELECTROMAGNETO-GASDYNAMICS

3.1. Approximate Equations of Flow

We deal with thin bodies placed in a uniform incoming stream of transonic speed in an electromagnetic field. In the approach below, we operate in the neighborhood of the point, line, or surface, at which the velocity of motion is equal to the speed of sound due only to the magnitude of the hydrodynamic pressure. Moreover, due to small perturbations, this requirement reduces the magnitude of the flow velocity to the vicinity of that of the incoming flow velocity, i.e. around $M_0 = 1$. We do not deal with the case where the actual speed of sound is equal to the actual velocity of the perturbed flow. The actual speed of sound is effected by both the electric and magnetic pressures. Nevertheless, because of small perturbations, the difference between the above two cases is small or negligibly small. We call this velocity (approximately equal to the transonic velocity in the incoming flow) the critical velocity. We assume that there are small deviations from free-stream conditions and that the fluid velocity is near the speed of sound in the incoming flow. From this viewpoint the appropriate perturbation velocity components are $(\alpha_{s*} - u_s)$, v_s and w_s. We also assume that:

$$\varrho_s = \varrho_{s0} + \varrho_{sp}; \quad \varrho_{s0} \cong \varrho_{s*}, \quad \varrho_s = \varrho_{s*} \quad \text{for} \quad u_0 = \alpha_{s*}, \quad \varrho_{s0} \gg \varrho_{sp}. \tag{3.1.1}$$

The equations to be approximated in the perturbation region are the z-component expression of Eq. (2.2.2), denoted by $(2.2.2)_3$, (2.2.10) and (2.4.16). We simplify coefficients of $u_{s,x}$, $v_{s,x}$, etc., in $(2.2.2)_3$ and (2.2.10)

The speed of sound is found from (2.4.9):

$$\alpha_s^2 = [\tfrac{1}{2}(K_s+1)\alpha_{s*}^2 - \tfrac{1}{2}(K_s-1)q_s^2][1+P_s-(K_s-1)\zeta_s]^{-1}, \quad (3.1.2)$$

and it is assumed that the singularity due to the value $[1-(K_s-1)\cdot(1+P_s)^{-1}\zeta_s] = 0$ is excluded. For this kind of singularity, a different approach must be used. Using (3.1.2) for α_s^2, we calculate the following:

$$\begin{aligned}\alpha_s^2-u_s^2 &= [\tfrac{1}{2}(K_s+1)\alpha_{s*}^2 - \tfrac{1}{2}(K_s-1)q_s^2][1+P_s-(K_s-1)\zeta_s]^{-1}-u_s^2\\ &\cong \{(K_s'+1)(1-2\beta^2\zeta_s)\alpha_{s*}(\alpha_{s*}-u_s)-\alpha_{s*}^2[P_s-(K_s-1)\zeta_s]\}\cdot\\ &\quad\cdot[1+P_s-(K_s-1)\zeta_s]^{-1}, \qquad (3.1.3)\end{aligned}$$

$$K_s' = K_s+2P_s; \quad \beta^2 = (K_s-1)(K_s'+1); \qquad (3.1.4)$$

$$u_s v_s = -(\alpha_{s*}-u_s)v_s+\alpha_{s*}v_s \cong \alpha_{s*}v_s; \qquad (3.1.5)$$

$$\begin{aligned}\alpha_s^2-v_s^2 &= [\tfrac{1}{2}(K_s+1)\alpha_{s*}^2 - \tfrac{1}{2}(K_s-1)q_s^2][1+P_s-(K_s-1)\zeta_s]^{-1}-v_s^2\\ &\cong [(K_s-1)\alpha_{s*}(\alpha_{s*}-u_s)+\alpha_{s*}^2][1+P_s-(K_s-1)\zeta_s]^{-1}; \qquad (3.1.6)\end{aligned}$$

$$\begin{aligned}&F_{sx}-X_{sx}+\sigma_s Z_{sx}-(K_s-1)(1+W_s)(1+A_s)^{-1}Q_{s,x}-L_{sx}\\ &= \varrho_{es}E_x-\varrho_{es}\mu_e w_s H_{0y}(1+H_{yp}H_{0y}^{-1}-f_1 H_{0y}^{-1}), \qquad (3.1.7)\end{aligned}$$

where f_1 refers to

$$\begin{aligned}f_1 = (\varrho_{es}\mu_e w_s)^{-1}\Big[&\sum_{r=1}^{n}\alpha(u_r-u_s)-X_s+\sigma_s Z_{sx}\\ &-(K_s-1)(1+W_s)(1+A_s)^{-1}Q_{s,x}-L_{s,x}\Big]; \qquad (3.1.8)\end{aligned}$$

$$\begin{aligned}&F_{sy}-X_{sy}+\sigma_s Z_{sy}-(K_s-1)(1+W_s)(1+A_s)^{-1}Q_{s,y}-L_{sy}\\ &= \varrho_{es}E_{0y}(1+E_{yp}E_{0y}^{-1})+\varrho_{es}\mu_e w_s H_{0x}(1+H_{xp}H_{0x}^{-1}+f_2 H_{0x}^{-1}), \qquad (3.1.9)\end{aligned}$$

$$\begin{aligned}f_2 = (\varrho_{es}\mu_e w_s)^{-1}\Big[&\sum_{r=1}^{n}\alpha(v_r-v_s)-X_{sy}+\sigma_s Z_{sy}\\ &-(K_s-1)(1+W_s)(1+A_s)^{-1}Q_{s,y}-L_{sy}\Big]. \qquad (3.1.10)\end{aligned}$$

The coefficient of $\sigma_s\varrho_s^{-1}$ is reduced to:

$$\alpha_s^2+u_s^2+v_s^2 \cong 2\alpha_{s*}^2. \qquad (3.1.11)$$

Relations (3.1.3), (3.1.5), (3.1.6), (3.1.7), (3.1.9) and (3.1.11) are substituted in (2.2.10) and the resulting equation, multiplied by $[1+P_s-(K_s-1)\zeta_s]\alpha_{s*}^{-2}$, becomes:

$$\begin{aligned}&\{(K_s'+1)(1-2\beta^2\zeta_s)(\alpha_{s*}-u_s)\alpha_{s*}^{-1}-[P_s-(K_s-1)\zeta_s]\}u_{s,x}\\ &-\alpha_{s*}^{-1}v_s[1+P_s-(K_s-1)\zeta_s](u_{s,y}+u_{s,x})+v_{s,y}+\gamma_s\alpha_{s*}^{-2}[1+P_s-(K_s-1)\zeta_s]E_x u_s\\ &-\gamma_s\mu_e w_s H_{0y}[1+P_s-(K_s-1)\zeta_s]\alpha_{s*}^{-2}u_s+\gamma_s(E_{0y}+\mu_e w_s H_{0x})\cdot\\ &\cdot[1+P_s-(K_s-1)\zeta_s]\alpha_s^{-2}v_s-2\sigma_s\varrho_s^{-1}[1+P_s-(K_s-1)\zeta_s]=0, \qquad (3.1.12)\end{aligned}$$

where the following quantities are assumed small and neglected:

$$(H_{yp}H_{0y}^{-1}-f_1H_{0y}^{-1})\alpha_{s*}^{-2} \ll 1; \quad E_{yp}E_{0y}^{-1}\alpha_{s*}^{-2} \ll 1,$$
$$(H_{xp}H_{0x}^{-1}+f_2H_{0x}^{-1})\alpha_{s*}^{-2} \ll 1; \quad (K_s-1)(\alpha_{s*}-u_s)\alpha_{s*}^{-1}v_{s,\,y} \ll 1.$$

Equation $(2.2.2)_3$ may be reduced to:

$$u_0w_{s,\,x} = \gamma_s\mu_e(u_{sp}H_{0y}+u_0H_{yp}-v_{sp}H_{0x})+\sigma_s\varrho_s^{-1}(Z_{sz}-w_s) - \varrho_s^{-1}X_{sz}+\varrho_s^{-1}\sum_{r=1}^{n}\alpha(w_r-w_s). \tag{3.1.13}$$

If we retain the first-order terms of perturbation in (2.4.16), the Crocco equation, after opening it in the present case, becomes:

$$\nabla^2H_s-\nabla^2Q_s+\gamma_s\nabla^2G_1-\nabla\cdot\mathbf{Y}_s' - \mu_e(H_{0x}w_{s,\,y}-H_{0y}w_{s,\,x}+H_{0z}\omega_{sz}-u_0H_{z,\,y})+(u_s\omega_{sz})_{,\,y} = 0. \tag{3.1.14}$$

Introducing a velocity function in the perturbed region $\varphi_s(x, y)$ such that:

$$u_s = A_1'+A_2'\varphi_{s,\,x}; \quad \varphi_{s,\,x} = \bar{\varphi}_{s,\,x}+A_2'^{-1}G_{s3}, \tag{3.1.15}$$
$$v_s = A_2'\varphi_{s,\,y}; \quad \varphi_{s,\,y} = \bar{\varphi}_{s,\,y}+A_2'^{-1}G_{s3}, \tag{3.1.16}$$

where A_1' is a constant to be determined; $A_2' = 1-M_1$, $M_1 = u_0\alpha_{s*}^{-1}$, $G_{s3}(x, y)$ is a function satisfying the following relation:

$$v_{s,\,x}-u_{s,\,y} = G_{s3,\,x}-G_{s3,\,y} = \omega_{sz}, \tag{3.1.17}$$

and the function $\bar{\varphi}_s$ is a velocity potential. Then we obtain:

$$u_{s,\,x} = A_2'\varphi_{s,\,xx}; \quad u_{s,\,y} = A_2'\varphi_{s,\,xy}; \quad v_{s,\,y} = A_2'\varphi_{s,\,yy}; \quad v_{s,\,x} = A_2'\varphi_{s,\,yx}. \tag{3.1.18}$$

Substituting (3.1.18) in (3.1.12), we obtain:

$$\begin{aligned}
&\{(K_s'+1)(1-2\beta^2\zeta_s)(\alpha_{s*}-A_1'-A_2'\varphi_{s,\,x})\alpha_{s*}^{-1}-[P_s-(K_s-1)\zeta_s]\}A_2'\varphi_{s,\,xx}\\
&\quad -\alpha_{s*}^{-1}[1+P_s-(K_s-1)\zeta_s]A_2'\varphi_{s,\,y}(\varphi_{s,\,xy}+\varphi_{s,\,yx})^{\dagger}+A_2'\varphi_{s,\,yy}\\
&+\gamma_s\alpha_{s*}^{-2}[1+P_s-(K_s-1)\zeta_s]E_xA'\varphi_{s,x}-\gamma_s\mu_e\alpha_{s*}^{-2}[1+P_s-(K_s-1)\zeta_s]w_sH_{0y}A_2'\varphi_{s,x}\\
&\quad +\gamma_s\alpha_{s*}^{-2}[1+P_s-(K_s-1)\zeta_s]E_{0y}A_2'\varphi_{s,\,y}\\
&+\gamma_s\mu_e\alpha_{s*}^{-2}[1+P_s-(K_s-1)\zeta_s]w_sH_{0x}A_2'\cdot\varphi_{s,\,y}-(\gamma_sE_x+\gamma_s\mu_ew_sH_{0y})[1+P_s\\
&\quad -(K_s-1)\zeta_s]\alpha_{s*}^{-2}A_1'-2[1+P_s-(K_s-1)\zeta_s]\sigma_s\varrho_s^{-1} = 0.
\end{aligned} \tag{3.1.19}$$

If we choose A_1' such that in the first term:

$$(K_s'+1)(1-2\beta^2\zeta_s)(\alpha_{s*}-A_1')\alpha_{s*}^{-1}-[P_s-(K_s-1)\zeta_s] = 0, \tag{3.1.20}$$

† φ_s is a function in which the order of differentiation cannot be interchanged. In the present case, Eq. (3.1.17) becomes $\varphi_{s,\,yx}-\varphi_{s,\,xy} = \omega_{sz} \neq 0$. Very little is known about such functions.

then A_1' takes the form:

$$A_1' = \alpha_{s*}\{(K_s'+1)(1-2\beta^2\zeta_s)-[P_s-(K_s-1)\zeta_s]\}(K_s'+1)^{-1}(1-2\beta^2\zeta_s)^{-1}. \tag{3.1.21}$$

It is seen that the value of A_1' given by (3.1.21) is not a constant but a function of position. However, we approximate P_s, ζ_s, by chosen constant mean values† $\bar{P}_s$, $\bar{\zeta}_s$, respectively. Thus, we have:

$$A_1' = \alpha_{s*}\{(K_s''+1)(1-2\beta^2\bar{\zeta}_s)-[\bar{P}_s-(K_s-1)\bar{\zeta}_s]\}(K''+1)^{-1}(1-2\beta^2\bar{\zeta}_s)^{-1}, \tag{3.1.22}$$

$$K_s'' = K_s+2\bar{P}_s. \tag{3.1.23}$$

Equation (3.1.19) now takes the form:

$$\begin{aligned} &-(K_s''+1)(1-2\beta^2\bar{\zeta}_s)(1-M_1)\alpha_{s*}^{-1}\varphi_{s,x}\varphi_{s,xx}-[1+\bar{P}_s \\ &-(K_s-1)\zeta_s](1-M_1)\alpha_{s*}^{-1}\varphi_{s,y}(\varphi_{s,yx}+\varphi_{s,xy})+\varphi_{s,yy}+\gamma_s(E_x-\mu_e w_s H_{0y}) \\ &[1+\bar{P}_s-(K_s-1)\bar{\zeta}_s]\alpha_{s*}^{-2}\varphi_{s,x}+\gamma_s(E_{0y}+\mu_e w_s H_{0x})[1+\bar{P}_s-(K_s-1)\bar{\zeta}_s]\alpha_{s*}^{-2}\varphi_{s,y} \\ &+\gamma_s(E_x-\mu_e w_s H_{0y})[1+\bar{P}_s-(K_s-1)\bar{\zeta}_s](1-M_1)^{-1}\{(K_s''+1)(1-2\beta^2\bar{\zeta}_s) \\ &-[\bar{P}_s-(K_s-1)\bar{\zeta}_s]\}(K_s''+1)^{-1}(1-2\beta^2\bar{\zeta}_s)^{-1}\alpha_{s*}^{-1} \\ &-2\sigma_s\varrho_s^{-1}(1-M_1)^{-1}[1+\bar{P}_s-(K_s-1)\bar{\zeta}_s] = 0. \end{aligned} \tag{3.1.24}$$

The boundary conditions are given at infinity and at the surface of the body. At infinity, there is a uniform stream:

$$u_s = u_0 = A_1'+A_2'\varphi_{s,x}\big|_\infty; \quad \varphi_{s,x} = (1-M_1)^{-1}(u_0-A_1'), \tag{3.1.25a}$$

$$\varphi_{s,y}\big|_\infty = 0. \tag{3.1.25b}$$

To consider the boundary condition at the surface of the thin body, we apply the following reasoning: Let l and t be respectively the chord and maximum thickness of the body and let $\delta \equiv tl^{-1}$ be the thickness ratio; also, let h represent the ratio of local surface slope to the thickness ratio; i.e.

$$h = \left(\frac{dy}{dx}\right)_{\text{surf}}\Big/\delta; \quad \left(\frac{dy}{dx}\right)_{\text{surf}} = \delta h, \tag{3.1.26}$$

where it is understood that h is a non-dimensional function of the dimensionless abscissa xl^{-1}, and y is a given function of position representing the body profile. The flow follows the contour, hence the boundary condition at the surface of the body, within the limits of small perturbation, is:

$$v_s u_s^{-1} \cong v_s \alpha_{s*}^{-1} = \delta h. \tag{3.1.27}$$

† This item is explained in Section 3.4.

Since the profile is assumed to be thin, the boundary condition may be evaluated on the horizontal x-axis:

$$y = 0, \quad 0 \leqq xl^{-1} \leqq 1; \quad A_2'\varphi_{s,\,y} = \alpha_{s*}\delta h; \quad \varphi_{s,\,y}\big|_{y=0} = \alpha_{s*}\delta A_2'^{-1}h. \tag{3.1.28}$$

If we use $(2.3.11)_2$ for H_{yp}, $(3.1.15)_1$, $(3.1.16)_1$, for u_{sp}, v_{sp}, respectively, in (3.1.13), we obtain:

$$u_0 w_{s,\,x} = \gamma_s \mu_e[(1-M_1)H_{0y}\varphi_{s,\,x} + u_0 G_{2,\,x} - (1-M_1)H_{0x}\varphi_{s,\,y}]$$
$$+\sigma_s \varrho_s^{-1}(Z_{sz} - w_s) + \varrho_s^{-1}X_{sz} + \varrho_s^{-1}\sum_{r=1}^{n}\alpha(w_r - w_s). \tag{3.1.29}$$

The system under consideration consists of $(2.3.6)_1$, $(2.3.12)_1$, (3.1.14), (3.1.24) with its boundary conditions, and (3.1.29) in unknowns G_1, G_2, ω_{sz}, φ_s and w_s.

3.2. Similarity Transformation

We now seek a transformation of the coordinate system whereby the system of the differential equations would furnish similarity relations. A similarity rule gives a solution for a whole class of related bodies under related conditions, if a solution for only one of these bodies, under one set of conditions, is known. The similarity rule would relate the pressure distributions on affinely related profiles of different thickness ratios and operating at different Mach numbers close to unity, and at different magnitudes of the intensities of electromagnetic fields. We seek a form of differential equations which would involve some constant coefficients composed of such parameters as the thickness ratio, Mach number, magnetic pressure numbers and magnetic Reynolds numbers.

The differential equations and boundary conditions are to be transformed from the (x, y)-coordinate system to the dimensionless (ξ, η)-coordinate system. The resultant system of differential equations, with the corresponding boundary conditions, is to contain constant coefficients satisfying a certain relation among themselves. Thus, we obtain an entire class of solutions which differ one from another in the magnitudes and values of the coefficients in the (x, y)-plane. This entire class represents a family of similar solutions. In accordance with [4], coefficients of equations of the system in the present case must be constant, invariant under the coordinate transformations regardless of variable properties characterizing the critical flow and the body placed in the flow.

We introduce:

$$\xi = xl^{-1}(A_6A_7)^{-n_1}, \quad \eta = yl^{-1}(A_1A_2)^{-n'_2}(A_5A_7)^{-n_2}\delta^{-m},$$
$$\bar{z} = zl^{-1}, \quad w_s = u_0(A_5A_6)^{-n_3}\bar{w}_s,$$
$$\varphi_s = \alpha_{s*}l(A_5A_6A_7)^{-n_4}f_s(\xi, \eta), \quad \omega_{sz} = \alpha_{s*}l^{-1}(A_5A_6A_7)^{-n_4}\bar{\omega}_{sz}(\xi, \eta), \tag{3.2.1}$$

where n_1, n'_2, n_2, n_3 and n_4 are to be determined and

$$A_1 = K_s+2\bar{P}_s+1, \quad A_2 = 1-2\beta^2\bar{\zeta}_s, \quad \beta^2 = (K_s-1)A_1^{-1},$$
$$A_5 = A_{0x}^2R_{msx}M_1^2, \quad A_6 = A_{0y}^2R_{msy}M_1^2, \quad A_7 = A_{0z}^2R_{msz}M_1^2,$$
$$A_{0i}^2 = \mu_eH_{0i}^2(\varrho_{s0}u_0^2)^{-1}, \quad R_{msi} = e_sv_0u_0lH_{0i}^{-1}, \quad i = x, y, z. \tag{3.2.2}^\dagger$$

Let:

$$\varrho_s = \varrho_{s0}A_{10}^{-n_{10}}\bar{\varrho}_s; \quad \sigma_s = \varrho_{s0}u_0l^{-1}A_{11}^{-n_{11}}\bar{\sigma}_s; \quad \mathbf{Z}_s = u_0A_{12}^{-n_{12}}\bar{Z}_s,$$
$$\mathbf{X}_s = \varrho_{s0}u_0^2l^{-1}A_{13}^{-n_{13}}\bar{\mathbf{X}}_s, \quad Q_s = u_0^2A_{14}^{-n_{14}}\bar{Q}_s, \quad \alpha_s = \alpha_{s*}\bar{\alpha}_s, \quad H_s = u_0^2A_{15}^{-n_{15}}\bar{H}_s, \tag{3.2.3}$$

where n_{10}, n_{11}, n_{12}, n_{13}, n_{14} and n_{15} are to be determined;

$$G_1 = E_{0y}lA_{16}^{-n_{16}}\bar{G}_1; \quad G_2 = H_{0x}lA_{17}^{-n_{17}}\bar{G}_2. \qquad (3.2.4), (3.2.5)$$

Then, we obtain the following derivatives:

$$\varphi_{s,x} = \varphi_{s,\xi}\xi_{,x} = \alpha_{s*}(A_6A_7)^{-n_1}(A_5A_6A_7)^{-n_4}f_{s,\xi}, \tag{3.2.6}$$
$$\varphi_{s,xx} = \varphi_{s,\xi\xi}(\xi_{,x})^2 = \alpha_{s*}l^{-1}(A_6A_7)^{-2n_1}(A_5A_6A_7)^{-n_4}f_{s,\xi\xi}, \tag{3.2.7}$$
$$\varphi_{s,xy} = \alpha_{s*}l^{-1}(A_6A_7)^{-n_1}(A_1A_2)^{-n'_2}(A_5A_7)^{-n_2}(A_5A_6A_7)^{-n_4}\delta^{-m}f_{s,\xi\eta}, \tag{3.2.8}$$
$$\varphi_{s,y} = \alpha_{s*}(A_1A_2)^{-n'_2}(A_5A_7)^{-n_2}(A_5A_6A_7)^{-n_4}\delta^{-m}f_{s,\eta}, \tag{3.2.9}$$
$$\varphi_{s,yy} = \alpha_{s*}l^{-1}(A_1A_2)^{-2n'_2}(A_5A_7)^{-2n_2}(A_5A_6A_7)^{-n_4}\delta^{-2m}f_{s,\eta\eta}. \tag{3.2.10}$$

Using these derivatives, we obtain the system of transformed equations

(i) *Hydrodynamics*

Equation for velocity function:

$$C_{s1}f_{s,\xi}f_{s,\xi\xi}+2C_{s2}f_{s,\eta}f_{s,\xi\eta}+2C_{s3}f_{s,\eta}\bar{\omega}_{sz}-f_{s,\eta\eta}-C_{s4}f_{s,\eta}-C_{s5}f_{s,\eta}\bar{w}_s$$
$$+C_{s6}f_{s,\xi}\bar{w}_s-C_{s7}\bar{G}_{1,\xi}f_{s,\xi}-C_{s8}\bar{G}_{1,\xi}+C_{s9}\bar{w}_s+2C_{s10}g_{s1} = 0, \tag{3.2.11}$$

$$C_{s1} = A_1^{2n'_2+1}A_2^{2n'_2+1}A_3A_5^{2n_2-n_4}A_6^{-3n_1-n_4}A_7^{-3n_1+2n_2-n_4}\,\delta^{2m}, \tag{3.2.12$_1$}$$
$$C_{s2} = A_1A_3A_4A_5^{-n_4}A_6^{-n_1-n_4}A_7^{-n_1-n_4},$$
$$C_{s3} = A_1^{n'_2+1}A_2^{n'_2}A_3A_4A_5^{n_2-n_4}A_6^{-n_4}A_7^{n_2-n_4}\,\delta^m,$$
$$C_{s4} = A_1^{n'_2+1}A_2^{n'_2}A_4A_5^{n_2}A_6A'_6A_7^{n_2}\delta^m,$$
$$C_{s5} = A_1^{n'_2+1}A_2^{n'_2}A_4A_5^{n_2-n_3+1}A_6^{-n_3}A_7^{n_2}\,\delta_m,$$
$$C_{s6} = A_1^{2n'_2+1}A_2^{2n'_2}A_4A_5^{2n_2-n_3}A_6^{-n_1-n_3+1}A_7^{-n_1+2n_2}\,\delta^{2m},$$

† A_{0i}^2 is called the magnetic pressure number [10] and A_{0i}^{-1} the Alfvén number [12, 16].

$$C_{s7} = A_1^{2n_2'+1}A_2^{2n_2'}A_4A_5^{2\dots}A_6^{-\dots 1+1}A_6'A_7^{-n_1+2n_2}A_{16}^{-n_{16}}\delta^{2m},$$

$$C_{s8} = A_1^{2n_2'+1}A_2^{2n_2'-1}A_3^{-1}A_4A_5^{2n_2+n_4}A_6^{-n_1+n_4+1}A_6'A_7^{2n_2-n_1+n_4}A_8A_{16}^{-n_{16}}\delta^{2m},$$

$$C_{s9} = A_1^{2n_2'}A_2^{2n_2'-1}A_3^{-1}A_4A_5^{2n_2-n_3+n_4+1}A_6^{-n_3+n_4}A_7^{2n_2+n_4}A_8\delta^{2m},$$

$$C_{s10} = A_1^{2n_2'+1}A_2^{2n_2'}A_3^{-1}A_4A_5^{2n_2+n_4}A_6^{n_4}A_7^{2n_2+n_4}A_{10}^{n_{10}}A_{11}^{-n_{11}}M_1\delta^{2m}, \quad (3.2.12)_{10}$$

$$g_{s1} = \bar{\sigma}_s(\bar{\varrho}_s)^{-1}, \quad (3.2.13)$$

$$A_3 = 1-M_1, \quad A_4 = 1-\gamma-\beta^2\bar{\zeta}_s, \quad \gamma = (K_s+\bar{P}_s)A_1^{-1}, \quad (3.2.14)$$

$$A_8 = A_1A_2-A_1A_4+1, \quad A_6' = R_{Ey} = E_{0y}(\mu_e u_0 H_{0y})^{-1}. \quad (3.2.15)$$

The boundary conditions for f_s are:

$$f_{s,\,\xi}\big|_\infty = C_{s11}; \quad f_{s,\,\eta}\big|_{\eta=0} = C_{s12}h(\xi), \quad (3.2.16)$$

$$C_{s11} = A_3^{-1}A_5^{n_4}A_6^{n_1+n_4}A_7^{n_1+n_4}M_1-(A_1A_2A_3)^{-1}A_5^{n_4}A_6^{n_1+n_4}A_7^{n_1+n_4}A_8,$$

$$C_{s12} = A_1^{n_2'}A_2^{n_2'}A_3^{-1}A_5^{n_2+n_4}A_6^{n_4}A_7^{n_2+n_4}\delta^{m+1}. \quad (3.2.17)_2$$

The momentum equation in the z-direction, (3.1.29), takes the form:

$$\bar{w}_{s,\,\xi} = C_{s13}f_{s,\xi},+C_{s14}\bar{G}_{2,\,\xi}-C_{s15}f_{s,\,\eta}+C_{s16}g_{s2z}$$
$$-C_{s17}g_{s3z}-C_{s18}g_{s4z}+C_{s19}g_{s5z}; \quad (3.2.18)$$

$$C_{s13} = A_3A_5^{n_3-n_4}A_6^{n_3-n_4+1}A_7^{-n_4}M_1^{-3},$$

$$C_{s14} = A_5^{n_3+1}A_6^{n_6}A_{17}^{-n_{17}}M_1^{-2},$$

$$C_{s15} = A_1^{-n_2'}A_2^{-n_2'}A_3A_5^{-n_2+n_3-n_4+1}A_6^{n_1+n_3-n_4}A_7^{n_1-n_2-n_4}M_1^{-3}\delta^{-m},$$

$$C_{s16} = A_{10}^{n_{10}}A_{11}^{-n_{11}}A_{12}^{-n_{12}}; \quad C_{s17} = A_6^{-n_1}A_7^{-n_1}A_{10}^{n_{10}}A_{11}^{-n_{11}},$$

$$C_{s18} = A_5^{n_3}A_6^{n_1+n_3}A_7^{n_1}A_{10}^{n_{10}}A_{13}^{-n_{13}}; \quad C_{s19} = A_6^{n_1+n_3}A_7^{n_1}A_5^{n_3}A_9A_{10}^{n_{10}},$$

$$A_9 = l\alpha(\varrho_{s0}u_0)^{-1}; \quad \alpha = e_s^2v_0^2\sigma^{-1},^\dagger \quad (3.2.19)$$

and Eq. (3.1.14) becomes:

$$C_{s20}\bar{H}_{s,\,\xi\xi}+C_{s21}\bar{H}_{s,\,\eta\eta}-C_{s22}\bar{Q}_{s,\,\xi\xi}-C_{s23}\bar{Q}_{s,\,\eta\eta}+C_{s24}\bar{G}_{1,\,\xi\xi}+C_{s25}\bar{G}_{1,\,\eta\eta}$$
$$-C_{s26}g_{s5\xi}-C_{s27}g_{s5\eta}-C_{s28}(g_{s2x})_{,\xi}+C_{s29}(g_{s1},)_\xi+C_{s30}(g_{s3x})_{,\xi}$$
$$-C_{s31}(g_{s2y})_{,\eta}+C_{s32}(g_{s3y})_{,\eta}-C_{s33}(g_{s4x})_{,\xi}-C_{s34}(g_{s4y})_{,\eta}$$
$$-C_{s35}\bar{w}_{s,\,\eta}+C_{s36}\bar{w}_{s,\,\xi}-C_{s37}\bar{\omega}_{sz}+C_{s38}g_{s6}$$
$$-C_{s39}g_{s7}+C_{s40}\bar{\omega}_{sz,\,\eta} = 0; \quad (3.2.20)$$

where we have used the following notation:

$$g_{s2i} = \bar{\sigma}_s(\bar{\varrho}_s)^{-1}\bar{Z}_{si}; \quad g_{s4i} = (\bar{\varrho}_s)^{-1}\bar{X}_{si}; \quad i = x, y, z,$$

$$g_{s3i} = \bar{\sigma}_s(\bar{\varrho}_s)^{-1}f_{s,\,i}; \quad i = \xi, \eta, \quad g_{s3z} = \bar{\sigma}_s(\bar{\varrho}_s)^{-1}\bar{w}_s,$$

† See ref. [10].

$$g_{s5\xi} = (\bar{\varrho}_s)^{-1}\sum_{r=1}^{n}(\alpha_{r*}\alpha_{s*}^{-1}A_3A_5^{-n_4}A_6^{-n_1-4}A_7^{-n_1-n_4}f_{r,\xi} - A_3A_5^{-n_4}A_6^{-n_1-n_4}A_7^{-n_1-n_4}f_{s,\xi}),$$

$$g_{s5\eta} = (\bar{\varrho}_s)^{-1}\sum_{r=1}^{n}(\alpha_{r*}\alpha_{s*}^{-1}A_1^{-n_2'}A_2^{-n_2'}A_3A_5^{-n_2-n_4}A_6^{-n_4}A_7^{-n_2-n_4}\delta^{-m}f_{r,\eta}$$

$$-A_1^{-n_2}A_2^{-n_2'}A_3A_5^{-n_2-n_4}A_6^{-n_4}A_7^{-n_2-n_4}\delta^{-m}f_{s,\eta}),$$

$$g_{s6} = \bar{\varrho}_s\bar{G}_{s4}; \quad \bar{G}_{s4} = 1+(\varrho_{es}u_s)^{-1}\sum_{r=1}^{n}\varrho_{er}u_r; \quad g_{s7} = \bar{\varrho}_s f_{s,\xi}\bar{G}_{s4}; \tag{3.2.21}$$

$$C_{s20} = A_6^{-2n_1}A_7^{-2n_1}A_{15}^{-n_{15}}; \quad C_{s21} = A_1^{-2n_2'}A_2^{-2n_2'}A_5^{-2n_2}A_7^{-2n_2}A_{15}^{-n_{15}}\delta^{-2m},$$

$$C_{s22} = A_6^{-2n_1}A_7^{-2n_1}A_{14}^{-n_{14}}; \quad C_{s23} = A_1^{-2n_2'}A_2^{-2n_2'}A_5^{-2n_2}A_7^{-2n_2}A_{14}^{-n_{14}}\delta^{-2m},$$

$$C_{s24} = A_6^{-2n_1+1}A_6'A_7^{-2n_1}A_{16}^{-n_{16}}M_1^{-2},$$

$$C_{s25} = A_1^{-2n_2'}A_2^{-2n_2'}A_5^{-2n_2}A_6A_6'A_7^{-2n_2}A_{16}^{-n_{16}}M_1^{-2}\delta^{-2m},$$

$$C_{s26} = A_6^{-n_1}A_7^{-n_1}A_9A_{10}^{n_{10}},$$

$$C_{s27} = A_1^{-n_2'}A_2^{-n_2'}A_5^{-n_2}A_7^{-n_2}A_9A_{10}^{n_{10}}\delta^{-m},$$

$$C_{s28} = A_6^{-n_1}A_7^{-n_1}A_{10}^{n_{10}}A_{11}^{-n_{11}}A_{12}^{-n_{12}},$$

$$C_{s29} = A_1^{-1}A_2^{-1}A_6^{-n_1}A_7^{-n_1}A_8A_{10}^{n_{10}}A_{11}^{-n_{11}}M_1^{-1},$$

$$C_{s30} = A_3A_5^{-n_4}A_6^{-2n_1-n_4}A_7^{-2n_1-n_4}A_{10}^{n_{10}}A_{11}^{-n_{11}}M_1^{-1},$$

$$C_{s31} = A_1^{-n_2'}A_2^{-n_2'}A_5^{-n_2}A_7^{-n_2}A_{10}^{n_{10}}A_{11}^{-n_{11}}A_{12}^{-n_{12}}\delta^{-m}.$$

$$C_{s32} = A_1^{-2n_2'}A_2^{-2n_2'}A_3A_5^{-2n_2-n_4}A_6^{-n_4}A_7^{-2n_2-n_4}A_{10}^{n_{10}}A_{11}^{-n_{11}}M_1^{-1}\delta^{-2m},$$

$$C_{s33} = A_6^{-n_1}A_7^{-n_1}A_{10}^{n_{10}}A_{13}^{-n_{13}}; \quad C_{s34} = A_1^{-n_2'}A_2^{-n_2'}A_5^{-n_2}A_7^{-n_2}A_{10}^{n_{10}}A_{13}^{-n_{13}}\delta^{-m},$$

$$C_{s35} = A_1^{-n_2'}A_2^{-n_2'}A_5^{-n_2-n_3+1}A_6^{-n_3}A_7^{-n_2}M_1^{-2}\delta^{-m},$$

$$C_{s36} = A_5^{-n_3}A_6^{-n_1-n_3+1}A_7^{-n_1}M_1^{-2}; \quad C_{s37} = A_5^{-n_4}A_6^{-n_4}A_7^{-n_4+1}M_1^{-3},$$

$$C_{s38} = A_1^{-1}A_2^{-1}A_7A_8R_{msz}M_1^{-3}; \quad C_{s39} = A_3A_5^{-n_4}A_6^{-n_1-n_4}A_7^{-n_1-n_4+1}R_{msz}M_1^{-3},$$

$$C_{s40} = A_1^{-n_2'}A_2^{-n_2'}A_5^{-n_2-n_4}A_6^{-n_4}M_1^{-1}\delta^{-m}. \tag{3.2.22}$$

(ii) *Electromagnetism*

Due to the change of coordinate systems, the components of the perturbed electric field, $(2.3.6)_3$, become:

$$E_{xp} = -G_{1,\xi}\xi_{,x} = -E_{0y}A_6^{-n_1}A_7^{-n_1}A_{16}^{-n_{16}}\bar{G}_{1,\xi}, \tag{3.2.23$_1$}$$

$$E_{yp} = -G_{1,\eta}\eta_{,y} = -E_{0y}A_1^{-n_2'}A^{-n_2'}A_5^{-n_2}A_7^{-n_2}A_{16}^{-n_{16}}\delta^{-m}\bar{G}_{1,\eta}. \tag{3.2.23$_2$}$$

Equation $(2.3.6)_1$ takes the form:

$$C_{s41}\bar{G}_{1,\xi\xi}+\bar{G}_{1,\eta\eta} = -C_{s42}\bar{\varrho}_s\bar{G}_{s5}, \tag{3.2.24}$$

with $C_{s41} = A_1^{2n_2'} A_2^{2n_2'} A_5^{2n_2} A_6^{-2n_1} A_7^{-2n_1+2n_2} \delta^{2m}$,

$$C_{s42} = A_1^{2n_2'} A_2^{2n_2'} A_5^{2n_2} A_6' A_7^{2n_2} A_{16}^{-n_{16}} R_R^{-1} R_{msy} \delta^{2m}$$

and where $\bar{G}_{s_5}$ is given by:

$$\bar{G}_{s5} = 1+(R_{msy}\bar{\varrho}_s)^{-1} \sum_{r=1}^{n} R_{mry}\bar{\varrho}_r, \tag{3.2.25}$$

$$R_R = \mu_e \varepsilon u_0^2 = (u_0 c^{-1})^2 \text{ (relativity parameter)}, \tag{3.2.26}$$

c being the velocity of light. Boundary conditions are:

$$\bar{G}_1 = \bar{G}_{1,\,\xi} = \bar{G}_{1,\,\eta} = 0, \quad \text{at} \quad \xi = \eta \to \infty. \tag{3.2.27}$$

The components of the perturbed magnetic field, $(2.3.10)_2$, $(2.3.11)_{1,\,2}$, are:

$$H_{xp} = -G_{2,\,\eta}\, \eta_{,y} = -H_{0x} A_1^{-n_2'} A_2^{-n_2'} A_5^{-n_2} A_7^{-n_2} A_{17}^{-n_{17}} \delta^{-m} \bar{G}_{2,\,\eta}, \quad (3.2.28)_1$$

$$H_{yp} = G_{2,\,\xi}\, \xi_{,x} = H_{0x} A_6^{-n_1} A_7^{-n_1} A_{17}^{-n_{17}} \bar{G}_{2,\,\xi}, \quad (3.2.28)_2$$

$$H_{zp} = H_{0z} u_0 \int_0^s \Bigg\{ \Bigg[\Big[\sum_{s=1}^{n} \bar{\varrho}_s (A_1^{-1} A_2^{-1} A_8 A_{10}^{-n_{10}} M_1^{-1} R_{msz})$$

$$- A_3 A_5^{-n_4} A_6^{-n_1-n_4} A_7^{-n_1-n_4} M_1^{-1} R_{msz} f_{s,\,\xi} \Big] A_1^{-n_2'} A_2^{-n_2'} A_5^{-n_2} A_7^{-n_2} \delta^{-m} d\eta$$

$$- \left[\sum_{s=1}^{n} A_1^{-n_2'} A_2^{-n_2'} A_3 A_5^{-n_2-n_4} A_6^{-n_2-n_4} A_7^{-n_4} \delta^{-m} M_1^{-1} \right] A_6^{-n_1} A_7^{-n_1} d\xi \Bigg\}. \quad (3.2.28)_3$$

The combined form $(2.3.12)_1$ obtained by differentiation of $(3.2.28)_{1,2}$ with respect to η and ξ respectively, takes the form:

$$C_{s43}\bar{G}_{2,\,\xi\xi} + \bar{G}_{2,\,\eta\eta} = C_{s44}\bar{\varrho}_s \bar{w}_s \bar{G}_{s6}; \tag{3.2.29}$$

$$C_{s43} = A_1^{2n_2'} A_2^{2n_2'} A_5^{2n_2} A_6^{-2n_1} A_7^{-2n_1+2n_2} \delta^{2m},$$

$$C_{s44} = A_1^{2n_2'} A_2^{2n_2'} A_5^{2n_2} A_7^{2n_2} A_{17}^{n_{17}} \delta^{2m} R_{msx},$$

$$\bar{G}_{s6} = 1+(\bar{\varrho}_s \bar{w}_s)^{-1} \sum_{r=1}^{n} R_{mrx} \bar{\varrho}_r \bar{w}_r, \tag{3.2.30}$$

with boundary conditions:

$$\bar{G}_2 = \bar{G}_{2,\xi} = \bar{G}_{2,\,\eta} = 0, \quad \text{at} \quad \xi = \eta \to \infty. \tag{3.2.31}$$

Equations (3.2.11) with B.C. (3.2.16), (3.2.18), (3.2.20), (3.2.24), and (3.2.29) constitute a fundamental system of equations.

3.3. Constant Coefficients

Under the transformation of coordinates, it is required that the coefficients of the dimensionless velocity function f_s, of $\bar{w}_s$, g_{si}'s, $\bar{G}_1$, $\bar{G}_2$, etc., remain unchanged. The task is to combine the coefficients C_{si}'s mentioned above into a certain number of composite constant coefficients such that the right-hand sides of the composite coefficients are in exponential forms raised to powers n_i's. Regardless of variable A_i's, we should have the left-hand sides constant. This implies that the powers of n_i's must vanish, thus furnishing an algebraic simultaneous system of equations in n_i's. To do this, we collect the coefficients of f_s, of $\bar{w}_s$, g_{si}'s, $\bar{G}_1$, $\bar{G}_2$, etc., of the differential system in Table 1. It is seen from Table 1 that A_i's with a power of unity should be eliminated. The procedure of elimination is in the order of A_3, A_4, A_8, A_9, A_5, A_6, A_7, and A_6'. The result is given in Table 2, which shows that, if we assume the powers of A_i's, δ to vanish, we have too many equations for the unknowns, n_i, $i = 1, 2, 2', 3, 4, 10$ to 17, and m. Therefore, we further simplify the system in Table 2 by direct and inverse multiplications of the composite constant coefficients. After a few manipulations, we get one equation only:

$$\frac{C_{s1}C_{s2}^2C_{s8}^3C_{s9}C_{s12}C_{s13}C_{s14}C_{s15}C_{s19}^2C_{s25}}{C_{s3}C_{s4}C_{s5}C_{s6}C_{s7}C_{s10}C_{s16}C_{s17}C_{s18}C_{s20}C_{s21}C_{s22}C_{s23}}\cdot$$

$$\frac{C_{s30}C_{s32}C_{s34}C_{s37}C_{s38}C_{s40}C_{s41}C_{s43}(R_{msx}R_{msy}R_{msz}R_R^{-1})}{C_{s24}C_{s26}C_{s27}C_{s28}C_{s29}^5C_{s31}C_{s33}C_{s35}^2C_{s36}C_{s39}^2C_{s42}C_{s44}}$$

$$= A_1^{3n_2'+1}A_2^{3n_2'+1}A_5^{3n_2+8n_3-n_4}A_6^{6n_1+8n_3-2n_4}A_7^{6n_1+3n_2-2n_4}A_{10}^{-6n_{10}}A_{11}^{8n_{11}}$$

$$\cdot A_{12}^{3n_{12}}A_{13}^{n_{13}}A_{14}^{2n_{14}}A_{15}^{2n_{15}}A_{16}^{-n_{16}}A_{17}^{-2n_{17}}\delta^{3m+1}. \tag{3.3.1}$$

Now, let us make the powers on the right-hand side of (3.3.1) vanish:

$$3n_2'+1 = 0; \quad n_2' = -1/3, \tag*{(3.3.2)$_1$}$$

$$3n_2+8n_3-n_4 = 0, \quad 6n_1+8n_3-2n_4 = 0, \quad 6n_1+3n_2-2n_4 = 0, \tag*{(3.3.2)$_2$}$$

$$n_{10} = n_{11} = n_{12} = n_{13} = n_{14} = n_{15} = n_{16} = n_{17} = 0, \tag*{(3.3.2)$_3$}$$

$$3m+1 = 0; \quad m = -1/3. \tag*{(3.3.2)$_4$}$$

The system of equations $(3.3.2)_2$ is indeterminate (four unknowns, three equations). Let us take an arbitrary value for n_4; the main point in such a choice is to obtain in the limiting case the solution corresponding to the classical gas-dynamic results:

$$n_4 = 0; \quad n_1 = n_2 = n_3 = 0. \tag{3.3.3}$$

Substituting values of $(3.3.2)_{1,4}$ and (3.3.3) in C_{si}'s in Table 1, we obtain appropriate constant coefficients C_{si}'s. When we use these coefficients in the left-hand side of (3.3.1), it is found that the equality of (3.3.1) holds true, i.e.

$$\frac{C_{s1}C_{s2}^2C_{s8}^3C_{s9}C_{s12}C_{s13}C_{s14}C_{s15}C_{s19}^2C_{s25}}{C_{s3}C_{s4}C_{s5}C_{s6}C_{s7}C_{s10}C_{s16}C_{s17}C_{s18}C_{s20}C_{s21}C_{s22}C_{s23}}\cdot$$

$$\cdot\frac{C_{s30}C_{s32}C_{s34}C_{s37}C_{s38}C_{s40}C_{s41}C_{s43}}{C_{s24}C_{s26}C_{s27}C_{s28}C_{s29}^5C_{s31}C_{s33}C_{s35}^2C_{s36}C_{s39}^2C_{s42}C_{s44}} =$$

$$= R_R(R_{msx}R_{msy}R_{msz})^{-1}. \tag{3.3.4}$$

TABLE 1. COEFFICIENTS OF DIFFERENTIAL SYSTEM

C_{s1}	$= A_1^{2n'_2+1}$	$A_2^{2n'_2+1}$	$A_3^1A_4^0A_5^{2n_2-n_4}$	$A_6^{-3n_1-n_4}$	$A_6'^0A_7^{-3n_1+2n_2-n_4}$	$A_8^0A_9^0A_{10}^0$	A_{11}^0	A_{12}^0	A_{13}^0	A_{14}^0	A_{15}^0	A_{16}^0	A_{17}^0	δ^{2m}	M_1^0
C_{s2}	$= A_1^1$	A_2^0	$A_3^1A_4^1A_5^{-n_4}$	$A_6^{-n_1-n_4}$	$A_6'^0A_7^{-n_1-n_4}$	$A_8^0A_9^0A_{10}^0$	A_{11}^0	A_{12}^0	A_{13}^0	A_{14}^0	A_{15}^0	A_{16}^0	A_{17}^0	δ^0	M_1^0
C_{s3}	$= A_1^{n'_2+1}$	$A_2^{n'_2}$	$A_3^1A_4^1A_5^{n_2-n_4}$	$A_6^{-n_4}$	$A_6'^0A_7^{n_2-n_4}$	$A_8^0A_9^0A_{10}^0$	A_{11}^0	A_{12}^0	A_{13}^0	A_{14}^0	A_{15}^0	A_{16}^0	A_{17}^0	δ^m	M_1^0
C_{s4}	$= A_1^{n'_2+1}$	$A_2^{n'_2}$	$A_3^0A_4^1A_5^{n_2}$	A_6^1	$A_6'^1A_7^{n_2}$	$A_8^0A_9^0A_{10}^0$	A_{11}^0	A_{12}^0	A_{13}^0	A_{14}^0	A_{15}^0	A_{16}^0	A_{17}^0	δ^m	M_1^0
C_{s5}	$= A_1^{n'_2+1}$	$A_2^{n'_2}$	$A_3^0A_4^1A_5^{n_2-n_3+1}$	$A_6^{-n_3}$	$A_6'^0A_7^{n_2}$	$A_8^0A_9^0A_{10}^0$	A_{11}^0	A_{12}^0	A_{13}^0	A_{14}^0	A_{15}^0	A_{16}^0	A_{17}^0	δ^m	M_1^0
C_{s6}	$= A_1^{2n'_2+1}$	$A_2^{2n'_2}$	$A_3^0A_4^1A_5^{2n_2-n_3}$	$A_6^{-n_1-n_3+1}$	$A_6'^0A_7^{-n_1+2n_2}$	$A_8^0A_9^0A_{10}^0$	A_{11}^0	A_{12}^0	A_{13}^0	A_{14}^0	A_{15}^0	A_{16}^0	A_{17}^0	δ^{2m}	M_1^0
C_{s7}	$= A_1^{2n'_2+1}$	$A_2^{2n'_2}$	$A_3^0A_4^1A_5^{2n_2}$	$A_6^{-n_1+1}$	$A_6'^1A_7^{-n_1+2n_2}$	$A_8^0A_9^0A_{10}^0$	A_{11}^0	A_{12}^0	A_{13}^0	A_{14}^0	A_{15}^0	$A_{16}^{-n_{16}}$	A_{17}^0	δ^{2m}	M_1^0
C_{s8}	$= A_1^{2n'_2}$	$A_2^{2n'_2-1}$	$A_3^{-1}A_2^1A_5^{2n_2+n_4}$	$A_6^{-n_1+n_4+1}$	$A_6'^1A_7^{-n_1+2n_2+n_4}$	$A_8^1A_9^0A_{10}^0$	A_{11}^0	A_{12}^0	A_{13}^0	A_{14}^0	A_{15}^0	$A_{16}^{-n_{16}}$	A_{17}^0	δ^{2m}	M_1^0
C_{s9}	$= A_1^{2n'_2}$	$A_2^{2n'_2-1}$	$A_3^{-1}A_4^1A_5^{2n_2-n_3+n_4+1}$	$A_6^{-n_3+n_4}$	$A_6'^0A_7^{2n_2+n_4}$	$A_8^1A_9^0A_{10}^0$	A_{11}^0	A_{12}^0	A_{13}^0	A_{14}^0	A_{15}^0	A_{16}^0	A_{17}^0	δ^{2m}	M_1^0
C_{s10}	$= A_1^{2n'_2+1}$	$A_2^{2n'_2}$	$A_3^{-1}A_4^1A_5^{2n_2+n_4}$	$A_6^{n_4}$	$A_6'^0A_7^{2n_2+n_4}$	$A_8^0A_9^0A_{10}^{n_{10}}$	$A_{11}^{-n_{11}}$	A_{12}^0	A_{13}^0	A_{14}^0	A_{15}^0	A_{16}^0	A_{17}^0	δ^{2m}	M_1^1
C_{s12}	$= A_1^{n'_2}$	$A_2^{n'_2}$	$A_3^{-1}A_4^0A_5^{n_2+n_4}$	$A_6^{n_4}$	$A_6'^0A_7^{n_2+n_4}$	$A_8^0A_9^0A_{10}^0$	A_{11}^0	A_{12}^0	A_{13}^0	A_{14}^0	A_{15}^0	A_{16}^0	A_{17}^0	δ^{m+1}	M_1^0
C_{s13}	$= A_1^0$	A_2^0	$A_3^1A_4^0A_5^{n_3-n_4}$	$A_6^{n_3-n_4+1}$	$A_6'^0A_7^{-n_4}$	$A_8^0A_9^0\ A_{10}^0$	A_{11}^0	A_{12}^0	A_{13}^0	A_{14}^0	A_{15}^0	A_{16}^0	A_{17}^0	δ^0	M_1^{-3}
C_{s14}	$= A_1^0$	A_2^0	$A_3^0A_4^0A_5^{n_3+1}$	$A_6^{n_3}$	$A_6'^0A_7^0$	$A_8^0A_9^0A_{10}^0$	A_{11}^0	A_{12}^0	A_{13}^0	A_{14}^0	A_{15}^0	A_{16}^0	$A_{17}^{-n_{17}}$	δ^0	M_1^{-2}
C_{s15}	$= A_1^{-n'_2}$	$A_2^{-n'_2}$	$A_3^1A_4^0A_5^{-n_2+n_3-n_4+1}$	$A_6^{n_1+n_3-n_4}$	$A_6'^0A_7^{n_1-n_2-n_4}$	$A_8^0A_9^0A_{10}^0$	A_{11}^0	A_{12}^0	A_{13}^0	A_{14}^0	A_{15}^0	A_{16}^0	A_{17}^0	δ^{-m}	M_1^{-3}
C_{s16}	$= A_1^0$	A_2^0	$A_3^0A_4^0A_5^0$	A_6^0	$A_6'^0A_7^0$	$A_8^0A_9^0A_{10}^{n_{10}}$	$A_{11}^{-n_{11}}$	$A_{12}^{-n_{12}}$	A_{13}^0	A_{14}^0	A_{15}^0	A_{16}^0	A_{17}^0	δ^0	M_1^0
C_{s17}	$= A_1^0$	A_2^0	$A_3^0A_4^0A_5^0$	$A_6^{n_1}$	$A_6'^0A_7^{n_1}$	$A_8^0A_9^0A_{10}^{n_{10}}$	$A_{11}^{-n_{11}}$	A_{12}^0	A_{13}^0	A_{14}^0	A_{15}^0	A_{16}^0	A_{17}^0	δ^0	M_1^0
C_{s18}	$= A_1^0$	A_2^0	$A_3^0A_4^0A_5^{n_3}$	$A_6^{n_1+n_3}$	$A_6'^0A_7^{n_1}$	$A_8^0A_9^0A_{10}^{10}$	A_{11}^0	A_{12}^0	$A_{13}^{-n_{13}}$	A_{14}^0	A_{15}^0	A_{16}^0	A_{17}^0	δ^0	M_1^0
C_{s19}	$= A_1^0$	A_2^0	$A_3^0A_4^0A_5^{n_3}$	$A_6^{n_1+n_3}$	$A_6'^0A_7^{n_1}$	$A_8^0A_9^0A_{10}^{n_{10}}$	A_{11}^0	A_{12}^0	A_{13}^0	A_{14}^0	A_{15}^0	A_{16}^0	A_{17}^0	δ^0	M_1^0
C_{s20}	$= A_1^0$	A_2^0	$A_3^0A_4^0A_5^0$	$A_6^{-2n_1}$	$A_6'^0A_7^{-2n_1}$	$A_8^0A_9^0A_{10}^0$	A_{11}^0	A_{12}^0	A_{13}^0	A_{14}^0	$A_{15}^{-n_{15}}$	A_{16}^0	A_{17}^0	δ^0	M_1^0
C_{s21}	$= A_1^{-2n'_2}$	$A_2^{-2n'_2}$	$A_3^0A_4^0A_5^{-2n_2}$	A_6^0	$A_6'^0A_7^{-2n_2}$	$A_8^0A_9^0A_{10}^0$	A_{11}^0	A_{12}^0	A_{13}^0	A_{14}^0	$A_{15}^{-n_{15}}$	A_{16}^0	A_{17}^0	δ^{-2m}	M_1^0
C_{s22}	$= A_1^0$	A_2^0	$A_3^0A_4^0A_5^0$	$A_6^{-2n_1}$	$A_6'^0A_7^{-2n_1}$	$A_8^0A_9^0A_{10}^0$	A_{11}^0	A_{12}^0	A_{13}^0	$A_{14}^{-n_{14}}$	A_{15}^0	A_{16}^0	A_{17}^0	δ^0	M_1^0
C_{s23}	$= A_1^{-2n'_2}$	$A_2^{-2n'_2}$	$A_3^0A_4^0A_5^{-2n_2}$	A_6^0	$A_6'^0A_7^{-2n_2}$	$A_8^0A_9^0A_{10}^0$	A_{11}^0	A_{12}^0	A_{13}^0	$A_{14}^{-n_{14}}$	A_{15}^0	A_{16}^0	A_{17}^0	δ^{-2m}	M_1^0

TABLE 1 (*cont.*)

$C_{s24} =$	A_1^0	A_2^0	$A_3^0A_4^0A_5^0$	$A_6^{-2n_1+1}$	$A_6'^1A_7^{-2n_1}$	$A_8^0A_9^0A_{10}^0$	A_{11}^0	A_{12}^0	A_{13}^0	A_{14}^0	A_{15}^0	$A_{16}^{-n_{16}}$	A_{17}^0	δ^0	M_1^{-2}
$C_{s25} =$	$A_1^{-2n'_2}$	$A_2^{-2n'_2}$	$A_3^0A_4^0A_5^{-2n_2}$	A_6^1	$A_6'^1A_7^{-2n_2}$	$A_8^0A_9^0A_{10}^0$	A_{11}^0	A_{12}^0	A_{13}^0	A_{14}^0	A_{15}^0	$A_{16}^{-n_{16}}$	A_{17}^0	δ^{-2m}	M_1^{-2}
$C_{s26} =$	A_1^0	A_2^0	$A_3^0A_4^0A_5^0$	$A_6^{-n_1}$	$A_6'^0A_7^{-n_1}$	$A_8^0A_9^1A_{10}^{n_{10}}$	A_{11}^0	A_{12}^0	A_{13}^0	A_{14}^0	A_{15}^0	A_{16}^0	A_{17}^0	δ^0	M_1^0
$C_{s27} =$	$A_1^{-n'_2}$	$A_2^{-n'_2}$	$A_3^0A_4^0A_5^{-n_2}$	A_6^0	$A_6'^0A_7^{-n_2}$	$A_8^0A_9^1A_{10}^{n_{10}}$	A_{11}^0	A_{12}^0	A_{13}^0	A_{14}^0	A_{15}^0	A_{16}^0	A_{17}^0	δ^{-m}	M_1^{-1}
$C_{s28} =$	A_1^0	A_2^0	$A_3^0A_4^0A_5^0$	$A_6^{-n_1}$	$A_6'^0A_7^{-n_1}$	$A_8^0A_9^0A_{10}^{n_{10}}$	$A_{11}^{-n_{11}}$	$A_{12}^{-n_{12}}$	A_{13}^0	A_{14}^0	A_{15}^0	A_{16}^0	A_{17}^0	δ^0	M_1^0
$C_{s29} =$	A_1^{-1}	A_2^{-1}	$A_3^0A_4^0A_5^0$	$A_6^{-n_1}$	$A_6'^0A_7^{-n_1}$	$A_8^1A_9^0A_{10}^{n_{10}}$	$A_{11}^{-n_{11}}$	A_{12}^0	A_{13}^0	A_{14}^0	A_{15}^0	A_{16}^0	A_{17}^0	δ^0	M_1^{-1}
$C_{s30} =$	A_1^0	A_2^0	$A_3^1A_4^0A_5^{-n_4}$	$A_6^{-2n_1-n_4}$	$A_6'^0A_7^{-2n_1-n_4}$	$A_8^0A_9^0A_{10}^{n_{10}}$	$A_{11}^{-n_{11}}$	A_{12}^0	A_{13}	A_{14}^0	A_{15}^0	A_{16}^0	A_{17}^0	δ^0	M_1^{-1}
$C_{s31} =$	$A_1^{-n'_2}$	$A_2^{-n'_2}$	$A_3^0A_4^0A_5^{-n_2}$	A_6^0	$A_6'^0A_7^{-n_2}$	$A_8^0A_9^0A_{10}^{n_{10}}$	$A_{11}^{-n_{11}}$	$A_{12}^{-n_{12}}$	A_{13}^0	A_{14}^0	A_{15}^0	A_{16}^0	A_{17}^0	δ^{-m}	M_1^0
$C_{s32} =$	$A_1^{-2n'_2}$	$A_2^{-2n'_2}$	$A_3^1A_4^0A_5^{-2n_2-n_4}$	$A_6^{-n_4}$	$A_6'^0A_7^{-2n_2-n_4}$	$A_8^0A_9^0A_{10}^{n_{10}}$	$A_{11}^{-n_{11}}$	A_{12}^0	A_{13}^0	A_{14}^0	A_{15}^0	A_{16}^0	A_{17}^0	δ^{-2m}	M_1^{-1}
$C_{s33} =$	A_1^0	A_2^0	$A_3^0A_4^0A_5^0$	$A_6^{-n_1}$	$A_6'^0A_7^{-n_1}$	$A_8^0A_9^0A_{10}^{n_{10}}$	A_{11}^0	A_{12}^0	$A_{13}^{-n_{13}}$	A_{14}^0	A_{15}^0	A_{16}^0	A_{17}^0	δ^0	M_1^0
$C_{s34} =$	$A_1^{-n'_2}$	$A_2^{-n'_2}$	$A_3^0A_4^0A_5^{-n_2}$	A_6^0	$A_6'^0A_7^{-n_2}$	$A_8^0A_9^0A_{10}^{n_{10}}$	A_{11}^0	A_{12}^0	$A_{13}^{-n_{13}}$	A_{14}^0	A_{15}^0	A_{16}^0	A_{17}^0	δ^{-m}	M_1^0
$C_{s35} =$	$A_1^{-n'_2}$	$A_2^{-n'_2}$	$A_3^0A_4^0A_5^{-n_2-n_3+1}$	$A_6^{-n_3}$	$A_6'^0A_7^{-n_2}$	$A_8^0A_9^0A_{10}^0$	A_{11}^0	A_{12}^0	A_{13}^0	A_{14}^0	A_{15}^0	A_{16}^0	A_{17}^0	δ^{-m}	M_1^{-2}
$C_{s36} =$	A_1^0	A_2^0	$A_3^0A_4^0A_5^{-n_3}$	$A_6^{-n_1-n_3+1}$	$A_6'^0A_7^{-n_1}$	$A_8^0A_9^0A_{10}^0$	A_{11}^0	A_{12}^0	A_{13}^0	A_{14}^0	A_{15}^0	A_{16}^0	A_{17}^0	δ^0	M_1^{-2}
$C_{s37} =$	A_1^0	A_2^0	$A_3^0A_4^0A_5^{-n_4}$	$A_6^{-n_4}$	$A_6'^0A_7^{-n_4+1}$	$A_8^0A_9^0A_{10}^0$	A_{11}^0	A_{12}^0	A_{13}^0	A_{14}^0	A_{15}^0	A_{16}^0	A_{17}^0	δ^0	M_1^{-3}
$C_{s38} =$	A_1^{-1}	A_2^{-1}	$A_3^0A_4^0A_5^0$	A_6^0	$A_6^{0'}A_7^1$	$A_8^1A_9^0A_{10}^0$	A_{11}^0	A_{12}^0	A_{13}^0	A_{14}^0	A_{15}^0	A_{16}^0	A_{17}^0	δ^0	$M_1^{-3}R_{msz}$
$C_{s39} =$	A_1^0	A_2^0	$A_3^1A_4^0A_5^{-n_4}$	$A_6^{-n_1-n_4}$	$A_6'^0A_7^{-n_1-n_4+1}$	$A_8^0A_9^0A_{10}^0$	A_{11}^0	A_{12}^0	A_{13}^0	A_{14}^0	A_{15}^0	A_{16}^0	A_{17}^0	δ^0	$M_1^{-3}R_{msz}$
$C_{s40} =$	$A_1^{-n'_2}$	$A_2^{-n'_2}$	$A_3^0A_4^0A_5^{-n_2-n_4}$	$A_6^{-n_4}$	$A_6'^0A_7^{-n_2-n_4}$	$A_8^0A_9^0A_{10}^0$	A_{11}^0	A_{12}^0	A_{13}^0	A_{14}^0	A_{15}^0	A_{16}^0	A_{17}^0	δ^{-m}	M_1^{-1}
$C_{s41} =$	$A_1^{2n'_2}$	$A_2^{2n'_2}$	$A_3^0A_4^0A_5^{2n_2}$	$A_6^{-2n_1}$	$A_6'^0A_7^{-2n_1+2n_2}$	$A_8^0A_9^0A_{10}^0$	A_{11}^0	A_{12}^0	A_{13}^0	A_{14}^0	A_{15}^0	A_{16}^0	A_{17}^0	δ^{2m}	M_1^0
$C_{s42} =$	$A_1^{2n'_2}$	$A_2^{2n'_2}$	$A_3^0A_4^0A_5^{2n_2}$	A_6^0	$A_6'^1A_7^{2n_2}$	$A_8^0A_9^0A_{10}^0$	A_{11}^0	A_{12}^0	A_{13}^0	A_{14}^0	A_{15}^0	$A_{16}^{+n_{16}}$	A_{17}^0	δ^{2m}	$M_1^0R_R^{-1}R_{msy}$
$C_{s43} =$	$A_1^{2n'_2}$	$A_2^{2n'_2}$	$A_3^0A_4^0A_5^{2n_2}$	$A_6^{-2n_1}$	$A_6'^0A_7^{-2n_1+2n_2}$	$A_8^0A_9^0A_{10}$	A_{11}^0	A_{12}^0	A_{13}^0	A_{14}^0	A_{15}^0	A_{16}^0	A_{17}^0	δ^{2m}	M_1^0
$C_{s44} =$	$A_1^{2n'_2}$	$A_2^{2n'_2}$	$A_3^0A_4^0A_5^{2n_2}$	A_6^0	$A_6'^0A_7^{2n_2}$	$A_8^0A_9^0A_{10}^0$	A_{11}^0	A_{12}^0	A_{13}^0	A_{14}^0	A_{15}^0	A_{16}^0	$A_{17}^{+n_{17}}$	δ^{2m}	$M_1^0R_{msx}$

Table 2. Composite Coefficients

$C_{s1}C_{s6}^{-1}C_{s8}C_{s9}^{2}C_{s23}^{-1}C_{s6}^{-1}C_{s27}^{-1}C_{s28}^{-1}C_{s29}^{-1}C_{s42}^{-1}$	$= A_1^{3n'_2+1}A_2^{3n'_2+1}A_5^{6n_2+3n_3}$	$A_6^{2n_1+3n_3}$	$A_7^{2n_1+3n_3}$	$A_{10}^{-n_{10}}$	$A_{11}^{2n_{11}}$	$A_{12}^{n_{12}}$	A_{13}^{0}	$A_{14}^{n_{14}}$	A_{15}^{0}	A_{16}^{0}	A_{17}^{0}	$\delta^{3m}M_1^2R_RR_{msy}^{-1}$
$C_{s1}^{2}C_{s2}^{-1}C_{s6}C_{s13}^{-1}$	$= A_1^{6n'_2+2}A_2^{6n'_2+2}A_5^{6n'_2-2n_3}$	$A_6^{-6n_1-2n_3}$	$A_7^{-6n_1+6n_2}$	A_{10}^{0}	A_{11}^{0}	A_{12}^{0}	A_{13}^{0}	A_{14}^{0}	A_{15}^{0}	A_{16}^{0}	A_{17}^{0}	$\delta^{6m}M_1^3$
$C_{s1}^{2}C_{s2}^{-1}C_{s6}C_{s29}^{-1}C_{s36}^{-1}C_{s38}C_{s39}^{-1}$	$= A_1^{6n'_2+2}A_2^{6n'_2+2}A_5^{6n_2}$	$A_6^{-3n_1}$	$A_7^{-3_1+6n_2}$	$A_{10}^{-n_{10}}$	$A_{11}^{n_{11}}$	A_{12}^{0}	A_{13}^{0}	A_{14}^{0}	A_{15}^{0}	A_{16}^{0}	A_{17}^{0}	$\delta^{6m}M_1^3$
$C_{s1}C_{s21}^{-1}C_{s22}^{-1}C_{s37}C_{s39}^{-1}$	$= A_1^{4n'_2+1}A_2^{4n'_2+1}A_5^{4n_2-n_4}$	$A_6^{-n_4}$	$A_7^{4n_2-n_4}$	A_{10}^{0}	A_{11}^{0}	A_{12}^{0}	A_{13}^{0}	$A_{14}^{n_{14}}$	$A_{15}^{n_{15}}$	A_{16}^{0}	A_{17}^{0}	$\delta^{4m}M_1^0R_{msz}^{-1}$
$C_{s1}^{2}C_{s2}^{-1}C_{s10}$	$= A_1^{6n'_2+2}A_2^{6n'_2+2}A_5^{6n_2}$	$A_6^{-5n_1}$	$A_7^{-5n_1+6n_2}$	$A_{10}^{n_{10}}$	$A_{11}^{-n_{11}}$	A_{12}^{0}	A_{13}^{0}	A_{14}^{0}	A_{15}^{0}	A_{16}^{0}	A_{17}^{0}	$\delta^{6m}M_1^1$
$C_{s1}C_{s12}$	$= A_1^{3n'_2+1}A_2^{3\ '_2+1}A_5^{3n_2}$	$A_6^{-3n_1}$	$A_7^{-3n_1+3n_2}$	A_{10}^{0}	A_{11}^{0}	A_{12}^{0}	A_{13}^{0}	A_{14}^{0}	A_{15}^{0}	A_{16}^{0}	A_{17}^{0}	$\delta^{3m+1}M_1^0$
$C_{s1}C_{s2}^{-1}C_{s3}C_{s30}^{-1}$	$= A_1^{3n'_2+1}A_2^{3n'_2+1}A_5^{3n_2}$	A_6^{0}	$A_7^{3n_2}$	$A_{10}^{-n_{10}}$	$A_{11}^{n_{11}}$	A_{1}^{0}	A_{13}^{0}	A_{14}^{0}	A_{15}^{0}	A_{16}^{0}	A_{17}^{0}	$\delta^{3m}M_1^1$
$C_{s1}C_{s16}C_{s17}C_{s18}C_{s20}C_{s32}^{-1}$	$= A_1^{4n'_2+1}A_2^{4n'_2+1}A_5^{4n_2+n_3}$	$A_6^{-3n_1+n_3}$	$A_7^{-3n_1+4n_2}$	$A_{10}^{2n_{10}}$	$A_{11}^{-n_{11}}$	$A_{12}^{-n_1}$	$A_{13}^{-n_{13}}$	A_{14}^{0}	$A_{15}^{-n_{15}}$	A_{16}^{0}	A_{17}^{0}	$\delta^{4m}M_1^1$
$C_{s1}C_{s7}^{-1}C_{s8}C_{s29}^{-1}C_{s31}^{-1}C_{s33}^{-1}$	$= A_1^{3n'_2+1}A_2^{3n'_2+1}A_5^{3n_2}$	$A_6^{-n_1}$	$A_7^{-n_1+3n_2}$	$A_{10}^{-3n_{10}}$	$A_{11}^{2n_{11}}$	$A_{12}^{-n_{12}}$	$A_{13}^{n_{13}}$	A_{14}^{0}	A_{15}^{0}	A_{16}^{0}	A_{17}^{0}	$\delta^{3m}M_1^1$
$C_{s1}C_{s4}^{-1}C_{s8}C_{s29}^{-1}$	$= A_1^{3n'_2+1}A_2^{3n'_2+1}A_5^{3n_2}$	$A_6^{-3n_1}$	$A_7^{-3n_1+3n_2}$	$A_{10}^{-n_{10}}$	$A_{11}^{n_{11}}$	A_{12}^{0}	A_{13}^{0}	A_{14}^{0}	A_{15}^{0}	$A_{16}^{-n_{16}}$	A_{17}^{0}	$\delta^{3m}M_1^1$
$C_{s1}^{2}C_{s2}^{-1}C_{s8}C_{s25}^{-1}C_{s29}^{-1}$	$= A_1^{8n'_2+1}A_2^{8n'_2+1}A_5^{8n_2}$	$A_6^{-5n_1}$	$A_7^{-7n_1+6n_2-n_4}$	$A_{10}^{-n_{10}}$	$A_{11}^{n_{11}}$	A_{12}^{0}	A_{13}^{0}	A_{14}^{0}	A_{15}^{0}	A_{16}^{0}	A_{17}^{0}	$\delta^{8m}M_1^3$
$C_{s1}^{2}C_{s2}^{-1}C_{s8}C_{s24}^{-1}C_{s29}^{-1}C_{s34}C_{s40}C_{s41}C_{s43}C_{s44}^{-1}$	$= A_1^{6n'_2+1}A_2^{6n'_2+1}A_5^{6n_2}$	$A_6^{-7n_1-n_4}$	$A_7^{-5n_1+8n_2}$	A_{10}^{0}	$A_{11}^{n_{11}}$	A_{12}^{0}	$A_{13}^{-n_{13}}$	A_{14}^{0}	A_{15}^{0}	A_{16}^{0}	$A_{17}^{-n_{17}}$	$\delta^{6m}M_1^2R_{msz}^{-1}$
$C_{s1}C_{s5}^{-1}C_{s9}C_{s29}^{-1}$	$= A_1^{3n'_2+1}A_2^{3n'_2+1}A_5^{3n_2}$	$A_6^{-2n_1}$	$A_7^{-2n_1+3n_2}$	$A_{10}^{-n_{10}}$	$A_{11}^{n_{11}}$	A_{12}^{0}	A_{13}^{0}	A_{14}^{0}	A_{15}^{0}	A_{16}^{0}	A_{17}^{0}	$\delta^{3m}M_1^1$
$C_{s1}^{3}C_{s2}^{-1}C_{s9}C_{s15}^{-1}C_{s29}^{-1}$	$= A_1^{9n'_2+3}A_2^{9n'_2+3}A_5^{9n_2-2n_3}$	$A_6^{-8n_1-2n_3}$	$A_7^{-8n_1+9n_2}$	$A_{10}^{-n_{10}}$	$A_{11}^{n_{11}}$	A_{12}^{0}	A_{13}^{0}	A_{14}^{0}	A_{15}^{0}	A_{16}^{0}	A_{17}^{0}	$\delta^{9m}M_1^4$
$C_{s1}^{2}C_{s2}^{-1}C_{s9}C_{s14}^{-1}C_{s29}^{-1}$	$= A_1^{6n'_2+2}A_2^{6n'_2+2}A_5^{6n_2-2n_3}$	$A_6^{-4n_1-2n_3}$	$A_7^{-4n_1+6n_2}$	$A_{10}^{-n_{10}}$	$A_{11}^{n_{11}}$	A_{12}^{0}	A_{13}^{0}	A_{14}^{0}	A_{15}^{0}	A_{16}^{0}	$A_{17}^{n_{17}}$	$\delta^{6m}M_1^3$
$C_{s1}^{2}C_{s2}^{-1}C_{s9}C_{s29}^{-1}C_{s35}^{-1}$	$= A_1^{7n'_2+2}A_2^{7n'_2+2}A_5^{7n_2}$	$A_6^{-4n_1}$	$A_7^{-4n_1+7n_2}$	$A_{10}^{-n_{10}}$	$A_{11}^{n_{11}}$	A_{12}^{0}	A_{13}^{0}	A_{14}^{0}	A_{15}^{0}	A_{16}^{0}	A_{17}^{0}	$\delta^{7m}M_1^3$

Obviously, we have proposed one of the many possible combinations of the magnitudes of the composite constant coefficients.

Note that if there are no electromagnetic fields and the flow is a perfect gas flow without vorticity, and if the motion reduces to a two-dimensional one, i.e. $A_1 = K_s+1$, $A_1A_4 = 1$, $\omega_{sz} = 0$, $w_s = 0$, $\overline{W}_s = \overline{A}_s = \overline{P}_s = 0$ g_{si}'s $= 0$, $A_5 = A_6 = A_7 = 0$, then the system of Eqs. (3.2.11), (3.2.18), (3.2.20), (3.2.24) and (3.2.29) reduces to v. Kármán's differential equation (1.1.3) for transonic flow.

The coefficients C_{si}'s are constant in the system of differential equations, and invariant under the coordinate transformations. The dimensionless velocity function f_s is a function of coordinates and coefficients C_{si}. If a series of bodies with the same dimensionless thickness distribution $h(\xi)$, is placed in flow patterns of various magnetic and electric fields, such that the above C_{si}'s remain constant and satisfy the condition (3.3.4), then the flow patterns are similar in the sense that the same function f_s describes the resultant different flow patterns.

3.4. Pressure Coefficient

The pressure coefficient is defined by:

$$C_{p_s} = -(p_s-p_{s0})(\tfrac{1}{2}\varrho_{s0}u_0^2)^{-1} = -(p_sp_{s0}^{-1}-1)(2K_s^{-1}M_0^{-2}). \quad (3.4.1)$$

To calculate the difference (p_s-p_{s0}), we perform the following calculation. The energy equation (2.4.9) is:

$$\tfrac{1}{2}(K_s-1)q_s^2+\alpha_s^2(1+P_s)-(K_s-1)\alpha_s^2\zeta_s = \tfrac{1}{2}(K_s-1)\alpha_{s*}^2+\alpha_{s*}^2, \quad (3.4.1a)$$

where ζ_s, given by (2.4.8), takes the form with $(2.3.6)_3$ and (2.4.7):

$$\zeta_s = \alpha_s^{-2}\left\{\int_0^s dQ_s+\int_0^s \varrho_s^{-1}[\sigma_s(\mathbf{Z}_s-\mathbf{q}_s)+\mathbf{X}_s+\sum_{r=1}^n \alpha(\mathbf{q}_r-\mathbf{q}_s)]\cdot d\mathbf{r}_s \right.$$
$$\left. +\gamma_s\int_0^s(\mathbf{E}_0+\mathbf{E}_p)\cdot d\mathbf{r}_s\right\}. \quad (3.4.2)$$

It is assumed that Q_s, σ_s, $\mathbf{Z}_s$, $\mathbf{X}_s$, and $\mathbf{E}_p$ are given functions of the coordinates and α_s is approximated as α_{s*}: $\alpha_s \cong \alpha_{s*}$. Then we can represent ζ_s approximately by the following method.

Let f be a function of s and let:

$$\int_0^s f(s_1)ds_1 = F(s); \quad F(0) = 0, \quad F'(s) = f(s). \quad (3.4.3)$$

We would like to find a constant parameter N which minimizes the difference $(N-F(s))$ on the entire interval $[0, a]$. This is equivalent to minimization of the integral,

$$\int_0^a (N-F(s))^2 \, ds = \text{min}, \tag{3.4.4}$$

where the upper limit a is a certain point in the perturbation region, and is representable by a characteristic length l, i.e. $a = ml$, m = integer. Let us divide the interval $[0, a]$ into n small equal sub-intervals Δs_i. Integral (3.4.4) can be written in the form:

$$\lim_{n\to\infty} \sum_{i=1}^{n} (N-F(s_i'))^2 \, \Delta s_i = \text{min}, \tag{3.4.5}$$

where

$$\Delta s_i = s_i - s_{i-1}, \quad 0 = s_0 < s_1 < s_2 \ldots < s_n = a,$$

s_i' is the mid-point of the ith interval with length an^{-1}. Let us minimize (3.4.5) with respect to N, i.e. for a finite n:

$$\sum_{i=1}^{n} \delta\{[N-F(s_i')]^2 \, an^{-1}\} = 0,$$

$$\sum_{i=1}^{n} \{+2[N-F(s_i')](an^{-1})\}\delta N = 0, \quad \sum_{i=1}^{n} [N-F(s_i')] = 0,$$

or

$$N = n^{-1}\sum_{i=1}^{n} F(s_i') = \text{arithmetic mean of } F(s_i')\text{'s}. \tag{3.4.6}$$

Applying (3.4.6) to integrals within the energy equation (3.4.2), we obtain a constant average value for ζ_s, say $\bar{\zeta}_s$. The same technique may be applied to find the average value for W_s, A_s, (2.1.16). Then we have a constant value for P_s, (2.4.5), denoted by $\bar{P}_s$. Our next problem is to simplify the expression for the speed of sound, α_s, Eq. (2.1.15). After the averaging process, it becomes:

$$\alpha_s^2 = p_s K_s \varrho_s^{-1}(1+\bar{N}_s), \quad M_s = 0, \quad \bar{W}_s, \bar{A}_s = \text{const}, \tag{3.4.7}$$

and Eq. (2.4.9) takes the form:

$$\tfrac{1}{2}(K_s-1)q_s^2 + p_s K_s \varrho_s^{-1}(1+\bar{N}_s)(1+\bar{P}_s)$$
$$-(K_s-1)p_s K_s \varrho_s^{-1}(1+\bar{N}_s)\bar{\zeta}_s = \tfrac{1}{2}(K_s-1)\alpha_{s*}^2 + \alpha_{s*}^2. \tag{3.4.8}$$

Rearranging terms, adding $-p_{s0}$ to both sides of (3.4.8), and neglecting perturbed velocity components of the kind $(u_s-\alpha_{s*})^2$, v_s^2, w_s^2 we obtain:

$$p_s - p_{s0} \cong -(K_s-1)\alpha_{s*}u_0[(A_1A_2A_1A_4)^{-1} - (A_1A_2)^{-1} + A_3(A_1A_4)^{-1}f_{s,\xi}]\cdot$$
$$\cdot K_s^{-1}\varrho_s(1+\bar{N}_s) + \alpha_{s*}^2 K_s^{-1}\varrho_s(1+\bar{N}_s)^{-1}(A_1A_4)^{-1} - p_{s0}. \tag{3.4.9}$$

Substitution of (3.4.9) in (3.4.1) gives:

$$C_{ps} = 2(K_s-1)K_s^{-1}(1+\bar{N}_s)\varrho_s\varrho_{s0}^{-1}[(A_1A_2A_1A_4)^{-1}-(A_1A_2)^{-1} +A_3(A_1A_4)^{-1}f_{s,\xi}]-2M_1^2K_s(1+\bar{N}_s)(A_1A_4)^{-1}\varrho_s\varrho_{s0}^{-1}+2M_0^2K_s^{-1}. \quad (3.4.10)$$

The free stream velocity is close to Mach number unity, and small perturbation refers to the disturbed region in the vicinity of the body. Hence, we approximate the component fluid density in the perturbation region ϱ_s by the one in the free stream ϱ_{s0}. Then (3.4.10) becomes:

$$C_{ps} = 2(K_s-1)K_s^{-1}(1+\bar{N}_s)[(A_1A_2A_1A_4)^{-1}-(A_1A_2)^{-1}+A_3(A_1A_4)^{-1}f_{s,\xi}] -2M_1^2K_s^{-1}(1+\bar{N}_s)(A_1A_4)^{-1}+2M_0^2K_s^{-1}, \quad (3.4.11)$$

where Mach numbers M_1 and M_0 have a relation given by (2.4.12). The pressure coefficient may be written in terms of C_{s1}, C_{s4}, C_{s5}, C_{s6}, A_1, A_2, A_4, and δ:

$$C_{ps} = 2K_{s1}[(A_1A_2)^{1/3}(A_1A_4)^{-1}\delta^{2/3}C_{s1}f_{s,\xi}+K_{s2}], \quad (3.4.12)$$

$$K_{s1} = (K_s-1)K_s^{-1}(1+\bar{N}_s); \quad (3.4.13)$$

$$K_{s2} = (A_1A_4)^{-9/4}(A_1^{-1}A_4^{-1}-1)(A_5A_6^2A_6')^{-3/4}\delta(C_{s4}C_{s5}C_{s6})^{3/4} -(A_1A_4)^{-1}M_1^2(K_s-1)^{-1}+M_0^2(1+\bar{N}_s)^{-1}(K_s-1)^{-1}. \quad (3.4.14)$$

For the gross fluid, the pressure coefficient is:

$$C_p = -2(p-p_0)(\varrho_0u_0^2)^{-1}. \quad (3.4.15)$$

Using (2.1.1) for ϱ_s and (2.1.2) for p_s, C_p is represented as:

$$C_p = -2\,[\sum_{s=1}^{n}(p_s-p_{s0})]\,u_0^{-2}(\sum_{s=1}^{n}\varrho_{s0})^{-1}$$

$$= [\sum_{s=1}^{n}-2(p_s-p_{s0})(\varrho_{s0}u_0^2)^{-1}\varrho_{s0}](\sum_{s=1}^{n}\varrho_{s0})^{-1} = (\sum_{s=1}^{n}C_{ps}\varrho_{s0})(\sum_{s=1}^{n}m_s\nu_{s0})^{-1}. \quad (3.4.16)$$

This completes the solution.

3.5. *Special Case*

A case where the magnetic field is parallel to the free-stream velocity in the absence of the electric field is analyzed below.

Substituting zero for H_{0y}, H_{0z}, E_{0y}, and E_{0z}, in (3.1.14), (3.1.24) and (3.1.29), we obtain a system of equations in which the boundary conditions of Eq. (3.1.24), $(2.3.6)_1$, $(2.3.12)_1$ remain unchanged. Instead of (3.2.1) and (3.2.5), we introduce:

$$\xi = xl^{-1}, \quad \eta = yl^{-1}(A_1A_2)^{-n_2'}A_5^{-n_2}\delta^{-m}, \quad \bar{z} = zl^{-1}, \quad w_s = u_0A_5^{-n_3}\bar{w}_s, \quad (3.5.1)$$

$$\varphi_s = \alpha_{s*}lA_5^{-n_4}f_s; \quad \omega_{sz} = \alpha_{s*}l^{-1}A_5^{-n_4}\bar{\omega}_{sz}; \quad G_1 = \mu_e u_0H_{0x}lA_{16}^{-n_{16}}\bar{G}_1. \quad (3.5.2)$$

Then a transformed system of equations is obtained:

$$C_{s1}f_{s,\,\xi}f_{s,\,\xi\xi}+2C_{s2}f_{s,\,\eta}f_{s,\,\xi\eta}+2C_{s3}f_{s,\,\eta}\bar{\omega}_{sz}-f_{s\,,\eta\eta}$$
$$-C_{s5}f_{s,\,\eta}\bar{w}_s-C_{s7}\bar{G}_{1,\,\xi}f_{s,\,\xi}-C_{s8}\bar{G}_{1,\,\xi}-2C_{s10}g_{s1}=0; \tag{3.5.3}$$

$$\bar{w}_{s,\,\xi}=C_{s14}\bar{G}_{2,\xi}-C_{s15}f_{s,\eta}+C_{s16}g_{s2z}-C_{s17}g_{s3z}-C_{s18}g_{s4z}+C_{s19}g_{s5z}; \tag{3.5.4}$$

$$C_{s20}\bar{H}_{s,\,\xi\xi}+C_{s21}\bar{H}_{s,\,\eta\eta}-C_{s22}\bar{Q}_{s,\,\xi\xi}-C_{s23}\bar{Q}_{s,\,\eta\eta}+C_{s24}\bar{G}_{1,\,\xi\xi}C_{s25}\bar{G}_{1,\,\eta\eta}$$
$$-C_{s26}g_{s5\xi}-C_{s27}g_{s5\eta}-C_{s28}(g_{s2x}),_{\xi}+C_{s29}(g_{s1}),_{\xi}+C_{s30}(g_{s3x}),_{\xi}$$
$$-C_{s31}(g_{s2y}),_{\eta}+C_{s32}(g_{s3y}),_{\eta}-C_{s33}(g_{s4x}),_{\xi}-C_{s34}(g_{s4y}),_{\eta}$$
$$-C_{s35}\bar{w}_{s,\,\eta}+C_{s38}g_{s6}-C_{s39}g_{s7}+C_{s40}\bar{\omega}_{sz,\,\eta}=0; \tag{3.5.5}$$

$$C_{s41}\bar{G}_{1,\,\xi\xi}+\bar{G}_{1,\,\eta\eta}=-C_{s42}\bar{\varrho}_s\bar{G}_{s5}; \tag{3.5.6}$$

$$C_{s43}\bar{G}_{2,\,\xi\xi}+\bar{G}_{2,\,\eta\eta}=C_{s\,44}\bar{\varrho}_s\bar{w}_s\bar{G}_{s6}. \tag{3.5.7}$$

Coefficients of the system are given in Table 3.

Since we cannot reduce the previous case to the present one, we follow the same technique as applied before to an arbitrary electromagnetic field, and obtain the following:

$$\frac{C_{s1}C_{s2}^2C_{s5}C_{s7}C_{s12}C_{s15}C_{s19}^2C_{s25}C_{s29}^2C_{s34}C_{s35}^2C_{s41}C_{s43}}{C_{s3}^2C_{s8}C_{s10}C_{s14}C_{s16}C_{s17}C_{s18}C_{s20}C_{s21}C_{s22}C_{s23}C_{s24}C_{s26}C_{s27}C_{s28}C_{s30}}\cdot$$
$$\cdot\frac{1}{C_{s31}C_{s32}C_{s33}C_{s38}C_{s39}C_{s40}C_{s42}C_{s44}}=A_1^{3n_2+1}A_2^{3n_2'+1}A_5^{3n_2-2n_3+n_4}A_{10}^{-4n_{10}}\cdot$$
$$\cdot A_{11}^{3n_{11}}A_{12}^{3n_{12}}A_{13}^{n_{13}}A_{14}^{2n_{14}}A_{15}^{2n_{15}}A_{16}^{3n_{16}}A_{17}^{2n_{17}}(M_1R_RR_{msx})^{-1}\delta^{3m+1}. \tag{3.5.8}$$

Assuming $n_3=n_4=0$, we obtain values of the exponents:

$$n_2'=-1/3,\quad m=-1/3,\quad n_2=n_{10}=n_{11}=n_{12}=n_{13}=n_{14}=n_{15}$$
$$=n_{16}=n_{17}=0. \tag{3.5.9}$$

The coefficients so obtained furnish a relation between constant coefficients (parameters):

$$\frac{C_{s1}C_{s2}^2C_{s5}C_{s7}C_{s12}C_{s15}C_{s25}C_{s29}^2C_{s34}C_{s35}^2C_{s41}C_{s42}}{C_{s3}^2C_{s8}C_{s10}C_{s14}C_{s21}C_{s23}C_{s24}C_{s27}C_{s30}C_{s31}C_{s32}C_{s38}C_{s39}C_{s40}C_{s42}C_{s44}}$$
$$=A_9^{-1}(M_1R_RR_{msx})^{-1}=M_1^{-1}R_R^{-1}A_{0x}^{-2}(R_{msx})^{-3}R_M. \tag{3.5.10}$$

Following the same procedure as in Section 3.4, we find the pressure coefficient:

$$C_{ps}=2K_{s1}[(A_1A_2)^{1/3}(A_1A_4)^{-1}\delta^{2/3}C_{s1}f_{s,\,\xi}+K_{s2}]; \tag{3.5.11}$$

$$K_{s1}=(K_s-1)K^{-1}(1+\bar{N}_s), \tag{3.5.12}$$

$$K_{s2}=(A_1^{-1}A_4^{-1}-1)(A_1A_4)^{-3}A_5^{-3}\delta C_{s5}^3-(A_1A_4)^{-1}M_1^2(K_s-1)^{-1}$$
$$+M_0^2(1+\bar{N}_s)(K_s-1)^{-1}. \tag{3.5.13}$$

Note that, in the absence of electromagnetic phenomena, momentum exchange, and mass sources in the fluid flow of perfect gas, we obtain:

$$C_{ps} = 2(K_s - 1)K_s^{-1}A_3 f_{s,\,\xi}, \tag{3.5.14}$$

which differs from the result [3, 13] by the factor $(K_s - 1)K_s^{-1}$ due to the different approximation method† used in calculating $(p_s - p_{s0})$. The pressure coefficient for the gross fluid is given by:

$$C_p = \sum_{s=1}^{n} \varrho_{s0} C_{ps} \left(\sum_{r=1}^{n} \varrho_{r0} \right)^{-1}. \tag{3.5.15}$$

3.6. Application

Above, we derived the system of differential equations in the (ξ, η)-plane which describes a family of similar flows in question. The coefficients satisfying condition (3.3.4) should have the same values for similar flows. The condition (3.3.4) may be remodelled as follows:

$$\frac{C_{s2}^2 C_{s8}^3 C_{s9} C_{s13} C_{s15} C_{s25} C_{s30}^2 C_{s38} C_{s40}}{C_{s3} C_{s4} C_{s5} C_{s6} C_{s7} C_{s10} C_{s21}^4 C_{s24} C_{s27} C_{s29} C_{s37} C_{s42} C_{s44}}$$

$$= R_R A_3^2 \mu_e^2 \varrho_{es0}^{-1} \varrho_{s0}^{-1} \sigma H_{0x}^2 H_{0y}^2 H_{0z}^{-1} u_0^{-2}; \tag{3.6.1}$$

$$C_{s1} = C_{s12}^{-\frac{1}{2}} = (A_1 A_2)^{\frac{1}{3}} A_3 \delta^{-\frac{2}{3}}. \tag{3.6.2}$$

The coefficients, C_{s14}, C_{s19}, C_{s26}, C_{s36}, and C_{s39} are transferred to the right-hand side of (3.6.1). The coefficient C_{s31} is of the form:

$$C_{s31} = (A_1 A_2)^{\frac{1}{3}} \delta^{\frac{1}{3}}; \quad C_{s21} = C_{s23} = C_{s31}^2 = C_{s34}^2 = C_{s41}^{-\frac{1}{2}} = C_{s43}^{-\frac{1}{2}}. \tag{3.6.3}$$

Relations (3.6.2) and (3.6.3) are combined:

$$C_{s1} C_{s31}^{-1} = A_3 \delta^{-1}. \tag{3.6.4}$$

The remaining coefficients are found to be unity:

$$C_{s16} = C_{s17} = C_{s18} = C_{s20} = C_{s22} = C_{s28} = C_{s33} = 1. \tag{3.6.5}$$

Expressions (3.6.1) to (3.6.4) provide a possibility of choosing arbitrarily a certain number of some parameters, i.e. Mach number, thickness ratio, specific-heat ratio, fluid density, electrical conductivity, and the magnitudes of magnetic and electric field intensities. The remaining must be calculated from the above system of equations. Suppose there are two flows, denoted by l and m, respectively, which are transonically similar. Then

† With an assumption that $(K_s - 1)K_s^{-1} \cong 1$, the classical result is obtained.

TABLE 3. COEFFICIENTS OF THE DIFFERENTIAL SYSTEM IN ALIGNED MAGNETIC FIELD CASE

$C_{s1} = A^{2n'_2+1}$	$A_2^{2n'_2+1}$	A_3^1	$A_4^0A_5^{2n_2-n_4}$	A_8^0	A_9^0	A_{10}^0	A_{11}^0	A_{12}^0	A_{13}^0	A_{14}^0	A_{15}^0	A_{16}^0	A_{17}^0	δ^{2m}	M_1^0
$C_{s2} = A_1^1$	A_2^0	A_3^1	$A_4^1A_5^{-n_4}$	A_8^0	A_9^0	A_{10}^0	A_{11}^0	A_{12}^0	A_{13}^0	A_{14}^0	A_{15}^0	A_{16}^0	A_{17}^0	δ^0	M_1^0
$C_{s3} = A_1^{n'_2+1}$	$A_2^{n'_2}$	A_3^1	$A_4^1A_5^{n_2-n_4}$	A_8^0	A_9^0	A_{10}^0	A_{11}^0	A_{12}^0	A_{13}^0	A_{14}^0	A_{15}^0	A_{16}^0	A_{17}^0	δ^m	M_1^0
$C_{s5} = A_1^{n'_2+1}$	$A_2^{n'_2}$	A_3^0	$A_4^1A_5^{n_2-n_3+1}$	A_8^0	A_9^0	A_{10}^0	A_{11}^0	A_{12}^0	A_{13}^0	A_{14}^0	A_{15}^0	A_{16}^0	A_{17}^0	δ^m	M_1^0
$C_{s7} = A_1^{n'_2+1}$	$A_2^{2n'_2}$	A_3^0	$A_4^1A_5^{2n_2+1}$	A_8^0	A_9^0	A_{10}^0	A_{11}^0	A_{12}^0	A_{13}^0	A_{14}^0	A_{15}^0	$A_{16}^{-n_{16}}$	A_{17}^0	δ^{2m}	M_1^0
$C_{s8} = A_1^{2n'_2}$	$A_2^{2n'_2-1}$	A_3^{-1}	$A_4^1A_5^{2n_2+n_4+1}$	A_8^0	A_9^0	A_{10}^0	A_{11}^0	A_{12}^0	A_{13}^0	A_{14}^0	A_{15}^0	$A_{16}^{-n_{16}}$	A_{17}^0	δ^{2m}	M_1^0
$C_{s10} = A_1^{2n'_2+1}$	$A_2^{2n'_2}$	A_3^{-1}	$A_4^1A_5^{2n_2+n_4}$	A_8^0	A_9^0	$A_{10}^{n_{10}}$	$A_{11}^{-n_{11}}$	A_{12}^0	A_{13}^0	A_{14}^0	A_{15}^0	A_{16}^0	A_{17}^0	δ^{2m}	M_1^1
$C_{s12} = A_1^{n'_2}$	$A_2^{n'_2}$	A_3^{-1}	$A_4^0A_5^{n_2+n_4}$	A_8^0	A_9^0	A_{10}^0	A_{11}^0	A_{12}^0	A_{13}^0	A_{14}^0	A_{15}^0	A_{16}^0	A_{17}^0	δ^{m+1}	M_1^0
$C_{s14} = A_1^0$	A_1^0	A_3^0	$A_4^0A_5^{n_2+1}$	A_8^0	A_9^0	A_{10}^0	A_{11}^0	A_{12}^0	A_{13}^0	A_{14}^0	A_{15}^0	A_{16}^0	$A_{17}^{-n_{27}}$	δ^0	M_1^{-2}
$C_{s15} = A_1^{-n'_2}$	$A_2^{-n'_2}$	A_3^1	$A_4^0A_5^{-n_2+n_3-n_4+1}$	A_8^0	A_9^0	A_{10}^0	A_{11}^0	A_{12}^0	A_{13}^0	A_{14}^0	A_{15}^0	A_{16}^0	A_{17}^0	δ^{-m}	M_1^{-3}
$C_{s16} = A_1^0$	A_2^0	A_3^0	$A_4^0A_5^0$	A_8^0	A_0^0	$A_{10}^{n_{10}}$	$A_{11}^{-n_{11}}$	$A_{12}^{-n_{12}}$	A_{13}^0	A_{14}^0	A_{15}^0	A_{16}^0	A_{17}^0	δ^0	M_1^0
$C_{s17} = A_1^0$	A_2^0	A_3^0	$A_4^0A_5^0$	A_8^0	A_9^0	$A_{10}^{n_{10}}$	$A_{11}^{-n_{11}}$	A_{12}^0	A_{13}^0	A_{14}^0	A_{15}^0	A_{16}^0	A_{17}^0	δ^0	M_1^0
$C_{s18} = A_1^0$	A_2^0	A_3^0	$A_4^0A_5^{n_3}$	A_8^0	A_9^0	$A_{10}^{n_{10}}$	A_{11}^0	A_{12}^0	A_{13}^{n-13}	A_{14}^0	A_{15}^0	A_{16}^0	A_{17}^0	δ^0	M_1^0
$C_{s19} = A_1^0$	A_2^0	A_3^0	$A_4^0A_5^{n_3}$	A_8^0	A_9^0	$A_{10}^{n_{10}}$	A_{11}^0	A_{12}^0	A_{13}^0	A_{14}^0	A_{15}^0	A_{16}^0	A_{17}^0	δ^0	M_1^0
$C_{s20} = A_1^0$	A_2^0	A_3^0	$A_4^0A_5^0$	A_8^0	A_9^0	A_{10}^0	A_{11}^0	A_{12}^0	A_{13}^0	A_{14}^0	$A_{15}^{-n_{15}}$	A_{16}^0	A_{17}^0	δ^0	M_1^0
$C_{s21} = A_1^{-2n'_2}$	$A_2^{-2n'_2}$	A_3^0	$A_4^0A_5^{-2n_2}$	A_8^0	A_9^0	A_{10}^0	A_{11}^0	A_{12}^0	A_{13}^0	A_{14}^0	$A_{15}^{-n_{15}}$	A_{16}^0	A_{17}^0	δ^{-2m}	M_1^0
$C_{s22} = A_1^0$	A_2^0	A_3^0	$A_4^0A_5^0$	A_8^0	A_9^0	A_{10}^0	A_{11}^0	A_{12}^0	A_{13}^0	$A_{14}^{-n_{14}}$	A_{15}^0	A_{16}^0	A_{17}^0	δ^0	M_1^0
$C_{s23} = A_1^{-2n'_2}$	$A_2^{-2n'_2}$	A_3^0	$A_4^0A_5^{-2n_1+1}$	A_8^0	A_9^0	A_{10}^0	A_{11}^0	A_{12}^0	A_{13}^0	$A_{14}^{-n_{14}}$	A_{15}^0	A_{16}^0	A_{17}^0	δ^{-2m}	M_1^{-2}

TABLE 3 (*cont.*)

$C_{s24} =$	A_1^0	A_2^0	A_3^0	$A_4^0A_5^1$	A_8^0	A_9^0	A_{10}^0	A_{11}^0	A_{12}^0	A_{13}^0	A_{14}^0	A_{15}^0	$A_{16}^{-n_{16}}$	A_{17}^0	δ^0	M_1^{-2}	
$C_{s25} =$	$A_1^{-2n'_2}$	$A_2^{-2n'_2}$	A_3^0	$A_4^0A_5^{-2n_2+1}$	A_8^0	A_9^0	A_{10}^0	A_{11}^0	A_{12}^0	A_{13}^0	A_{14}^0	A_{15}^0	$A_{16}^{-n_{16}}$	A_{17}^0	δ^{-2m}	M_1^{-2}	
$C_{s26} =$	A_1^0	A_2^0	A_3^0	$A_4^0A_5^0$	A_8^0	A_9^1	$A_{10}^{n_{10}}$	A_{11}^0	A_{12}^0	A_{13}^0	A_{14}^0	A_{15}^0	A_{16}^0	A_{17}^0	δ^0	M_1^0	
$C_{s27} =$	$A_1^{-n'_2}$	$A_2^{-n'_2}$	A_3^0	$A_4^0A_5^{-n_2}$	A_8^0	A_9^1	$A_{10}^{n_{10}}$	A_{11}^0	A_{12}^0	A_{13}^0	A_{14}^0	A_{15}^0	A_{16}^0	A_{17}^0	δ^{-m}	M^0	
$C_{s28} =$	A_1^0	A_2^0	A_3^0	$A_4^0A_5^0$	A_8^0	A_9^0	$A_{10}^{n_{10}}$	$A_{11}^{-n_{11}}$	$A_{12}^{-n_{12}}$	A_{13}^0	A_{14}^0	A_{15}^0	A_{16}^0	A_{17}^0	δ^0	M_1^0	
$C_{s29} =$	A_1^{-1}	A_2^{-1}	A_3^0	$A_4^0A_5^0$	A_8^1	A_9^0	$A_{10}^{n_{10}}$	$A_{11}^{-n_{11}}$	A_{12}^0	A_{13}^0	A_{14}^0	A_{15}^0	A_{16}^0	A_{17}^0	δ^0	M_1^{-1}	
$C_{s30} =$	A_1^0	A_2^0	A_3^1	$A_4^0A_5^{-n_4}$	A_8^0	A_9^0	$A_{10}^{n_{10}}$	$A_{11}^{-n_{11}}$	A_{12}^0	A_{13}^0	A_{14}^0	A_{15}^0	A_{16}^0	A_{17}^0	δ^0	M_1^{-1}	
$C_{s31} =$	$A_1^{-n'_2}$	$A_2^{-n'_2}$	A_3^0	$A_4^0A_5^{-n_2}$	A_8^0	A_9^0	$A_{10}^{n_{10}}$	$A_{11}^{-n_{11}}$	$A_{12}^{-n_{12}}$	A_{13}^0	A_{14}^0	A_{15}^0	A_{16}^0	A_{17}^0	δ^{-m}	M_1^0	
$C_{s32} =$	$A_1^{-2n'_2}$	$A_2^{-2n'_2}$	A_3^1	$A_4^0A_5^{-2n_2-n_4}$	A_8^0	A_9^0	$A_{10}^{n_{10}}$	$A_{11}^{-n_{11}}$	A_{12}^0	A_{13}^0	A_{14}^0	A_{15}^0	A_{16}^0	A_{17}^0	δ^{-2m}	M_1^{-1}	
$C_{s33} =$	A_1^0	A_2^0	A_3^0	$A_4^0A_5^0$	A_8^0	A_9^0	$A_{10}^{n_{10}}$	A_{11}^0	A_{12}^0	$A_{13}^{-n_{13}}$	A_{14}^0	A_{15}^0	A_{16}^0	A_{17}^0	δ^0	M_1^0	
$C_{s34} =$	$A_1^{-n'_2}$	$A_2^{-n'_2}$	A_3^0	$A_4^0A_5^{-n_2-n_3+1}$	A_8^0	A_9^0	A_{10}^0	A_{11}^0	A_{12}^0	A_{13}^0	A_{14}^0	A_{15}^0	A_{16}^0	A_{17}^0	δ^{-m}	M_1^{-2}	
$C_{s38} =$	A_1^{-1}	A_2^{-1}	A_3^0	$A_4^0A_5^1$	A_8^1	A_9^0	A_{10}^0	A_{11}^0	A_{12}^0	A_{13}^0	A_{14}^0	A_{15}^0	A_{16}^0	A_{17}^0	δ^0	M_1^{-3}	R_{msx}
$C_{s39} =$	A_1^0	A_2^0	A_3^1	$A_4^0A_5^{-n_4+1}$	A_8^0	A_9^0	A_{10}^0	A_{11}^0	A_{12}^0	A_{13}^0	A_{14}^0	A_{15}^0	A_{16}^0	A_{17}^0	δ^0	M_1^{-3}	R_{msx}
$C_{s40} =$	$A_1^{-n'_2}$	$A_2^{-n'_2}$	A_3^0	$A_4^0A_5^{-n_2-n_4}$	A_8^0	A_9^0	A_{10}^0	A_{11}^0	A_{12}^0	A_{13}^0	A_{14}^0	A_{15}^0	A_{16}^0	A_{17}^0	δ^{-m}	M_1^{-1}	
$C_{s41} =$	$A_1^{2n'_2}$	$A_2^{2n'_2}$	A_3^0	$A_4^0A_5^{2n_2}$	A_8^0	A_9^0	A_{10}^0	A_{11}^0	A_{12}^0	A_{13}^0	A_{14}^0	A_{15}^0	A_{16}^0	A_{17}^0	δ^{2m}	M_1^0	
$C_{s42} =$	$A_1^{2n'_2}$	$A_2^{2n'_2}$	A_3^0	$A_4^0A_5^{2n_2+1}$	A_8^0	A_9^0	A_{10}^0	A_{11}^0	A_{12}^0	A_{13}^0	A_{14}^0	A_{15}^0	$A_{16}^{-n_{16}}$	A_{17}^0	δ^{2m}	M_1^0	R_R^{-1}
$C_{s43} =$	$A^{2n'_2}$	$A_2^{2n'_2}$	A_3^0	$A_4^0A_5^{2n_2}$	A_8^0	A_9^0	A_{10}^0	A_{11}^0	A_{12}^0	A_{13}^0	A_{14}^0	A_{15}^0	A_{16}^0	A_{17}^0	δ^{2m}	M_1^0	
$C_{s44} =$	$A_1^{2n'_2}$	$A_2^{2n'_2}$	A_3^0	$A_4^0A_5^{2n_2}$	A_8^0	A_9^0	A_{10}^0	A_{11}^0	A_{12}^0	A_{13}^0	A_{14}^0	A_{15}^0	A_{16}^0	$A_{17}^{-n_{17}}$	δ^{2m}	M_1^0	R_{msx}

we have the following relations:

$$R_R\mu_e\varrho_{es0}^{-1}\varrho_{s0}^{-1}\sigma H_{0x}^2 H_{0y}^2 H_{0z}^{-1}(1-M_1)^2 M_1^{-2}\alpha_{s*}^{-2}\big|_l$$
$$= R_R\mu_e\varrho_{es0}^{-1}\varrho_{s0}^{-1}\sigma H_{0x}^2 H_{0y}^2 H_{0z}^{-1}(1-M_1)^2 M_1^{-2}\alpha_{s*}^{-2}\big|_m, \qquad (3.6.6)_1$$

$$(A_1A_2)^{\frac{1}{3}}A_3\delta^{-\frac{2}{3}}\big|_l = (A_1A_2)^{\frac{1}{3}}A_3\delta^{-\frac{2}{3}}\big|_m, \qquad (3.6.6)_2$$

$$A_3\delta^{-1}\big|_l = A_3\delta^{-1}\big|_m. \qquad (3.6.6)_3$$

Let us present an example for illustrative purposes. Let us vary the parameters, M_1, H_{0x}, H_{0y}, H_{0z}. For simplicity's sake, we assume $W_s \equiv \overline{W}_s \equiv A_s \equiv \bar{A}_s \equiv N_s \equiv \bar{N}_s \equiv P_s \equiv \bar{P}_s \equiv 0$, $s = (i, e, n)$, referring to ion (i), electron (e) and neutral (n) fluid flows, respectively, for both flows. Then (3.6.6) becomes:

$$H_{0x}^2 H_{0y}^2 H_{0z}^{-1}(1-M_1)^2(M_1\alpha_{s*})^{-2}\big|_l = H_{0x}^2 H_{0y}^2 H_{0z}^{-1}(1-M_1)^2(M_1\alpha_{s*})^{-2}\big|_m, \qquad (3.6.7)_1$$

$$A_2^{\frac{1}{3}}A_3\delta^{-\frac{2}{3}}\big|_l = A_2^{\frac{1}{3}}A_3\delta^{-\frac{2}{3}}\big|_m, \qquad (3.6.7)_2$$

$$A_3\delta^{-1}\big|_l = A_3\delta^{-1}\big|_m. \qquad (3.6.7)_3$$

The pressure coefficient, (3.4.16), in the flow "l", "m", takes the form:

$$C_p = (m\nu_0+m_e\nu_0+m\nu_{n0})^{-1}(C_{pi}m\nu_0+C_{pe}m_e\nu_0+C_{pn}m\nu_{n0}$$
$$\cong \nu_0(\nu_0+\nu_{n0})^{-1}2(K_i-1)K_i^{-1}\{[K_i+1-2(K_i-1)\bar{\zeta}_i]^{\frac{1}{3}}[1-(K_i-1)\bar{\zeta}_i]^{-1}\cdot$$
$$\cdot\delta^{\frac{2}{3}}C_{i1}f_{i,\xi}+[1-(K_i-1)\bar{\zeta}_i]^{-\frac{13}{4}}(K_i-1)\bar{\zeta}_i(A_5A_6^2A_6')^{-\frac{3}{4}}\delta(C_{i4}C_{i5}C_{i6})^{\frac{3}{4}}$$
$$-[1-(K_i-1)\bar{\zeta}_i]^{-1}M_1^2(K_i-1)^{-1}+M_0^2(K_i-1)^{-1}\}$$
$$+\nu_{n0}(\nu_0+\nu_{n0})^{-1}2(K_n-1)K_n^{-1}\{[K_n+1-2(K_n-1)\bar{\zeta}_n]^{\frac{1}{3}}[1-(K_n-1)\bar{\zeta}_n]^{-1}\delta^{\frac{2}{3}}$$
$$\cdot C_{n1}f_{n,\xi}+[1-(K_n-1)\bar{\zeta}_n]^{-\frac{13}{4}}(K_n-1)\bar{\zeta}_n(A_5A_6^2A_6')^{-\frac{3}{4}}\delta(C_{n4}C_{n5}C_{n6})^{\frac{3}{4}}$$
$$-[1-(K_n-1)\bar{\zeta}_n]^{-1}M_1^2(K_n-1)^{-1}+M_0^2(K_n-1)^{-1}\}, \qquad (3.6.8)$$

where we have assumed $m_i \cong m_n = m$, $M_{1i} \cong M_{1n} = M_1$ and neglected $m_e m^{-1}\nu_0$, $m_e m^{-1}C_{pe}$. To evaluate $\bar{\zeta}_i$, $\bar{\zeta}_n$, we further assume that the second integral in (3.4.2) is much smaller than the remaining terms, and that $d\bar{Q}_i \equiv d\bar{Q}_n \equiv 0$, $\bar{Q}$ being energy other than that of the electromagnetic origin. Then we approximate $\bar{\zeta}_i$, $\bar{\zeta}_n$, for ion flow retaining only the first term (Joule heat [51]) and the last term in (3.4.2); in neutral flow all terms vanish:

$$\bar{\zeta}_i \cong \alpha_{s*}^{-2}(J_{i0})^2(\sigma\varrho_{s0}u_0)^{-1}\int_0^s dr_s+A_6A_6'\int_0^{s'}dy+A_7A_7'\int_0^{s''}dz; \quad J_{i0} = e_i\nu_0u_0; \qquad (3.6.9)$$

$$\bar{\zeta}_n \cong 0. \qquad (3.6.10)$$

We also assume:

$$\int_0^s dr_s \cong \int_0^s dx = 1, \quad \int_0^{s'} dy = \delta, \quad \int_0^{s''} dz = \delta. \tag{3.6.11}$$

With (3.6.11), (3.6.9) becomes:

$$\bar{\zeta}_i = e_i^2 \nu_0 u_0 (m_i \sigma \alpha_{s*}^2)^{-1} + A_6 A_6' \delta + A_7 A_7' \delta. \tag{3.6.12}$$

Using (3.6.10) in (3.6.8), C_p becomes:

$$C_p = \nu_0(\nu_0+\nu_{n0})^{-1}(C_{pi}^{(1)}+C_{p_i}^{(2)})+\nu_{n0}(\nu_0+\nu_{n0})^{-1}C_{p_n}; \tag{3.6.13}$$

$$C_{p_i}^{(1)} = 2(K_i-1)\,K_i^{-1}[(K_i+1-2(K_i-1)\,\bar{\zeta}_i]^{\frac{1}{3}}[1-(K_i-1)\,\bar{\zeta}_i]^{-1}\,\delta^{\frac{2}{3}} C_{i1} f_{i,\xi}\,; \tag{3.6.14}$$

$$C_{p_i}^{(2)} = 2(K_i-1)\,K_i^{-1}\Big\{[1-(K_i-1)\bar{\zeta}_i]^{-\frac{13}{4}}(K_i-1)\,\bar{\zeta}_i(A_5 A_6^2 A_6')^{-\frac{3}{4}}\,\delta(C_{i4}C_{i5}C_{i6})^{\frac{3}{4}}$$
$$-[1-(K_i-1)\,\bar{\zeta}_i]^{-1}(K_i-1)^{-1}M_1^2+(K_i-1)^{-1}M_0^2\Big\}; \tag{3.6.15}$$

$$C_{p_n} = +2(K_n-1)\,K_n^{-1}(K_n+1)^{-\frac{1}{3}}\delta^{\frac{2}{3}}C_{n1}f_{n,\xi}\,. \tag{3.6.16}$$

The pressure coefficient C_{p_i} is resolved into component coefficients $C_{p_i}^{(k)}$, $k = 1, 2$. The component coefficient $C_{p_i}^{(1)}$ contains the function f_i. For this coefficient, we shall present the "influence lines", i.e. we assume $f_{i,\xi} = 1$. As representative medium we take nitrogen. The molecular mass of the nitrogen ion is practically the same as that of neutral nitrogen. The values below refer approximately to the altitude of 60 km, where the amount of oxygen is negligibly small in comparison with that of nitrogen. We choose (denoting the properties of this flow by the subscript l):†

$$6000°\text{K}, \quad 0.001 \text{ atm}, \quad \alpha_{i*} = 2000 \text{ m/sec}, \quad \nu_i = \nu_0 \cong 10^{21}/\text{m}^3,$$
$$K_i = 1.4; \quad B_{0x_l} = 10 \text{ gauss}, \quad B_{0y_l} = 0.75\,\text{B}_{0x_l}, \quad B_{0z_l} = 0.5\,\text{B}_{0x_l},$$
$$M_{1_l} = 0.8, \quad \delta = 0.1, \quad l = 1\,m. \tag{3.6.17}$$

The magnetic permeability is that of free space, $\mu_e = 4\pi\times10^{-7}$ henry/m. The electrical conductivity is calculated by means of the formula given in [14]:

$$\sigma = \eta^{-1}; \quad \eta = 6.53\times10^3 T^{-\frac{3}{2}} \log \Lambda \text{ ohm-cm}. \tag{3.6.18}$$

Using (3.6.17) and the chosen values, we obtain the numerical value $\bar{\zeta}_i$, (3.6.12), and consequently get:

$$C_{pi_l}^{(1)} = 0.15 f_{i,\xi} = 0.15; \quad C_{pi_l}^{(2)} = 0.394. \tag{3.6.19}$$

† E. V. Stupochenko, I. P. Stakhanov, E. V. Samuilov, A. S. Pleshanov, and I. B. Rozhdestvenskii, "Thermodynamic properties of air in the temperature interval from 1000 to 12,000 K and the pressure intervals, 0.001 to 1000 atm", *ARS Journal*, **30**, 98–112 (1960).

We choose the following patterns denoted by m_1, m_2, etc.:

(1) $H_{0x_{m_1}} = H_{0x_l}$, $H_{0y_{m_1}} = iH_{0y_l}$, $i = 2, 3, 4, \ldots$, with other quantities kept the same as in the flow "l". Then we calculate $H_{0z_{m_1}}$ from $(3.6.7)_1$. Next, we obtain $\bar{\zeta}_i$ by means of (3.6.14) and $C^{(1)}_{pi_{m_1}}$, and $C^{(2)}_{pi_{m_1}}$ from (3.6.14) and (3.6.15), respectively;

(2) $H_{0x_{m_2}} = H_{0x_l}$, $H_{0z_{m_2}} = iH_{0z_l}$;

(3) $H_{0z_{m_3}} = H_{0z_l}$, $H_{0y_{m_3}} = iH_{0y_l}$; $i = 2, 3, \ldots$,

(4) $H_{0x_{m_4}} = H_{0x_l}$, $H_{0y_{m_4}} = iH_{0y_l}$, $i = 2, 3, \ldots$; $M_{im_4} = 0.9$ (or $\delta_{m_4} = 0.075$).

Following the procedure in (1), we obtain pressure coefficients $C^{(k)}_{pim_j}$, $k = 1, 2$, $j = 1, 2, 3, 4$. The ratios of the pressure coefficients of chosen patterns to that of the flow "l" are plotted in Fig. 1. In diagrams (a), (b),

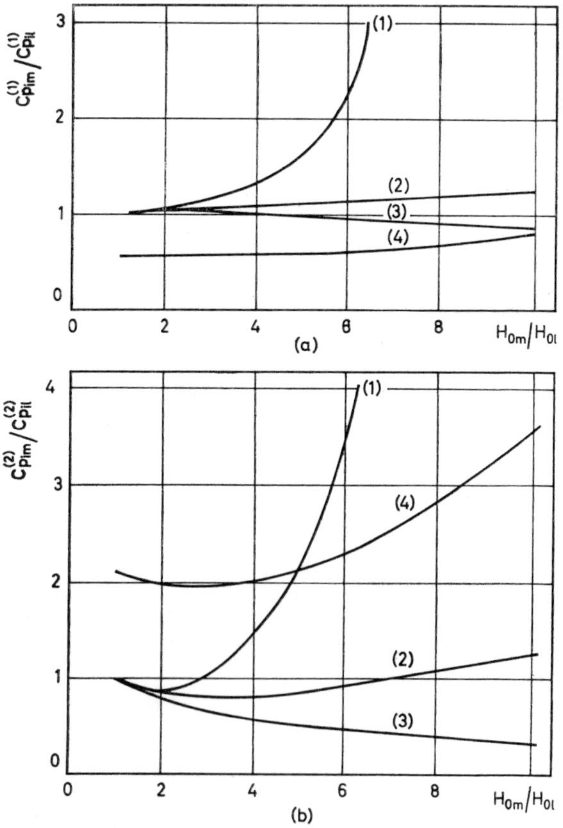

FIG. 1. Ion pressure coefficients $C^{(k)}_{p_i}$, $k = 1, 2$ for flow patterns (1), (2), (3) and (4).

we measure along the horizontal axes the ratio H_{0m}/H_{0l}, where for various patterns the ratio is equal to:

$$(1)\ H_{0m}/H_{0l} \equiv H_{0y_{m_1}}/H_{0yl}, \qquad (2)\ H_{0m}/H_{0l} \equiv H_{0z_{m_2}}/H_{0zl},$$

$$(3)\ H_{0m}/H_{0l} \equiv H_{0y_{m_3}}/H_{0yl}, \qquad (4)\ H_{0m}/H_{0l} \equiv H_{0y_{m_4}}/H_{0yl}.$$

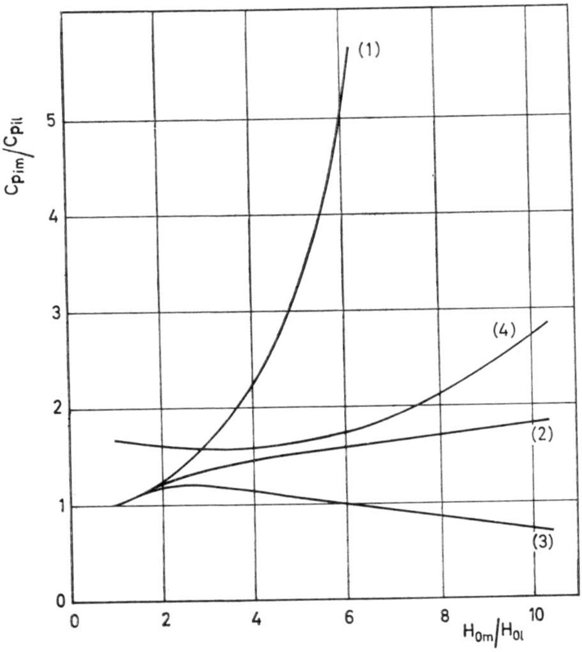

FIG. 2. Ion pressure coefficients C_{p_i} for flow patterns (1), (2), (3) and (4).

In pattern (1), the pressure coefficients increase very deeply, whereas in pattern (3) both decrease. In pattern (4), $C_{p_i}^{(2)}$ increases in comparison with patterns (2), (3), maintaining $C_{p_i}^{(1)}$ almost unchanged with increasing *y*-components of the magnetic field. In pattern (3) pressure coefficients decrease with increasing *y*-components of the magnetic field. The curves $C_{p_i}^{(1)}$ are "influence curves," i.e. $f_{i,\xi} = 1$. Figure 2 represents the sum of $C_{p_i}^{(1)}$ and $C_{p_i}^{(2)}$. Figs. 1 and 2 show the general tendency of the actual pressure coefficients with $f_{i,\xi} \neq 1$ in relation to varying parameters: the electromagnetic field, Mach number and thickness ratio.

Regarding the case of the magnetic field parallel to the free-stream velocity, the relation corresponding to (3.6.1) derived from (3.5.8) becomes:

$$\frac{C_{s2}^2 C_{s5} C_{s7} C_{s15} C_{s25} C_{s29}^2 C_{s35}^2}{C_{s3}^2 C_{s8} C_{s10} C_{s21}^5 C_{s24} C_{s27} C_{s30}^2 C_{s38} C_{s39} C_{s40} C_{s42} C_{s44}}$$

$$= R_{\bar{R}}^{-1} \mu_e \sigma (e_s^2 \nu_0^2 l)^{-1} M_1^{-1} H_{0x}^2 . \tag{3.6.20}$$

Relations (3.6.2), (3.6.3), (3.6.4) and (3.6.5) remain valid in this case. For two similar flows "l" and "m" with the same fluid properties, electrical conductivity, density and charge, we have:

$$H_{0x}^2 M_1^{-1}|_l = H_{0x}^2 M_1^{-1}|_m; \quad A_3 \delta^{-1}|_l = A_3 \delta^{-1}|_m . \tag{3.6.21}$$

Using the values from (3.6.17), (3.6.18) and assumptions described above (3.6.9), we obtain $\bar{\zeta}_i$, $C_{p_i}^{(1)}$, and $C_{p_i}^{(2)}$ from (3.5.11), (3.5.12), (3.6.12), respectively:

$$\bar{\zeta}_i = e_i^2 \nu_0 u_0 (m_i \sigma \alpha_{i*}^2)^{-1}, \tag{3.6.22}$$

$$C_{p_i}^{(1)} = 2(K_i - 1) K_i^{-1} [K_i + 1 - 2(K_i - 1)\bar{\zeta}_i]^{\frac{1}{3}} [1 - (K_i - 1)\bar{\zeta}_i]^{-1} \delta^{\frac{2}{3}} C_{i1} f_{i,\xi} , \tag{3.6.23}$$

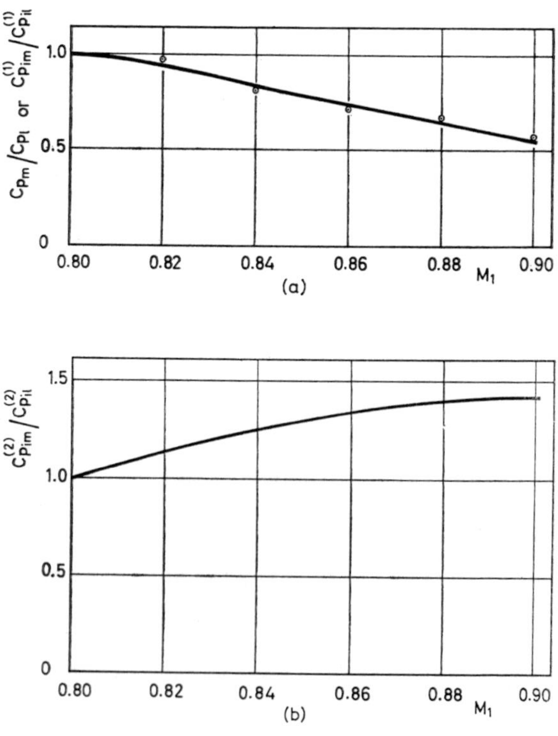

FIG. 3. Ion pressure coefficients $C_{p_i}^{(k)}$, $k = 1, 2$ for aligned field case.

$$C_{p_i}^{(2)} = 2(K_i-1)^2 K_i^{-1}[1-(K_i-1)\bar{\zeta}_i]^{-4}\bar{\zeta}_i A_5^{-3}\delta C_{s5}^3 - [1+(K_i-1)\bar{\zeta}_i]^{-1}\bar{\zeta}_i(K_i-1)^2 M_1^2, \quad (3.6.24)$$

$$C_{p_i}^{(1)} = 0.1675, \quad C_{p_i}^{(2)} = 0.580. \quad (3.6.25)$$

We use the flow "l" given in (3.6.17) with $B_{0yl} = B_{0zl} = 0$. Then we obtain δ_m, H_{0xm}, from (3.6.21). The values (3.6.22), (3.6.23) and (3.6.24) are calculated. The Mach numbers are chosen in the interval $M_{1m} = 0.82$, $0.84, \ldots, 0.90$, and the resulting values are plotted in Fig. 3.

The pressure coefficient C_{pl} given by Tamada's expressions, (1.2.5), (1.2.6), with the values used in (3.6.25), is found to be:

$$C_{pl} = 0.216. \quad (3.6.26)$$

We choose the calculated values of M_{1m}, δ_m given above, and determine the magnitude of the magnetic field by means of (1.2.5) to obtain the pressure coefficient given by (1.2.6). The ratio C_{p_m}/C_{pl} using Tamada's formula (1.2.6) for C_{p_m} is shown in Fig. 3a. It is seen that values of $C_{p_{im}}^{(1)}/C_{p_{il}}^{(1)}$ with $C_{p_{im}}^{(1)}$ calculated according to the procedure proposed in the present work, and C_{p_m}/C_{pl} with C_{p_m} as in Tamada's work marked in Fig. 3a by small circles are very close. This seems to suggest that were we to choose an appropriate value for $f_{i,\xi}$, we would have pressure coefficients obtainable by solving the Tricomi equation (1.2.4), as in Tamada's approach. This seems to indicate the validity and correctness of the present approach, where no attempt was made to solve equations. The differential system in the present work may be solved by some sort of iteration or successive-approximation procedure. In doing so, the solution for the Tricomi equation in the perturbed potential φ, which is similar to (1.2.4), may serve as first approximation for the velocity function f_i in the differential system of the present work.

REFERENCES

1. N. Geffen, "Magneto-gasdynamic flows with shock waves", *Physics of Fluids*, **6**, No. 4, 567–71 (1962).
2. H. Grad, "General fluid equations", *Notes on Magnetohydrodynamics*, No. 1, Inst. of Math. Sciences, New York University, NYO-6486 (1956).
3. G. K. Guderley and H. Yoshihara, The flow over a wedge profile at Mach number one", *Journal of the Aeronautical Sciences* **17**, 723–35 (1950).
4. T. v. Kármán, "Similarity law of transonic flow", *Journal of Math. and Physics* **25**, 182–90 (1947).
5. M. Z. v. Krzywoblocki and J. Nutant, "On the similarity rule in magneto-gas-dynamics", *Acta Physica Austriaca* **13**, No. 1, 1–18 (1960).
6. J. E. McCune and E. L. Resler, "Compressible effects in magneto-aerodynamic flows past thin bodies", *Journal of the Aero/Space Sciences* **27**, 493–503 (1960).

7. R. v. MISES, *Mathematical Theory of Compressible Fluid Flow*, Academic Press Inc. (1958).
8. K. OSWATITSCH, *Gas Dynamics*, Academic Press Inc. (1956).
9. K. OSWATITSCH, "Similarity and equivalence in compressible flow", *Advances in Applied Mechanics*, vol. VI, pp. 153–271, Academic Press Inc. (1960).
10. S. I. PAI, *Magnetogasdynamics and Plasma Dynamics*, Springer-Verlag, Vienna, and Prentice-Hall, Inc., Englewood Cliffs, New Jersey (1962).
11. S. I. PAI, "Gasdynamic effects on electric current density in magnetogasdynamics", *Journal of the Aero/Space Sciences* **29**, 483–4 (1962).
12. R. SEEBASS, "On 'transcritical' and 'hypercritical' flows in magnetogasdynamics", *Quart. Applied Math.* **19**, 231–7 (1961).
13. A. H. SHAPIRO, *The Dynamics and Thermodynamics of Compressible Fluid Flow*, vol. II, The Ronald Press Company, New York (1954).
14. L. SPITZER, Jr., *Physics of Fully Ionized Gases*, Interscience Publishers, Inc., New York (1956).
15. J. R. SPREITER, "Alternative forms for the basic equations of transonic flow theory", *Journal of the Aero-Space Sciences* **21**, No. 1, 70–72 (1954).
16. K. TAMADA, "Transonic flow of a perfectly conducting gas with aligned magnetic field", *Physics of Fluids* **5**, No. 8, 871–8 (1962).

ON THE PROTECTIVE ACTION OF HIP SAFETY BELTS IN VEHICLES

Franz Ollendorff

Faculty of Electrical Engineering, Technion—Israel Institute of Technology, Haifa

INTRODUCTION

In modern vehicles the human body is subjected to effects of inertia due to sudden changes in speed. In passenger aircraft, international regulations prescribe safety belts attached to the seats, in order to provide maximum protection against the psychological and physical consequences, and similar safety regulations are being introduced with regard to motor vehicles.

The heaviest impact on the human body is probably that due to the collision of the vehicle with a more or less solid object. The present article is a preliminary attempt at quantitative study of the impact mechanism, and more extensive treatment is called for in the future.

BASIC EQUATIONS

As shown in Fig. 1, a rectangular block represents the human body strapped to the seat around the hips and at rest relative to the vehicle. The dimensions of the block (including the neck and the head, but not the arms) are its width (a), depth (b), and height (h). The internal structure of the body is disregarded, and a uniform modulus of elasticity E is assumed.

The seat is so designed that the passenger strapped to it looks straight ahead in the direction of motion. The centre of the plane through the belt (assumed to be substantially parallel to the seat) is the origin of a Cartesian system of coordinates rigidly attached to the vehicle, with the x-axis pointing left to right, the y-axis forwards, and the z-axis upwards. The instant of impact is assumed as $t = 0$; prior to it ($t < 0$) the vehicle had been moving at a uniform speed, V, parallel to the x'-axis of a "geographical" Cartesian system of coordinates (x'; y'; z') in turn at rest relative to the earth. V is small enough compared with the velocity of light for sufficiently accurate results to be obtained by Newtonian mechanics. If, in addition, the influence of the earth's rotation is disregarded throughout—in other words, the system $(x'; y'; z')$ is treated as inertial—the primed and unprimed coordinates are related by the Galilean transformation:

$$x' = x + Vt; \quad y' = y; \quad z' = z. \tag{1}$$

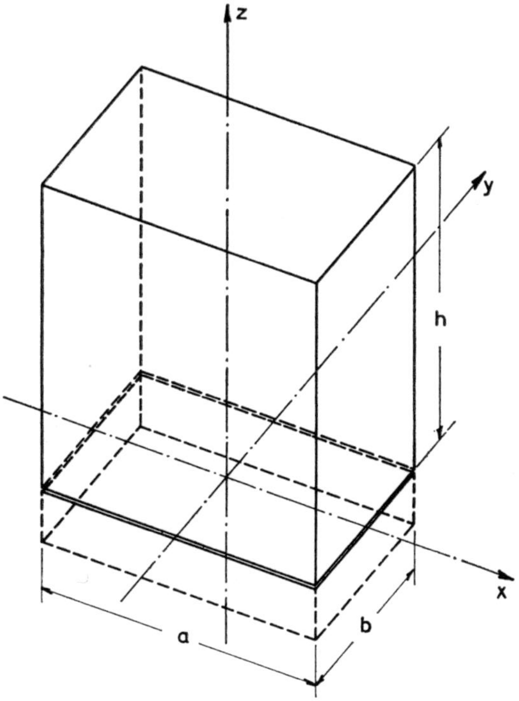

FIG. 1. Model of upper body and system of reference.

The forces of the earth's gravitational field may (in fact should) be disregarded compared with d'Alembert's forces of inertia. This being so, the z-axis of the upper part of the body, initially assumed to be straight, becomes a curve s, whose line elements ds derive from the axial elements dz through dilatation-free lateral displacement $u = u(z; t)$.

With the origin of s suitably located, the equality

$$s = z \tag{2}$$

is of general validity (Fig. 2).

In accordance with Saint-Venant's theory of the bending of homogeneous beams, the plane

$$y = y_0 = u(z; t); \quad |x| \leqq \tfrac{1}{2}a; \quad 0 < z \leqq h \tag{3}$$

is regarded as the neutral plane of the deformed body, free of tensile stresses. Moreover, it is assumed that the planes originally characterized by their axial coordinate z remain plane at any fixed time $t > 0$. Denoting

$$\bar{y} = y_0 + y \tag{4}$$

and

$$\pm\sigma_{\max} = \pm\sigma_{\max}(s; t) \quad (\bar{y} = \mp\tfrac{1}{2}b) \tag{5}$$

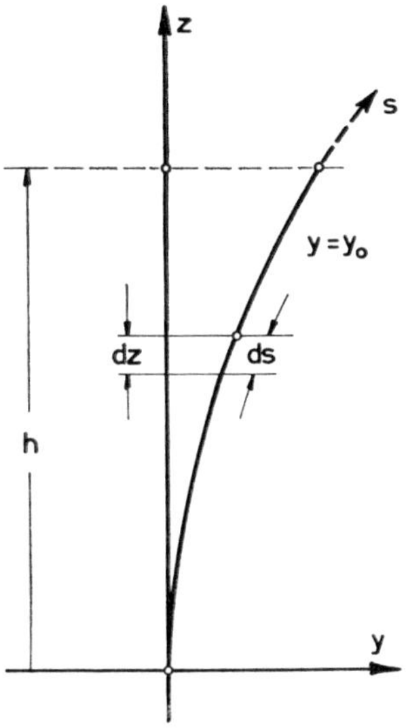

FIG. 2. Reference system in deformed body.

being the tensile stresses in the plane $s = z$ acting simultaneously at time t on the breast and back, the so-called "law of straight lines"

$$\sigma(\bar{y}; s; t) = -\sigma_{\max}(s; t)\cdot 2\bar{y} \tag{6}$$

represents the spatial distribution of σ in each of the planes $0 < s \leqq h$. The elements of the body bounded by the infinitesimally close planes s and $(s+ds)$ are thus affected by the right-hand moment

$$M(s+ds; t) = a\int_{-\frac{1}{2}b}^{\frac{1}{2}b} \sigma(\bar{y}; s+ds; t)\,\bar{y}\,d\bar{y} = \frac{\theta}{b}\sigma_{\max}(s+ds; t) \tag{7}$$

and by the left-hand moment

$$M(s; t) = a\int_{-\frac{1}{2}b}^{\frac{1}{2}b} \sigma(\bar{y}; s; t)\,\bar{y}\,d\bar{y} = \frac{\theta}{b}\,\sigma_{\max}(s; t), \tag{8}$$

the moment of inertia of the cross section of the body being

$$\theta = \tfrac{1}{12}ab^3. \tag{9}$$

For sufficiently small deformation, the kinematic relationship

$$\frac{\partial^2 u}{\partial z^2} = \frac{\sigma_{max}}{Eb} = \frac{M}{E\theta} \tag{10}$$

follows from Hooke's law.

Let $F_y = F_y(s+ds;\, t)$ be the lateral force caused by the shear stresses τ, which in turn act in the plane $(s+ds)$ of the considered element of the body in the direction of the positive y-axis. Then the condition of rotational equilibrium

$$F_y + \frac{\partial M}{\partial z} = 0 \tag{11}$$

applies at any moment, regardless of the lateral displacement of the body. That displacement, however, induces a force of inertia $-\gamma\,(\partial^2 u/\partial t^2)$ in accordance with the (mean) density of the upper part of the body γ so that virtual equilibrium relative to the y-axis is only ensured through the equation:

$$-\gamma\frac{\partial^2 u}{\partial t^2} + \frac{1}{ab}\frac{\partial F_y}{\partial z} = 0. \tag{12}$$

Substituting Eq. (11), we have

$$\gamma\frac{\partial^2 u}{\partial t^2} = -\frac{1}{ab}\frac{\partial^2 M}{\partial z^2} \tag{13}$$

and by recourse to Eq. (10), it is seen that the displacement u of the partial differential equation for rod vibrations

$$\frac{\partial^4 u}{\partial z^4} + \frac{\gamma ab}{E\theta}\frac{\partial^2 u}{\partial t^2} = 0 \tag{14}$$

satisfies our requirements.

INITIAL AND BOUNDARY CONDITIONS

The heaviest impact occurs when the vehicle is brought to an instantaneous standstill. If the motion relative to the geographical system of reference prior to the impact is referred to the system $(x;\, y;\, z)$ for times $t > 0$, then the virtually rigid state of the human body strapped to the seat is described by

$$\left|\frac{\partial u}{\partial t}\right|_{t=0} = \begin{matrix} 0 & \text{for} & t < 0 \\ -V & \text{for} & t > 0 \end{matrix} \tag{15}$$

while the belt ensures that

$$\left.\frac{\partial u}{\partial z}\right|_{z=0} = 0. \tag{16}$$

Since, moreover, it had been assumed that the upper part of the body was at rest relative to the vehicle before the impact, we have

$$u = 0 \quad \text{for} \quad t < 0. \tag{17}$$

Unlike the situation in space craft, the breast and head are assumed to be completely free. Accordingly, there is neither a torque M, nor a lateral force T_y, especially in the plane through the apex, $z = h$. The boundary condition

$$\left.\frac{\partial^2 u}{\partial z^2}\right|_{z=h} = 0 \tag{18}$$

follows from Eq. (10), and furthermore, by Eq. (11),

$$\left.\frac{\partial^3 u}{\partial z^3}\right|_{z=h} = 0. \tag{19}$$

INTEGRAL REPRESENTATION OF THE SOLUTION

In order to represent the conditions of Eq. (15) in analytical form, Gauss' plane of complex numbers

$$p = q + ir \quad (i = \sqrt{-1}) \tag{20}$$

is resorted to. Then

$$\left.\frac{\partial u}{\partial t}\right|_{z=0} = -\frac{V}{2\pi i}\int_{-i\infty}^{i\infty} \frac{e^{pt}}{p}\, dp, \tag{21}$$

the path of integration shown in Fig. 3 coinciding with the imaginary axis and the origin being by-passed by means of an arc in the half-plane $q > 0$.

On the strength of the linearity of Eq. (14), u is now substituted in synchronous form with Eq. (21), namely:

$$u = -\frac{V}{2\pi i}\int_{-i\infty}^{i\infty} w(p)\frac{e^{pt}}{p}\, dp. \tag{22}$$

Accordingly, the complex amplitude $w(p)$, dependent on z, obeys the ordinary differential equation

$$\frac{d^4 w}{dz^4} = \alpha^4 w \tag{23}$$

with

$$\alpha^4 = -\frac{\gamma ab}{E\theta} p^2 \tag{24}$$

substituted for simplicity. Selecting, at random for the moment, three constant A, B, and C, we have,

$$w = A\,[\sin \alpha z - \sinh \alpha z] + B \cos \alpha z + C \cosh \alpha z, \tag{25}$$

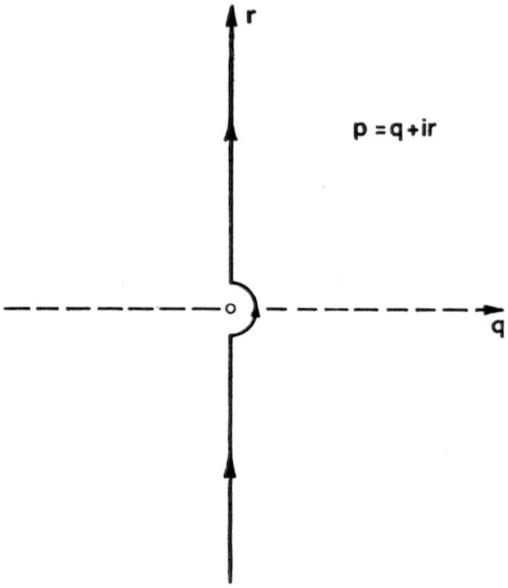

FIG. 3. Path of integration in Eq. (21).

which is a particular integral of Eq. (23). It complies with the "belt condition" of Eq. (16) because

$$\frac{dw}{dz} = A\alpha[\cos \alpha z - \cosh \alpha z] - B \sin \alpha z + C \sinh \alpha z \tag{26}$$

is identical in A, B and C. To satisfy Eqs. (18) and (19) as well, we proceed to:

$$\frac{d^2w}{dz^2} = A\alpha^2[-\sin \alpha z - \sinh \alpha z] - B\alpha^2 \cos \alpha z + C\alpha^2 \cosh \alpha z \tag{27}$$

and

$$\frac{d^3w}{dz^3} = A\alpha^3[-\cos \alpha z - \cosh \alpha z] + B\alpha^3 \sin \alpha z + C\alpha^3 \sinh \alpha z \tag{28}$$

stipulating

$$A[\sin \alpha h + \sinh \alpha h] = -B \cos \alpha h + C \cosh \alpha h \tag{29}$$

and

$$\alpha[\cos \alpha h + \cosh \alpha h] = B \sin \alpha h + C \sinh \alpha h. \tag{30}$$

These equations yield

$$B = A\cdot\frac{1+\cos\alpha h\cosh\alpha h-\sin\alpha h\sinh\alpha h}{\sin\alpha h\cosh\alpha h+\cos\alpha h\sinh\alpha h} \tag{31}$$

and

$$C = A\cdot\frac{1+\cos\alpha h\cosh\alpha h+\sin\alpha h\sinh\alpha h}{\sin\alpha h\cosh\alpha h+\cos\alpha h\sinh\alpha h}. \tag{32}$$

With

$$\left|\frac{\partial u}{\partial t}\right|_{z=0} = -\frac{V}{2\pi i}\int_{-i_\infty}^{i_\infty}(B+C)e^{pt}\,dp \tag{33}$$

evaluated from Eqs. (22) and (25),

$$B+C = 2A\cdot\frac{1+\cos\alpha h\cosh\alpha h}{\sin\alpha h\cosh\alpha h+\cos\alpha h\sinh\alpha h} = \frac{1}{p} \tag{34}$$

satisfies Eq. (21). Hence, finally,

$$A = \frac{1}{2p}\cdot\frac{\sin\alpha h\cosh\alpha h+\cos\alpha h\sinh\alpha h}{1+\cos\alpha h\cosh\alpha h}, \tag{35}$$

$$B = \frac{1}{2p}\cdot\frac{1+\cosh\alpha h\cosh\alpha h-\sin\alpha h\sinh\alpha h}{1+\cos\alpha h\cosh\alpha h}, \tag{36}$$

$$C = \frac{1}{2p}\cdot\frac{1+\cos\alpha h\cosh\alpha h+\sin\alpha h\sinh\alpha h}{1+\cos\alpha h\cosh\alpha h}. \tag{37}$$

It is seen that the function of Eq. (25) has a simple pole at the points of the x-axis obtained from the solutions $\varkappa_n$ of the transcendental equation:

$$\cos\varkappa_n\cdot\cosh\varkappa_n+1 = 0 \tag{38}$$

(according to Fig. 4 and the accompanying table)† as

$$p_n = \pm i\sqrt{\left(\frac{E\theta}{\gamma ab}\right)}\cdot\frac{\varkappa_n^2}{h^2}. \tag{39}$$

† Solution of the transcendental equation (38):

n	1	2	3	4	5	6
$\varkappa_n$	1·8751	4·6941	7·8548	10·9955	14·1372	17·2788

[K. Hayashi, *Fünfstellige Funktionentafeln*, p. 52, Springer, Berlin, 1930.]

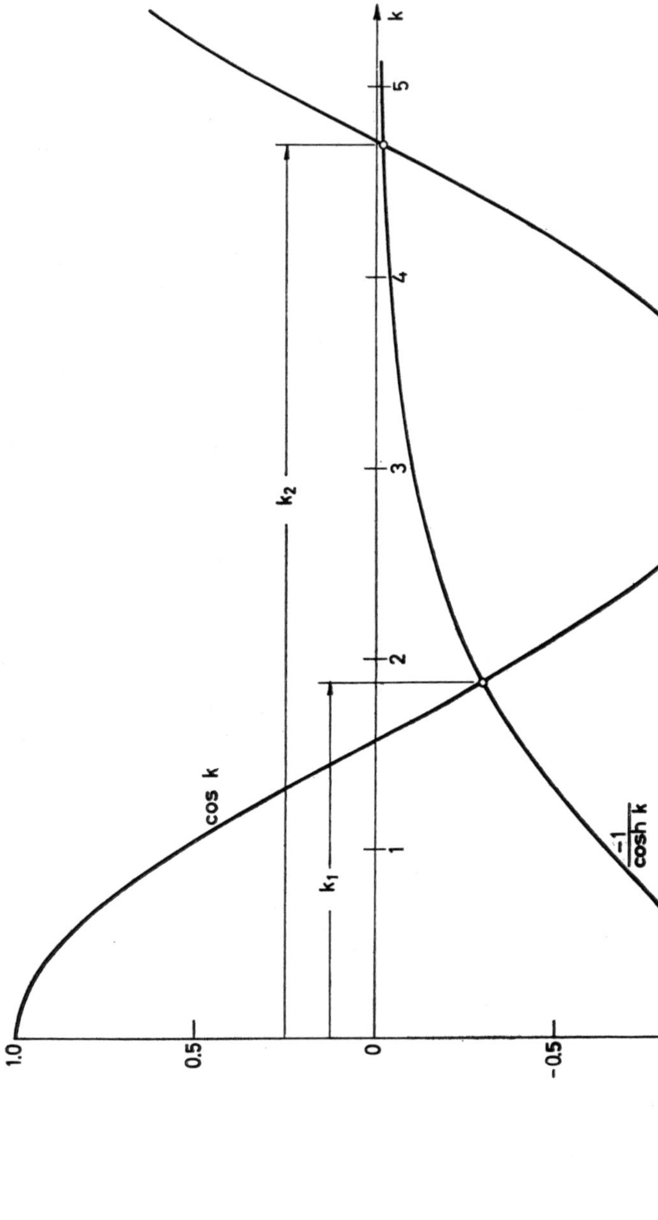

FIG. 4. Solution of the transcendental equation (38).

Each pair of conjugate complex poles represents a natural frequency. If the time-constant of the body is defined as

$$T = h^2 \sqrt{\left(\frac{\gamma ab}{E\theta}\right)}, \tag{40}$$

the absolute value ω_n of the appropriate angular frequency is

$$\omega_n = \frac{\varkappa_n^2}{T}. \tag{41}$$

Specifically, the fundamental frequency is given by:

$$f_1 = \frac{1}{2\pi}\frac{\varkappa_1^2}{T} = \frac{1}{2\pi}\frac{1{\cdot}875^2}{T}. \tag{42}$$

This formula will serve for numerical evaluation of the modulus of elasticity of the upper part of the body. In practice the following figures may be assumed:

$$f_1 = \frac{4}{3}\,\text{cps}; \quad \gamma = 1000\,\frac{\text{kg}}{\text{m}^3}; \quad a = 0{\cdot}4\ \text{m}; \quad b = 0{\cdot}3\ \text{m}; \quad h = 0{\cdot}8\ \text{m}. \tag{43}$$

Hence, by Eq. (9),

$$\theta = \frac{1}{12}\times 0{\cdot}4\times 0{\cdot}3^3 = 9\times 10^{-4}\ \text{m}^4. \tag{44}$$

By Eqs. (42) and (43), the time constant is

$$T = \frac{1{\cdot}875^2}{2\pi}\times\frac{3}{4} = 0{\cdot}42\ \text{sec}, \tag{45}$$

and finally, by Eq. (40), the modulus of elasticity is

$$E = \frac{h^4}{T^2}\frac{\gamma ab}{\theta} = \frac{0{\cdot}8^4}{0{\cdot}42^2}\times\frac{1000\times 0{\cdot}4\times 0{\cdot}3}{9{\cdot}10^{-4}} = 0{\cdot}312\times 10^6\,\frac{\text{Nt}}{\text{m}^2}. \tag{46}$$

In order to satisfy the initial conditions, Eq. (17), the path of integration given previously must be modified—having regard to Eq. (22)—so as to by-pass the singularities, Eq. (39), in each case by an arc in the half-plane $q > 0$ (see Fig. 5).

MOTION OF THE HEAD

For $z = h$, Eqs. (25), (35), (36) and (37) yield

$$w = \frac{1}{p}\,\frac{\cos\alpha h + \cosh\alpha h}{1 + \cos\alpha h\cdot\cosh\alpha h} \tag{47}$$

so that, in accordance with Eq. (22),

$$[u]_{z=h} = -\frac{V}{2\pi i}\int_{-i_\infty}^{i_\infty}\frac{e^{pt}}{p^2}\,\frac{\cos\alpha h+\cosh\alpha h}{1+\cos\alpha h\cosh\alpha h}\,dp \tag{48}$$

describes the displacement of the head, whose velocity is in turn given by

$$[v]_{z=h} = -\frac{V}{2\pi i}\int_{-i_\infty}^{i_\infty}\frac{e^{pt}}{p}\,\frac{\cos\alpha h+\cosh\alpha h}{1+\cos\alpha h\cosh\alpha h}\,dp. \tag{49}$$

Since, accordingly, the component

$$[v_0]_{z=h} = -V \tag{50}$$

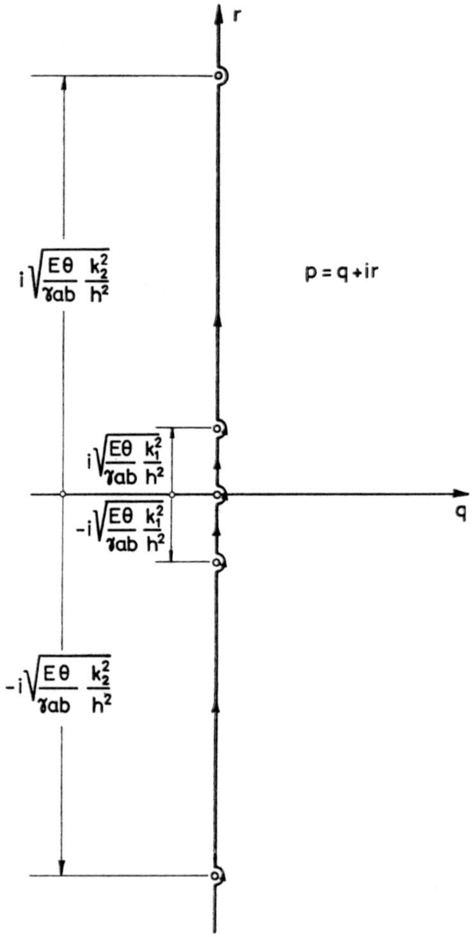

FIG. 5. The path of integration as adapted to the initial conditions.

corresponds to the pole $p = 0$ for $t > 0$, the head, following the impact, merely undergoes non-harmonic oscillation relative to the geographical system of reference. In the vicinity of the poles, Eq. (39), and having regard to Eq. (40), the Taylor series

$$1+\cos \varkappa h \cosh \varkappa h \tag{51}$$
$$= (\sinh \varkappa_n \cos \varkappa_n - \cosh \varkappa_n \sin \varkappa_n) \left[\frac{d(\varkappa h)}{dp}\right]_{p=p_n} \cdot \Delta p + \dots$$
$$= \pm iT \times \frac{1}{2\varkappa_n} \times (\sinh \varkappa_n \cos \varkappa_n - \cosh \varkappa_n \sin \varkappa_n)\, \Delta p + \dots$$

is valid, so that for $t > 0$ the residue theorem yields

$$[u]_{z=h} = -Vt + VT \sum_{n=1}^{\infty} \frac{4}{\varkappa_n^3} \frac{\cos \varkappa_n + \cosh \varkappa_n}{\cosh \varkappa_n \sin \varkappa_n - \sinh \varkappa_n \cos \varkappa_n} \sin \omega_n t. \tag{52}$$

This formula is, however, valid only as long as $|u'| = |u+Vt| < h$, since otherwise the premises of Saint-Venant's theory do not apply. In addition, allowance should then be made for the fact that the head is forcibly flung against a solid object within the vehicle shortly after the impact. Allowing for these restrictions, Eq. (41) yields the velocity of the head following the impact:

$$[v]_{z=h} = -V + V \sum_{n=1}^{\infty} \frac{4}{n} \frac{\cos \varkappa_n + \cosh \varkappa_n}{\cosh \varkappa_n \sin \varkappa_n - \sinh \varkappa_n \cos_n \varkappa_n} \cos \omega_n t. \tag{53}$$

As an example the oscillation of the head will now be calculated, using the same numerical data as above. For $n = 1$,

$$\varkappa_1 = 1{\cdot}875; \quad \cosh \varkappa_1 = 3{\cdot}316; \quad \sinh \varkappa_1 = 3{\cdot}183; \quad \cos \varkappa_1 = -0{\cdot}299;$$
$$\sin \varkappa_1 = 0{\cdot}953, \tag{54}$$

hence

$$\frac{4}{\varkappa_1} \frac{\cosh \varkappa_1 + \cos \varkappa_1}{\cosh \varkappa_1 \sin \varkappa_1 - \sinh \varkappa_1 \cos \varkappa_1} = 1{\cdot}56,$$
$$\frac{4}{\varkappa_1^3} \frac{\cosh \varkappa_1 + \cos \varkappa_1}{\cosh \varkappa_1 \sin \varkappa_1 - \sinh \varkappa_1 \cos \varkappa_1} = 0{\cdot}444. \tag{55}$$

For $n = 2$,

$$\varkappa_2 = 4{\cdot}69; \cosh \varkappa_2 = 54{\cdot}43; \quad \sinh \varkappa_2 = 54{\cdot}42; \quad \cos \varkappa_2 = -0{\cdot}022;$$
$$\sin \varkappa_2 = -0{\cdot}999 \tag{56}$$

and

$$\frac{4}{\varkappa_2} \frac{\cosh \varkappa_2 + \cos \varkappa_2}{\cosh \varkappa_2 \sin \varkappa_2 - \sinh \varkappa_2 \cos \varkappa_2} = -0{\cdot}852,$$
$$\frac{4}{\varkappa_2^3} \frac{\cosh \varkappa_2 + \cos \varkappa_2}{\cosh \varkappa_2 \sin \varkappa_2 - \sinh \varkappa_2 \cos \varkappa_2} = -0{\cdot}0388. \tag{57}$$

For $n > 2$,

$$\varkappa_n \simeq (2n-1)\times\frac{\pi}{2} \tag{58}$$

is sufficiently accurate, hence

$$\frac{4}{\varkappa_n}\,\frac{\cosh\varkappa_n+\cos\varkappa_n}{\cosh\varkappa_n\sin\varkappa_n-\sinh\varkappa_n\cos\varkappa_n} \simeq \frac{2}{\pi(2n-1)}\times(-1)^{n-1}, \tag{59}$$

$$\frac{4}{\varkappa_n^3}\,\frac{\cosh\varkappa_n+\cos\varkappa_n}{\cosh\varkappa_n\sin\varkappa_n-\sinh\varkappa_n\cos\varkappa_n} \simeq \frac{8}{\pi^3(2n-1)^3}\times(-1)^{n-1}. \tag{60}$$

Since the series of Eq. (52) converges rapidly, the term for $n = 1$ suffices, so that Eq. (45) yields

$$[u']_{z=h} = [u+Vt]_{z=h} = V\times 0{\cdot}42\times 0{\cdot}44\sin\omega_1 t = V\times 0{\cdot}185\sin\omega_1 t. \tag{61}$$

Stipulating that the amplitude of this oscillation, u' max, should not exceed

$$u'_{\max} = 0{\cdot}75\ \mathrm{m} \tag{62}$$

the speed of the vehicle may not exceed

$$V_{\max} = \frac{0{\cdot}75}{0{\cdot}185}\,\frac{\mathrm{m}}{\mathrm{sec}} = 4{\cdot}06\,\frac{\mathrm{m}}{\mathrm{sec}} = 14{\cdot}6\,\frac{\mathrm{km}}{\mathrm{h}}. \tag{63}$$

This value is surprisingly low, but it should be borne in mind that it is based on the extreme assumption of instantaneous braking. In practice, it may be assumed that the vehicle is braked within a *finite* time Δt at a uniform deceleration:

$$a = -\frac{V}{\Delta t}. \tag{64}$$

Using the convolution integral, Eq. (52) is replaced, for $t > \Delta t$, by the more moderate

$$[u]_{z=h} = -V\left(t-\frac{\Delta t}{2}\right)$$

$$+VT\sum_{n=1}^{\infty}\frac{4}{\varkappa_n^3}\,\frac{\cos\varkappa_n+\cosh\varkappa_n}{\cosh\varkappa_n\sin\varkappa_n-\sinh\varkappa_n\cos\varkappa_n}\,\frac{\sin(\omega_n\Delta t/2)}{(\omega_n\Delta t/2)}\sin\left[\omega_n\left(t-\frac{\Delta t}{2}\right)\right] \tag{65}$$

Then, if a time $\Delta t = \frac{1}{2}$ sec elapses until the final standstill of the vehicle, the amplitude of the fundamental oscillation following the "braked" impact will be reduced in relation to that of the instantaneous impact by the attenuation factor:

$$\left|\frac{\sin\dfrac{\omega_1\Delta t}{2}}{\dfrac{\omega_1\Delta t}{2}}\right| = \left|\frac{\sin\pi\dfrac{4}{3}\dfrac{1}{2}}{\pi\dfrac{4}{3}\dfrac{1}{2}}\right| = 0{\cdot}413 \tag{66}$$

so that, other conditions being equal, the permissible maximum speed may be increased to

$$V_{\max} = \frac{4{\cdot}06}{0{\cdot}413}\frac{\text{m}}{\text{sec}} = 9{\cdot}85\frac{\text{m}}{\text{sec}} = 35{\cdot}6\frac{\text{km}}{\text{h}}. \tag{67}$$

For determining the acceleration of the head due to the impact, a finite braking time must be assumed. This acceleration is of extreme importance in view of its physiological effect on the brain, as an impact of the kind described frequently results in severe concussion.

For $t > \Delta t$, Eqs. (41) and (65) yield:

$$\left[\frac{\partial^2 u}{\partial t^2}\right] = -\frac{V}{T}\sum_{n=1}^{\infty} 4\varkappa_n \frac{\cos\varkappa_n + \cosh\varkappa_n}{\cosh\varkappa_n \sin\varkappa_n - \sinh\varkappa_n \cos\varkappa_n}\,\frac{\sin(\omega_n \Delta t/2)}{\omega_n \Delta t/2} \times \sin\left[\omega_n\left(t - \frac{\Delta t}{2}\right)\right]. \tag{68}$$

Disregarding the poor convergence of this series and making do with its first term, we have:

$$\left[\frac{\partial^2 u_{1,\max}}{\partial t^2}\right]_{z=h} = \frac{V}{T}\times 4\varkappa_1 \frac{\cos\varkappa_1 + \cosh\varkappa_1}{\cosh\varkappa_1 \sin\varkappa_1 - \sinh\varkappa_1 \cos\varkappa_1}\left|\frac{\sin(\omega_1 \Delta t/2)}{\omega_1 \Delta t/2}\right| \tag{69}$$

for the amplitude of the fundamental oscillation of the acceleration of the head. Substituting the numerical data as above, $\Delta t = \frac{1}{2}$ sec, and the maximum speed by Eq. (67), we have the amplitude

$$\left[\frac{\partial^2 u_{1,\max}}{\partial t^2}\right]_{z=h} = \frac{9{\cdot}85}{0{\cdot}42}\times 1{\cdot}875^4 \times 0{\cdot}444 \times 0{\cdot}413 = 53{\cdot}5\frac{\text{m}}{\text{sec}^2} \tag{70}$$

which exceeds gravitational acceleration fivefold and demonstrates the extreme danger involved.

THE STRAPPING-IN MOMENT

In the plane $z = 0$,

$$\left[\frac{d^2 w}{dz^2}\right]_{z=0} = \alpha^2[C-B] = \frac{\alpha^2}{p}\,\frac{\sin\alpha h \sinh\alpha h}{1+\cos\alpha h \cosh\alpha h} \tag{71}$$

is valid by Eqs. (25), (31) and (32). Hence, based on Eqs. (10), (22) and (24), the "strapping-in moment" M_0, which raises the body, is given by:

$$M_0 = \gamma abh^2 \frac{V}{2\pi i}\int_{-i\infty}^{i\infty} \frac{1}{(\alpha h)^2}\,\frac{\sin\alpha h \sinh\alpha h}{1+\cos\alpha h \cosh\alpha h}\, e^{pt}\, dp, \tag{72}$$

indicating absence of a permanent moment after the impact. Substituting Eq. (51), the divergent series:

$$M_0 = \frac{\gamma abh^2V}{T} \sum_{n=1}^{\infty} 4\varkappa_n \frac{\sin \varkappa_n \sinh \varkappa_n}{\cosh \varkappa_n \sin \varkappa_n - \sinh \varkappa_n \cos \varkappa_n} \sin \omega_n t \tag{73}$$

is obtained for $t > 0$, representing solely the (non-harmonic) natural oscillations of the body. A series useful for $t > \Delta t$ is only obtainable for a finite braking time:

$$M_0 = \frac{\gamma abh^2V}{T} \sum_{n=1}^{\infty} 4\varkappa_n \frac{\sin \varkappa_n \sinh \varkappa_n}{\cosh \varkappa_n \sin \varkappa_n - \sinh \varkappa_n \cos \varkappa_n} \frac{\sin (\omega_n \Delta t/2)}{\omega_n \Delta t/2} \times \sin \left[\omega_n \left(t - \frac{\Delta t}{2} \right) \right] \tag{74}$$

so that

$$M_{0,1_{\max}} = \frac{\gamma abh^2V}{T} \times 4\varkappa_1 \times \frac{\sin \varkappa_1 \sinh \varkappa_1}{\cosh \varkappa_1 \sin \varkappa_1 - \sinh \varkappa_1 \cos \varkappa_1} \left| \frac{\sin (\omega_1 \Delta t/2)}{\omega_1 \Delta t/2} \right| \tag{75}$$

is the amplitude of the fundamental frequency of the strapping-in moment. By Eq. (8), this induces in the back a tensile stress with amplitude:

$$\sigma_{\max} = \frac{b}{\theta} M_{0,1_{\max}} . \tag{76}$$

Substituting the relevant data, and $V = V_{\max}$ by Eq. (67) for $\Delta t = \frac{1}{2}$ sec, we have

$$M_{0,1_{\max}} = \frac{1000 \times 0{\cdot}4 \times 0{\cdot}3 \times 0{\cdot}64 \times 9{\cdot}85}{0{\cdot}42} \times 4 \times 1{\cdot}875 \times \frac{3{\cdot}183 \times 0{\cdot}953}{3{\cdot}316 \times 0{\cdot}953 + 3{\cdot}183 \times 0{\cdot}299} \times 0{\cdot}413 = 5620 \text{ mNt} = 573 \text{ m kgf} \tag{77}$$

and

$$\sigma_{\max} = \frac{0{\cdot}3}{9 \times 10^{-4}} \times 5620 \frac{\text{Nt}}{\text{m}^2} = 1{\cdot}6 \times 10^6 \frac{\text{Nt}}{\text{m}^2} = 16{\cdot}3 \frac{\text{kgf}}{\text{cm}^2} . \tag{78}$$

A stress of the same order of magnitude acts on the kidneys, and the question is whether they can survive it.

BELT FORCE

The lateral force $[F_y]_{z=0}$ in the plane $z = 0$ must be taken up in equal parts by either side of the belt—disregarding, in this instance, the seat friction. Consequently, the force

$$G = \tfrac{1}{2}\,[F_y]_{z=0} \tag{79}$$

induces a stress. By Eqs. (10) and (11)

$$[F_y]_{z=0} = -E\theta\left[\frac{\partial^3 u}{\partial z^3}\right]_{z=0} \tag{80}$$

is valid. Since, by Eqs. (25) and (35),

$$\left[\frac{d^3 w}{dz^3}\right]_{z=0} = -2A\alpha^3 = -\frac{\alpha^3}{p}\,\frac{\sin\alpha h\cosh\alpha h+\cos\alpha h\sinh\alpha h}{1+\cos\alpha h\cosh\alpha h} \tag{81}$$

the belt force is given by:

$$G = \frac{\gamma abhV}{2}\,\frac{1}{2\pi i}\int_{-i_\infty}^{i_\infty}\frac{1}{\alpha h}\,\frac{\sin\alpha h\cosh\alpha h+\cos\alpha h\sinh\alpha h}{1+\cos\alpha h\cosh\alpha h}\,e^{pt}\,dp. \tag{82}$$

Substituting Eq. (51), we again have a divergent series for the natural oscillations at $t > 0$:

$$G = \frac{\gamma abhV}{T}\sum_{n=1}^{\infty}2\varkappa_n^2\times\frac{\cosh\varkappa_n\sin\varkappa_n+\sinh\varkappa_n\cos\varkappa_n}{\cosh\varkappa_n\sin\varkappa_n-\sinh\varkappa_n\cos\varkappa_n}\cos\omega_n t, \tag{83}$$

but assuming a finite braking time we obtain a series useful for all times $t > \Delta t$, namely:

$$G = \frac{\gamma abhV}{T}\sum_{n=1}^{\infty}2\varkappa_n^2\,\frac{\cosh\varkappa_n\sin\varkappa_n+\sinh\varkappa_n\cos\varkappa_n}{\cosh\varkappa_n\sin\varkappa_n-\sinh\varkappa_n\cos\varkappa_n}\,\frac{\sin(\omega_n\Delta t/2)}{\omega_n\Delta t/2}\times\cos\left[\omega_n\left(t-\frac{\Delta t}{2}\right)\right] \tag{84}$$

the amplitude of whose fundamental oscillation is given by:

$$G_{1,\,\max} = \frac{\gamma abhV}{T}\times 2\varkappa_1^2\times\frac{\cosh\varkappa_1\sin\varkappa_1+\sinh\varkappa_1\cos\varkappa_1}{\cosh\varkappa_1\sin\varkappa_1-\sinh\varkappa_1\cos\varkappa_1}\left|\frac{\sin(\omega_1\,\Delta t/2)}{\omega_1\Delta t/2}\right| \tag{85}$$

and substituting the data as before we have:

$$G_{1,\,\max} = \frac{1000\times 0{\cdot}4\times 0{\cdot}3\times 0{\cdot}8\times 9{\cdot}85}{0{\cdot}42}\times 2\times 1{\cdot}875$$

$$\times\frac{3{\cdot}316\times 0{\cdot}953-3{\cdot}183\times 0{\cdot}299}{3{\cdot}316\times 0{\cdot}953+3{\cdot}183\times 0{\cdot}299}\times 0{\cdot}413\text{ Nt} = 3540\text{ Nt} = 360\text{ kgf}. \tag{86}$$

This means that the lateral force tending to detach the upper part of the body is doubled! However, in practice, the finite width of the belt should be allowed for, and the resulting pressure per unit area of the belt is relatively low for the usual dimensions of safety belts.

CONCLUSIONS

The behaviour of the human body—replaced here by a highly idealized model for the purpose of this analysis—strapped to its seat by means of a hip belt is represented by the partial differential equation for rod vibrations. This equation was integrated by means of the theory of functions, allowing for the initial and boundary conditions dictated by the preceding equilibrium. The resultant motion was derived for the natural oscillations of the upper part of the body. On the basis of certain plausible assumptions as to the elasticity parameters involved it was found that a speed as low as 15 km/h leads to dangerous displacement of the head in the case of instantaneous braking, and even an assumption of a braking time of $\frac{1}{2}$ sec does not raise the permissible maximum speed above 36 km/h. At the same time the brain is subjected to acceleration exceeding that of gravity about fivefold, and the internal organs undergo heavy mechanical stresses. In the light of these results, the protection afforded by a hip belt must be considered inadequate.

The method developed herein is readily applicable to other protective devices. In particular, a belt around one or both shoulders promises to be more effective. The quantitative aspect of such a device is the subject of a separate study.

MATERIAL PROPERTIES

THE RONAY EFFECT IN MEDIA SHOWING CREEP

Alfred M. Freudenthal

Department of Civil Engineering and Engineering Mechanics, Columbia University, New York, U.S.A.

ABSTRACT

The complete constitutive equation of a viscous medium is used to illustrate the accumulation of second-order axial extension under cyclic axial load and cyclic torsion. Theoretical "times to failure" are established on a similar basis as the time to (creep) failure under sustained axial load.

NOTATION

a_1, a_2	first- and second-order parameters
A, A_0	cross section, initial cross section
C_0, C_1, C_2	zero, first- and second-order compliances
d	diameter
e_{ij}	components of strain deviation
J_2, J_2'	second invariant of stress tensor, stress deviation
L, L_0	length, initial length
M_t	torque
p, P	hydrostatic pressure, axial force
t, t_F	time, time to failure
x	time variable
y, z	variables denoting area contraction
δ_{ij}	Kronecker delta
ε_{ij}, ε	strain component, axial strain
η	coefficient of shear viscosity
λ	coefficient of viscous traction
σ_{ij}, σ	stress component, axial stress
τ	shear stress
ω	frequency

1. INTRODUCTION

In recent experiments in which small cylindrical specimens of superpure polycrystalline aluminium at room temperature were subjected to reversed cyclic torsion producing small surface shear strains ($\pm 7.5\times 10^{-4}$ to $\pm 60\times 10^{-4}$) the accumulation of irreversible longitudinal strain increments at double the frequency of the torsion cycle was demonstrated [1]. This accumulation, which for large numbers of torsion cycles produced large permanent extensions and severe lateral contraction in solid specimens and collapse (buckling) characteristic of instability under external pressure in thin-walled tubular specimens could be related to the existence, in the constitutive equation of strain-hardening media, of second-order terms [2] which give rise to a hydrostatic pressure and to normal stresses in the plane of the motion.

In elastic media the effects of these second-order stresses are known as the Kelvin and the Poynting effects, respectively [3]. In incompressible polymeric fluids the effect of the normal stresses [4] is known as the Weissenberg effect [5], while the volume change associated with shear in granular media is known as dilatancy or the Reynolds effect [6]. The theory of these second-order effects under conditions of unidirectional straining has been developed by Reiner in two classical papers [7].

The recently discovered second-order strain accumulation in strain-hardening media under reversed cyclic straining [1], to be referred to as the Ronay effect, differs from the second-order effects in elastic solids and polymeric fluids by the fact that by increasing the number of strain cycles it can be magnified to such an extent that it becomes technically significant and easily measurable. The extension of the experimental investigation on superpure aluminium to elevated temperatures between 100°C and 400°C has shown a substantial increase of the axial strain increment with increasing temperature [8] which might be the result of the increase with temperature of the significance of the viscous component in the mechanical response of the polycrystalline metal (Fig. 1). If this were the case, the implication would be that the second-order axial strain accumulation under reversed torsion at elevated temperatures is the result of a combination of strain-hardening (plastic) and of viscous second-order parameters in the constitutive equation.

In view of the considerable complexity of constitutive equations for combined strain-hardening–viscous media as well as in view of the difficulty involved in the analytic solution of even simple strain-hardening problems it may be of interest to illustrate the implications of the Ronay effect by the analysis of the problem of reversed cyclic torsion or tension of a thin cylindrical tube of a visco-elastic medium that shows creep under a sustained load. It has been shown that within the range of significant creep the effect of the elastic component is insignificant [9]. It has also been shown that the second-order effects due to tensorial non-linearity of a visco-elastic constitutive equation are independent of an existing parametric non-linearity [10]. Consequently the use of a parametrically linear viscous medium can be justified as a first approximation.

2. CONSTITUTIVE RELATIONS

The constitutive equation for a viscous medium is of the general form [11]

$$\dot{\varepsilon}_{ij} = C_0\delta_{ij}+C_1\sigma_{ij}+C_2\sigma_{ik}\sigma_{kj}; \tag{2.1}$$

assuming the medium to be incompressible the condition $\varepsilon_{ii} = 0$ has to be satisfied for $\sigma_{ii} = 3p$. Hence $3C_0+3pC_1+C_2\sigma_{ik}\sigma_{ik} = 0$ and therefore $-C_0 = pC_1+2C_2J_2/3$ since $\sigma_{ik}\sigma_{ik} = 2J_2$. Hence

$$\dot{\varepsilon}_{ij}= \dot{e}_{ij} = C_1' s_{ij}+C_2(s_{ik}\, s_{ki}-\tfrac{2}{3}\, J_2'\delta_{ij}), \tag{2.2}$$

where $s_{ij} = \sigma_{ij}-p\delta_{ij}$, $2J_2' = s_{ij}s_{ij}$ and $C_1' = C_1+2C_2p$.

For uniaxial tension of a bar of cross section A $\sigma_{ij} = 0$ for all values i and j except $\sigma_{33} = \sigma = P/A$. Hence $s_{11} = s_{22} = -\sigma/3 = -p$, $s_{22} = 2\sigma/3$, $J_2' = \sigma^2/3$ and therefore

$$[\dot{e}_{ij}] = C_1\sigma \begin{bmatrix} -\frac{1}{3} & 0 & 0 \\ 0 & -\frac{1}{3} & 0 \\ 0 & 0 & \frac{2}{3} \end{bmatrix} + \tfrac{1}{9}\, C_2\sigma^2 \begin{bmatrix} -1 & 0 & 0 \\ 0 & -1 & 0 \\ 0 & 0 & 2 \end{bmatrix} \tag{2.3}$$

Thus the onedimensional equation in the direction of the axis

$$\dot{\varepsilon}_{33} = \dot{\varepsilon} = \tfrac{2}{3}\, C_1\sigma+\tfrac{2}{9}\, C_2\sigma^2. \tag{2.4}$$

Comparison with the uniaxial equation for the classical viscous medium $\dot{\varepsilon} = \sigma/\lambda = \sigma/3\eta$ shows that $C_1 = \frac{1}{2}\eta = 3/2\lambda$.

In torsion about the axis $\sigma_{23} = \sigma_{32} = \sigma_{13} = \sigma_{31} = \tau = \tau_{max}$ while all other stress components are zero. Hence with $p = 0$ and $J_2' = \tau^2$

$$[\dot{e}_{ij}] = C_1\tau \begin{bmatrix} 0 & 0 & 1 \\ 0 & 0 & 1 \\ 1 & 1 & 0 \end{bmatrix} + C_2\tau^2 \begin{bmatrix} -\frac{1}{3} & 0 & 0 \\ 0 & -\frac{1}{3} & 0 \\ 0 & 0 & +\frac{2}{3} \end{bmatrix}. \tag{2.5}$$

Thus the one-dimensional relation in the direction of the axis

$$\dot{\varepsilon}_{33} = \dot{\varepsilon} = \tfrac{2}{3}\, C_2 \tau_{max}^2 , \tag{2.6}$$

where for a tubular cross-section of diameter d subject to a torque M_t: $\tau_{max} = 2M_t/\pi d^2 \delta$. The strain-rate in axial direction for finite deformation

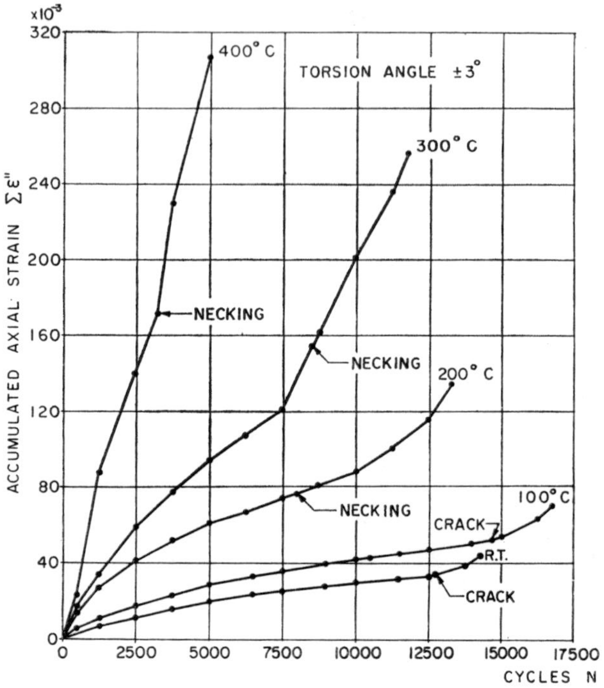

FIG. 1

of an incompressible cylindrical bar is defined by

$$\dot{\varepsilon} = \frac{1}{L}\frac{dL}{dt} = -\frac{1}{A}\frac{dA}{dt} . \tag{2.7}$$

3. OSCILLATING AXIAL FORCE

The differential equation for the (massless) cylindrical bar subject to an oscillating axial force $P = P_0 . \sin \omega t$ is obtained by introducing

$$\sigma = \frac{P}{A} = \frac{P_0}{A} \sin \omega t \tag{3.1}$$

into Eq. (2.4) in which $\dot{\varepsilon}$ is expressed by Eq. (2.7). Hence

$$-\frac{1}{A}\frac{dA}{dt} = \frac{2}{3}C_1\frac{P_0}{A}\sin\omega t+\frac{2}{9}C_2\left(\frac{P_0}{A}\right)^2\sin^2\omega t. \tag{3.2}$$

With the abbreviations

$$\sigma_0 = \frac{P_0}{A_0}, \quad \omega t = x, \quad \frac{A}{A_0} = z; \quad \frac{2}{3}\frac{C_1\sigma_0}{\omega} = a_1; \quad \frac{2}{9}\frac{C_2\sigma_0^2}{\omega} = a_2 \tag{3.3}$$

Eq. (3.2) takes the form

$$z\frac{dz}{dx} = -a_1 z\sin x - a_2\sin^2 x \tag{3.4}$$

and, with $v = z^2$, is transformed into

$$\frac{dy}{dx}+2a_1\sqrt{v}\sin x = -2a_2\sin^2 x. \tag{3.5}$$

It is obvious that the solution of the homogeneous equation ($a_2 = 0$) with the initial condition $v = z^2 = 1$ for $x = 0$

$$\sqrt{v} = z = 1-a_1(1-\cos x) \tag{3.6}$$

represents the linear viscous response. Since $z^{-1} = A_0/A = L/L_0 = (L_0+\Delta L)/L_0$, Eq. (3.6) can be written in the form

$$\frac{L_0}{L_0+\Delta L} \doteq 1-\frac{\Delta L}{L_0} = 1-a_1(1-\cos x)$$

or

$$\frac{\Delta L}{L_0} = a_1(1-\cos x) = \frac{\sigma_0}{\lambda\omega}(1-\cos\omega t) \tag{3.7}$$

valid for values of $\Delta L/L_0 \ll 1$.

The particular integral of the non-homogeneous Eq. (3.5) represents therefore the second-order effect. While such an integral is not readily obtainable, an approximation in the vicinity of $v \to 0$ or $z \to 0$ ($A \to 0$) is obtained by neglecting the second term on the left-hand side of Eq. (3.5). This approximation is of the form

$$v = 1-a_2(x-\tfrac{1}{2}\sin 2x) \tag{3.8}$$

or

$$\frac{A}{A_0} = \frac{L_0}{L} = \left(1+\frac{\Delta L}{L_0}\right)^{-1} = \left[1-\frac{2}{9}C_2\sigma_0^2 t\left(1-\frac{\sin 2\omega t}{2\omega t}\right)\right]^{\frac{1}{2}}. \tag{3.9}$$

If the time to failure t_F is defined as the time at which the maximum stress amplitude $\sigma = P_0/A$ attains the fracture stress [9], Eq. (3.9) can be solved for t_F for $A = A_F = P_0/\sigma_F$. Hence

$$\frac{2}{9} C_2\sigma_0^2 t_F \left(1-\frac{\sin 2\omega t_F}{2\omega t_F}\right) = 1-\left(\frac{\sigma_0^2}{\sigma_F}\right). \tag{3.10}$$

For sufficiently long times t_F the time to fracture is independent of frequency and can be expressed in the form

$$t_F = \left(\frac{1}{9} C_2\sigma_0^2\right)^{-1}\left[1-\left(\frac{\sigma_0}{\sigma_F}\right)^2\right] = t_0\left[1-\left(\frac{\sigma_0}{\sigma_F}\right)^2\right], \tag{3.11}$$

where t_0 is the failure time defined by $A \to 0$.

In Fig. 2 the functions (L/L_0) versus t according to Eqs. (3.9) and (3.10) have been plotted. They describe the behavior within the range of values (L/L_0) close to unity and the asymptotic behavior in the vicinity of $t = t_0$. For the intermediate range an exact solution of Eq. (3.4) would have to be obtained.

4. OSCILLATING TORQUE

The differential equation for the (massless) cylindrical bar subject to an oscillating torque $M_t = M_{t0} \sin \omega t$ acting at one end of the bar while the other is fixed is obtained by introducing

$$\tau_{\max} = \tau_0 \left(\frac{A_0}{A}\right)^{\frac{3}{2}} \sin \omega t \tag{4.1}$$

and Eq. (2.7) into Eq. (2.6). Hence

$$-\frac{1}{A}\frac{dA}{dt} = \frac{1}{3} C_2\tau_0^2 \left(\frac{A_0}{A}\right)^3 \sin^2 \omega t \tag{4.2}$$

with the abbreviations

$$\omega t = x, \quad \frac{A}{A_0} = z, \quad \frac{1}{3}\frac{C_2\tau_0^2}{\omega} = a_3 \tag{4.3}$$

Eq. (4.2) takes the form

$$z^2\frac{dz}{dx} = -a_3 \sin^2 x, \tag{4.4}$$

which can be directly integrated under the initial condition $z = 1$ for $x = 0$:

$$z^3 = 1-\frac{3}{2} a_3 \left(x-\frac{1}{2} \sin 2x\right) \tag{4.5}$$

or

$$\frac{A}{A_0} = \frac{L_0}{L} = \left(1+\frac{\Delta L}{L_0}\right)^{-1} = \left[1-\frac{1}{2}\,C_2\tau_0^2 t\left(1-\frac{\sin 2\omega t}{2\omega t}\right)\right]^{\frac{1}{3}}. \tag{4.6}$$

Since no force is acting in the direction of the extension, failure is defined by the condition $A \to 0$. Hence the time to failure is obtained from the equation

$$C_2\tau_0^2 t_F\left(1-\frac{\sin 2\omega t_F}{2\omega t_F}\right) = 2. \tag{4.7}$$

For sufficiently long times the time to failure is approximately independent of frequency, tending towards

$$t_F = (C_2\tau_0^2)^{-1}. \tag{4.8}$$

In Fig. 2 the function (L/L_0) versus t according to Eq. (4.6) is also plotted.

It should be noted that for the sake of simplicity the equations have been derived without considering that the distribution of the shear stress is non-uniform over solid sections and that, therefore, the distribution of the second-order axial strain is also non-uniform. However, the use of $\tau_{\max}$ instead of an average value of τ affects the magnitude of the total axial elongation only slightly.

5. INTERACTION OF OSCILLATING TORQUE WITH CONSTANT AXIAL FORCE

For interaction of torsion and tension $\sigma_{23} = \sigma_{32} = \sigma_{13} = \sigma_{31} = \tau$ and $\sigma_{33} = \sigma$ while all other stress-components are zero. Hence the matrix equation

$$[e_{ij}] = C_1\begin{bmatrix} -\frac{1}{3}\sigma & 0 & \tau \\ 0 & -\frac{1}{3}\sigma & \tau \\ \tau & \tau & \frac{2}{3}\sigma \end{bmatrix} + C_2\begin{bmatrix} -\frac{1}{9}(\sigma^2+6\tau^2) & \tau & \sigma\tau \\ \tau^2 & -\frac{1}{9}(\sigma^2-3\tau^2) & \sigma\tau \\ \sigma\tau & \sigma\tau & \frac{2}{9}(\sigma^2+\frac{3}{2}\tau^2) \end{bmatrix}. \tag{5.1}$$

The one-dimensional relation in the direction of the axis is therefore

$$\dot{\varepsilon}_{33} = \dot{\varepsilon} = \tfrac{2}{3}C_1\sigma + \tfrac{2}{9}C_2(\sigma^2+\tfrac{3}{2}\tau^2). \tag{5.2}$$

Introducing $P = P_0 =$ const and Eqs. (2.7) and (4.1), Eq. (5.2) takes the form

$$-\frac{1}{A}\,\frac{dA}{dt} = \frac{2}{3}\,C_1\sigma_0\left(\frac{A_0}{A}\right) + \frac{2}{9}\,C_2\left[\sigma_0^2\left(\frac{A_0}{A}\right)^2 + \frac{3}{2}\tau_0^2\left(\frac{A_0}{A}\right)^3 \sin^2\omega t\right]. \tag{5.3}$$

With the abbreviations

$$\omega t = x, \frac{A}{A_0} = z, \quad \frac{2}{3}\frac{C_1\sigma_0}{\omega} = a_1, \quad \frac{2}{9}\frac{C_2\sigma_0^2}{\omega} = a_2 \tag{5.4}$$

it is transformed into

$$z\frac{dx}{dx} = -a_1 z - a_2 - \frac{3}{2}a_2\left(\frac{\tau_0}{\sigma_0}\right) z^{-1}\sin^2 x \tag{5.5}$$

and, with $v = z^2$, into

$$\frac{dv}{dx} = -2a_1\sqrt{v} - 2a_2 - 3a_2\left(\frac{\tau_0}{\sigma_0}\right)^2 \frac{1}{\sqrt{v}}\sin^2 z. \tag{5.6}$$

For values of the variable v close to unity the principal effect is that of the first term on the right-hand side. Hence Eq. (5.6) degenerates into

$$\frac{dv}{dx} + 2a_1\sqrt{v} = 0 \tag{5.7}$$

the solution of which for $v = 1$ at $x = 0$ is

$$\sqrt{v} = 1 - a_1 x = z = \frac{A}{A_0}. \tag{5.8}$$

Introducing a fracture stress $\sigma = \sigma_F$ this equation predicts the time to fracture under sustained stress alone [9]

$$x = a_1^{-1}\left(1 - \frac{\sigma_0}{\sigma_F}\right) \quad \text{or} \quad t_F = (\lambda/\sigma_0)\left(1 - \frac{\sigma_0}{\sigma_F}\right). \tag{5.9}$$

However, it can be seen from Eq. (5.6) that as v decreases, the last term on the right-hand side tends to become of a magnitude comparable with that of the first term in spite of the fact that the ratio (a_2/a_1) is rather small since the ratio C_2/C_1 is very small; in all cases the constant term $2a_2$ can be neglected. Thus the interaction Eq. (5.6) can be approximated by

$$\frac{dv}{dx} + 2a_1\sqrt{v} + 3a_2\left(\frac{\tau_0}{\sigma_0}\right)^2 \sin^2 x \frac{1}{\sqrt{v}} = 0. \tag{5.10}$$

While a solution of this equation is not readily obtainable an approximation of the time to failure due to the interaction can be obtained by considering that the approach to the condition of instability failure due to cyclic torsion alone is extremely slow over most of the time (Fig. 2); moreover, it can be seen from the form of Eq. (5.10) that as the instability condition is approached ($v \to 0$) the significance of the second term decreases while that of the third term increases. Hence it seems reasonable to consider that the cross section ratio $(A/A_0) = \sqrt{v}$ used in the determination of the shear stress τ_{max} is a function of time of the form suggested by the solution of Eq. (5.8). Hence $\sqrt{v}$ in the third term of Eq. (5.10) is replaced by $\sqrt{v}\,(1-a_1x)$ while the second is eliminated, producing the equation

$$\frac{dv}{dx}+3a_2\left(\frac{\tau_0}{\sigma_0}\right)^2\frac{1}{\sqrt{v}}\,\frac{\sin^2 x}{(1-a_1x)}=0, \tag{5.11}$$

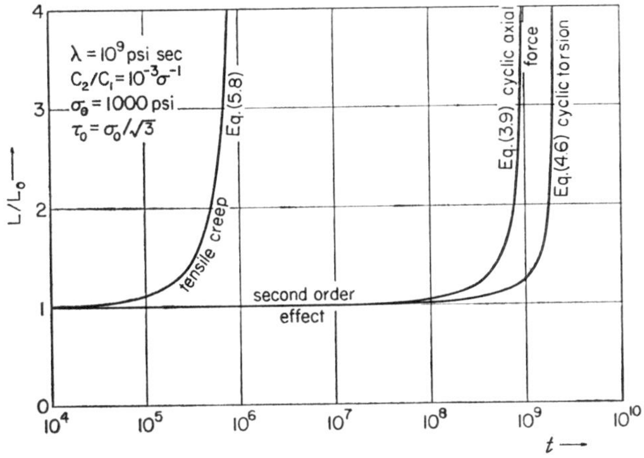

FIG. 2.

the solution of which, for the initial condition $v = 0$ for $x = 0$, can be written in the form

$$v^{3/2} = 1-\frac{9}{2}a_2\left(\frac{\tau_0}{\sigma_0}\right)^2\int\frac{\sin^2 x}{1-a_1x}\,dx. \tag{5.12}$$

If failure is defined by $\sqrt{v} \to 0$, the time to failure $t_F = x_F\omega^{-1}$ is obtained by solving the equation

$$\int\frac{\sin^2 x_F}{(1-a_1x_F)}\,dx = \frac{2}{9a_2}\left(\frac{\sigma_0}{\tau_0}\right)^2, \tag{5.13}$$

which, in first rough approximation, can be replaced by the expression

$$x_F \sim \frac{1}{a_1}\left[1-\exp\left(-\frac{4a_1}{9a_2}\cdot\frac{\sigma_0^2}{\tau_0^2}\right)\right] \sim \frac{1}{a_1}, \tag{5.14}$$

which illustrates the insignificance of the reduction of the time to failure due to the axial load resulting from the second-order effect of cyclic torsion. Equation (5.8) has been plotted in Fig. 2 in order to illustrate the difference in the rate of approach to failure instability due to tension.

An interesting second-order interaction phenomenon arises when a cyclic axial force is superimposed on a sustained torque. This follows directly if the relation between the shear strain rate $\dot{\varepsilon}_{13} = \dot{\varepsilon}_{23} = \dot{\gamma}$ and the shear stress is deduced from Eq. (5.1) under the assumption $\sigma_{33} = \sigma_0 \sin \omega t$ and $\sigma_{13} = \sigma_{23} = \tau_0 = \text{const}$:

$$\dot{\gamma} = C_1\tau + C_2\sigma_0\tau_0 \sin \omega t \tag{5.15}$$

while $\dot{\varepsilon}_{33} = \dot{\varepsilon}$ is governed by Eq. (2.4) or (3.2). Thus, the applied cyclic tension is associated with a secondorder cyclic shear component which is superimposed on the steady creep produced by the applied constant torque. The cyclic shear is proportional to the product $(\sigma \cdot \tau)$, suggesting the existence of a linear second-order interaction effect between cyclic tension and sustained torsion.

In certain polycrystalline metals in which a superimposed cyclic stress is known to accelerate the steady creep under constant stress in the same direction [12], the superposition of a second-order cyclic shear stress component associated with cyclic tension would lead to an acceleration of the creep in shear under the sustained torque [13].

This is the case whether or not the response of the material is viscous, elastic or strain-hardening or a combination of these responses, since in first approximation an equation of a form similar to Eq. (5.15) but involving the strain γ or strain increment $d\gamma$ can be derived for any type of isotropic material.

REFERENCES

1. M. Ronay, *British J. Appl. Phys.* **16,** 727 (1965).
2. A. M. Freudenthal and M. Ronay, *Proc. Royal Soc.* A **292,** 14 (1966).
3. M. Reiner, *Deformation, Strain and Flow,* H. K. Lewis, London, pp. 309, 317 (1960).
4. A. S. Lodge, *Elastic Fluids,* Academic Press, New York, pp. 231–46 (1964).
5. K. Weissenberg, *Nature* **159,** 310 (1947).
6. Reference 3, p. 307.
7. M. Reiner, *Amer. J. Math.* **67,** 350 (1945); **70,** 433 (1948).
8. M. Ronay, *Internat. J. Solid and Structures,* **3,** 167 (1967).
9. M. Reiner and A. M. Freudenthal, *Proc. 5th Int. Congr. Appl. Mech.,* Cambridge, Mass., 1938, J. Wiley, New York, p. 228 (1939).
10. E. M. Lenoe, R. A. Heller, and A. M. Freudenthal, *Trans. Soc. Rheol.,* **9,** Part 2, p. 97 (1965).
11. A.M. Freudenthal and H. Geiringer, "Math. theories of the inelastic continuum", *Handbuch d. Physik* (S. Fluegge, Ed.), vol. 6, p. 258.
12. A. J. Kennedy, *Processes of Creep and Fatigue in Metals,* J. Wiley & Sons, New York, p. 453 (1963).
13. L. F. Coffin, *J. Basic Eng.* (*ASME*), **82,** 671 (1960).

THE "YIELD POINT" OF CONCRETE

JOSEPH GLUCKLICH

Department of Mechanics, Technion—Israel Institute of Technology, Haifa

ABSTRACT

In formulating a strength theory for concrete, as well as for other purposes, a definition of concrete fracture, common to all states of stress and to all conditions, is required. Fracture-mechanics considerations are used in distinguishing two phases in the fracture process, viz. slow and fast crack propagation. It is proposed to define fracture as the onset of fast propagation. In view of the similarity in behaviour to plastic materials, this is termed the "yield point" of concrete, although it is realized that concrete has no true plastic deformation.

Three alternative methods are proposed for identifying this "yield point" in the course of a compression test. These are: (a) strain-controlled loading; (b) creep tests at varying stress levels; and (c) volume-change measurements.

1. INTRODUCTION

A strength criterion has not been determined to-date for concrete. For ductile metals (such as mild steel), it has long been established that the Mises–Hencky criterion of distortional strain-energy is applicable. This conclusion could be arrived at in the case of ductile metals through both theoretical and empirical approaches. In the case of concrete, the theoretical difficulties are almost insurmountable, concrete being a heterogeneous, porous and strain-rate-sensitive material. In view of the second-named property, it could be expected that, in addition to the second invariant of the stress deviator, as in metals, the first invariant of strain would also be involved in the strength criterion, but very little is known beyond that.

As for the empirical approach, the difficulties are similarly considerable, although it is obviously the only promising way. The course of attack consists in subjecting specimens to various combinations of principal stresses and determining the failure stresses. The limiting surface (in the three-dimensional case) of strength is then constructed, and various functions of the stress and strain invariants are applied to the experimental curve. This has been attempted at several laboratories throughout the world, but with little success so far. The lack of success is attributable to the following causes. (a) No single laboratory has covered the entire range of stress combinations; since concretes vary in ingredients, mix proportions, *w/c* ratio, etc., and test rooms vary in humidity and temperature, there is little hope of a unified curve. (b) No attention has been paid to the rate of straining or stressing. (c) The definition of "fracture" varies not only between laboratories, but also, within the same laboratory, between fractures caused by different modes of stressing. The last-named circumstance is the main cause of difficulties in solving the problem.

Reverting again to mild steel, no difficulty is encountered in identifying the yield point, be it the result of tension, compression, shear or any other stress. It is evident on the specimen itself, and reflected even more distinctly in the stress–strain behaviour. Concrete, as all researchers in the field know, fails in completely different ways under different states of stress. While in the case of simple tension it is perhaps legitimate to equate fracture with separation into two parts as well as with the point of the highest stress, in the case of compression these events do not coincide and in fact clear-cut separation is rarely realized. What should then be defined as fracture in compression? Appearance of the first crack? The peak load level in the loading machine? Or complete disintegration, which may set in appreciably beyond the peak load point? One cannot avoid the feeling that the loading method and machine play an all-too-important role in determining the "strength" value. The stress–strain curve is also of little help here, as it lacks any distinctive features. It is agreed that disintegration originates fairly early in the course of loading, and that this process is reflected on the stress–strain curve in the gradual decrease in modulus. The end-point of this process, or what is known as the ultimate load in stress-controlled tests, is an arbitrary value, largely dependent on the ability of the machine to hold the separated parts together.

Obviously, what is needed for concrete is a parameter analogous to the yield value of plastic materials. It should be common to all states of stress and define fracture under all such conditions. Also, ideally, this value should be easily obtainable from phenomenological behaviour, as is the yield point from the stress–strain curve of mild steel. This definition would then be used in constructing the limiting surface of strength. A start in this direction was made by Newman [1] and associates, who provided both ultimate-strength and "discontinuity" strength curves. Their definition of "discontinuity" was, however, a little ambiguous.

To be able to define strength, one should have reasonable insight into the mechanism leading to ultimate disintegration. It is now almost universally agreed that this mechanism consists mainly in initiation and propagation of cracks. For a detailed treatment of this subject, the reader ought to refer to Griffith's theory of crack propagation. Griffith [2] developed his theory for ideally brittle materials; Irwin [3] and Orowan [4] extended it to plastic materials, Rivlin and Thomas [5] included rubbery materials, and Kaplan [6] and the author [7] included concrete. In principle, the differences between all these types of materials lie in the mechanism of energy dissipation.

The treatment of concrete fracture from Griffith's viewpoint, although reported before [7, 8, 9], will be presented again here, as it is still far from common knowledge. This will be followed by a new definition of the "yielding" of concrete, by the phenomenological expressions of this "yielding", and finally by methods of identifying the onset of this event.

2. GRIFFITH'S CONCEPT APPLIED TO CONCRETE

The essential of Griffith's analysis, concerning the stability of an elliptical crack normal to a tension field in an elastic material, will be briefly outlined here. The geometry is presented in Fig. 1, which deals with a

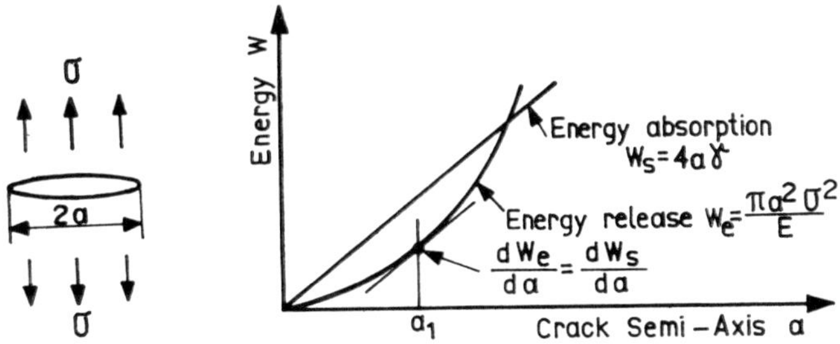

FIG. 1. The Griffith case.

plane-stress case. Based on the principle of minimum energy, the theory states that the crack will grow when the rate (crack-length rate) of elastic energy release caused by this growth equals or exceeds that of energy absorption. Griffith dealt with ideally brittle materials, and his sole energy-absorption mechanism was that of new-surface formation. Thus, on the credit side of his energy balance he had the elastic energy release (after Inglis [10])

$$W_e = \frac{\pi a^2 \sigma^2}{E}, \tag{1}$$

and on the debit side—the new surface energy

$$W_s = 4a\gamma, \tag{2}$$

where γ is the surface tension of the material.

Since for crack instability

$$\frac{dw_e}{da} = \frac{dw_s}{da} \tag{3}$$

it follows that

$$\frac{2\pi\sigma^2 a}{E} = 4\gamma$$

and

$$\sigma = (2E\gamma/\pi a)^{\frac{1}{2}} . \tag{4}$$

Therefore, for a crack $2a$ to grow, a stress expressed by Eq. (4) is required. The process is shown graphically in Fig. 1, in which the point of crack instability a_1 lies where the two energy curves are parallel.

In an ideally elastic brittle material this is all that needs be known, and the quantities E and γ determine the crack toughness of the material. In real materials, however, most of the elastic energy is not transformed into surface energy but dissipated in some other way. In ductile metals almost all of the energy is dissipated by plastic deformations in the highly stressed zones near the tips of the crack. Accordingly, the crack will not grow unless the rate of elastic energy release equals or exceeds the rate of energy absorption by plastic strains. Even then, growth will still not be spontaneous, as the plastic zone grows with the crack, so that the energy requirement per unit crack length increases. Hence the need for gradual increase in stress to force the crack to propagate. This is illustrated in Fig. 2(a), where the energy absorption curve has an increasing slope, in contrast to the constant slope of the Griffith case. An initial crack a_0 will not start propagating until the stress is raised to σ_0, so that

$$\frac{d}{da}\left(\frac{\pi a^2 \sigma_0^2}{E}\right) = \frac{dW_D}{da} \quad \text{for} \quad a = a_0, \tag{5}$$

implying that the release curve is parallel to the absorption curve at that point. This growth will, however, be limited, since the energy-release rate soon lags behind the demand rate. Assuming that growth terminates at a_1, the stress should be further raised to σ_1 (where the release and absorption curves are again parallel) for the crack to repropagate, this time to a_2. A further increase in stress to σ_2 extends the crack to a_3. This process continues until at a_4 the stress is raised to σ_4 and the resulting propagation proceeds unchecked, as from now on the slope of the release curve either equals or exceeds that of the absorption curve. (The two curves become convergent.)

In Fig. 2(b) the derivatives of the two energies are presented. The energy-release rate is termed the "driving force", as the factor "propelling" the

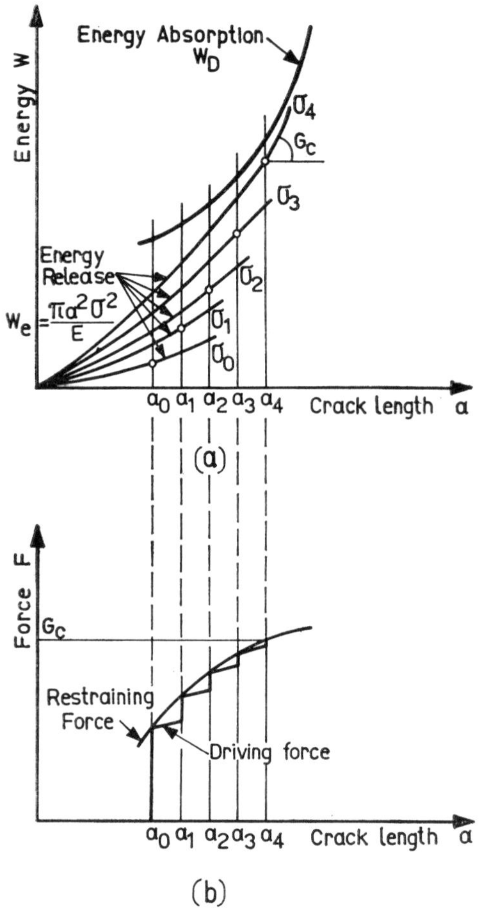

FIG. 2. The forced growth of a crack presented in terms of: (a) energies, (b) forces.

crack ahead; the energy-absorption rate is termed the "restraining force". The driving force must be increased by raising the stress σ, to keep up with the increase in the restraining force, in order to promote crack propagation. This is shown in Fig. 2 as a stepped process for the sake of simplicity; in reality, the growth of both stress and crack length is continuous. As mentioned, beyond point a_4 no further increase of σ is necessary for the crack to continue spontaneously. a_4 is in this case the critical crack length and σ_4 the breaking stress. If the strain-energy release rate is measured at this point of onset of instability, it will be found to be a material constant. The strain-energy release rate is usually denoted by G and its critical value by G_c. The latter may be regarded either as the driving force at onset of instability, or as the energy required for formation of unit new surface at that instant. It thus includes the overall requirement of energy (irrespective of its dissipative mechanism) but excludes kinetic energy (which need not be considered below the point of instability, since velocities are then negligible).

G_c replaces E in Griffith's equation, and it can be seen from Fig. 2 that in the case of tension

$$G_c = \frac{2\pi a_c \sigma^2}{E} \tag{6}$$

where a_c is the critical crack length. For compression, when propagation is of the "shear mode", Irwin [11] has shown that

$$G_c = \frac{\pi a_c \sigma^2 \sin^2 \varphi \cos^2 \varphi}{E}, \tag{7}$$

where φ is the angle between the crack plane and the direction of compression.

For compression, when propagation is of the "opening mode", the author [7, 9] has shown that

$$G_c = \frac{\pi b \sigma^2}{2E} \tag{8}$$

where $2b$ is the maximum projection of the crack (which may have any orientation) on an axis normal to the compression field.

In plane-strain cases the expressions in Eqs. (6), (7) and (8) should be multiplied by $(1-\nu^2)$, where ν is Poisson's ratio.

So far only materials whose dissipative mechanism is by plastic flow have been discussed. No such mechanism exists in concrete which, although quite often referred to colloquially as "plastic" (because of phenomenological similarity), is in fact, rheologically, a typical non-plastic material very similar to glass, to which it is chemically akin. On the other hand, a certain analogy to metals may be seen in the common effect of energy dissipation which in the case of concrete is the result of discrete microcracking in the highly stressed zone ahead of the crack. These microcracks do not necessarily merge with the main crack, so that (as far as fracturing is concerned) the energy expended in their formation is wasted. This means that in order to extend the main crack, the strain energy should provide for more surfaces than those required by the main crack alone. Kaplan [6] has shown that the total energy release in concrete is approximately twelve times the newly formed surface energy in the main crack, and it might be concluded that the total new surface area is twelve times that of the main crack. Thus, phenomenologically, the behaviour resembles that of metals, the energy expended on microcracking being the counterpart of that expended on plastic deformation. Figure 2 and Eqs. (6), (7) and (8) are therefore valid for concrete. Indeed, concrete also has the two distinct stages of fracturing: up to the point of instability crack growth is slow and proceeds in stages of equilibrium; above this point it is spontaneous and rapid. G_c is a material constant for concrete as it is for metals, and Kaplan and the author have found it to be of the order of 0·1 lb/in. The fact that concrete is heterogeneous complicates matters somewhat, but changes nothing in principle: Crack growth is often arrested by an obstruction, such as an aggregate particle. In most aggregates surface tension is higher than in the cement paste, so that when the crack penetrates the particle the energy demand is suddenly increased; a similar increase occurs if the crack is diverted to by-pass the obstacle, for then the actual surface formed is larger than the effective one. In this manner aggregates act as crack arresters and their combined effect consists, statistically, in adding to the energy-absorption mechanism described earlier. The heterogeneity of concrete is treated in more detail elsewhere [9].

3. "YIELDING" OF CONCRETE

The internal processes accompanying the loading of concrete can now be understood.

When the stress is such that its predominating component is tensile, cracks develop in a direction normal to it. In most cases a large enough crack exists prior to load application, so that on loading it alone grows and eventually causes the specimen to break. With better homogeneity, several cracks may grow concurrently. There is no difference in principle between the two cases, the important point being that it is a stage of slow crack growth, with the driving and restraining forces in equilibrium. Since energy is wasted in providing for microcracking, the stress–strain curve deviates from the straight line.† Next comes a stage when the energy-release rate exceeds all likely energy-absorption rates (as in point a_4, Fig. 2) and

† Viscoelasticity also contributes to this non-linearity; this will be discussed later.

the crack grows spontaneously at almost acoustic speed and traverses the entire specimen. This is the "breaking" point, and the stress–strain curve terminates abruptly at the stress level at which this fast propagation occurred (σ_4 in Fig. 2).

When the stress is such that its predominating component is compressive, cracksdevelop parallel to the direction of compression. (This includes cracks which were initially non-parallel.) The author [7] has shown that this growth is much better controlled than that of tension cracks. As a result, the process is not confined to one or few cracks as in tension, but rather involves almost the entire bulk of the specimen. Certain cracks propagate until they are stabilized; other cracks follow suit and similarly become stable, with the process continuously spreading throughout the speci-

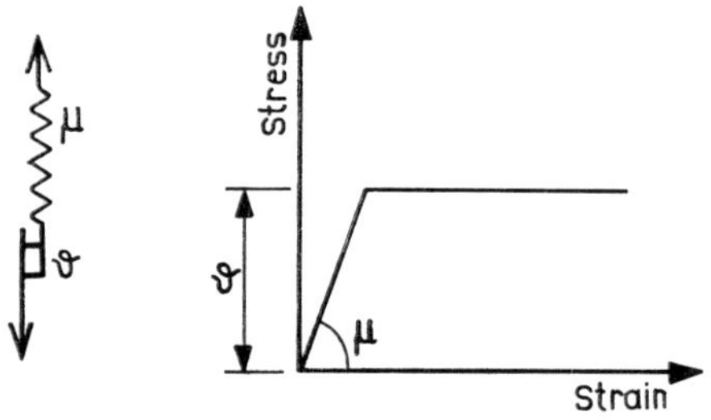

FIG. 3. Stress–strain behaviour of a Prandtl body.

men. Consequently more energy is dissipated, and deviation of the stress–strain curve from the straight line is more pronounced than in tension. This is the slow stage, much more prolonged in compression than in tension, as while in the latter the damage is largely confined to a single cross-section, in the former it involves the entire bulk. (It should be noted that the term "damage" is not quite appropriate in the case of compression, as the effect consists in gradual elimination of potential fracture nuclei and actually enables the specimen to accommodate higher loads.) This is again eventually followed by the stage when the strain-energy release rate assumes its critical value (G_c) and fast propagation sets in. In contrast to its tensile counterpart, however, the specimen in this case is still capable of sustaining loads and thus a certain amount of further loading is possible. Moreover, since failure is not immediate, other parts of the specimen are allowed to reach their G_c value. Thus a whole series of fast crack growths, as distinct from the single extension in tension, follows the phase of slow growth.

The onset of fast crack propagation in concrete is regarded in this work as the counterpart to the yield point in plastic materials. More precisely, the analogy refers to an ideal Prandtl body with a stress–strain characteristic as in Fig. 3. In the course of loading of such a material, equilibrium prevails up to the stress level ϑ. Once ϑ has been reached, equilibrium is disrupted and deformations set in spontaneously. The mechanism in the Prandtl body is obviously completely different from that of crack extension, but there is similarity in behaviour which may in fact be carried one step further: as in metals yielding first sets in at points of high stress concentration, relieving the stress at these points and permitting eventual plastic flow in the entire body—so in concrete crack growth originates in isolated flaws and continues until they are stabilized, permitting increase of the external load to the point of final crack extension.

According to this definition of "yielding" in concrete, its determination for a specimen in tension is no problem: it coincides with complete separation and with the end-point of the stress-strain curve. For a specimen in compression, on the other hand, identification of the onset of fast extension, whether by symptoms on the specimen itself or from the stress–strain curve, is quite a problem. In view of its importance, however, in formulating a strength theory for concrete, an attempt in this direction is certainly worth while.

4. PHENOMENOLOGICAL MANIFESTATION OF "YIELDING" IN CONCRETE

Now that the onset of fast crack propagation has been adopted as the unambiguous definition of fracture, all that is needed is an external manifestation of this event, so that its occurrence can be spotted in a test. The conventional stress–strain curve obtained under a constant rate of loading is useless. As is well known, this is a monotonically rising curve which deviates from the straight line at fairly low loads (it is customary to assume linearity up to about 50%, but recent evidence [12, 13] points to non-linearity almost from zero load) and increases in curvature gradually up to its termination at, or near, a point of maximum. Many investigators have come to the conclusion that the shape of this curve is related to cracking as well as to the viscoelastic properties, but no attempt has been made at quantitative analysis from the fracture-mechanics viewpoint. This is partly because (as mentioned earlier) application of fracture mechanics to concrete is very recent; but mainly because this curve lacks any distinctive features indicating the onset or termination of events. Some investigators [14, 15, 16] have resorted to acoustic, ultrasonic and similar techniques, all aimed at direct or indirect recording of the effects of crack formation, and actually succeeded in relating the cracking process to the shape of the curve. In particular, it was thus confirmed that a change in the process of cracking takes place around 50% of the ultimate load. At this point the sonic apparatus registered an appreciable increase in emitted sound, the ultrasonic apparatus—a sudden decrease in pulse velocity, and lateral strain measurements—an increase in Poisson's ratio. On the basis of the findings of Hsu *et al.* [12], the author [13] has interpreted this as the stage at which cracking, hitherto confined to matrix-aggregate interfaces, shifts to the tougher medium of the matrix proper. The stress–strain curve reflects this event by an increase in curvature or, as some claim, by the onset of curvature. However, it is certainly still a stage of slow crack propagation in which cracks are made to propagate by the increasing load. Beyond the 50% point the stress–strain curve lacks any feature that might identify the onset of fast propagation, now defined as the "yield point".

Methods for determining this point in the compression test will now be proposed.

(a) Method of Strain-controlled Loading

The answer to the question—why not associate the "yield point" in compression with the ultimate load as in the case of simple tension?—is that in compression the specimen can still sustain loads after major cracking has taken place, since this cracking is parallel to the direction of compression. Therefore it is reasonable to assume that this "yield point" is slightly below the ultimate load. How much so depends on several factors connected with test conditions, mainly on the friction capacity at the contact faces of the specimen and on the rate of loading. As these two factors increase, the capacity of the specimen to sustain load above the "yield point" also increases. The second factor is the more instructive of the two in the present context, as it is seen that at a very slow rate the load will never exceed the "yield point". This indicates a possible solution to the problem of identifying this point: if loading is strain- rather than stress-controlled, and if the rate of straining is kept constant, the stress will never exceed the "yield point" for it will have to be reduced as soon as spontaneous straining has set in, to prevent a change in the rate. The stress at onset of spontaneous propagation thus determines the peak of the curve, not an arbitrary load level depending on test conditions and on the rate of loading

Figure 4 shows two stress–strain curves for identical compressive specimens, with curve (a) obtained at a constant rate of stressing and curve (b)

at a constant rate of straining.† If the failure stress in curve (a) is taken as 100%, the point of onset of spontaneous yielding in curve (b) is about 90%, which is the "yield point" in the case in question.

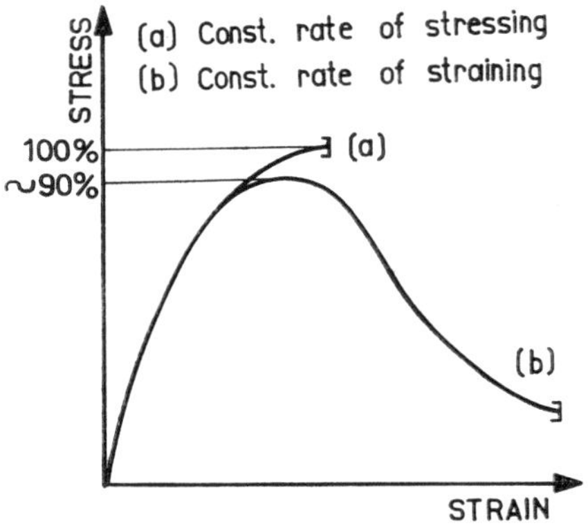

FIG. 4. Diagrammatic representation of the stress–strain relationship of concrete.

(b) Method of Creep under Constant Load

Another approach to the problem of identifying the "yield point" of concrete in compression is as follows.

Assuming an energy criterion of strength, a concrete specimen can approach fracture in either of two ways: (i) monotonic loading to the limit of its energy content; (ii) loading up to a certain level (not below approximately 80% of the static breaking load), at which level it then remains. Creep sets in, followed by fracture after a certain interval. The author [17] has shown that this creep fracture is partly due to the presence of a Kelvin body within its rheological structure, with strain rather than stress as limiting value (Reiner [18]). The author has also shown that

$$W_e = \frac{\sigma^2}{4\mu}(1-e^{-(\mu/\eta)t}) \tag{9}$$

where μ and η are, respectively, the elastic and viscous constants of the Kelvin body, e the base of natural logarithms and t time. Equation (9) implies that under a constant stress σ, the elastic energy content increases exponentially with time. This is, apparently, why a specimen under a sustained load of, say, 80% of the static strength ultimately fractures. A specimen so loaded first creeps in a normal manner while at the same time its

† In reality the processes were force- and displacement-controlled, but changes in specimen cross-section and length were negligible.

energy content increases and slow cracking develops. Eventually the strain energy reaches the level where its release rate (the "driving force") equals the critical value G_c and the slow cracking becomes spontaneous. At this point the creep curve passes through a point of inflection beyond which its rate increases progressively, terminating in fracture. This point of inflection is the sought "yield point"; repeating the procedure for specimens under different constant loads (say, 75–95% of the static ultimate) and extrapolating the points of inflection to the regular short-term test, the "yield point" for this no-creep test will be obtained. As an example of this procedure, attention is called to the excellent results reported by Rüsch [19] in 1960 (Figs. 5, 6), which he used for determining the "sustained-load strength"—overlooking the possibility of utilizing them for separating the visco-elastic from the cracking strains. Figure 5 shows the creep curves for compressive specimens loaded to varying stress levels. Two families of curves are distinguishable: below $\alpha = 0{\cdot}77$ (which Rüsch termed the "sustained-load strength") the deformations become stabilized, tending to a limiting value of deformation; above $\alpha = 0{\cdot}77$ the creep curves eventually depart from the visco-elastic creep behaviour, increase their rate and terminate in failure. As the curves are drawn to a logarithmic scale, the points of inflection are not easily discernible; they would stand out better in a linear-scale graph. It is at these points that spontaneous crack propagation sets in. The same results are presented in Fig. 6 in stress–strain coordinates. The curve on the left is the short-term (20 min) stress–strain relationship. The creep curves of Fig. 5, terminating in failure above $\alpha = 0{\cdot}77$ and reaching a creep limit below it, are presented as horizontal straight lines. The points of inflection A, B and C are also transferred to this figure, indicating on the horizontal creep lines the end of visco-elastic creep and the onset of spontaneous cracking. Curve A-B-C-D is thus the "yield curve" and point D, obtained by extrapolation, the "yield point" for short-term loading. Rüsch assumes that in the course of a regular load-increasing compression test a drastic change in the cracking behaviour of the material takes place at $\alpha = 0{\cdot}77$ leading to eventual failure; in reality the present analysis shows that nothing happens at this point as far as cracking is concerned—it is merely the lower limit of a range of sustained loads which cause, with time, sufficient accumulation of elastic energy for G_c to be attained. (Creep has two components, the delayed-elastic and the non-recoverable; their development with time is shown in Fig. 7, reproduced from a work by the author and Ishai [20]. As shown, the elastic component terminates relatively early, while the non-recoverable continues. Therefore, although creep continues below $\alpha = 0{\cdot}77$, it is not elastic and

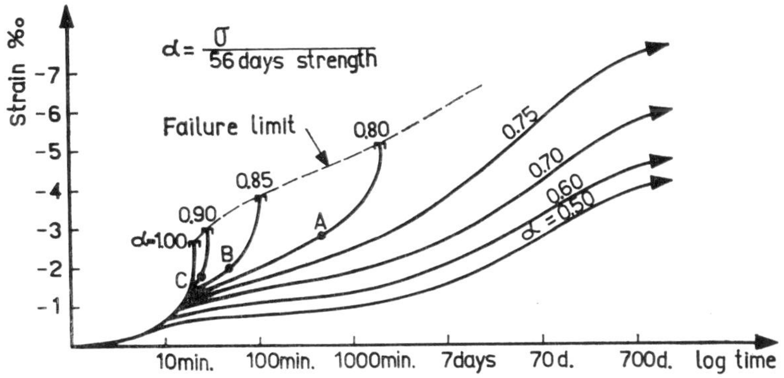

FIG. 5. Strains of concrete prisms under sustained compressive loads of varying levels (after Rüsch [19]).

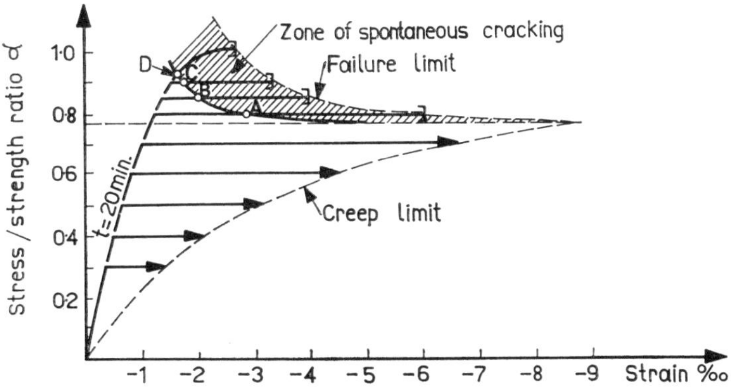

FIG. 6. The strains of Fig. 5 as a function of the stress level (after Rüsch [19]).

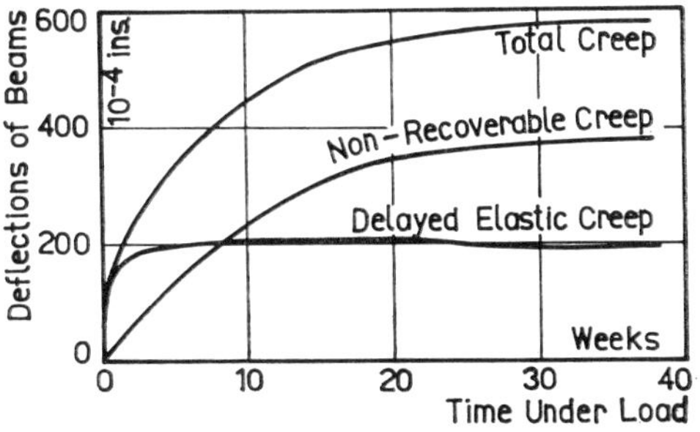

FIG. 7. The two components of the creep of concrete. Deflections of slender beams (after Glucklich and Ishai [20]).

does not involve storage of energy.) Above the 0·77 level, G_c is attained after varying time intervals under load.

It is also seen that $\alpha \approx 0{\cdot}92$ is a significant point in the cracking history of the short-term compression specimen, at which cracking becomes catastrophic. This value is in very close agreement with that obtained under strain-controlled loading as described earlier, but it should be clear by now that the "yield" value is not absolute, and that the 0·92 figure is only valid for a loading duration of about 20 min; it generally varies between 77% and 92% of the ultimate, depending on the loading rate.

(c) Method of Volume Changes

Many investigators have observed the volume changes during loading of a concrete specimen. These observations never involved direct volume measurement, only that of axial and lateral strains. The increase in Poisson's ratio around 50% of the ultimate load, already mentioned and interpreted as the onset of matrix cracking,† is one result of these observations. The dissimilarity between the lateral-strain and longitudinal-strain curves, as shown schematically in Fig. 8, was noted by Richart, Brandtzaeg and

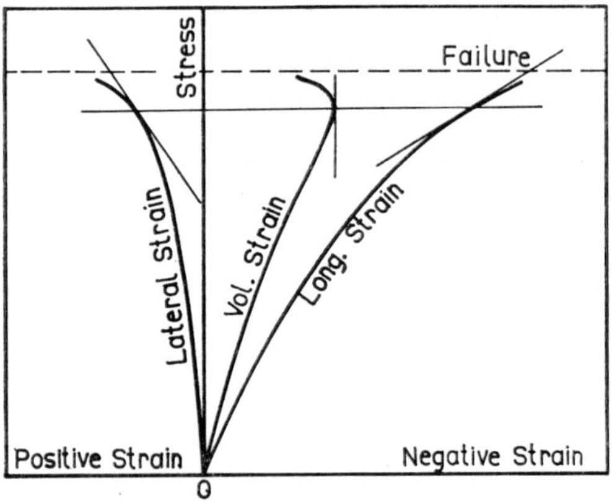

FIG. 8. Typical stress–strain diagrams for concrete specimens in compression, presenting longitudinal, lateral and volumetric strains.

Brown [21] as early as 1929; while the longitudinal-strain curve shows a gradual increase in curvature up to maximum, the lateral-strain curve is nearly linear to about 80% of the maximum, followed by a steep increase

† Accordingly, this "Poisson's ratio" should be regarded merely as an apparent value.

in curvature. As a result, the volumetric strain (which is the sum of the longitudinal and the two lateral strains) presents a curve of unusual type: usually negative, it is initially slightly convex with respect to the stress axis and then becomes, almost abruptly, strongly concave while rapidly tending to zero strain; in some cases it even becomes positive (i.e. "Poissons's ratio" exceeds 0·5) before fracture. The point of curvature reversal is obviously of high significance in the context of cracking. Rüsch describes it as the onset of "disruption of the structure of the material" and relates the corresponding stress level to the sustained-load strength. In the light of the above considerations, there may be grounds for stating that this point is the onset of fast crack propagation. Rüsch, as already mentioned, found his sustained-load strength at 77% of the short-term strength, and his results show the point of reversal in the vicinity of that level. Other investigators located this point, on the basis of more careful lateral-strain measurements at several cross-sections along the height of the specimen (with resulting improved accuracy in the volume derivations), at around 90% of the ultimate short-term strength. It will be recalled that for a short-term stress-controlled loading the author had found the onset of fast crack propagation at around 92% of the ultimate. This may be regarded as good agreement. Weigler and Becker's [22] results confirm this 90% figure for the point of reversal in three different concretes; their data substantiate this for uniaxial as well as for biaxial compression with different principal-stress ratios. Barnard [23], on the basis of highly ingenious and accurate volume-change measurements, found the point of reversal to coincide, in all cases, with the maximum stress attained by him in strain-controlled tests. His results provide, therefore, further confirmation of the proposed theory.

5. SUMMARY

The proposed theory is based on experimental results from various sources. Briefly, it can be summarized by the following statements:

1. The process of concrete fracture has two distinctive phases: slow crack propagation during which equilibrium prevails; fast (spontaneous) propagation, terminating in disintegration.
2. The onset of fast crack propagation is the true strength of the material; in view of the similarity in behaviour to plastic materials, it is here termed the "yield point" of concrete.
3. Under tensile loading, this "yielding" coincides with complete separation.

4. Under compressive loading, identification of the onset of "yielding" is a problem. Three alternative methods are proposed:
   (a) When loading is strain-controlled, the maximum stress attained would be the "yield stress".
   (b) In creep tests at varying stress levels, the "yield points" are identifiable as the points of inflection in the creep curves; the yield point for a short-term test is obtainable by extrapolation to a stress level of 100%.
   (c) The onset of volume increase coincides with the onset of "yielding"; it is obtainable by measuring volume changes during compression.
5. For formulation of a strength theory for concrete, loading tests under varying stress combinations are carried out. The resulting strengths, used in the limiting strength curves, should correspond to the stresses at the onset of fast crack extension.

REFERENCES

1. K. Newman, Concrete Systems, Chapter 8 in *Complex and Heterophase Materials*, to be published by Elsevier Pub. Co.
2. A. A. Griffith, *Phil. Trans. Roy. Soc.* A **221**, 163 (1921).
3. G. R. Irwin, *Fracturing of Metals*, Am. Soc. Metals, Cleveland, Ohio (1948).
4. E. Orowan, *MIT Symp. on Fatigue and Fracture of Metals*, Wiley, New York (1950).
5. R. S. Rivlin and A. G. Thomas, *J. of Polymer Sc.* **10** (3), 291 (1953).
6. M. F. Kaplan, *Proc. Am. Conc. Inst.* **58** (5), 591–610 (Nov. 1961).
7. J. Glucklich, *T. & A. M. Report* No. 215, Eng. Exp. Stn., Univ. of Ill. (1962).
8. J. Glucklich, T. *& A. M. Report* No. 622, Eng. Exp. Stn., Univ. of Ill. (1962).
9. J. Glucklich, *J. Eng. Mech. Div.*, Proc. A.S.C.E., vol. **89**, No. EM 6, 127–38 (1963).
10. C. E. Inglis, *Inst. Naval Arch. London* **60**, p. 219 (1913).
11. G. R. Irwin, Private communication (1963).
12. T. T. C. Hsu, F. O. Slate, G. M. Sturman and G. Winter, *Proc. Am. Conc. Inst.* **60** (2), 209 (Feb. 1963).
13. J. Glucklich, *Proc. Int. Conf. Struct. Conc.*, Imperial College, London (1965).
14. R. L'Hermite, *Rilem Bull.* No. 1 (New Series) (March 1959).
15. H. Rüsch, *C. & C. A. Translation*, No. 86, London (1960).
16. R. Jones, *Brit. Jour. App. Phys.* **3** (7), 229–32 (July 1952).
17. J. Glucklich, *Rilem Bull.* No. 5 (New Series) (Dec. 1959).
18. M. Reiner, *Deformation and Flow*, H. K. Lewis, London (1949).
19. H. Rüsch, *Proc. Am. Conc. Inst.* **57** (1), 1 (July 1960).
20. J. Glucklich and O. Ishai, *Proc. Am. Conc. Inst.* **57,** (8) 947 (Feb. 1961).
21. F. E. Richart, A. Brandtzaeg and R. L. Brown, *Univ. of Ill. Bull. No.* 190 (1929).
22. H. Weigler and G. Becker, *Deutscher Ausschuss für Stahlbeton*, Heft 157, Berlin, (1963).
23. P. R. Barnard, *Mag. Conc. Res.* **16,** (49) 203–10 (Dec. 1964).

EFFECT OF ORIENTATION ON THE FATIGUE BEHAVIOR OF SURFACE GRAINS IN POLYCRYSTALLINE ALUMINUM†

WILLIAM DAVID HANNA

McDonnell Douglas Corporation, Santa Monica, California, U.S.A.

and

D. ROSENTHAL

Department of Engineering, University of California, Los Angeles, U.S.A.

ABSTRACT

The fatigue behavior of surface grains in aluminum is strongly dependent on their orientation. At small strain amplitudes the orientation which causes the *least* strain hardening of the grain is the *most prone* to fatigue failure. Thus, strain hardening in surface fatigue does not appear to be a necessary condition for failure. The dependence of fatigue on grain orientation suggests a definite influence of texture.

1. INTRODUCTION

Fatigue has been an important problem in the study of metals for many years. A tremendous amount of research has been conducted in an effort to understand its basic nature. Although considerable progress has been made, there are still many unresolved questions in fatigue. Among the most important are the role of strain hardening and the effect of the free surface of the metal.

Unlike static fracture, which may be caused by a single application of stress, fatigue fracture is brought about by repeated application of stresses, none of which need be great enough to produce appreciable plastic deformation of the body. However, the plastic strain which occurs on each cycle, although small, does produce changes in the structure of the metal. These changes first occur on the atomic level, and are believed to progress through

† Based on a thesis by the first author in the Department of Engineering, University of California at Los Angeles.

several stages on their way to becoming a macroscopic phenomenon, but their ultimate result is cracking and fracture of the metal.

Most metals of interest deform plastically by slipping of crystallographic planes. Experimental evidence [1] has confirmed that slips produced by the tensile part of a cycle do not eliminate those originating in the compressive part, but tend to occur on adjacent planes. Therefore, the slip from each part of the cycle is cumulative, causing the crystal lattice to undergo a large plastic strain during fatigue, although the body as a whole may suffer practically no permanent deformation.

The plastic behavior of a metal, which may include strain hardening, is certainly an important consideration in the study of fatigue. Equally important, however, is the role played by the free surface of the metal. Many studies [1] have shown that the free surface is much more susceptible to fatigue damage than the interior. In fact, if the surface of a metal has received no prior treatment to increase its fatigue resistance, such as shot peening or nitriding, then fatigue cracks almost always originate there.

Early studies of fatigue did not take the importance of the surface into account. Theories such as those proposed by Gough and Hanson [2] and by Orowan [3] assume that fatigue is a consequence of localized strain hardening, and that fatigue failure of a ductile metal eventually takes place by brittle fracture. According to this view, failure occurs because cyclic

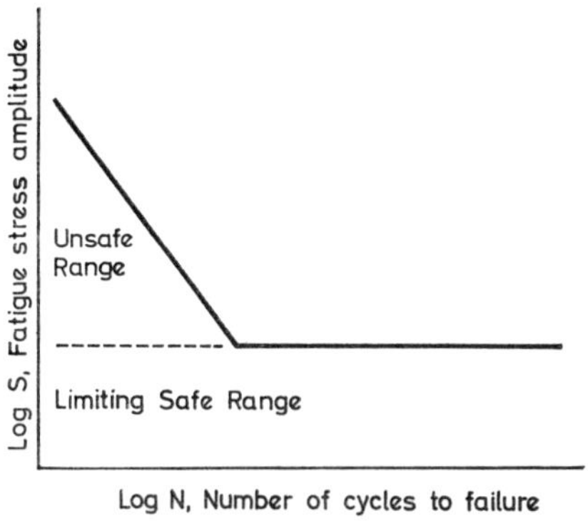

FIG. 1. Typical S–N fatigue curve for iron and iron alloys.

straining uses up the capacity for plastic deformation at critical regions inside the metal. Orowan points out that the typical curve of stress amplitude versus number of cycles to failure, shown in Fig. 1, exhibits two essentially linear regions, which he calls the unsafe range and the limiting safe range. The line representing the safe range is very nearly horizontal for some metals, notably for iron and steel. Using this criterion Orowan developed a quantitative theory which gives good predictions when applied to metals exhibiting this "safe" range. However, in most metals the limiting safe range does not exist, and the second portion of the curve continues downward.

More recent studies, mainly on face-centered cubic metals, have demonstrated that the surface plays an important role in metal fatigue. While investigating fatigue in aluminium alloys in 1953, Forsyth [4] observed that there were thin ribbons of metal extruded from the slip bands on the surface which became more pronounced as the fatigue test progressed. He speculated that the crevices which they left behind would form stress raisers in the material which would contribute to fatigue failure.

Thompson, Wadsworth and Louat [5] reported observing extrusions and intrusions of metal along the slip bands on the surface of copper specimens which were fatigued at low stresses. They noted that the crack which eventually caused failure of their specimens originated at these surface disturbances. They further demonstrated the role of these surface disturbances in fatigue by showing that the fatigue life of their specimens could be extended indefinitely by periodically polishing away the damaged surface during the course of the fatigue test.

Wood [6], by examining cross sections of the surface of fatigued copper and brass, also showed that fatigue cycling causes progressive surface deterioration which eventually leads to cracking and fracture. However, he noted a distinction between fatigue at high and low amplitudes of cyclic stress. Referring to Fig. 1, he pointed out that fatigue fracture along the high stress portion of the curve is actually delayed static fracture, accompanied by high internal stresses and a relatively small cumulative plastic strain, which is built up by coarse slip. At the smaller fatigue stresses, on the other hand, he found that internal stresses in the metal were low, and that extremely large totals of plastic strain could build up by a process of fine slip.

In more recent work, Wood [7] and his associates used the electron microscope to study fatigue in copper and brass. They reported that in all ranges of alternating stress, fatigue in these metals begins with the formation of microscopic voids inside the metal, continued cycling causes these voids to coalesce, forming minute cracks which eventually link together to bring about failure of the specimen. In the low-stress range, however, where the increased lifetime of the part is of greater practical interest, they found that surface deterioration was prevalent.

Evidence that the surface is important in static, plastic deformation was presented by Rosenthal and Grupen [8]. They investigated the static behavior of polycrystalline aluminum at small strains, and showed that a distinctive layer existed in which plastic behavior was different from that in the interior of the metal. In the early stages of plastic deformation a thin surface layer of each grain, possibly a few microns thick, strain hardened less than the bulk of the material. By measuring lattice strains at the surface using an X-ray technique, they showed that the degree of strain hardening of this soft layer depended upon the relative orientation of the grain with respect to the surface. A theoretical model based on the motion and interaction of dislocations near the free surface was proposed. It was based on the fact that because there are fewer constraints on displacements at the free surface, a single slip mechanism can operate unimpeded by and in conjunction with multislip, if it contributes to shear displacements normal to the surface. However, in the interior of a grain, constraints would preclude such a mechanism becoming operative.

Using this model, Rosenthal and Grupen were able to predict the plastic behavior of the surface under static loading. The present study is an attempt to extend their theory to fatigue; more specifically to determine the relationship between the small amplitude fatigue behavior of the surface grains and their orientations with respect to the surface.

2. EXPERIMENTAL PROCEDURE

(*a*) *Material*

The fatigue specimens were prepared from 0.063-in. sheet of 1100-0 aluminum alloy. This material was selected for the following reasons:

1. The 1100, and other aluminum alloys, had already been used in the previous study [8] which showed the existence of a distinctive surface layer in a polycrystalline aggregate.
2. The elastic behavior of aluminum exhibits a high degree of isotropy, which was desirable in this investigation for reasons stated below (specimen design).
3. Aluminum has a face-centered cubic structure and the plastic behavior is well understood in terms of accepted theories of slip.

The properties of this material, listed in the *Alcoa Aluminum Handbook*, 1962, are given in Table 1.

TABLE 1. NOMINAL COMPOSITION AND MECHANICAL PROPERTIES OF THE ALUMINUM ALLOY 1100

Nominal composition	Percent
Al	99.00 min
Si+Fe	1.00 max
Cu	0.20 max
Mn	0.05 max
Zn	0.10 max
Other	0.15 max
Mechanical properties	1100–0
Ultimate strength	13,000 psi
Yield strength	5000 psi
Young's modulus	10×10^6 psi

(*b*) *Specimen Preparation*

It was necessary, in the course of the investigation, to observe individual grains on the surface of the aggregate as the fatigue test progressed. Under the experimental conditions imposed, it was decided that a grain size of about 0.25 in. would be acceptable for the observations, and that it would provide twenty to thirty usable grains in the test area of each fatigue specimen.

The grain size of the sheet, as received from the vendor, was much too small for the above purpose. The strain anneal method was therefore used to obtain the correct size. The procedure consisted of a 6% plastic strain in the transverse direction of the sheet followed by a 2-hr recrystallization anneal at 1080°F.

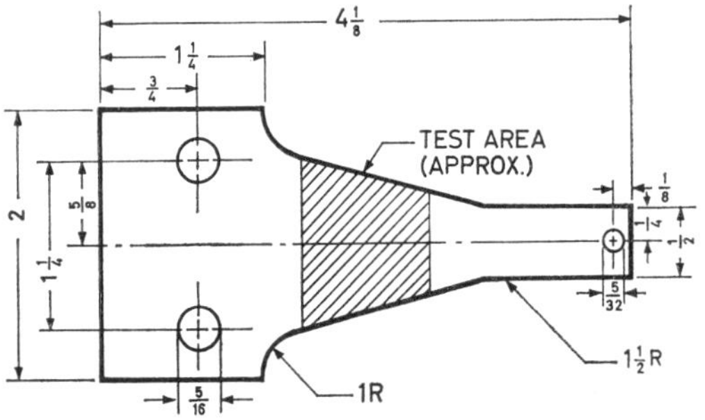

FIG. 2. Detail drawing of fatigue specimen. All dimensions in inches. Thickness of part is 0.059 ± 0.001.

The fatigue specimens, shown in Figs. 2 and 3, were very carefully cut from the prepared material and finished by hand filing in order to avoid residual stresses in the material. Each specimen was mounted to a block, sanded on a plane table until the surface was flat, and then annealed at 600°F for 30 min. After being etched and photographed to provide a record of the surface, they were sanded smooth again with No. 600 paper and then electrolytically polished in a phosphoric acid solution. The electrolyte consisted of:

H_3PO_4 (conc)	817 milliliters
H_2SO_4 (conc)	134 milliliters
CrO_3 (crystals)	156 grams
H_2O (distilled)	40 milliliters

(c) *Specimen Design*

The specimens were fatigued by alternate bending. They were mounted as cantilever beams with one end fastened to a rigid support and the other end to a rod connected to an adjustable, eccentric, rotating shaft. Loaded in this manner, they were subjected to a surface stress, σ, given by the known expression

$$\sigma = \frac{6P}{h^2}\frac{x}{b} \tag{1}$$

where P is the applied load, h is the thickness of the beam, x is the distance of the cross section from the point of application of the load, b is the width of the beam. In order to have a uniform stress over the area of interest on the surface, the specimen was designed in such a way that the ratio of the width of the cross section to its distance from the point of application of the load, b/x, was constant.

Strictly speaking, Eq. (1) is valid only for an isotropic material in which the elastic constants are independent of direction. In a face-centered cubic metal E_{max} is in the [111] direction and E_{min} is in the [100] direction. In general, Young's modulus in any direction, **r**, in a cubic lattice can be found from the relation

$$\frac{1}{E_r} = \frac{1}{E_{100}} - 3\left(\frac{1}{E_{100}} - \frac{1}{E_{111}}\right)(a_{xr}^2 a_{xy}^2 + a_{xr}^2 a_{zr}^2 + a_{yr}^2 a_{zr}^2) \tag{2}$$

where x, y and z are coordinate axes coincident with the cube axes, and a_{xr}, a_{yr} and a_{zr} are direction cosines of the vector **r**. For aluminum,

$$E_{max} = E_{111} = 10.95 \times 10^6 \text{ psi and}$$
$$E_{min} = E_{100} = 9.10 \times 10^6 \text{ psi}$$

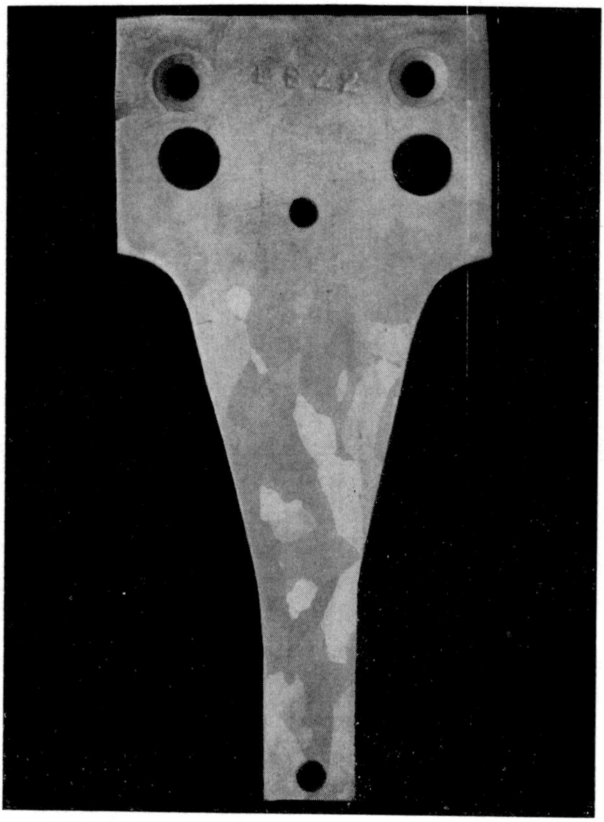

FIG. 3. Photograph of a fatigue specimen. Orig. mag. 1×.

FIG. 4. Test setup. A, Fatigue specimen. B, Electric contact. C, Connecting rod. D, Eccentric shaft. E, Vertical adjustment. F, Steel support plate. G, Vise.

The difference between the extreme values of Young's modulus for aluminum is only 18%, and so is the maximum error introduced by assuming the material to be isotropic.

(*d*) *Test Setup*

The experiments were performed on a constant-speed sheet-fatigue-testing machine manufactured by Krouse Testing Machine Inc., Columbus, Ohio. The machine was equipped with a device to mount the specimen and an electric motor with an adjustable eccentric crank to control the degree of bending. The test setup is shown in Fig. 4.

The testing machine operated at 1750 c/s. To make certain that the specimens would undergo no undesirable dynamic modes of vibration at that speed during the fatigue test, a foil strain gage was mounted to the surface of a specimen and attached to a Sanborn 301 Carrier Amplifier–Recorder and its associated equipment. This specimen was cycled at various deflections, and the curves of strain versus time traced out by the recorder showed no disturbing modes. The dynamic behavior was also checked with a stroboscopic light synchronized with the cycling specimen, with the same negative result.

(*e*) *Testing Procedure*

To discriminate between the fatigue behavior of grains according to their orientations with respect to the surface, the specimens were cycled a certain number of times in each of a series of constant strain amplitudes, beginning with a very small strain and progressing to larger values. The surface was observed periodically at each strain amplitude with a 60-power microscope. Data were taken to record the onset of fatigue as indicated by the appearance of slip lines and the orientation of the slip lines with respect to the center line of the specimen in each grain in the test area. The onset of fatigue in a grain was assumed to coincide with the appearance of the first observable slip lines under the microscope. Using this criterion, grains were labeled as belonging to group I, group II, etc., depending upon the amplitude of strain at which slip lines first appeared on the surface.

The first strain level was set just at the elastic limit of the material to insure that the initial stress would not cause all of the grains to slip. Generally, this strain was too low to cause observable slip in any grain in the test area and the slip lines did not appear until the second or third level.

The strain amplitude for each succeeding level was obtained by increasing the preceding deflection about 25%. In this way it was possible to separate the grains in each specimen into three different groups. The test was continued for each specimen until fracture occurred, which generally required a series of four to five levels of strain amplitudes and a total of about 5×10^6 cycles.

(*f*) *Determination of Grain Orientation*

A back-reflection Laue X-ray photograph of each grain was made to determine its orientation with respect to the surface of the specimen. This method was chosen because it was convenient, non-destructive and accurate to within about 2 degrees.

3. RESULTS

Figures 5 a, b and c show the orientations of the surface normals of grains in the standard stereographic triangle for three fatigue specimens. The orientation of the applied macroscopic stress for the same specimens is given in Figs. 6 a, b, and c. In these figures the Roman numerals represent the group to which a grain belongs and the small Arabic number refers to the particular grain. An asterisk after the number indicates that the grain developed an observable crack on the surface.

It was found that in almost all grains only one set of parallel slip lines appeared on the surface, indicating that single slip was occurring at the surface of these grains. The X-ray photographs revealed that in all cases the slip lines were the traces of a set of octahedral planes. Because of the near isotropy of the material and the small strains imposed in the test, it was assumed that the strain was homogeneous at the surface of the specimen. Under this condition, calculations showed that in all grains the crystallographic planes which slipped were those in which the resolved shear stress was the highest of any slip plane in the crystal. Since it was not possible to determine the slip direction from the X-ray data, it cannot be stated with certainty that two slip systems on the same plane were not operating in these grains. However, there are only three possible slip directions in each of the four octahedral planes. Of the twelve possible slip systems, the probability is only 2/11, or about 18%, that in a randomly oriented grain the second system which slips will be in the same plane as the first. In this experiment about 80% of the grains had only one set of

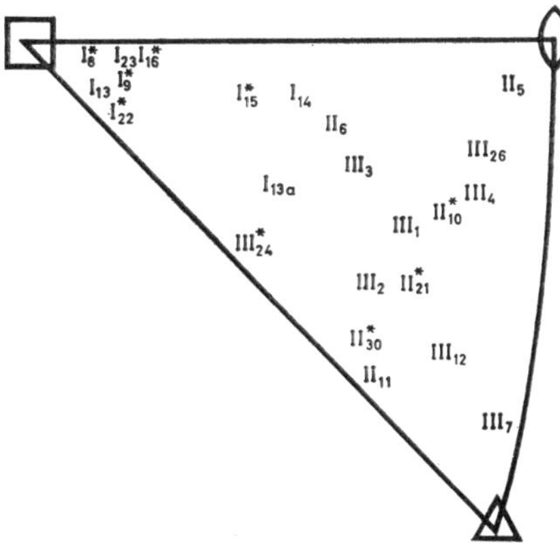

FIG. 5a. Orientation of surface normals of grains in a standard (001) projection for fatigue specimen 23.

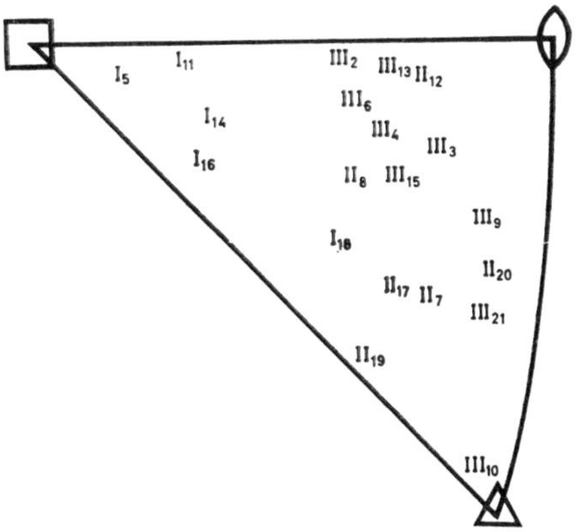

FIG. 5b. Orientation of surface normals of grains in a standard (001) projection for fatigue specimen 24.

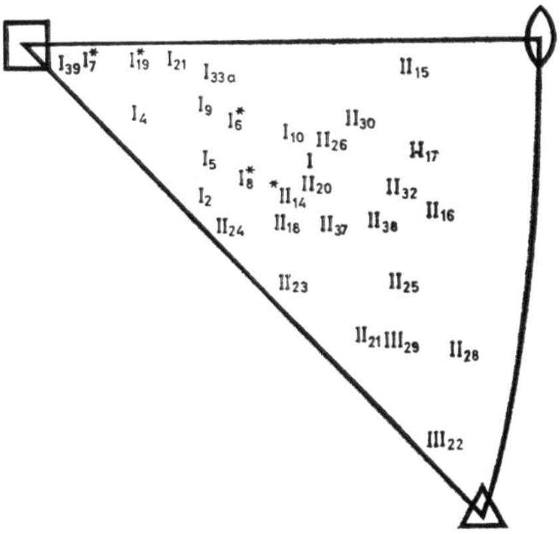

FIG. 5c. Orientation of surface normals of grains in a standard (001) projection for fatigue specimen 26.

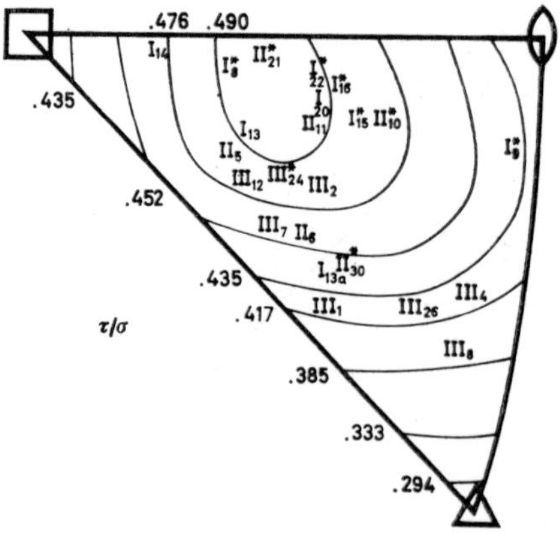

FIG. 6a. Orientation of applied stress on grains in a standard (001) projection for fatigue specimen 23.

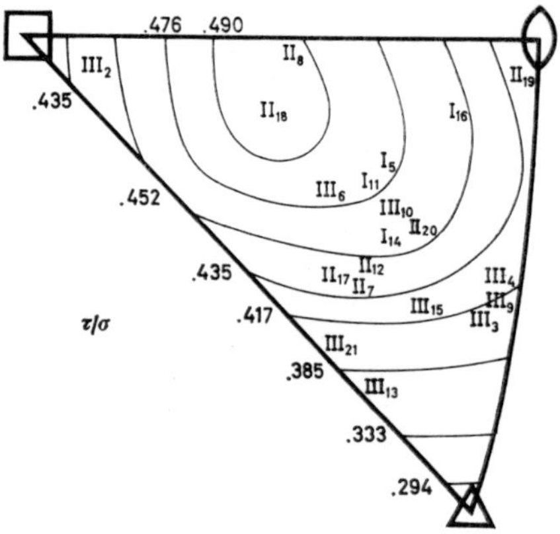

FIG. 6b. Orientation of applied stress on grains in a standard (001) projection for fatigue specimen 24.

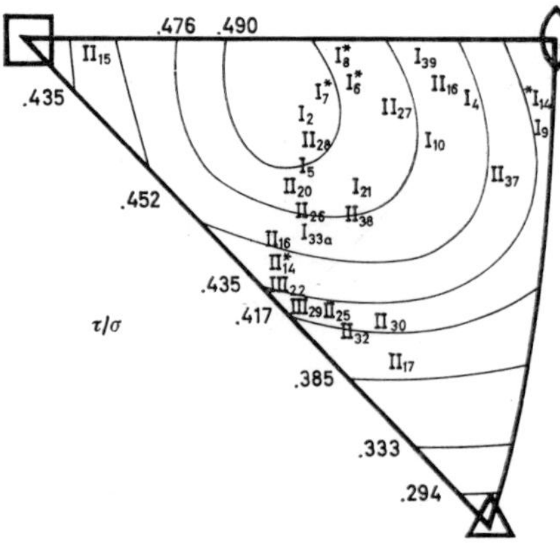

FIG. 6c. Orientation of applied stress on grains in a standard (001) projection for fatigue specimen 26.

2.6×10^6 cycles 4.1×10^6 cycles 5.7×10^6 cycles

FIG. 7. Progressive fatigue damage on the surface of a typical group I grain. Orig. mag. 35×.

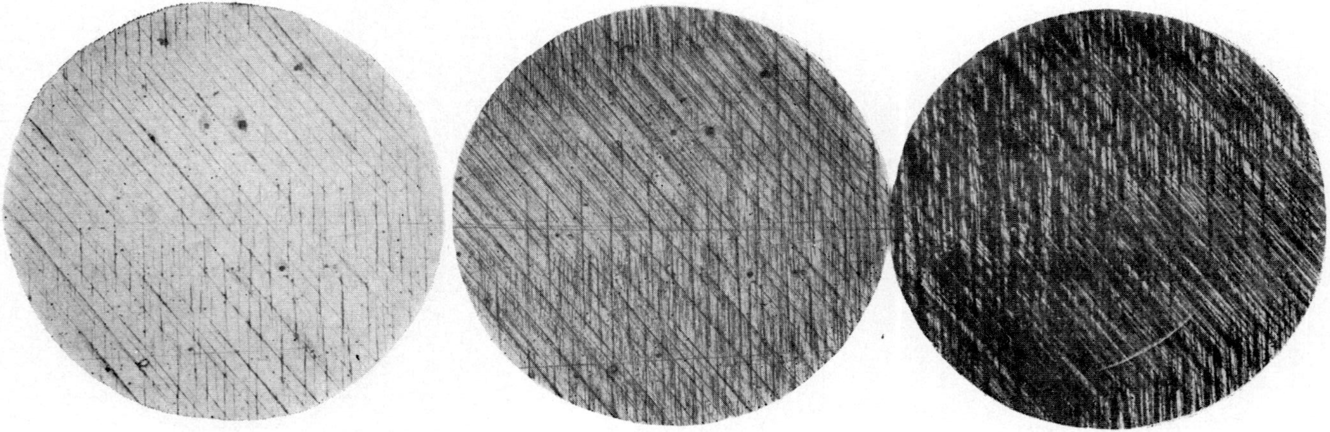

2.6×10^6 cycles　　4.1×10^6 cycles　　5.7×10^6 cycles

FIG. 8. Progressive fatigue damage on the surface of a group II grain having two active slip systems. Orig. mag. 35×.

FIG. 9. Surface of a fatigue specimen after fatigue cycling at a high stress amplitude. Orig. mag. 35×.

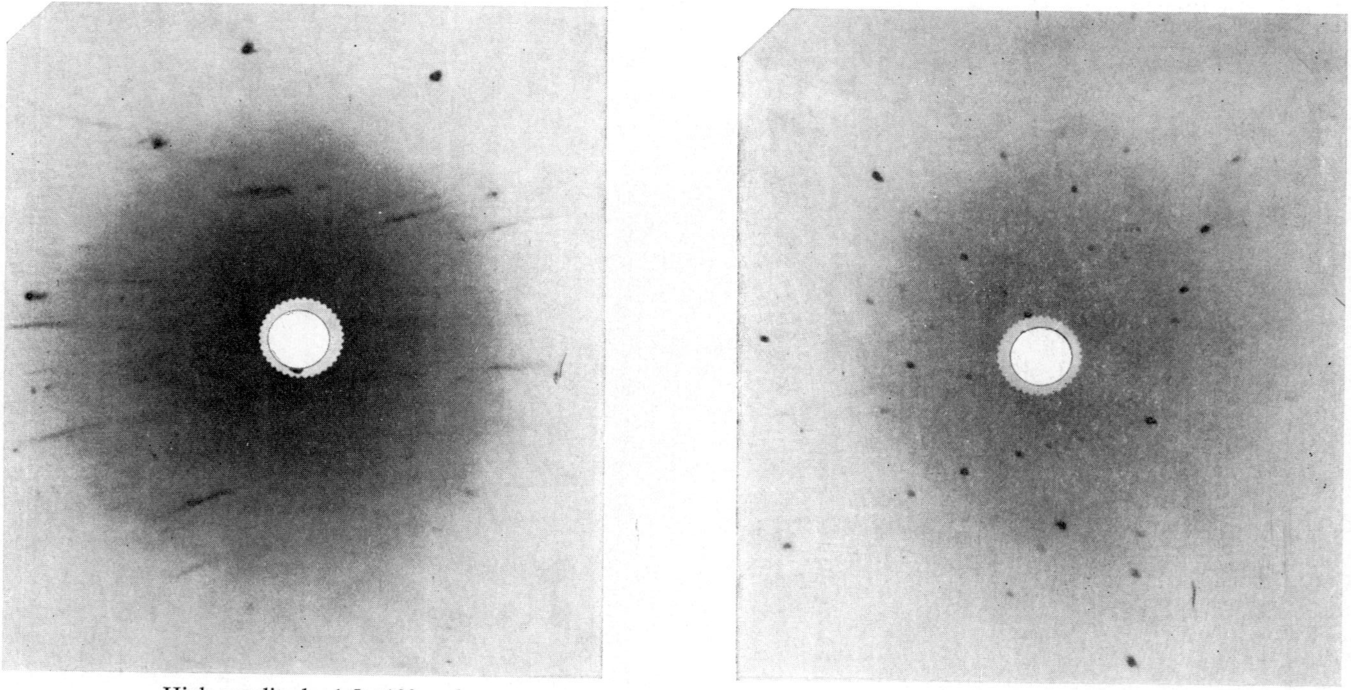

High amplitude, 1.5×10^6 cycles   Low amplitude, 5.2×10^6 cycles

FIG. 10. Back-reflection Laue X-ray photographs of grains subjected to high and low fatigue stress amplitudes.

slip lines, indicating that single slip must have occurred in many grains; the remaining grains had either two or three sets of slip lines.

Figures 7 and 8 are photomicrographs of selected areas of a specimen taken at intervals during the fatigue test. Figure 7 shows a typical group I grain while Fig. 8 is a group II grain, which in this instance exhibited multiple slip.

4. DISCUSSION

Slip lines on the surface of a grain during fatigue are a definite indication that plastic deformation is taking place. However, if the extension and contraction portions of the cycle are equal, the slip occurs with no resultant macroscopic deformation of the grain. The individual slips build up as a series of parallel lines or bands which cause the originally polished surface to take on a roughened appearance as shown in Fig. 7. This appearance, characteristic of low amplitude cycling, has been noted by several workers [4], [6], [11].

It was found in this investigation that continued cycling eventually caused the slip lines on the surface of some grains to open up into cracks extending straight across the grain. These cracks essentially cleaved the crystal lattice between the slip planes. This phenomenon occurred, however, only for the case of very small strains. Large plastic strains caused severe damage to the entire crystal lattice which resulted in coarse slip, large-scale deformation bands and irregular cracks that were not crystallographically oriented. Rumpling of the surface occurred along the high stress directions in the metal similar to that encountered in static loading. The type of deformation was basically the same as that observed by Wood [6] at large strain amplitudes, Fig. 9. Further evidence to support his observation is presented in Fig. 10. This is a back reflection Laue X-ray photograph taken of a grain in the same specimen, which fractured after 1.5×10^6 cycles. The other photograph in the figure is from a similar specimen which underwent over 5×10^6 small amplitude cycles before failure. The asterism of the reflection spots in the former shows that the crystal lattice was highly distorted, while the latter shows no such distortion even though the grain underwent more than three times the number of fatigue cycles. The X-ray and microscopic observations both attest that the damage is due to a large, rather than small strain amplitude fatigue. For this reason it was not taken into account in our study.

As for the small amplitude fatigue, the effect of orientation of surface crystals may be presented in a number of ways. Because of the previous work by Rosenthal and Grupen [8], it was desirable in the present study to determine the effect of grain orientation with respect to the surface. However, since the resolved shear stress criterion appeared to be valid for the tests, it did not seem unreasonable to expect a correlation between this criterion and the fatigue behavior of each grain. The data for the specimens are presented both ways in Figs. 5 a, b, c and 6 a, b, c.

Figures 5 a, b, and c, which give the orientation of surface normals of grains, show that there was a definite correlation between the orientation and the onset of fatigue. The surface normals of group I grains, which were the first to develop slip lines, were all oriented near the {001} pole of the standard triangle. The other grains were all definitely further from the {001} pole than the group I grains, although the distinction between the group II and group III is not too sharp.

Using the resolved shear stress criterion, where τ is the magnitude of the resolved shear stress on the most favorably oriented slip system and σ is the applied tensile or compressive stress, grains with higher values of τ/σ should be the first to slip. Figures 6 a, b and c, show that this was not the case. It can be seen that many of the first grains to slip had smaller values of resolved shear stress than the group II and III grains.

It appears then, from the limited amount of data available, that the orientation of a grain with respect to the surface has more bearing on the plastic behavior at the surface during fatigue than has the magnitude of the shear stress on the most favorably oriented slip system. This result correlates well with the work of Rosenthal and Grupen who found that the plastic behavior of their samples also depended strongly on the surface orientations. Their results are reproduced schematically in Figs. 11 a and b. These graphs show the lattice strain at the free surface versus applied stress in the bulk of the sample for polycrystalline aluminum strained statically in compression. The {*hkl*} value shown on each graph refers to the family of planes parallel to the surface of the sample, using the notation employed in the present work to designate orientations of grains in the fatigue specimens. For comparison, the standard stress–strain curve for the macroscopic behavior of their samples is plotted in Fig. 12.

Figures 11 a and b show that there was an absence of strain hardening over a fairly wide range of stress on the surface of grains oriented with {001} planes parallel to the surface. In contrast to this, grains oriented with {111} planes parallel to the surface of the sample did not show this "soft", non-strain-hardening behavior, but continued to exhibit elastic strain proportional to the applied stress over the range of stress investigated. Grains with orientations other than these extremes displayed behavior intermediate between the two.

This leads to a very interesting result when applied to the fatigue data from the two samples which failed under small amplitude strains in this investigation. In both cases the grains in which cracking was initiated were the group I grains which were oriented near the {001} pole. In these grains there were numerous straight cracks parallel to the slip lines, in addition to

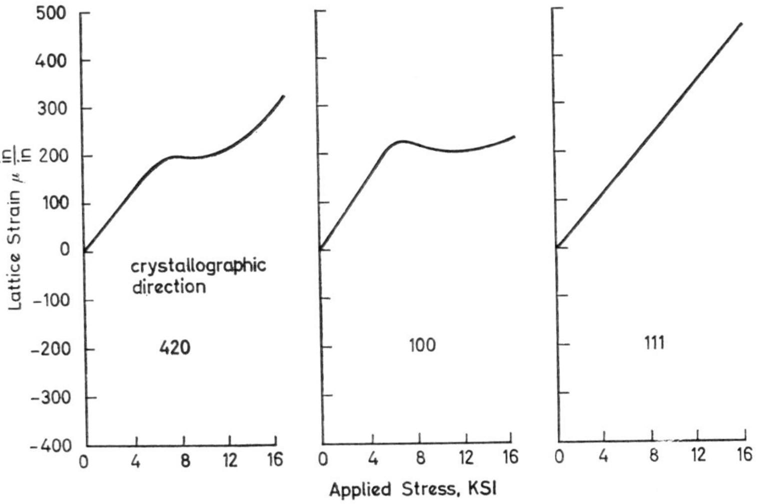

FIG. 11a. Dependence of lattice strain on applied stress on the surface of an aluminum alloy strained statically in compression [8].

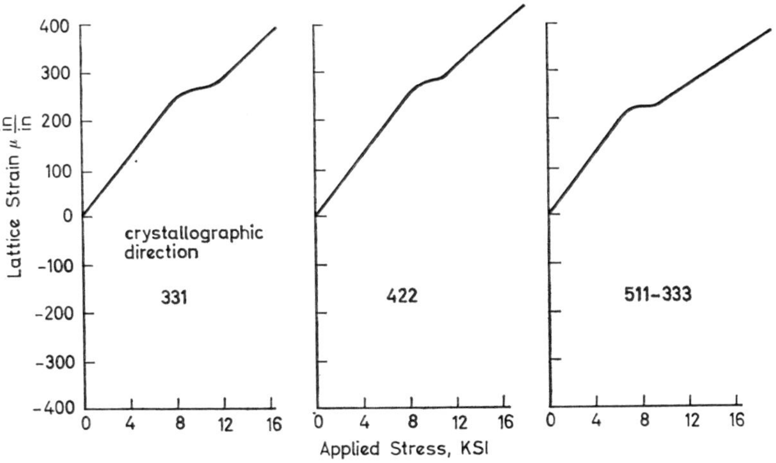

FIG. 11b. Dependence of lattice strain on applied stress on the surface of an aluminum alloy strained statically in compression [8].

the main crack which propagated across the specimen. In all instances where a group II or group III grain cracked, it was either due to propagation of the already fatal crack from a group I grain or to the increase of stress after the group I grains had cracked and could no longer support the load. All cracks in the group II and III grains were jagged, irregular and definitely non-crystallographic, while cracks in the group I grains were all parallel to the slip lines.

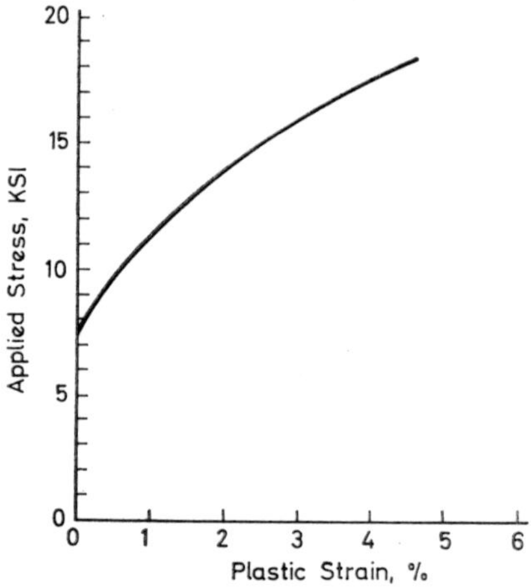

FIG. 12. Stress–strain curve for the aluminum alloy of Fig. 11.

Since the group I grains were those in which the surface did not strain harden appreciably at small strains, the indication is that strain hardening was not a necessary condition for fatigue in these tests. Instead, it appears more likely that the failure was brought about by progressive deterioration of the surface until macroscopic cracks developed.

Strain hardening of the thin layer at the free surface is practically nil for small plastic strains, but this does not preclude the possibility that the bulk of the material will strain harden. It was, in fact, observed that when cycling a specimen at a strain amplitude which initially caused plastic deformation, the macroscopic strain quickly became elastic due to work hardening. This result was indeed expected, since Rosenthal and Grupen showed that the surface layer in question is only a few microns thick. Thus, it could have little effect on the macroscopic stress–strain characteristics of a sample of the thickness used in this experiment.

5. CONCLUSIONS

It was shown in this study that the onset of fatigue in individual grains of a polycrystalline aggregate, as revealed by the appearance of slip lines on the surface, was highly dependent on the orientation of the grain with respect to the surface of the specimen. The magnitude of the resolved shear stress on the most favorably oriented slip system was found to be a much less important factor.

Fatigue failure was shown to be brought about by progressive deterioration of the structure of the metal at the surface. This condition occurred more strongly in grains that began exhibiting fatigue damage early in the test, and which had the orientations shown by Rosenthal and Grupen to exhibit little or no strain hardening of the surface layers. Thus, it appears that strain hardening was not a necessary condition for fatigue failure in the material used and under the conditions imposed in this investigation.

It was also confirmed that there were definitely two different mechanisms of fatigue damage: one which occurred at stresses of large amplitude and another at very small amplitudes. The latter, which was the subject of this investigation, is perhaps more interesting from the standpoint of the engineer, since most structures in which fatigue life is a consideration are not designed to undergo reversed plastic strains of appreciable amplitudes.

Future work should encompass b.c.c. metals, in which the mechanism of fatigue so far has been little investigated. It should also be extended to metals with texture, such as titanium, with particular attention to the recently observed phenomenon of texture hardening.

REFERENCES

1. GEORGE SINES and J. L. WAISMAN, *Metal Fatigue*, McGraw-Hill, New York (1959).
2. H. J. GOUGH and D. HANSON, "The behavior of metals subjected to repeated stresses", *Proc. Roy. Soc.* **104** (A), 538–65 (Nov. 1923).
3. E. OROWAN, "Theory of the fatigue of metals", *Proc. Roy. Soc.* **171** (A), 79–106 (May 1939).
4. P. J. E. FORSYTH, "Exudation of material from slip bands at the surface of an aluminium–copper alloy", *Nature* **171**, 172–3 (24 Jan. 1953).
5. N. THOMPSON, N. WADSWORTH and N. LOUAT, "The origin of fatigue fracture in copper", *Phil. Mag.* **1** (8), 113–26 (Feb. 1956).
6. W. A. WOOD, "Some basic studies of fatigue in metals", *International Conference on the Atomic Mechanisms of Fracture*, ed. by B. L. AVERBACH *et al.*, Nation Science Foundation, Office of Naval Research, Air Force Office of Scientific Research and Ship Structure Committee, Swampscott, Massachusetts, April 12–16, 1959. Wiley & Sons, New York, pp. 412–34 (1959).
7. W. A. WOOD, S. MCK. COUSLAND and K. R. SARGANT, "Systematic microstructural changes peculiar to fatigue deformation", *Acta Metal.* **11**, 643–52 (July 1963).

8. D. Rosenthal and W. B. Grupen, "Second order effect in crystal plasticity: Deformation of surface layers in face-centered cubic aggregates", *Proceedings of the International Symposium on Second-Order Effects in Elasticity, Plasticity and Fluid Dynamics*, ed. by M. Reiner and D. Abir, International Union of Theoretical and Applied Mechanics, Israel Academy of Sciences and Humanities and the Technion–Israel Institute of Technology, Haifa, Israel, April 23–27, 1962. Pergamon Press, Oxford, pp. 391–415 (1964).

POLYMERS IN MATERIAL SCIENCE

H. F. MARK

Polytechnic Institute of Brooklyn, New York, U.S.A.

INTRODUCTION

A new branch of Chemistry developed rapidly during the last three decades—the science and technology of *high polymers*. In general these materials consist of the elements C, H, N and O and are, therefore, conventionally classified as *organic polymers*; but in some instances other elements (such as B, Si, P, S, F and Cl) are present in certain proportions and have a more or less important influence on the ultimate properties of the products. Nevertheless, it is customary to refer to this group of compounds as organic polymers or organic macromolecules. Together with the large family of metallic compounds and of ceramic systems, these organic polymers represent essential engineering materials in the construction of buildings, vehicles, engines, appliances, textiles, packaging and writing sheets, plastics, rubber goods and household articles of all kinds.

The rapid growth of these relatively new engineering materials in the recent past is due to several factors:

(a) The basic raw materials for their production are readily available in large quantities and are, in general, inexpensive. Natural organic polymers such as cellulose (paper, textiles), proteins (wool, silk, leather), starch (food, adhesives) and rubber (tires and many other goods) are mainly available as products of farming and forestry activities, whereas the raw materials for the production of synthetic polymers come essentially from the industries of oil and coal. The simplest building units are called *monomers*, some of which (like ethylene, propylene, isobutylene, butadiene and styrene) are by-products of gasoline and luboils, of very low cost (5c per lb and less) and are available in very large quantities. Many others are simple derivatives of ethylene, benzene, formaldehyde, phenol, urea and other basic organic chemicals; they range in cost from 5c to 25c (at the most) per lb, and are also, in general, large-scale industrial products. Thus the basic building units for organic polymers, as defined above, represent a wide variety of compounds, readily available and of low or moderate cost [1].

(b) Intense research activities in many laboratories have succeeded, during the last 30 years, in elucidating the mechanism of the reactions with the aid of which long-chain molecules are formed from the above basic units. They are called "polymerization reactions" and represent either typical chain reactions of highly exothermic character, or step reactions in which chains of systematically repeating units are formed. The results of this research have provided a sound fundamental background for understanding these reactions. During the same period, systematic engineering efforts yielded a number of relatively simple unit-type processes which permit polymerization and poly-condensation reaction to be translated into large-scale industrial operations. Several of them—such as polymerization in the gas phase at high pressures, in solution, suspension, emulsion and even in solid state—are today well-developed standard procedures which allow monomers to be converted into polymers rapidly, conveniently and at low cost. In fact, the actual conversion cost from monomer to polymer can, in many cases, be as low as 5c/lb, a condition which has greatly contributed to the rapid expansion of this field [2].

(c) Thirty years ago there existed few well-developed processes and machines for converting organic polymers from the state of a latex—a sheet or a molding powder—into the ultimate commercial product such as fibers, bristles, films, plates, rods, tubes, bottles, cups. combs and other marketable commodities. Today, numerous continuous automatic, rapid and inexpensive methods are available for spinning, casting, blow-molding, injection and compression molding, stamping and vacuum forming, which open, for each polymer with attractive properties, an almost immediate chance of being converted into useful and marketable consumer goods. Obviously, the existence of manifold applications, and the availability of standardized and automated methods for leading a new polymer into many different channels, stimulate synthesis and development of such new members of the organic polymer family [3].

(d) The large number of available monomers, and the even larger number of polymers and copolymers made from them, has provided an almost continuous spectrum of composition and structure of organic macromolecules. On the other hand, systematic exploration of their mechanical, optical, electrical and thermal behavior has provided an equally dense spectrum of characteristic practical properties, and the study of their correspondence has led to relatively profound and dependable understanding of structure–property relationships. This has the great advantage that new polymers or copolymers with desired and prescribed properties need not be looked for by an empirical, more or less random, system of synthetic efforts, but can be designed on paper, and successful elimination of many possibilities can be effected before work is actually started in the laboratory. This approach has been so successful in many instances, that one can speak of a *molecular engineering* approach in the synthesis and development of new polymeric materials [4].

All fundamental and applied efforts of monomer syntheses, polymerization techniques, and manufacturing processes can eventually be condensed in a few guiding principles which represent, so to speak, the essence of our present understanding and know-how. Such principles are, of course, only qualitative generalizations, and in each individual case have to be supplemented by quantitative considerations and numerical refinements; but they give a convenient and clarifying "helicopter view" of the present state of our knowledge and its practical applicability. As a consequence they are good working hypotheses or guide posts, provided they are used with caution and with realization of their character as approximations and illustrations.

In the following sections, we shall try to enumerate and discuss the most important principles of this type and to indicate their most prominent applications.

MOLECULAR WEIGHT

A factor of great importance in the synthesis and application of organic polymers is their molecular weight [5]. The words "macromolecule", "giant molecule", "high polymer" and "polymer" already indicate that the molecules of this class of compounds are large, and hence consist of many parts. In fact, all existing experience indicates that valuable and interesting properties of natural and synthetic polymers can only be obtained if their molecular weight (MW) is sufficiently high. Since many important materials consist of chain molecules with repeating units, one can also introduce the concept of *degree of polymerization* (DP) representing the number of basic units in a given macromolecule. In general, "molecular weight" and "degree of polymerization" are used interchangeably in the sense that they are related by the equation:

$$\mathrm{MW} = \mathrm{DP} \times (\mathrm{MW})_u \tag{1}$$

where $(\mathrm{MW})_u$ is the molecular weight of the repeating unit or monomer.

It has also been established by many contributors, over the last 30 years, that polymeric materials do not consist of strictly identical molecules, but always represent a mixture of species, each of which has a different molecular weight or DP. The character of a given material is indicated by a *molecular weight distribution function*, which can be conveniently represented by a curve in which the frequency or percentage of each species is plotted against the molecular weight or the DP of this particular species. The narrower the distribution range of a given polymer, the more homogeneous is the material. As a consequence of this polymolecularity of all polymeric compounds, one cannot simply speak of "a" molecular weight or of "a" DP, but has to operate with an *average* molecular weight or average degree of polymerization [6].

Many tests have established that all important mechanical properties of polymers, particularly tensile strength, elongation to break, impact strength and reversible elasticity depend in a very definite manner on the average molecular weight, or the DP, in the sense that up to a certain (relatively low) DP level no strength at all is developed. From then on, there is a steep increase in mechanical performance with DP until, at still higher molecular weights, the curve flattens out and one enters a domain of diminishing returns of strength on further DP increments. Figure 1 shows a

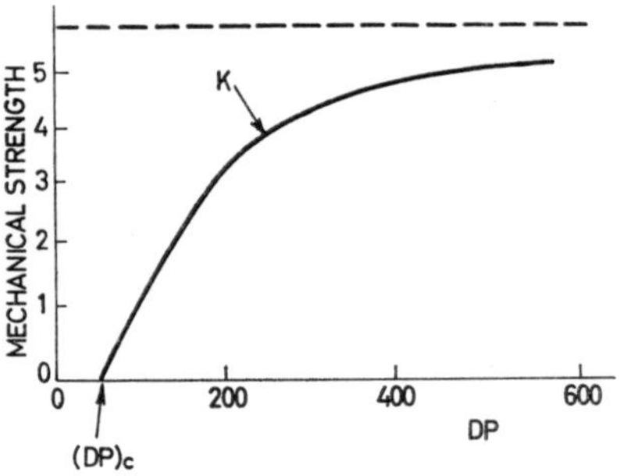

FIG. 1. Tensile properties as a function of molecular weight of DP.

characteristic shape of this curve, typical of all polymers and differing for each individual material only in numerical details. Each polymer has, for instance, a critical DP value, DP_c, below which it is essentially a friable powder, but for each polymer its numerical value is different; polyamides start to develop strength already at DP's around 40, whereas cellulose needs values around 60 and many vinyl polymers need still higher values (around 100). The "knee" K of the curve occurs at DP 150 for polyamides, 250 for cellulose and 400 for many vinyls. However, all polymers exhibit no strength below DP 30 and approach limiting strength above 600. Even if there are still small gains in the higher molecular-weight range of the curve, they are difficult to attain in practice because very long chains have high viscosities in the dissolved and molten state and become increasingly difficult and impracticable to process. The characteristic shape of the curve in Fig. 1 has the consequence that virtually all practically useful polymers fall within the DP range of 200 to 2000, which, in general, corresponds to molecular weights of 20,000 to 200,000.

Evidently the curve in Fig. 1 is of great interest and importance for anybody who wants to prepare and launch a new and useful polymer. If, at a certain point of his research, he finds that the molecular weight of his present samples is around 5000, his attention and efforts will be mainly directed at modifying the polymerization conditions, in order to penetrate a higher molecular weight range. If, on the other hand, he establishes that his present samples already have molecular weights around 60,000, he will focus his interest on factors such as molecular-weight distribution, branching, influence of reactive groups or stereoregulation, but not on a further increase in molecular weight or chain length.

The curve of Fig. 1 can be represented by an equation of the type:

$$\mathrm{MS} = (\mathrm{MS})_i - A/\mathrm{DP} \qquad (2)$$

where $(\mathrm{MS})_i$ = mechanical strength of infinitely long chains,
MS = mechanical strength measured for a given DP, and
A = a constant.

The parameters A and $(\mathrm{MS})_i$ are characteristic for any given polymer, and are connected with $(\mathrm{DP})_c$, the critical DP, through

$$(\mathrm{DP})_c = A/(\mathrm{MS})_i.$$

Hence, Eq. (2) may also be written as:

$$\mathrm{MS} = (\mathrm{MS})_i[1-(\mathrm{DP})_c/\mathrm{DP}]. \qquad (3)$$

It should be added that Eqs. (2) and (3) lose their significance when DP drops below $(\mathrm{DP})_c$; it is also clear that, at least at this stage, they are only empirical, but it was shown by Flory and others [5] that they can be rationalized by considering that the bonds along each individual chain (chemical bonds) are much stronger than those between chains (van der Waals bonds), which has the consequence that short chains slip very easily along each other and offer little or no stress-transfer action. This explains the existence and significance of $(\mathrm{DP})_c$. On the other hand, if the chains become very long, the accumulated resistance of many van der Waals bonds to slippage becomes so high that eventually chemical bonds begin to break. It is clear that, from this point on, further lengthening of the chains has little influence on improved mechanical performance. Formulating these arguments in equations, it was possible not only to interpret Eq. (3) rationally, but also to arrive at improved expressions representing the conditions for mechanical failure of chain polymers in greater detail.

Thus far, we have only discussed the influence of the length of the chain molecules; let us now consider other chain properties, also important for the mechanical and thermal behavior of polymeric systems.

CRYSTALLIZATION

One of the first important notions on the basic properties of chain molecules was that they exhibit a tendency to form crystallite bundles or aggregates. When using the terms "crystallite" and "crystallize" in connection with polymers, it must be understood that one wishes to emphasize that certain volume elements of the polymeric system have reached a state of three-dimensional order, which in certain respects approaches the crystals of normal materials such as sugar, naphthalene, stearic acid or maleic anhydride. However, crystalline domains in a polymeric material do not have the regular shape of normal crystals; they are much smaller in size, contain many more imperfections, and are connected with the disordered amorphous areas by through-going polymer chains, so that there are no sharp boundaries between the laterally ordered (crystalline) and disordered (amorphous) parts of the system. On the other hand, polymeric crystals have relatively sharp melting points, high densities, and high moduli of rigidity; they resist dissolution and swelling and are virtually impervious to diffusion of small molecules. They yield relatively sharp X-ray and electron-diffraction patterns, and show characteristic absorption-peak splitting in the infrared absorption spectrum. In some instances (cellulose, polyvinyl alcohol and certain proteins) there are reasonably sharp boundaries between the crystalline and amorphous portions of a polymeric material; in other cases (linear polyethylene, polyesters and polyacetals) it appears more appropriate to consider the entire system as crystalline, but with the understanding that flaws and imperfections (twisted or folded chain segments) are more or less uniformly spread out over its length and width, and are responsible for the deviation of its behavior from that of a perfect crystal [7].

Whatever the best mental picture for incomplete three-dimensional order in any special case may be, there can be no doubt that the tendency to crystallize plays a very important role in the thermal and mechanical behavior of polymers, and it will be useful to relate this tendency to the chemical composition and structural details of the individual macromolecules. There are, of course, many contributing factors to so complicated a process as three-dimensional order and, in this paper, it will be appropriate to discuss only the most important of them. They seem to be the following:

(a) Structural regularity of the chains, which can readily lead to establishment of identity periods.
(b) Free vibrational and rotational motion in the chains, so that different conformations can be assumed without surmounting high energy barriers.
(c) Presence of specific groups producing strong lateral intermolecular (van der Waals) bonds and regular, periodic arrangement of such groups.
(d) Absence of bulky irregularly spaced substituents which inhibit the chain segments from fitting into a crystal lattice and prevent the laterally bonding groups from approaching each other to the distance of best interaction.

Let us now analyze these factors, one after the other, and estimate how much each of them contributes to the crystallizability of a given material.

(a) *Structural regularity*. The simplest polymer molecules are linear polyethylene and polyformaldehyde; their chains readily assume a planar zigzag conformation characterized by a sequence of *trans* bonds and can, therefore, produce a very short identity period along the chain. This, evidently, favors establishment of lateral order, particularly if the macromolecules are oriented by stress or shear; in fact, both polymers are easily orientable and can attain very high degrees of crystallinity. The polyethylene chains are non-polar, and all intermolecular attraction is due to dispersion forces; rotation about the C—C bond is inhibited by an energy barrier of about 2.7 kg cal/mole of bonds. This limited flexibility, and the dispersion forces between adjacent chains, are responsible for the high melting point, high rigidity and low solubility of this material. In the case of polyformaldehyde, rotation about the C—O bond is less inhibited than about the C—C bond, but the dipole character of the  group causes polar forces between adjacent chains, which act over longer ranges and are stronger than the dispersion forces. As a consequence, polyformaldehyde has a higher rigidity and a higher melting point than polyethylene; it is also less soluble.

If a substituent such as CH_3, Cl or CN is attached to the chain of linear polyethylene in a 1,3 sequence, there exist two simple ways of establishing regular geometric placement of the substituents along the chain: the isotactic placement, in which all substituents (or at least a long row of them) have the same configurational position (either "*d*" or "*l*"), and the syndiotactic placement characterized by regular alternation of "*d*" and "*l*" over the entire molecule, or at least over longer stretches of it. Any deviation from these two cases, or any combination of them, is called "atactic" and refers to more or less random geometric positions of the substituents along the length of the chain. Natta [8], who first succeeded in preparing pure (or almost pure) representatives of these cases with polypropylene and other *alpha*-olefinic polymers, demonstrated that the stereoregulated or stereospecific species are rigid, crystallizable, high-melting and poorly soluble materials, whereas the atactic or irregular species are comparatively soft, low-melting and readily soluble polymers, which do not crystallize under any conditions. The spectacular influence of this structural regularity on crystallization, and with it on most mechanical and thermal properties, was established by many investigators for numerous vinylic, acrylic and allylic polymers, and has become an important principle in designing and "tailoring" new polymers, in particular in view of the possibility of aiming definite prevalence of one of the three structures by appropriate choice of experimental conditions—such as catalyst, solvent, temperature and additives [9].

Another, equally important, influence of structural regularity on ultimate properties has been found in polymerization of conjugated dienes. If a butadiene molecule becomes the unit of a long chain, the following three structures are possible:

1,4–*cis*

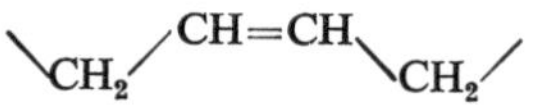

(I)

1,4–*trans*

(II)

1,2

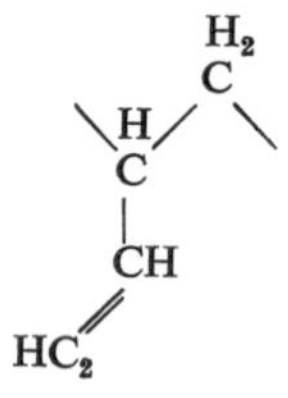

(III)

Using appropriate initiators, solvents, temperatures and additives, each of these structures has been synthesized in almost pure form. Even though the same chemical monomer is used, the three different structural arrangements shown above exihibit noticeably different behavior [10]:

Structure (I) is a soft, readily soluble elastomer with a glass point (T_g) around -60°C and a high retractive force; it crystallizes on stretching over 200%. (Hevea rubber belongs to this species, with isoprene as the monomer instead of butadiene.)

Structure (II) is a hard, poorly soluble polymer which crystallizes readily without elongation and has a melting point around 70°C. (Balata and guttapercha belong to this species, with isoprene as the monomer instead of butadiene.)

Structure (III) exists in isotactic, syndiotactic and atactic forms, all of which were prepared by Natta and his collaborators with the aid of Ziegler catalysts. The stereoregulated forms are rigid, crystalline and poorly soluble materials; the atactic species are soft elastomers with slow and sluggish recovery characteristics. Many other vinylic and acrylic polymers have been obtained in atactic and stereoregulated forms, and in all cases there is pronounced influence of the structural character on the ultimate properties—in the sense that regularity favors crystallizability and with it rigidity, high melting point and resistance to dissolution [9].

Extensive work on copolymers of all kinds fully confirms the influence of structural regularity on crystallization tendency and, consequently, on properties. Copolymers characterized by regular alternation of the two components A and B, which can be represented as

—A B A B A B A B A B A B A B A B A B—,

show a distinct tendency to crystallize, whereas the corresponding one-to-one copolymers with random geometric distribution of the two components

—A B B A A A B A B B A A B A B B B A A B A B B A A B—

are intrinsically amorphous and represent non-rigid, soluble and low-softening resinous materials.

(b) *Chain flexibility*. The term "flexibility" in the context of linear macromolecules refers to the activation energies required to initiate vibrational and rotational motions about single bonds in a macromolecule, as a consequence of which different conformations of the chain can be assumed at moderate temperatures in relatively short times. The energy barriers separating different individual conformations in organic molecules have been thoroughly studied on many ordinary, low molecular-weight organic compounds with the aid of specific heats, infrared absorption and magnetic resonance. These studies resulted in fairly complete knowledge of the stability of different conformational isomers and of the rate at which equilibrium between them is established. This information has been applied, with the necessary reservations to macromolecules, and has led to the following general conclusions:

(1) Linear polymers containing only, or mainly, single bonds between C and C, C and O, or C and N, are capable of rapid conformational changes. If regularly built, and/or if there exist considerable intermolecular forces—the materials are crystallizable, relatively high-melting, rigid and poorly soluble; if irregularly built, they are amorphous, soft and rubbery materials.

(2) Ether- and imine bonds or double bonds in the *cis* form depress the energy barrier to rotation of the adjacent bonds and "soften" the chain, in the sense that the polymers are less rigid, more rubbery and more readily soluble than the corresponding chains of consecutive carbon–carbon bonds. This is particularly true if the "plasticizing" bonds are irregularly distributed along the chains, so that they do not favor but rather inhibit crystallization.

(3) Cyclic structures in the backbone of a chain inhibit conformational changes very drastically, and lead to laborious or slow crystallization. This can go so far, that under most practical conditions the polymers remain amorphous.

(c) *Groups producing intermolecular attraction*. If flexible chains with structural regularity are allowed to form aggregates or bundles, the stability of these supermolecular entities depends on firm cohesion between adjoining chains, and it is obvious that specific groups producing strong intermolecular bonds between such chains have a favorable effect on crystallization. This is particularly true if these groups are arranged along the macromolecules at regular distances, so that they can draw close without causing any valence strain in the chains themselves. In fact, it has been found that all groups carrying dipoles or highly polarizable groups, or those which permit development of interchain hydrogen bonds, favor crystallinity and with it all the valuable properties which are a consequence of the presence of crystalline areas [5]. Thus in polyvinyl chloride and polyvinylidene chloride the C—Cl dipoles increase the lateral cohesive energy density of the system and with it the rigidity, softening temperature and resistance to dissolution and swelling. This effect is enhanced in syndiotactic polyvinyl chloride because of the spatial regularity of the C—Cl bonds. An example of the beneficial influence of polarizable groups is found in polyethylene terephthalate, where the phenylene rings are polarized by the C = O dipoles, resulting in a firm and rigid lattice structure. Very pronounced is the consequence of interchain hydrogen bonding for crystallinity, rigidity, high melting and poor solubility. A well-known example is polyvinyl alcohol, with hydroxyl groups at alternate carbon atoms. Even atactic chains can establish frequent hydrogen bonds if they are first parallelized by stretch or shear; the polymer is, therefore a high-melting and rigid fiber-former. The syndiotactic species is even more rigid, higher-softening and less soluble in water. Another case of strong lateral hydrogen bonding is cellulose, where the effect is enhanced by the rigidity of the gluco-pyranose rings in *beta*-d-junction; as a consequence, cellulose is highly crystalline, non-fusible and insoluble (or very poorly soluble), and has a modulus of rigidity unusually high for an organic polymer. Perhaps the most striking effect of lateral hydrogen bonding between regularly spaced groups is demonstrated by the linear polyamides, where the —CO—NH— groups are responsible for the establishment of these bonds. Since these groups can be spaced at different distances by interposing paraffinic —CH_2— chains of different length between them, it can be established that small and regular spacing of successive amide groups along the chains leads to rigid, high-melting and poorly soluble types, whereas reduction of lateral hydrogen bonding by large and/or irregular spacing of the amide groups produces low-melting and even rubbery types, easily soluble in many organic solvents.

Combination of pronounced chain rigidity with lateral hydrogen bonding would lead to polymers of the structure

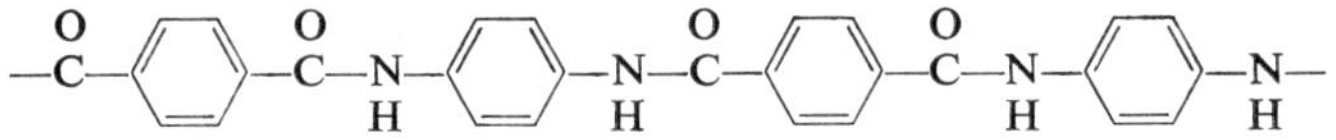

These have actually been prepared, and represent polyamides which are extremely rigid, high-melting and resistant to heat, dissolution and swelling.

(d) *Bulky substituents.* The vibrational and rotational mobility of an intrinsically flexible chain can be inhibited by bulky substituents; the degree of stiffening depends on the size, shape and mutual interaction of the latter. Methyl-, carboxymethyl- and phenyl groups have, in general, a noticeably inhibitory influence on the segmental mobility of linear macromolecules, as shown by the relatively high glass-transition and heat-distortion points of polymethylmethacrylate or of polystyrene compared with polyethylene. Larger aromatic groups have an even stronger influence, and polyvinyl naphthalene, -anthracene and -carbazol are amorphous polymers with remarkably high heat-distortion points and unusual rigidity; they are also poorly soluble. Substituents from ethyl- to hexyl generally exhibit a softening influence, in view of the increased average spacing of the main chains, preventing their dipole groups from drawing close enough for favorable interaction. These substituents are open chains with considerable internal mobility of their own, and act as ingrown (or chemical) plasticizers rather than as stiffeners. Striking examples of this behavior are the polyacrylic and methacrylic esters from ethyl to hexyl, and the polyvinyl esters from propionate to hexoate, in which rigidity, softening temperature and resistance to dissolution and swelling decrease with increasing number of carbon atoms in the chain of the substituent. If these chains become still longer—twelve to eighteen carbon atoms—and remain unbranched (normal paraffinic alcohols or acids), a new phenomenon sets in, namely a tendency of the side chains to form crystalline domains of their own, in which the side chains of adjoining macromolecules arrange themselves in bundles of laterally ordered units [5]. The result is the same as when two brushes are pushed into each other, their bristles being forced to form tightly packed bundles, and producing a firm and resilient system. In all these cases the softening range of the material is close to the melting point of the side-chain crystallites. If the side chains are not homopolar, but themselves contain polar or even hydrogen bonding groups, rather rigid, high-softening and solvent-resistant polymers are formed—as, for instance, in the case of polyphenyl-methacrylamides or polytrichlorostyrene. Thus it can be seen that a proper choice of substituents, and their influence on crystallizability, permit preparation of a large variety of polymers capable of useful applications in the synthesis of rubbery and fibrous materials.

MOLECULAR ENGINEERING OF ELASTOMERS

A particularly important class of polymeric compounds consists of the soft, tough, stretchable and resilient elastomers. They represent a combination of properties which cannot be matched by inorganic materials such as metals or ceramics. In fact, it is now recognized that rubber elasticity can only be exhibited by systems consisting of long, and flexible chain molecules, having weak intermolecular forces and interconnected by primary valence bonds at suitable intervals, so as to form a three-dimensional network. In addition, it is also desirable that these chain molecules be capable of reversible partial crystallization induced by stretching. In these terms, the phenomenon of rubber elasticity—namely, the capacity for rapid and complete recovery from imposed large strains—is ascribed to uncoiling and recoiling of the long and flexible chain molecules. As for the phenomenon of partial crystallization on stretching, this has considerable bearing on the stress–strain curve and on the rupture and tear phenomena exhibited by the elastomer.

In order to treat rubber elasticity in a quantitative manner, it is necessary to define the statistical segment as that portion of the chain molecule which lies between cross-links. Its mass is often referred to as the molecular weight between network junctures (M_n). It is obvious that the elastic behavior of the material is governed by the length of these segments. For instance, it is readily seen that the force required to obtain a given extension is an inverse function of this length, i.e. it increases with the number of cross-links per unit volume of the elastomer.

Thermodynamic Considerations

As in all molecular phenomena, rubber elasticity can be treated both by means of a thermodynamic and a kinetic approach. In either case, the network segment takes the place of the individual molecule, and an analogy may be drawn between it and the molecule in a gas or in a solution. Linear extension of such a chain can thus be compared to volume compression of a gas. It must be realized from the outset that the flexibility of an elastic chain is due to the possibility of rotational movement about its single carbon–carbon bonds, which permit the chain to be extended by uncoiling and, conversely, to retract by recoiling to its original conformation. The latter phenomenon is caused by the kinetic energy of the chain atoms, whose transverse motions would tend to force the extended chain to form a random coil.

The thermodynamic treatment of this phenomenon proposes that the unperturbed, randomly coiled network chain is in a state of highest probability or entropy, and that this entropy decreases when the chain is extended under an external force. In these terms, rubber elasticity can be considered purely as an entropy manifestation, unless the intermolecular forces between chains are large enough for additional energy changes during extension and contraction, in close analogy to the behavior of a non-ideal gas under the influence of van der Waals forces. This postulate can be expressed by the equation:

$$F = (\partial E/\partial L)_{T,P} - T(\partial S/\partial L)_{T,P} \tag{4}$$

where F = external force of extension,
L = length of rubber specimen,
E = internal energy,
S = entropy, and,
P and T = pressure and temperature respectively.

Equation (4) can also be written as:

$$F = (\partial E/\partial L)_{T,P} + T(\partial S/\partial T)_{T,P} \quad (5)$$

which is readily seen to be analogous to the equation of state for a gas:

$$P = -(\partial E/\partial V)_T + T(\partial P/\partial T)_v. \quad (6)$$

In the absence of intermolecular forces, the first right-hand terms of Eqs. (5) and (6) vanish, and the equations simplify to the ideal expressions:

$$P = T(\partial P/\partial T)_v \quad (7)$$

and

$$F = T(\partial F/\partial T)_{P,L}. \quad (8)$$

The best information concerning the extent to which changes in internal energy (E) and entropy (S) contribute to the force of retraction, is obtained from a study of force–temperature relationships. These have been repeatedly investigated, and except at very small elongations, force was found to increase with temperature—in other words, stretched rubber tends to retract on heating, as would be expected from entropy considerations. Figure 2 represents the modulus of an elastomer as a function of temperature.

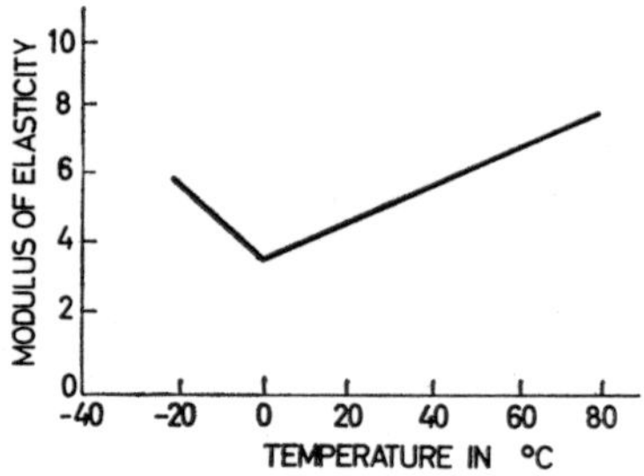

FIG. 2. Modulus of rubber as a function of temperature.

ture. The negative slopes at very small elongations are due to normal thermal expansion of the material, which predominates under these conditions. The elongation at which the two effects balance is known as the thermoelastic inversion point; it occurs at about 10% elongation.

Extrapolation of the force–temperature plots to zero temperature yields the appropriate values of $(\partial E/\partial L)_{T,P}$, permitting appreciation of the influence of intermolecular forces. It was found experimentally that by far the major contribution to the elastic force is due to the entropy change, with only a minor effect of the internal energy due to intermolecular forces. Hence the entropy theory of rubberlike elasticity appears to be well justified.

The Kinetic Approach

The relationship between the applied force F and the extension of a chain network can also be treated by means of kinetic considerations, where the extension is related to the restriction on possible spatial distributions of the freely rotating statistical chain segments; this leads to the following equation for rubber elasticity:

$$F/A = NRT(\alpha - 1/\alpha^2), \tag{9}$$

where N = number of moles of network segments per unit volume,
A = original cross-sectional area of sample,
α = ratio of extended to original length.

It can be seen that Eq. (9) is again fully analogous to the equation of state for an ideal gas:

$$P = nRT/V.$$

An attempt at experimental verification of Eq. (9) has shown some deviations from theory associated with energy contributions, in the range of small elongations, but much more serious descrepancies can be observed

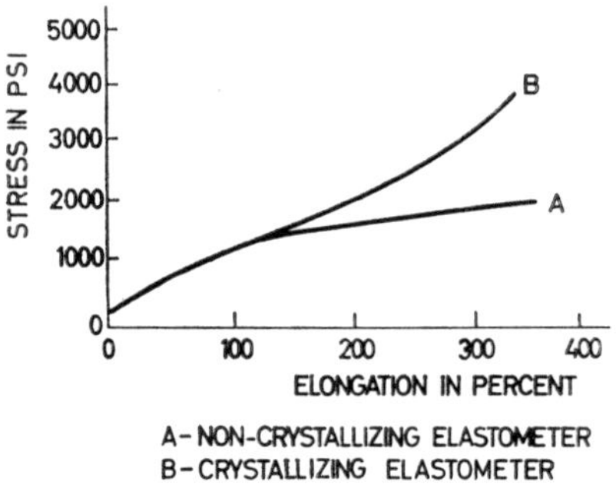

FIG. 3. Stress–strain curves of non-crystallizing and crystallizing elastomers.

in the range of large elongations such as 200% and more. This is shown in Fig. 3 where a theoretical stress–strain curve for rubber is compared with an experimental one.

Influence of Crystallization

The marked upward deviation of the stress–strain curve from the theoretical force values, observed at higher elongations, has been interpreted as a consequence of crystallization. Some elastomers, such as natural rubber, high *cis*-polybutadiene, polyisobutylene and polychloroprene, which possess relatively regular chain structures, exhibit reversible partial crystallization induced either by stretching or cooling. Hence, at sufficiently high elongations, some portions of the chains become parallelized and are in a favorable position to form crystallites. These crystalline domains are, of course, inextensible and cause a sharp rise in the stress–strain curve; upon extension, rubber produces its own reinforcing filler, namely rubber crystallites.

The occurrence of crystallization on stretching of all sufficiently regular elastomers has been proved experimentally by the use of X-ray and electron diffraction and by IR absorption. While only diffuse halo patterns are obtained from stretched non-crystallizing amorphous elastomers such as SBR EPR, distinct diffraction spots are observed in the case of natural rubber and other crystallizing elastomers. It has also been found that sharpness and intensity of the interference pattern are greatly enhanced at higher degrees of extension, because the crystalline areas become more frequent and larger. Those elastomers which crystallize on stretching also generally crystallize on cooling in the unstretched state, and exhibit the phenomenon of freezing and melting at an appropriate temperature at which the network segments lose sufficient kinetic energy to participate in the crystallization process. In view of the long-chain character of the rubber molecules, this process is imperfect and the material can be only partly crystalline at best, leading to rather diffuse melting or freezing points, compared with those of ordinary organic compounds. The phase transformations which a material such as natural rubber undergoes with changes in temperature can be readily demonstrated by the volume–temperature relationships.

The imperfect crystallization process in elastomers is reflected in the sensitivity of the crystalline melting-point to the conditions of crystallization. Thus, crystallites formed more rapidly at lower temperatures are also more imperfect and exhibit a lower melting point. This was strikingly demonstrated by the work of Roberts and Mandelkern, as shown in Table 1 [9]. Thus, by careful cooling at 14°C (and slow heating), the most accurate T_m value for natural rubber is found to be as high as 28–29°C.

Since the crystallites formed on stretching can also help in distributing the stresses to which a number of chains are subjected, they play an important role in rupture phenomena, such as tensile or tear strengths. It is thus not surprising that elastomers showing ability to crystallize on stretching also exhibit high tensile strengths in gum vulcanizates. On the other hand, elastomers which cannot crystallize show relatively poor tensile strength, and require the use of reinforcing fillers (such as carbon black) in order to attain sufficient tenacity for practical use.

TABLE 1. EFFECT OF CRYSTALLIZATION TEMPERATURE ON THE MELTING POINT OF NATURAL RUBBER

Crystallization temperature (°C)	Melting temperature (°C)
−18	21–22
0	22.5–23.5
+8	22.5–23
+14	28–29

TABLE 2. IMPORTANT ELASTOMERS AND THEIR CHARACTERISTICS

Name	Chemical name	Vulcanization agent	Crystallization upon stretching	Gum strength
Natural rubber	Cis-1,4-polyisoprene	Sulfur	High	Excellent
SBR	Poly (butadiene-co-styrene)	Sulfur	none	Poor
Butyl rubber	Poly (isobutylene-co-isoprene)	Sulfur	High	Good
Neoprene	Polychloroprene	MgO or ZnO	Fair	Good
Nitrile rubber	Poly (butadiene-co-acrylonitrile	Sulfur	None	Poor
Silicone	Polydimethylsiloxane	Peroxides	Small	Poor
Thiokol	Polyalkylene-di-sulfides	Zinc oxide	Fair	Moderate
Urethane rubber	Polyester urethanes	Di-isocyanates	Med. rate	Good
Polybutadiene rubber	Polybutadiene	Sulfur	Good for high *cis* structure	Poor to fair
EPR	Poly (ethylene-co-propylene)	Peroxides and sulfur	None	Poor

Table 2 shows a list of the more important elastomers, from the viewpoint of their ability to crystallize on stretching; it demonstrates the importance of regular structure on the strength of soft vulcanizates.

MOLECULAR ENGINEERING OF FIBER FORMERS

The manifold industrial uses of fibers necessitate an extremely wide variety of basic materials. In the domain of natural products we use as fibers such diverse substances as flax, cotton, silk, wool, rubber, metals, asbestos and glass. The range of just one property, for example stiffness, exhibited by commercial fibers of these materials is shown in Fig. 4, which presents the initial stress–strain properties of fibrous materials derived from natural sources. One can see that the initial slope of the stress–strain curve traverses the range from rubber, with a modulus of only about 100 psi, to steel, with a modulus of 30 million psi.

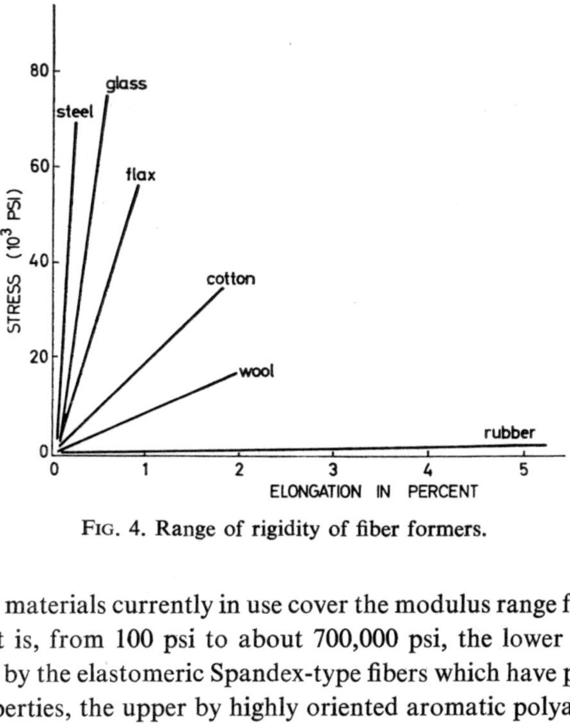

FIG. 4. Range of rigidity of fiber formers.

Synthetic materials currently in use cover the modulus range from rubber to flax, that is, from 100 psi to about 700,000 psi, the lower limit being represented by the elastomeric Spandex-type fibers which have pronounced rubber properties, the upper by highly oriented aromatic polyamides such as Nomex. Most classical synthetic fibers, particularly rayon, cellulose acetate, the conventional nylons, polyesters, vinyls, acrylics and polyolefins fall between these extremes.

Rigidity (or modulus) is only one of many properties which are important for the successful end use of a fiber; demand with respect to other properties are equally diverse. Experience has shown that many significant adjustments and changes of fiber properties can be effected by technology, i.e. through mechanical or thermal treatment during the fiber-manufacturing process. However, all these physical processes are only effective up to certain limits; for really fundamental changes in fiber properties one has to go back to chemistry and synthesize a new polymer.

The following particularly important fiber properties—namely, melting point, modulus, recovery power, tensile strength, moisture absorption and dyeability—shall be briefly discussed in the light of the above, in order to show how the general principles can be put to immediate practical use.

Considering first the melting point, it should be borne in mind that all melting points, both of polymers and of low molecular-weight materials, depend on the ratio of the heat of fusion ΔH to the entropy of fusion ΔS:

$$T_m = \Delta H/\Delta S\,. \tag{10}$$

Specifically, Fig. 5 shows that the melting point of a polymer is related to the strength and regularity of the sites of intermolecular attraction (which in turn affects the heat of fusion) and the stiffness of the polymer chains (which in turn affects the entropy of fusion).

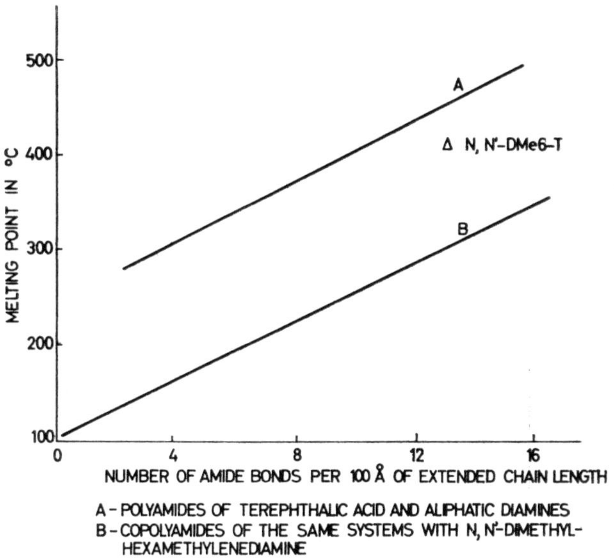

FIG. 5. Melting points of polyamides as functions of spacing of the CO-NH groups along the backbone chain.

Figure 5 demonstrates clearly the effect of intermolecular forces provided by hydrogen bonds in polyamides by plotting the melting point against the concentration of amide groups per 100 Å of extended chain length. It can first be seen that the melting point is essentially a linear function of the amide group concentration. The heat of fusion increases in direct proportion to the frequency of hydrogen bonds produced by the increase in amide concentration. One might expect the melting point to extrapolate to that of linear polyethylene at zero amide concentration, as in this case there are only relatively weak van der Waals forces which determine the heat of fusion of this polymer. In reality, the extrapolated melting point is somewhat lower, probably owing to the polymolecularity of the samples used for determining the melting points.

The effect of chain stiffness is well known in the case of Terylene or Dacron, which are polyesters with terephthalic acid as component; this results in paraphenylene rings along the main chain. Since macromolecules of this type do not coil as flexibly in their amorphous state, they have a lower entropy of fusion and, consequently, a higher melting point.

The melting point of a polymer is also affected by the reduction of the regularity at which the monomer groups are spaced along the backbone chain, such as, for example, by random copolymerization. This is shown in Fig. 6, where the melting points of random copolymers of 6/10–6/6 nylon

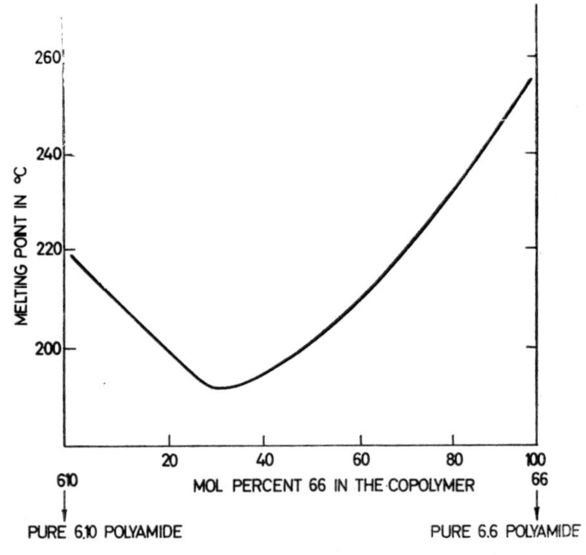

FIG. 6. Influence of random copolymerization on melting and softening of linear polyesters.

are plotted against the composition of the material. The lowest melting point (190°C) is reached in the neighborhood of a 70/30 composition. Disturbance of the regularity of the amide group spacings impairs the efficiency of hydrogen bonding, reduces ΔH in the basic melting point Eq. (10) and, thereby, lowers the temperature of fusion. Monomer unit rigidity and regularity also have a pronounced influence on the melting points of polyesters. Compounds based on ethylene glycol and sebacic acid are amorphous and soften at about 70°C, whereas polyesters made of ethylene glycol and terephthalic acid are crystalline and melt around 250°C. Thus, a 180°C increase in melting point is contributed by the stiff *para*-phenylene ring in the main chain.

The melting point–composition diagram for the random copolymer system of these two polymers shows very clearly the progressive softening effect of the reduction in stiff-segment content and regularity. All factors affecting the melting point also affect the modulus for much the same reasons but, in addition, there are now other contributing structural characteristics, namely orientation and degree of crystallinity. Thus the modulus of rigidity of a perfect 66-nylon or Dacron crystal has been estimated to be in the range of 20 million psi. However, the density of commercial nylons and polyesters shows them to be only about 50% crystalline, which accounts for the fact that observed moduli are only a small fraction of the theoretical value, since at small deformations it is the amorphous parts of the fiber which yield under load and produce the elongation. Thus, for high fiber modulus one needs regularly spaced crystallizable groups along the chains, highly oriented parallel to the fiber axis [5].

In discussing the fiber modulus, we limited ourselves to values referring to room temperature. Let us now consider how the modulus of rigidity of a fiber changes with temperature. Figure 7 shows the characteristic behavior, essentially valid for all polymers. Since the initial modulus is chiefly a property of the amorphous regions in the fiber, it is relatively high at low tem-

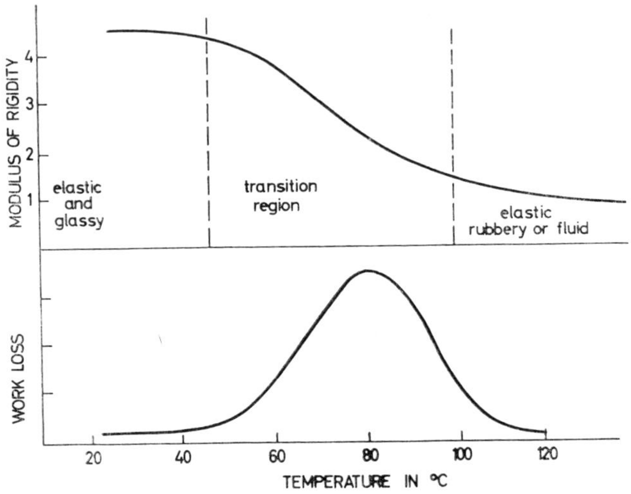

FIG. 7. Dynamic modulus and work loss of polymers versus temperature.

peratures and does not show much change with temperature in this part of the modulus–temperature curve. This region finds the amorphous domains of the polymer in the glassy state; and at relatively small deformations, the restoring force results from an increase in the energy of the bonds undergoing deformation. At the high-temperature end of the curve, although the modulus is considerably reduced, it is again relatively temperature-independent. In this low-modulus region, the amorphous portions of the fiber are fluid and quite readily deformable, and the restoring force produced by extension results from the decrease in entropy of the uncoiled polymer chains. At this end of the curve, both restoring force and modulus are proportional to the concentration of cross-links between polymer chains, as well as to the amount of essentially non-deformable crystalline regions. The cross-links may be either individual hydrogen bonds, or covalent bonds, or small crystallites. There is little hysteresis on cyclic loading of the fiber in either of these two regions, and recovery from deformation is good in both. By contrast, in the intermediate range (known as the "glass" or "second-order transition" region), the onset of sluggish responses in the partially melted amorphous regions results in high hysteresis and relatively poor recovery. The temperature of maximum work loss is close to the glass-transition temperature T_g.

As the degree of crystallinity of a polymer is reduced, the curve tends to move to the left, as well as to a decrease in modulus level. If, finally, T_g is well below room temperature, the modulus becomes quite low, and if the cross-linking efficiency through crystallites or through covalent cross-linking is adequate, we have an elastomer with rapid response to deformation and good recovery. For maximum recovery, we do not want stress-sensitive cross-links such as hydrogen bonds because they rupture and eventually reform in new conformations which impair recovery. Without cross-links of any kind, we would have a viscous liquid with no recovery at all.

It is not surprising that the structure characteristics which influence the fiber modulus also affect the glass transition temperature; it follows that, ideally, the glass transition temperature of a good fiber former should not be too close to the normal end-use temperature. Table 3 illustrates the effects of changing structural polymer features by reducing crystallizability through disrupted chain regularity and reduced hydrogen bonding. Fifty percent substitution of n-alkylated 610 polyamide in the regular 610 polymer, and 60% substitution of ethylene–glycol sebacate in enthylene-glycol terephthalate, produce a marked reduction in fiber modulus, reduce the glass transition temperature below 20°C, and yield for the polyamide and the polyester a typically elastomeric fiber.

TABLE 3. REDUCTION OF REGULARITY AND HYDROGEN BONDING DECREASES MODULUS AND T_g AND INCREASES EXTENSIBILITY

	Properties at 25°C		
	Modulus (in 10^3 psi)	Elongation to break (in %)	T_g (in °C)
610	280	25	~50
610/n-alkylated 610 (50/50)	6	400	~ 0
ethylene-glycol terephtalate/ ethylene-glycol sebacate (40/60)	7	300	~ −20

Elastic properties can be improved if, instead of preparing a random copolymer, two homopolymers are blended in the melt; for instance, polyethylene–glycol terephthalate and polyethylene–glycol sebacate in a 40/60 ratio, allowing limited ester interchange. This results in a block copolymer (compare Table 4) which is also an elastomer, but whose melting point is *ca.* 50°C higher than that of the random copolymer (170°C compared with 120°C).

TABLE 4. COMPARISON OF THE PROPERTIES OF A RANDOM COPOLYMER WITH THOSE OF A BLOCK COPOLYMER OF THE SAME CHEMICAL COMPOSITION

Random copolymer made of 40% ethylene-glycol terephthalate and 60% ethylene-glycol sebacate:

M.P. = 120°C
Elongation to break = 300%
Modulus = 7000 psi

Block copolymer made of the same components in the same percentage:

M.P. = 170°C
Elongation to break = 200%
Modulus = 1000 psi

Still further improvements in properties of elastomeric fibers have been achieved in the block-polymer framework by partly equilibrating relatively long chains (4000 molecular weight) of polyethylene glycol with polyethylene–glycol terephthalate. Here, because of the chemical inertness of the ether links to interchange reactions, one obtains a substantial weight-percent modification at a relatively small mole-percent modification and melting point depression. Thus, with a 60 weight-percent polyethylene glycol modification, the melting point–elasticity relationship is even better than for the partially interchanged block copolymer. This effect is shown in

Table 5 and compared with the properties of the two other polyesters discussed above.

TABLE 5. MELTING POINT, EXTENSIBILITY AND STRESS DECAY FOR VARIOUS COPOLYESTERS

Polymer	Melting point (°C)	Elongation to break (%)	Short-term stress decay (%)
A	120	300	20
B	170	200	13
C	200	350	8

A = random 40/60 polyethylene–glycol terephthalate-co-sebacate.
B = block 40/60 same composition.
C = block 40/60 polyethylene–glycol terephthalate–co-polyethyleneoxide.

While discussing structural conditions which influence modulus and melting point, it is apparent that we have been forced to consider factors which also govern elasticity and deformational recovery. For a typically elastomeric fiber with high elongation and complete strain recovery, for example "Lycra", we need certain segments with low modulus and a T_g below the use temperature. On the other hand, all important natural and synthetic apparel fibers are characterized by a balance of structural design with a view to much higher moduli and T_g values above the normal end-use temperature.

Turning now to the factors which govern tensile strength, we already know that molecular weight is of prime importance. Most linear polymers with T_g above room temperature have high tensile strength if their molecular weight is above about 10,000 and the chains are highly oriented parallel to the fiber axes. 66 nylon of molecular weight 18,000 can yield fibers, with well above 100,000 psi tensile strength, and there are many other linear polymers which give similar values at normal temperatures. However, on tensile testing at different temperatures or rates of loading other details of the polymer structure come into play.

Up to now we mainly discussed the behavior of homopolymers, random copolymers and block copolymers, in terms of structure–property relationships. In the latter two types the chemical modification of the polymer is in the chain. Other types of copolymers are obtainable by creating free radical sites along the trunk polymer chain and initiating vinyl polymerization in the presence of a desired monomer. The products of such an operation are graft copolymers, in which the polymer backbone is unchanged and the chemical modification has the form of branches. High-energy radiation,

TABLE 6. PROPERTY COMPARISON OF GRAFTED FIBERS VERSUS THE UNGRAFTED SYSTEMS AND RANDOM COPOLYMERS (66-NYLON BASE)

Fiber	Moisture regain (%) (50% RH)	Static propensity (log R)	Fiber stick temp. (°C)	Wet crease recovery (%)	Dye absorption rate
66-nylon	2.5	13.3	240	70	normal
20% acrylic acid graft on 66-nylon (Na salt)	7.5	3	380	94	rapid
20% acrylic acid on 66-nylon (Ca salt)	5	13	420	70	normal
66/6 (80/20)	3.5	13	200	65	rapid

chemical initiation of free radicals and even mechanical action and heat can be utilized to promote grafting processes. Grafts on preformed fibers may be uniformly distributed throughout the cross section of a fiber, or confined to surface regions, by utilizing knowledge of the diffusion and solubility characteristics of the various monomers in the substrate. Proper selection of the modifying monomer, in the light of its own polymer properties, permits the fiber modulus and recovery properties to be increased or reduced, and other properties—such as adhesivity, water absorption and increased dye receptivity—to be imparted. In grafting a monomer onto a performed fiber, the crystal structure of the backbone polymer remains essentially unaffected, as acrylic acid grafts on preformed 66 nylon fibers demonstrates. Table 6 shows how the balance of some key fiber properties is altered by grafting 20% acrylic acid onto 66-nylon fibers and converting the graft into the sodium or calcium salt of polyacrylic acid. These properties are compared in the table with those of a random copolymer of 66- and 6-nylon. It is readily seen that:

(1) A significant increase in moisture regain is obtained with the hydrophilic graft.
(2) The static propensity, as indicated by the log electrical resistance of a fabric, is markedly reduced in the sodium salt of the graft.
(3) The sodium salt of the graft also shows improved wet-crease recovery because of the highly swollen nature of the graft in water.
(4) The dye-absorption rate is considerably higher in the graft, as it is also in the random copolymer.
(5) The divalent calcium ion produces ionic cross-links in the fiber, yielding a marked increase in stick temperature and melting point.

By contrast, the random copolymer contributed improvement in dye absorption rate only, and that at a marked sacrifice in melting point.

The data presented, and the examples chosen for illustrating characteristic behavior, refer mainly to condensation polymers and their fibers, but the principles discussed and the conclusions drawn hold for fibers made from all classes of linear polymers. In vinyls and acrylics, steric isomerism is important in governing intermolecular forces. Random-, block-, or graft copolymerization affects them as it does condensation polymers; chain stiffening is preferably carried out with the aid of side groups, since it is difficult to insert rings in the backbone chain of addition polymers.

MOLECULAR ENGINEERING OF POLYMERS FOR BUILDING CONSTRUCTION

There are many indications that the conventional fields of application of organic polymers are becoming amply supplied with various naturally competing materials offering little prospect of substantial further expansion. These fields are essentially textiles, packaging materials, rubbers, plastic coatings and adhesives, which together represent a very large volume of synthetic polymer consumption—in 1963 more than 20 billion 1b (10 million tons) for the U.S.A. alone. The approaching saturation of these markets raises the question whether other large areas of industrial activity could be found which would be able to absorb substantial quantities of organic polymers, if the properties and costs could be so adjusted that their systematic and large-scale application would present a technical and economic advantage for these industries.

It appears that building construction of all kinds—small and large homes, office buildings, factories and laboratories—offers a very promising chance for the consumption of rather large quantities of many types of organic polymers, if certain properties (or, even better, combinations of properties) could be conveniently built into these materials.

For a rough estimate of the scale involved, it should be mentioned that the house- and home-building activities in the U.S.A. in 1963 amounted to a total investment of about $50 billion of which about 20% were spent on construction materials such as metals, concrete, bricks, glass and wood. If plastics are to be used in the building trade they can serve either as structural (load-bearing) units or as finishing materials, servicing a building with water, air, gas, and electricity and providing for the finishing of floors, ceilings and walls. Until now organic polymers have been mainly used in the finishing sector in the form of floor tiles, coatings, insulating foams, window- and door frames, and roofing. About 20% of all plastics and coatings are already being used in these capacities and this volume is likely to keep expanding during the next years. More important expansion, however, can be seen in the use of organic polymers as structural units, provided certain improvements in properties can be effected (see Table 7).

TABLE 7. KEY PROPERTIES OF POLYMERS FOR BUILDING CONSTRUCTION

Required property	Currently available materials	Outlook
Reasonable price	improvement needed	fair
Light weight	excellent	fair
Structural strength	generally satisfactory	good
Fire proof	needs improvement	good
Long life	almost satisfactory	very good
Easily transported	excellent	very good
Corrosion-resistant	excellent	very good
Moisture-resistant	satisfactory	very good
Solvent-resistant	excellent	very good

It is obvious that a successful invasion of this field cannot be brought about by the development of one single property—for example, tensile strength—to an extravagantly high level, but that one must aim at a combination of properties which would make a polymeric material particularly valuable and attractive for a special use. For rubber tubing, for example, one would require softness, flexibility, elasticity, abrasion resistance, impermeability to air and resistance to temperatures up to about 150°C.

For use as structural units in the building industry, one may characterize a favorable compromise of properties as follows:

(a) High modulus of rigidity, preferably above 700,000 psi.
(b) High softening or melting point, preferably above 500°C.
(c) High tensile strength, preferably above 100,000 psi.
(d) High elongation to break, preferably above 10%.

Favorable values of (c) and (d) give a large area under the stress–strain curve, which means a high energy requirement for breaking or tearing a piece of the material. This, in turn, manifests itself in high impact strength and high abrasion resistance.

(e) High resistance to solvents and swelling agents, even at elevated temperatures.
(f) High resistance to deterioration through heat (flame-proofing), radiation and aggressive chemicals.

Every material of low specific gravity and cost is bound to find numerous profitable applications, if one succeeds in incorporating into it a favorable combination of the above properties. It might therefore be useful to recapitulate very briefly the principles, with the aid of which it has already proved possible to achieve well-balanced compromises, and explore whether and how still superior combinations could be arrived at. In a general and simplified sense, there are three main principles which have been very useful in the past and should be good working hypotheses for future efforts: (a) crystallization, (b) cross-linking, (c) chain stiffening.

Let us, therefore, evaluate each of these principles for the present purpose and, in conclusion, consider how they could be combined to yield improved products.

(*a*) *Crystallization*

The principle of crystallization has already been discussed in some detail; it has long been known to be a very valuable property of linear, flexible macromolecules, whenever one wants a good combination of thermal and mechanical properties. Even linear polyethylene, a completely non-polar material with weak interchain bonding, is rigid, relatively high-melting (130°C), strong, tough, abrasion-resistant and insoluble in anything at room temperature—only because it possesses a strong tendency to crystallize, which means, in this context, to form bundles or domains of high lateral order. The same is true of isotactic polypropylene (melting point 170°C) and isotactic polystyrene (around 230°C). If the macromolecules contain polar groups and are of regular architecture, even better combinations of mechanical and thermal properties can be realized, such as in polyvinyl alcohol, polyvinylidene chloride, polyoxymethylene and many aliphatic polyesters and polyamides such as polycaprolactam and 66- or 610-nylon. In all these cases we have intrinsically flexible chain molecules of regular architecture with a distinct tendency to form crystalline domains, within which a systematic accumulation of interchain forces reinforces the entire structure to such an extent that the system becomes hard, high softening and insoluble.

Crystallization of linear flexible macromolecules is a phenomenon which has not only found numerous practical applications, but also stimulated detailed statistical analysis of the thermodynamics of these systems, in turn permitting rationalization of many empirical facts and providing for rather satisfactory basic understanding of the behavior of such systems.

(*b*) *Cross-linking*

It has long been known that a combination of favorable properties is also obtainable by chemical cross-linking of long flexible chain molecules. Rubber is a convenient example here. As one reduces the original segmental mobility of the individual chains by means of localized but strong carbon–sulfur and sulfur–sulfur cross-links, the material becomes more rigid, higher-softening and less soluble. As more and more cross-links are introduced, their average spacing along the flexible chains decreases, the system is progressively stiffened and one finally winds up with hard rubber or ebonite, a material which is very rigid, has an extremely high softening range and is completely insoluble and non-swelling.

The length of the cross-linking element is of importance, because very short cross-links, such as direct bonds between the carbon atoms of two chains,

methylene (—CH_2—) bridges or —S—, —S—S— and $\underset{\displaystyle \mathrm{S}}{\overset{|}{\text{—S—}}}$ bonds,

create stiffness and high softening sooner than longer cross-links, as has been established with the aid of diamines, diepoxides, diolefins, dialdehydes and diisocyanates. Apparently the relatively flexible chains of such cross-links permit segmental mobility to be maintained even if the tie-points are relatively close together, and build up three-dimensional networks of remarkable resilience, toughness and recovery power.

The chemical nature of the cross-links is not closely connected with their mechanical and thermal effect, but can be very important for their resistance to elevated temperatures and to chemical reagents. Thus —C—C— cross-links are very resistant to both influences: sulfide (—S—), disulfide (—S—S—) and ester (—CO—O—) cross-links are sensitive to heat and alkali; acetalic (—O—CH_2—O—) tie-points are sensitive to acids, and urethane-type (—NH—CO—O—) bonds are cleaved by elevated temperatures. Depending on the nature of the base polymer, there is a rather wide choice of cross-linking agents available, permitting a three-dimensional network with the desired properties to be produced.

It should be added here that an effect similar to that of cross-linking is also obtainable by incorporation of a reinforcing filler into a polymer. The term "reinforcing" refers to the fact that there is very intimate contact in molecular dimensions, between the filler and the polymer chains, and that strong adsorption forces immobilize the polymer chains at the surface of the filler particles.

Although, at the first glance, the effect of cross-linking is very similar to that of crystallization, there are several important differences:

(1) In a crystalline system, rigidity is the result of numerous regularly spaced lateral bonds between the oriented chains; each of these bonds is weak, and the ultimate effect is due to their large number and regularity. In a cross-linked system, by contrast, the bonds between the long flexible chains are strictly localized, individually strong, and arranged at random within the system. As a result, crystallization is a reversible phenomenon, whereas cross-linking is irreversible; it is the preferred technique for the production of thermoset resins.

(2) Crystallization is a physical effect which takes place at all temperatures and is strongly influenced by physical processes such as orientation and swelling. Cross-linking, on the other hand, is a chemical phenomenon, requiring the presence of certain special reagents; it is strongly accelerated by elevated temperatures, but little affected by orientation or swelling.

Many important hard, infusible and insoluble products are obtained by cross-linking—such as all hard rubbers, urea-, melamine- and phenol formaldehyde condensation products, polyesters using glycerol, trimethylol propane or pentaerythritol as components, and resins hardened by grafting of styrene onto a polyester backbone containing aliphatic double bonds. While there are obvious and significant differences in the way in which crystallization and cross-linking act, it was very useful to have two different principles with which to achieve a desirable combination of valuable properties. Even more encouraging is the availability of a third independent principle, namely the use of stiff linear chain molecules.

(c) *Chain Stiffening*

Crystallization and cross-linking produce stiffness, high softening and difficult solubility by establishing firm lateral connections between instrinsically flexible chains; one can, however, attain the same result by incorporating the stiffness in the individual chains and designing them in such a manner that their segmental motion is intrinsically restricted; the hardening effect of bulky substituents has been already mentioned for polystyrene which is amorphous, has no cross-links, but still is a hard, relatively high-softening (90°C) polymer. The absence of crystallinity results in complete transparency, and the absence of cross-linking in reversible moldability and easy flow characteristics. Similar conditions prevail in the case of polymethyl methacrylate, which is linear, amorphous and has intrinsically flexible backbone chains stiffened by the two substituents

$(CH_3$ and $COOCH_3)$ at alternate carbon atoms; as a result, it has a hard, brilliantly transparent, relatively high-softening (95°C) thermoplastic polymer, which has found many useful and valuable applications. The only weakness of both materials is their low resistance to swelling and dissolution; it seems that the bulky and eventually polar substituents are capable of producing favorable mechanical and thermal effect associated with the overall mobility and flexibility of the chain segments, but not of imparting sufficient resistance to penetration of the system by solvents or swelling agents, because this process is a strictly localized phenomenon and depends on the affinity of the substituents for the particular solvent molecules.

Similar effects are also produced if the backbone chains themselves are rigid, and if their substituents are so arranged as to prevent crystallization. Classical examples of such polymers are cellulose acetate and cellulose nitrate, where the glucosidic backbone chains represent considerable intrinsic stiffness, and the irregularly arranged acetyl and nitrate groups of incompletely substituted samples preclude formation of a crystalline order. In fact, both materials are hard, transparent, high-melting and amorphous thermoplastic resins, widely and successfully applied for many years; cellulose acetate is still rather sensitive to solvents and swelling agents (just like polystyrene and polymethyl methacrylate), whereas cellulose nitrate is only soluble in a few selected systems.

Recently, the principle of using intrinsically rigid chains has been studied in more detail and found several new and interesting embodiments. Based on rigid monomeric units, such as:

Bisphenol

$$HO-\bigcirc-\overset{CH_3}{\underset{CH_3}{C}}-\bigcirc-OH$$

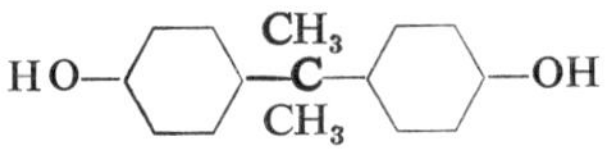

Tetraminodiphenyl

$$H_2N-\bigcirc-\bigcirc\overset{NH_2}{\diagup}-NH_2$$
$$\diagdown NH_2$$

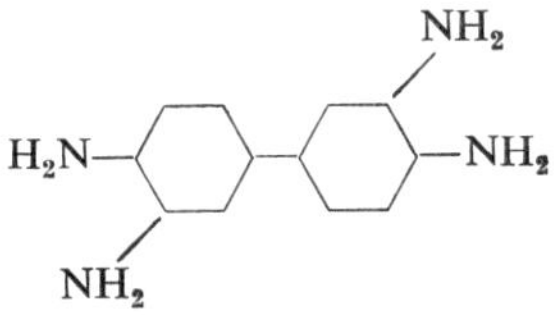

Terephthalic acid

$$HOOC-\bigcirc-COOH$$

Methylene bis-phenyl-isocyanate

$$OCN-\bigcirc-CH_2-\bigcirc-NCO$$

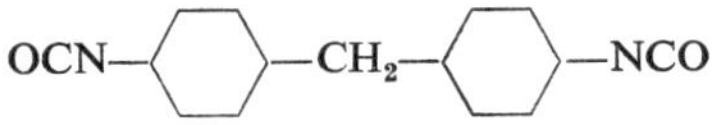

Para-phenylene diamine

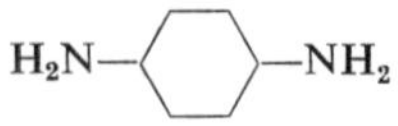

Pyromellitic anhydride

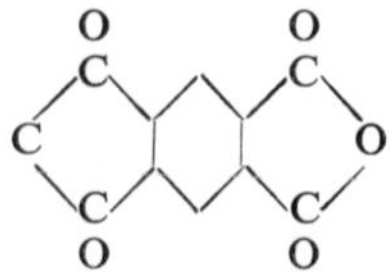

and others, a series of polymers have been synthesized which are substantially amorphous and non-cross-linked, but represent very hard, high-softening and solvent-resistant materials. Earlier examples of this type are the polycarbonates and the linear epoxy resins, both based on bisphenol; more recent representatives are the polybenzimidazoles, polyamides, and polyphenyl oxazoles, which exhibit unusually high resistance to softening, swelling and decomposition. In fact, some of these newer materials can stand temperatures up to 500°C for long periods without softening and deterioration, and are completely insoluble in all organic solvents up to 300°C.

Other rigid molecules currently under study for possible application in the field of high-temperature resistant materials are based on other aromatic chains such as polyphenylene

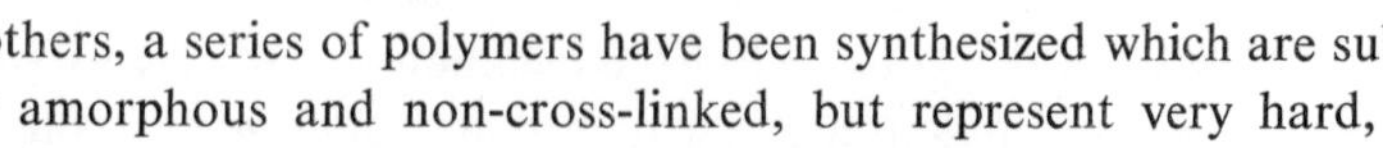

which cannot fold even at rather high temperatures, because rotation about the carbon–carbon single bond between the *para*-combined phenylene rings can only lead to different angles between the planes of consecutive rings, but not to a kink or bend in the main chain. In fact, representatives of this species are very rigid and high-melting, possess a pronounced tendency to crystallize, and are highly insoluble. This combination of valuable properties has not yet been fully brought to fruition, because currently known polyphenylenes are in the relatively low molecular weight range.

Para-polyphenylene oxide and polyphenylene sulfide are other cases showing the chain stiffening action of a *para*-phenylene unit:

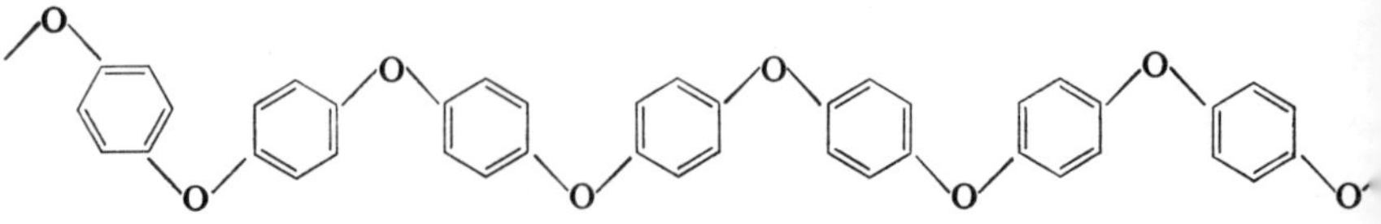

Rotation about the bonds between an ether oxygen atom and the adjacent carbon atoms of the rings both changes the angle between the planes of consecutive units, and causes bends and kinks in the chain, but rotational freedom is noticeably inhibited by the presence of the aromatic rings on each side of the oxygen or sulfur atom. As a consequence, it has been found that chains of this type represent high-melting, rigid and poorly soluble materials.

Another interesting way to arrive at chains made up of condensed rings is the synthesis of so-called ladder polymers. The first case of such a structure was prepared by exposing polyacrylonitrile to elevated temperatures, leading to formation of rows of six-membered rings by an electron-pair displacement:

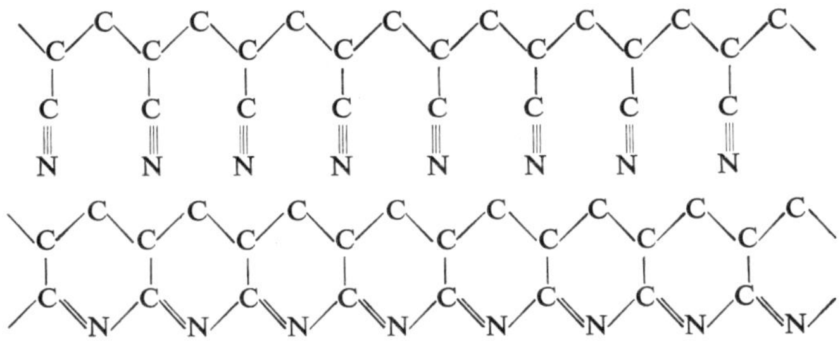

which involves stiffening, insolubility and discoloration. Further heating leads to evolution of H_2 and to aromatization,

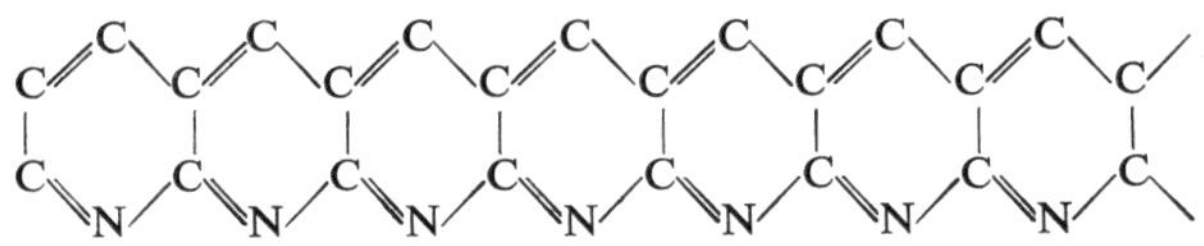

yielding a black, completely non-fusible and insoluble material corresponding in structure to linear graphite with one carbon atom in every ring replaced by nitrogen.

These examples clearly indicate the numerous possibilities for formation of long stiff chains; in all cases, the properties of the resulting materials confirm the expectations.

The existence of three different and independent ways of establishing favorable compromises of valuable properties stimulates the attempt to explore combinations of these principles with a view to even better results. For a convenient survey of such combinations, let us consider a triangle (Fig. 8) in which the three principles of crystallization, cross-linking and chain stiffening are represented respectively by the three vertices *A*, *B* and *C*. Vertex *A* represents a large number of crystallizable thermoplastic polymers with flexible chains which have proved particularly successful as fiber and film formers. Representative materials are polyethylene, polypropylene, polyoxymethylene, polyvinyl alcohol, polyvinyl chloride, polyvinylidene chloride and such polyamides as 6- and 66-nylon. *B* represents the typical thermoset highly cross-linked systems such as hard rubbers, urea-, melamine-, and phenol-formaldehyde condensates, highly reticulated polyesters, polyepoxides and polyurethanes. Finally, *C* represents the amorphous, thermoplastic resins with relatively high rigidity and high softening range such as polystyrene, polymethylmethacrylate, *ABS* resins, polystyrene derivatives and more recently polycarbonates, linear polyepoxides, polyethers and polycondensation products with inflexible chains.

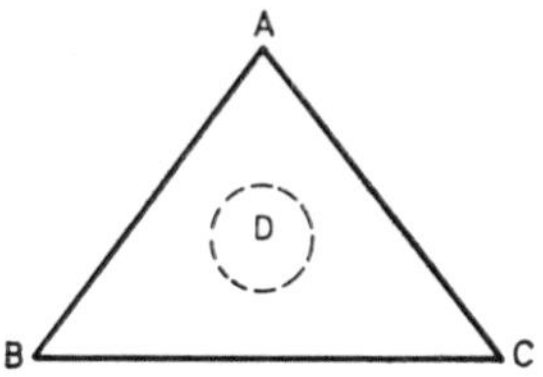

A - CRYSTALLIZATION
B - CROSS-LINKING
C - CHAIN STIFFENING

FIG. 8. Illustration of the three principles influencing the properties of high polymers.

To see how combinations of the principles have resulted in more attractive and valuable product properties, let us consider the sides of the triangle. Side *AC* accommodates several interesting fiber- and film formers. One of them is polyethylene–glycol terephthalate (Terylene or Dacron), in which the *para*-phenylenic unit of the acid introduces enough chain stiffening to raise the melting point of this polymer to about 260°C, as high as that of 66-nylon, although a polyester has no lateral hydrogen bonding available for stiffening its solid crystalline phase. Thus chain stiffening combines with crystallization to produce attractive properties without bringing either of these two principles to an extremely high value. Cellulose is another case in which excellent fiber- and film-forming properties are obtained through a combination of chain stiffening and crystallinity; the resulting polymer is extremely rigid, non-melting and soluble only in a very small number of particularly potent liquid systems. The presence of substantially rigid chains has the favorable consequence that high tensile strength and high softening characteristics become already apparent at relatively low degrees of crystallinity, thereby placing cellulose somewhere half-way between *A* and *C*. Many of the recent spectacular improvements of cellulosic filaments are based on the dual origin of its fiber-forming potential. Another example of a beneficial combination of *A* and *C* is cellulose triacetate, in which the capacity for crystallization superimposes several favorable properties on those of normal cellulose acetate, particularly insolubility in many organic liquids and heat settability through additional crystallization.

On *AB* are located all slightly or moderately cross-linked rubbers, which crystallize on progressive stretching; examples are natural rubber, high *cis*-polybutadiene and polyisoprene, butyl rubber and neoprene. Depending on the degree of cross-linking, they are more or less close to *B*.

Until now we have only considered single-component systems, namely a specific polymer or copolymer; in elastomer technology, however, it is customary to produce stiffening and temperature resistance by inclusion of a hard, finely divided solid filler (crystalline or pseudocrystalline) such as carbon black, silica or alumina. The inclusion is analogous to crystallization in that the flexible chains of the original polymeric matrix are restricted in their segmental mobility by the presence of the very small and hard particles of the filler to whose surface they are attached by strong adsorptive forces. Thus the presence of a filler simulates and replaces crystallization and brings the system closer to point *A*.

Side *CB* represents certain useful polymers characterized by increased rigidity, high softening and insolubility of stiff chain systems obtained through additional cross-linking. Well-known examples of combination of these two principles are elevation of the heat-distortion point of acrylic- and methacrylic polymers by incorporation of allyl methacrylate or ethylene–glycol dimethacrylate and curing of epoxy polymers based on such stiff chain elements as bisphenol and cyclic acetals of pentaerythritol. Additional attempts are now under way to improve the properties of more recent amorphous systems with intrinsically stiff chains by cautious cross-linking, shifting these materials away from point *C* in the direction of *B*. The principal object of this approach is improvement of the dissolution and swelling resistance at elevated temperatures.

The success in combining the above principles in pairs naturally leads to the question whether or not combination of all three could lead to still further improvements. This is a field in which much exploratory work is done at present and certain interesting results have already been obtained. One such successful application is after-treatment of cotton with certain cross-linking agents, or the spinning of rayon in the presence of such agents. The result is fibers of satisfactory strength, elongation and dyeing characteristics, but of insufficient recovery power. Introduction of a cautiously controlled system of cross-links with the aid of bifunctional reagents leads to substantial improvement of the recovery power and wrinkle resistance while leaving all other desirable properties unchanged. Similarly promising results have been obtained with mildly reticulated amorphous stiff chain systems of the epoxy and urethane type, in which crystallization has been replaced by a reinforcing filler.

The triple combinations are situated somewhere in the interior of the triangle (around point *D*) and one can expect that thorough and systematic exploration of this area will lead to many new and interesting polymeric systems with properties superior to those at our disposal today.

REFERENCES

1. K. H. Meyer, *Natural and Synthetic High Polymers*, 2nd ed., Interscience, New York (1950).
2. F. W. Billmeyer, Jr., *Textbook of Polymer Chemistry*, Interscience, New York (1957).
3. Brage Golding, *Polymers and Resins*, Van Nostrand, Princeton, New Jersey (1959).
4. H. F. Mark, *Scientific American* **197**, No. 3, 81–89 (1957).
5. P. J. Flory, *Principles of Polymer Chemistry*, Cornell University Press, Ithaca (1953).
6. H. F. Mark and A. V. Tobolsky, *Physical Chemistry of High Polymeric Systems*, 2nd ed., Interscience, New York (1950).
7. L. E. Nielsen, *Mechanical Properties of Polymers*, Reinhold, New York (1962).
8. G. Natta, *Advances in Catalysis*, vol. XI, Academic Press, New York (1959).
9. N. G. Gaylord and H. F. Mark, *Linear and Stereoregular Addition Polymers*, Interscience, New York (1959).
10. G. Natta *et al.*, *Chim. e Ind. (Milan)* **41**, 398 (1959).

RHEOLOGY

RHEOLOGY—ITS STRUCTURE AND ITS POSITION AMONG THE NATURAL SCIENCES

HANSWALTER GIESEKUS

Abteilung Angewandte Physik, Farbenfabriken Bayer AG., Leverkusen, Germany

INTRODUCTION

It is not yet 40 years since Eugene Cook Bingham and Markus Reiner collaborated in 1928 in laying the foundation of rheology as a natural science in its own right. It appears justified, therefore, to ask whether this still so young discipline has already acquired a fully developed structure and gained an established place in the field of natural sciences. This cannot be taken for granted, because the first "heroic" stage of a science is known to be characterized by a group of new problems coming into view for the first time and methods for their solution being perfected, whereas precise definition of the subject-matter (and thus, at the same time, demarcation of its boundaries with adjoining sciences) is reserved for a second stage, that of consolidation.

Up to a few years ago, the general impression was that rheology is still in its first stage, but anyone who has carefully been following its development during the last decade will have found so many indications of progressive consolidation, that it may safely be maintained today that the structure of rheology, and its relationship to adjoining disciplines, have at least been laid down in their basic features. Accordingly, a scheme will be drawn up and discussed below, showing the breakdown of rheology into its component fields within the circle of adjoining sciences.

DEFINITION

Rheology is the science which seeks to describe, explain, measure and apply the phenomena occurring in bodies on being deformed.†

† As restriction to mere processes of *flow*—in accordance with a literal interpretation of the term "rheology" (i.e. "science of flow")—would entail an unnatural boundary, there is a general consensus that deformable bodies capable of flow only to a limited extent, or not at all, are also to be included in the sphere of rheology.

By this definition, rheology is a branch of *physics;* on the other hand, it is seen to be closely linked with both *chemistry* and *engineering sciences.*†

Since it is mechanical stresses, first and foremost, that constitute the phenomena associated with deformation, there is primarily a close relationship to *mechanics.* Still, rheology cannot be considered a mere component field of mechanics, as deformation causes also *electromagnetic* and *optic* phenomena (electro- and magnetostriction, stress and flow birefringence, etc.). Furthermore, *thermodynamic* viewpoints play such an important part in the deformation of bodies that it would be inappropriate merely to subordinate them to the mechanical ones.

However, quite apart from these facts, there is a considerable difference between the traditional theories of mechanics (elastomechanics, hydrodynamics, gas dynamics, etc.) and rheology—as regards the aspects under which problems are considered. The theories of mechanics take the material for granted (e.g. the elastic solid, the inviscid or viscous fluid) and are primarily interested in the static and dynamic phenomena (vibrations, flow fields and drag in them, stability criteria, etc.). By contrast, the interest of rheology is primarily directed at the material and at characterization of its properties. Its central subject-matter is the *rheological equations of state (constitutive equations)*, and it considers processes associated with the material primarily with a view to its sought properties. It is only in the field of application that interest is diverted from the properties of the material to the processes themselves.

CLASSIFICATION

In accordance with the four terms used in the above definition: *description, explanation, measurement, and application*, rheology comprises four principal fields. The first is frequently called *macrorheology* but should more appropriately be termed *phenomenological rheology* because in it, irrespective of the structure of the material, phenomena are described as such and considered in their mutual relationship. The field associated with the second term is the so-called *microrheology*, in which properties are derived from the structure of bodies; it should therefore be preferably

† Our definition thus largely coincides with Bingham's original intuition, which he formulated to Reiner in the following words: "Here you, a civil engineer, and I, a chemist, are working together at joint problems. With the development of colloid chemistry, such a situation will be more and more common. We therefore must establish a branch of physics, where such problems will be dealt with" [1].

termed *structural rheology.*† The third field is *rheometry,*‡ which develops, on the basis of the phenomenological theory, measuring techniques for the various classes of bodies with a view to characterizing the material as fully as possible. The final field is *applied rheology*, primarily concerned with the influence of the rheological properties in industrial processing; it takes them into consideration in the design of processing equipment, conveyor lines, pumps, etc., and utilizes them with a view to desired effects.

Across this classification, rheology may also be subdivided according to the nature of the phenomena investigated. Thus, frequently only the mechanical aspect of deformational behaviour is associated with rheology (in a narrower sense), whereas the term *rheo-optics* is used for the optical aspects. Similarly, electric and magnetic deformation phenomena are appropriately grouped together as *electro-rheology* and *magneto-rheology.*§

As the subject-matter of rheology is deformational behaviour of bodies in general, all types of materials fall in principle into its sphere of application. It is naturally advisable to exclude such materials as are treated in the traditional theories of continuum mechanics (infinitesimal theory of elasticity, hydromechanics, gas dynamics), unless interest is actually focused on their deviations from the predictions of these theories. Thus, in practice, it is preferably certain classes of materials that are objects of rheological research in so far as their behaviour is specially informative (rubber and polymers, their solutions and melts, colloid systems, glass, concrete, clay suspensions, etc.). The question whether the deformational behaviour of metals—largely treated under plasticity—should in practice be included under rheology is still unsettled, but in principle this should be recommended, and many research workers engaged in these problems are of this opinion.

Materials investigated to date include two classes whose rheological treatment entails so many specific problems that each of them has been assigned a field of its own: the deformational processes in the earth's crust are treated under *geo-rheology*, while the flow processes in organisms are the subject-matter of *bio-rheology*.

The close association of rheology with mechanics and thermodynamics

† Because of the analogy with "statistical mechanics", this field is sometimes referred to as "statistical rheology"; occasionally, the term "structural mechanics" is used.

‡ For this field we also find the term "rheogoniometry". In accordance with the wider use of the term "rheology" as mentioned under footnote (†), p. 341, "rheometry" by analogy covers measuring techniques concerning the deformational behaviour of materials incapable of free flow (e.g. elastometry and plastometry).

§ The related terms "electroviscosity" and "magnetoviscosity" are frequently used.

has already been pointed out. This applies particularly to phenomenological rheology, frequently dealt with within the framework of *rational mechanics* or *thermodynamics of irreversible processes*.† Structural rheology, on the other hand, undoubtedly closely adjoins *statistical mechanics* and *thermodynamics (the theory of transport processes)*, whereas the more complex materials (e.g. macromolecular solutions) are even more closely related to *physical chemistry* and to the relevant research of special categories of materials within its framework (colloids, polymers, rubber, wood, concrete, brick, glass, metals, etc.).

Rheometry primarily utilizes the measuring techniques of mechanics, as well as calorimetric, electrical and optical methods. As regards its aim, it is closely related to what is known as *materials testing*, as well as to the metrology of *chemical* and *process engineering*.

In the same manner applied rheology, as a constructive discipline, overlaps the fields of chemical and process engineering, and bears on other engineering sciences. In order to illustrate this, the breakdown of rheology into its various subdivisions, as described here, is demonstrated in a scheme indicating the relationship to adjoining sciences. The dashed line enclosing the rheological subjects indicates that separation is not real and that in all cases there are close links and even overlapping.

		Applied Mathematics		
	Mechanics		Thermodynamics	
Materials Testing	Rheometry	Phenomenological Rheology (Macrorheology)	Electro- and Magneto-rheology	Electro- and Magneto-dynamics
Chemical and Process Engineering	Applied Rheology		Rheo-optics	Optics
Biology and Medicine	Bio-rheology	Structural Rheology (Microrheology)	Geo-rheology	Geophysics
		Colloid, Polymer, Rubber, Wood, Concrete, Brick, Glass, Metals Research		
	Physical Chemistry		Statistical Thermodynamics	

† In the former case, it is mostly treated under the title of "general (non-linear) theory of continua", while the latter is usually restricted to the linear theory (e.g. of viscoelastic materials) as a special case of the theory of linear dissipative (passive) systems.

ORGANIZATION

As intended by its founders, rheology has from the very outset been characterized by a tendency to overlap and to link up with adjoining disciplines, and to the present day this has been one of its most peculiar features: rheology conventions are attended by mathematicians, physicists, engineers, chemists, biologists, and physicians, and rheological research is carried on in both industrial and research laboratories of governmental and commercial institutions as well as at various departments of universities and institutes of technology.† The real task of rheology cannot, however, be defined as mere provision of a discussion forum, for the various scientists interested in rheological problems, by means of national or regional rheological societies, conventions or periodicals, with individual research left exclusively to existing disciplines of natural and engineering sciences; specifically *rheological research* is also called for. This follows from the above analysis of the structure of rheology, which indicates no single scientific discipline with which rheology *as a whole* might be associated. Although phenomenological rheology may still be considered, under the loosest possible classification, as part of mechanics, and structural rheology as part of physical chemistry, no common platform is possible for both phenomenological and structural rheological research to be pursued *simultaneously*.‡ The same applies to the link between theoretical rheology and rheometry or applied rheology. Unreasonable as it were even to attempt to detach rheological work from its present setting and concentrate it in future within a separate framework called "rheology", special rheological research centres, established as an *addition* (at least

† The first volume of *Rheologica Acta* (1958–61) contains publications from laboratories of general chemical industry, of the petroleum, synthetic-fibre, and paper industries, from governmental institutions such as the National Bureau of Standards (Washington), the Physikalisch-technische Bundesanstalt (Brunswick), and the Bundesamt für Materialprüfung (Berlin), and from research institutes for fibres, ceramics, mining, peat and wood, biophysics, medicine, sweets and dairying. Work in universities and institutes of technology was carried on, *inter alia*, at departments of applied mathematics, theoretical and experimental physics, mechanics, fluid dynamics, engineering sciences, process engineering, industrial and colloid chemistry, polymer research, mining, metallurgy, foundry and pathology.

‡As an illustration how unnatural such a situation would be, let us visualize phenomenological and statistical thermodynamics, or Maxwell's theory and the electron theory, assigned to different disciplines. Thermodynamics is known to owe many of its most important advances to research workers who were at home in both branches. Similarly, a number of scientists working in the field of rheology hold that it will be impossible in future to achieve important results without, above all, close interlinking of the macro- and micro-rheological viewpoints.

in some places), would certainly be desirable. Such centres would permit qualified research on a sufficiently broad and solid theoretical and experimental basis, and at the same time improve the efficiency of rheological work done elsewhere, by means of consultation and suggestions.

The earlier statement, to the effect that rheology as a scientific discipline is already in the stage of consolidation, did not refer to organization of rheological research, which is still in its infancy. It is, however, making increasing progress in various quarters in view of growing awareness, particularly during recent years, of its need for intensive promotion combined with closer coordination.† Similar remarks also apply to rheology as a subject of a university course.‡ Thus, it would not be unrealistic if we wish Markus Reiner, on his eightieth birthday, that he may live to see rheology, of which he was one of the founders, assume its established place also within the framework of research and teaching institutions, so as to be able to realize its tasks even more effectively—not least that of bridging between the various branches of science.

REFERENCES

1. M. Reiner, "The Deborah number", *Physics Today* **62** (Jan. 1964).
2. A. B. Metzner, "The mechanics of non-Newtonian fluids: American research and research needs", *Rheol. Acta* **5,** 65 (1966).
3. A. Y. McLean and A. T. J. Hayward, *Rheological Research in Great Britain*, publ. by National Engineering Laboratory, East Kilbride, Glasgow (1964).
4. A. S. Lodge, "The promotion of research in rheology", *Rheol. Acta* **5,** 63 (1966).
5. J. H. C. Vernon, "The teaching of rheology", *Bull. Inst. Physics* (London), (Mar. 1965), pp. 95; (abridged) *Rheol. Acta,* **5,** 61 (1966).

† Reference is in order here, for example, to the activities of the A.I.Ch.E. in the United States [2] or to the National Engineering Laboratory [3], and the "Exploratory Committee" of the British Society of Rheology [4] in Great Britain.

‡ Cf. lectures and discussions at the annual meeting of the British Society of Rheology (Exeter, 1964) [5].

VISCOSITY OF RIGID PARTICLE SUSPENSIONS†

ZVI HASHIN

Towne School of Civil and Mechanical Engineering, University of Pennsylvania, Philadelphia, Pennsylvania, U.S.A.

1. INTRODUCTION

The problem of the theoretical determination of the viscosity coefficient of a suspension consisting of Newtonian incompressible fluid and rigid particles is one of long standing. The pioneering work in the subject is Einstein's [1] well-known investigation in which it was assumed that the particles are spherical and that their fractional volume is very small compared with unity. The effective viscosity coefficient of the suspension, η^*, is then given by

$$\eta^* = \eta(1+2{\cdot}5c), \tag{1.1}$$

where η is the viscosity coefficient of the fluid and c the particle fractional volume. Experiments show that with increasing c, the suspension viscosity increases very much faster than (1.1). This is not surprising, since in the derivation of (1.1) interaction of spheres was neglected. In general, (1.1) does describe correctly the slope of experimental η^* versus c curves at $c = 0$.

There have been numerous attempts to predict η^* for arbitrary c. A recent extensive review article by Rutgers [2] lists close to 300 papers and close to a hundred different results for η^*. These results are generally based on various simplifying assumptions and approximations, while some are also semi-empirical, containing an undetermined parameter to be hopefully determined by experiment.

This discouraging state of affairs is really not surprising in view of the enormous inherent difficulties. As will be explained further below, in order to calculate η^* it is necessary to find the detailed flow field in the suspension, which for many interacting particles is an intractable problem, even for spherical shapes. To complicate matters even further, the position of

† Supported by the U.S. Army Research Office-DURHAM under Contract DA-31-124-ARO-D-194 with the University of Pennsylvania.

particles is not really known at any time. It is clear that one is faced with a statistical problem, and that suspension geometry should be characterized in statistical terms. As in the theory of random processes, this can be done in terms of N-point probability functions specifying the joint probability that each point of a given N-point system is within the fluid or particle phase. In general, η^* will depend upon the infinite system of these probability functions. It is important to realize that volume fractions in a statistically homogeneous suspension are one-point probabilities, i.e. the probability that a point thrown at random into the suspension is within one or the other phase. Therefore, it is by no means to be expected that volume fractions alone will be sufficient geometrical information for determination of η^*, as has unfortunately been assumed by a number of writers.

The goal of the statistical approach to the problem is to use statistical information in the form of joint-probability functions to predict η^*. At present, this approach is confined to two- and three-point probability functions. Since such probability functions are only partial information, the only possibility is to construct bounds on η^* in terms of this information. Lower bounds in terms of such probability functions have been derived by Prager [3], and Weissberg and Prager [4, 5]. Upper bounds in terms of statistical information are apparently not to be found in literature.

Judging by the results obtained so far it does not, unfortunately, appear that the probability functions up to order 3 can provide sufficient information for construction of lower bounds close to experimental results. It thus seems that higher-order probability functions have to be taken into account. At present, they have to be regarded as information to be obtained by experiment. However, experimental determination of these is very difficult, indeed much more difficult than measurement of η^* itself. It must therefore be concluded that until methods are found for determining such probability functions theoretically on the basis of some plausible statistical assumptions, the statistical approach must remain chiefly of academic interest.

It should also be pointed out that while N-point probability functions are undoubtedly convenient for description of random geometry, it is entirely possible that there exists a better way of description, although the author is not aware of it. One is brought to this thought by the fact that such probability functions seem to be incapable of describing the simple topological restriction that one phase is in the form of particles. Thus there is, to-date, no statistical criterion for the definition of a suspension. This seems to be the reason that it has not proved possible to derive finite upper bounds in terms of statistical information, for the case of a rigid phase. For if the possibility of a *continuous* rigid phase is not excluded, then the upper bound must also apply for such a possibility and thus must be infinite.

The approach adopted in this work is motivated by the inherent difficulties described above. Here geometrical information in addition to the volume fractions is provided artificially, by restricting the analysis to suspensions of very special geometry. These suspension models, described further below, can be rigorously treated and this is the chief motivation for their choice.

2. GENERAL THEORY OF EFFECTIVE VISCOSITY

It is assumed that the suspension geometry is statistically homogeneous.† Two cases of statistical symmetry will be here considered. In the first case the suspension is also assumed to be statistically isotropic and in the second case only transverse isotropy is assumed. The first case is an appropriate description when the particle shapes do not show any preferred orientation, while the second case applies when the particles are very elongated and all oriented in one direction of homogeneous flow.

It is now assumed that the suspension is subjected to statistically homogeneous strain rate and stress fields. A necessary (but not sufficient) condition for such statistical homogeneity is that averages of strain-rate and stress components, taken over local representative volumes of the suspension, do not vary with location of the representative volume.

The effective viscosity coefficient η^* is then defined by the relation

$$\bar{s}_{ij} = 2\eta^*\bar{e}_{ij}, \tag{2.1}$$

where overbar denotes average, s_{ij} is the stress deviator and e_{ij} is the strain-rate deviator.

In the case of statistical isotropy, (2.1) remains valid with the same η^* regardless of orientation of the coordinate system to which the stresses and strain rates are referred. In the transversely isotropic case, however, (2.1) has a more restricted interpretation, as will be explained further below.

To produce statistically homogeneous fields of strain rate and stress, a very large suspension control volume is subjected to boundary conditions which would produce homogeneous states of strain rate and stress

† For details here and in the rest of this section, see ref. [6].

if the fluid in the control volume were homogeneous. A suitable velocity boundary condition is

$$v_i(S) = e^0_{ij} x_j, \tag{2.2}$$

where S is the surface of the control volume, the range of subscripts is 1, 2, 3, a repeated subscript denotes summation, v_i are velocity components, e^0_{ij} are constant strain-rate components, and x_i are cartesian surface coordinates. It is rigorously true that for (2.2) prescribed, the strain rates averaged over the entire control volume are e^0_{ij}. It follows from statistical homogeneity that they are also the local strain rate averages.

A suitable stress boundary condition is

$$T_i(S) = -p^0 n_i + s^0_{ij} n_j, \tag{2.3}$$

where T_i are the surface fractions, n_i the components of outward normal to S, and p^0 and s^0_{ij} constant pressure and deviatoric stresses respectively. A boundary condition of type (2.3) is physically realizable only for an extremely viscous fluid, but it is certainly mathematically appropriate for any viscous fluid. For (2.3) prescribed, it is rigorously true that average control volume pressure and deviatoric stresses are p^0 and s^0_{ij}, respectively. Again, by statistical homogeneity, they are also the local averages.

The mathematical problem with which one is faced in calculating η^* is now clear. Prescribing (2.1) on the control volume boundary defines $\bar{e}_{ij}$ in (2.1) as e^0_{ij}. If $\bar{s}_{ij}$ are then calculated, η^* becomes known. However, calculation of $\bar{s}_{ij}$ requires determination of the detailed s_{ij} field, which is in general an intractable problem. Conversely, (2.3) may be prescribed, whence $\bar{s}_{ij}$ in (2.1) are given by s^0_{ij}. Now $\bar{e}_{ij}$ has to be calculated, which is just as difficult.

An alternative definition of η^* is in terms of energy dissipation. If (2.2) is prescribed, it is rigorously true that the energy dissipated per unit time in the control volume V is given by

$$D_e = 2\eta^* e^0_{ij} e^0_{ij} V. \tag{2.4}$$

On the other hand, when (2.3) is prescribed the dissipation is

$$D_s = \frac{1}{2\eta^*} s^0_{ij} s^0_{ij} V. \tag{2.5}$$

Equation (2.4) is the definition originally adopted by Einstein [1]. The definition (2.5) has been used by Prager [7]. Both are equivalent to (2.1). Expressions (2.4) and (2.5) are of fundamental importance. Together with the extremum principles for creeping viscous flow theory, they provide the basis for the construction of bounds for the effective viscosity coefficient η^*.

Consider first the Helmholtz–Korteweg extremum principle (compare, for example, Lamb [8], p. 617). Define an admissible velocity field $\tilde{v}_i(\mathbf{x})$, within the suspension volume, by the following requirements:

(a) $\tilde{v}_i(\mathbf{x})$ vanishes within the particles and on their surfaces, apart from rigid body motions which are immaterial to the theorem;
(b) $\tilde{v}_i(\mathbf{x})$ satisfies the boundary conditions (2.2);
(c) $\tilde{v}_{i,i} = 0$;
(d) $\tilde{v}_i(\mathbf{x})$ is continuous throughout the fluid volume V_2.

Define the strain rate dissipation integral $\tilde{D}_e$ by

$$\tilde{D}_e = 2\eta \int_{(V_2)} \tilde{e}_{ij}\tilde{e}_{ij}\, dV, \tag{2.6}$$

where $\tilde{e}_{ij}$ are the strain rates derived from $\tilde{v}_i$. Then, according to the Helmholtz–Korteweg principle

$$\tilde{D}_e \geqslant D_e, \tag{2.7}$$

where D_e is given by (2.4). Thus (2.7) provides an upper bound for η^*, for any $\tilde{v}_i$.

Now consider the second extremum principle, in terms of viscous stresses (compare, for example, Prager [7]). Here it is necessary to define an admissible pressure $\tilde{p}$ and an admissible stress deviator field $\tilde{s}_{ij}$ which fulfil the following requirements, within the fluid volume V_2 only:

(a) The tractions $\tilde{T}_i = -\tilde{p}n_i + \tilde{s}_{ij}n_j$ satisfy the boundary conditions (2.3);
(b) $\tilde{s}_{ij,j} = \tilde{p}_i$;
(c) the tractions $\tilde{T}_i$ are continuous.

Define the stress dissipation integral $\tilde{D}_s$ by

$$\tilde{D}_s = \frac{1}{2\eta} \int_{(V_2)} \tilde{s}_{ij}\tilde{s}_{ij}\, dV. \tag{2.8}$$

Then the extremum principle is expressed by the inequality

$$\tilde{D}_s \geqslant D_s, \tag{2.9}$$

where D_s is given by (2.5). Thus (2.9) provides an upper bound for $1/\eta^*$ for any s_{ij}, and hence a lower bound for η^*.

The extremum principles used here are completely analogous to the well-known linear elasticity principles of minimum potential energy and minimum complementary energy, for the special case of an incompressible elastic body. The analogous quantities are given by the following scheme:

Creeping viscous flow	*Incompressible elasticity*
viscosity coefficient	shear modulus
velocity	displacement
strain rate	small strain
pressure	negative mean normal stress
stress deviator	stress deviator
dissipation	twice the strain energy density

The analogy is well known for solutions of boundary value problems in creeping viscous flow and incompressible elasticity, with similar geometries and boundary conditions. In this context the analogy dates back to Rayleigh [10] and has been given in more detail by Goodier [11]. The wider validity of the analogy for admissible fields entering into the extremum principles has been pointed out by Hill and Power [9]. It is also interesting to note that the extremum principles in terms of polarization tensors derived by Hashin and Shtrikman [12] have their analogous counterparts for creeping viscous flow (Hashin [13]).

The elasticity extremum principles have been used to great advantage to bound effective elastic moduli of heterogeneous materials.

It follows, by the analogy, that bounds for the elastic shear modulus of a two-phase medium, in which one phase is rigid and the other incompressible elastic, can be immediately translated into bounds for the effective viscosity coefficient of a rigid particle suspension. Such bounds will be given and discussed in what follows.

3. ISOTROPIC SUSPENSIONS

Consider first the case of a statistically homogeneous and isotropic suspension in which the particles are of arbitrary shapes. For the corresponding elastic problem of a two phase-material, Paul [14] gave a variational bounding method according to which the effective shear modulus μ^* is bracketed as follows,

$$\frac{1}{\dfrac{v_1}{\mu_1}+\dfrac{v_2}{\mu_1}} \leqslant \mu^* \leqslant v_1\mu_1+v_2\mu_2, \tag{3.1}$$

where 1, 2 denote the phases and v denotes volume fraction.

Let the first phase be deformable and the second perfectly rigid. Then $\mu_2 \to \infty$. Denote

$$v_2 = c. \tag{3.2}$$

Then, using the analogy, it is found that

$$\frac{\eta}{1-c} \leqslant \eta^*, \tag{3.3}$$

where η is the fluid viscosity coefficient. The upper bound for η^* becomes infinite.

A better lower bound *for arbitrary particle shapes* has been obtained by Hashin [13]. This result is

$$\eta^*_{(-)} = \eta\left(1+\frac{2{\cdot}5c}{1-c}\right) \leqslant \eta^*. \tag{3.4}$$

The result (3.4) has been derived on the basis of a new extremum principle for creeping viscous flow. It can, however, be easily derived on the basis of the viscous flow-elasticity analogy and bounds for the elastic shear modulus of a two-phase material given by Hashin and Shtrikman [15]. To do this, the phase shear moduli in the elasticity bound are replaced by viscosity coefficients, one of the coefficients being infinite to account for rigid particles, and Poisson's ratios are assigned the value $\frac{1}{2}$ to satisfy incompressibility. Then, the shear-modulus lower bound transforms into (3.4), while the upper bound becomes infinite. The infinite value of the upper bound is not surprising; since the theory is based on knowledge of volume fractions only, the bounds cover all possible geometries with same volume fractions. Therefore, they also apply for the case when the rigid phase is in the form of a continuous matrix. However, in that case no flow is possible and the effective viscosity is infinite. Therefore, in order to improve upon the result (3.4), it is necessary to take account of the fact that in a suspension the rigid phase is in the form of particles. How to do this for arbitrary particle shapes is apparently not known at present. However, useful information may be obtained by analysis of a very idealized model of a spherical particle suspension.

To describe this suspension, consider a control volume containing non-overlapping composite spheres. Each composite sphere consists of a rigid spherical particle and a concentric viscous fluid matrix shell with viscosity coefficient η. In each composite sphere the particle-to-shell volume ratio has the same value c. Let the volume of all composite spheres be denoted by V'. The remaining volume V'' can be made infinitely small by progressive addition of composite spheres of indefinitely diminishing sizes.

Consider the case when (2.2) is prescribed on the surface of the suspension volume. To construct an admissible displacement field for the Helmholtz–Korteweg principle, (2.2) is chosen as the field in V''. Then each composite sphere is subjected to (2.2) on its boundary (apart from a rigid-body motion). This defines a creeping viscous flow boundary-value problem for the composite sphere, with zero velocities on the particle surface. The velocity fields in all spherical shells, together with (2.2) in V'', obviously satisfy all requirements for $\tilde{v}_i$. If these are used in (2.6) together with (2.4), an upper bound for η^* of the idealized suspension is obtained. Since the volume V'' is vanishingly small, its contribution to the bound also becomes vanishingly small. The upper bound then follows from the fields in the fluid shells, only. Indeed, it suffices to compute the bound for one composite sphere, because of the equal particle-to-shell volume ratios.

An entirely analogous procedure can be used to construct an admissible stress field for (2.3) prescribed on the surface, leading to a lower bound by (2.5) and (2.7).

Fortunately, the necessary calculations do not have to be performed. The same suspension model as described here has been analyzed for the linear elastic case (Hashin [16]). Therefore, by the variational creeping viscous flow–incompressible elasticity analogy, the results can be immediately transcribed to the present case. The upper bound for the elastic case is defined by equations (50), (84–86) in [16]. Specializing to rigid particles and incompressible elasticity, the following upper bound $\eta^*_{(+)}$ is obtained:

$$\eta^*_{(+)} = \eta \left\{ 1 + \frac{2{\cdot}5c}{1 - c\left[1 + \dfrac{21}{4}\,\dfrac{\left(1 - c^{\frac{2}{3}}\right)^2}{1 - c^{\frac{7}{3}}}\right]} \right\}. \qquad (3.5)$$

The lower bound for η^* may be similarly obtained. It may, however be proved that this lower bound is always below the lower bound (3.4). Since (3.4) applies for arbitrary particle shapes, it is certainly valid for the present idealized suspension. Accordingly, η^* for the suspension considered is bounded by (3.5) from above and by (3.4) from below. Note that (3.4), (3.5) also satisfy the Einstein result (1.1) for small c.

It is of great interest to compare the present theoretical results with experimental findings. Two widely quoted experimental investigations of suspension viscosity are due to Eilers [17] and Vand [18]. Eilers' experiments were performed on a suspension of bitumen particles in water, whereas Vand measured the viscosity of suspensions of glass spheres in a solution of ZnI_2 in water–glycerol mixture. The results are shown in the table (p. 356) as a function of particles fractional volume. Also shown are the upper bound results according to Eq. (3.5). A graphical comparison of

Eilers' and Vand's results with the bound (3.4) and (3.5) is shown in Fig. 1. The two different columns for Vand's experiments indicate measurements in which the suspension was not stirred (N) or stirred (S) before measurement of η^*. Note the considerable increase in Vand's η^* at high concentration due to preliminary stirring. In the author's opinion, this clearly

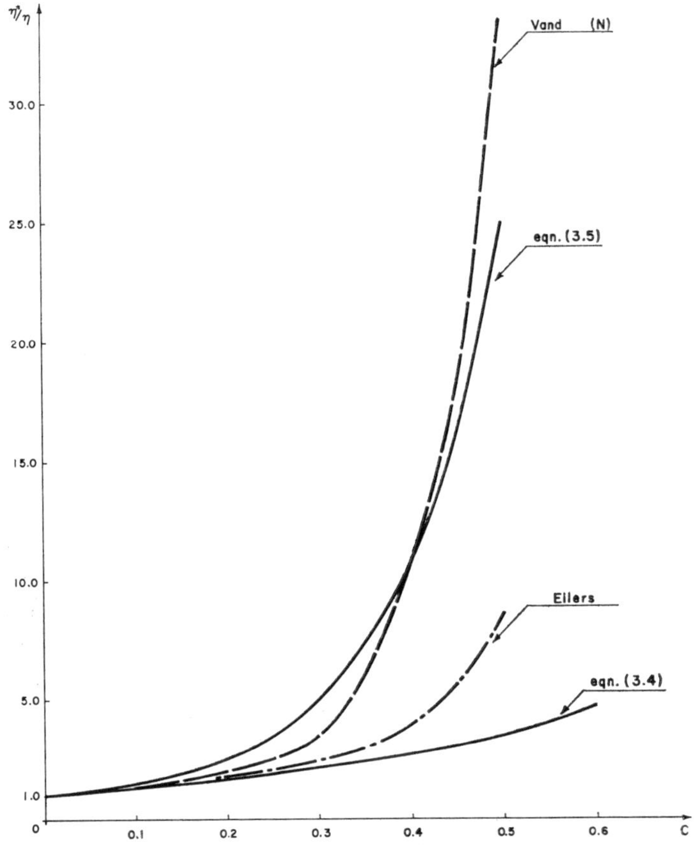

FIG. 1. Effective viscosity coefficient; bounds and experimental results.

indicates the strong influence of the statistical distribution of the spheres, and the futility of regarding η^* as a function of c only. Also note the considerable difference between Eilers' and Vand's experimental results, which is probably due to the same cause.

Comparison of the present upper-bound results with Vand's experiments shows that from about $c = 0{\cdot}40$ even the (N) results increase beyond the upper bound. This phenomenon may be explained by the nature

c	Vand		Eilers	Eq. (3.5)
	η^*/η (N)	η^*/η (S)	η^*/η	$\eta^*_{(+)}/\eta$
0·10	1·342	1·342	1·25	1·449
0·20	2·024	2·024	1·84	2·492
0·30	3·556	3·636	2·55	4·958
0·40	10·53	11·77	4·0	10·73
0·45	18·13	33·33		16·31
0·50	33·33	200·00	7·6	25·20

of the composite sphere assemblage model. It is there assumed that each particle is always surrounded by fluid. Thus the model cannot take account of sphere clustering, mutual friction and collisons which are liable to increase the resistance to flow and thus the suspension viscosity at high fractional volume of the particles. On the other hand, such effects should also have been expected in Eilers' experiments, and there remains the open question why his results are so much lower.

4. TRANSVERSELY ISOTROPIC SUSPENSION

Unlike an isotropic suspension in which any direction is an axis of material symmetry, a transversely isotropic suspension has only one such axis. Such symmetry may be assumed to exist in a suspension of needle-like particles. Under overall homogeneous shear flow (Couette flow) the particles will ultimately orient themselves in the flow direction which may then be assumed to be an axis of transverse material isotropy. This kind of symmetry is also found in fiber-reinforced material where the fibers are all parallel and otherwise randomly located. For a statistical description of this kind of symmetry see ref. [19].

Consider statistically homogeneous shear flow in the x_1-direction with velocity gradient in the x_2-direction. The effective viscosity coefficient η^* is then defined by a relation of type (2.1), which for the present case assumes the form

$$\bar{s}_{12} = 2\eta^* \bar{e}_{12}. \tag{4.1}$$

A convenient control volume for the present case is a large slab whose lower and upper faces are parallel to the flow direction (Fig. 2). A suitable velocity boundary condition of type (2.2) is given by

$$v_1(S) = 2e^0_{12} x_2 , \tag{4.2a}$$

$$v_2(S) = v_3(S) = 0. \tag{4.2b}$$

Under these conditions

$$\bar{e}_{12} = \bar{e}_{21} = e^0_{12}. \tag{4.3}$$

(These correspond to pure shear flow, whereas (4.2) also contains a homogeneous rotatory part.) All other $\bar{e}_{ij}$ vanish and (2.4) becomes

$$D_e = 4\eta^* e^{0\,2}_{12} V. \tag{4.4}$$

A suitable traction boundary condition of type (2.3) is given by constant shear stresses $s^0_{21} = s^0_{12}$ on the faces normal to the x_1- and x_2-axes respectively, while the pressure and all other stress components, such as may act on the surface, vanish. Then

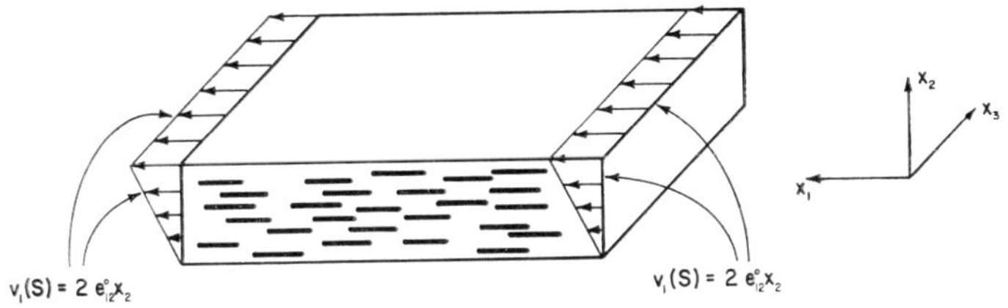

FIG. 2. Control volume for transversely isotropic suspension.

$$\bar{s}_{12} = \bar{s}_{21} = s^0_{12}. \tag{4.5}$$

All other $\bar{s}_{ij}$ vanish, and (2.5) becomes

$$D_s = \frac{1}{\eta^*} s^{0\,2}_{12}{}^2 V. \tag{4.6}$$

It is now assumed that all particles can be adequately described by cylinders whose length-to-diameter ratio is very large. In that case the end effects are negligible, and it may be assumed with sufficient accuracy that the cylinders extend continuously through the entire control volume. The situation is then entirely analogous to that found in a fiber-reinforced material, and the results obtained for elastic properties of such materials may be transcribed to the present case by analogy. The pertinent elastic modulus is the axial shear modulus denoted by G_1^* in [20] and by μ in [19].

For the case of arbitrarily shaped cylinders, Hashin [19] (equations (5.12–13)) has given bounds for this elastic modulus. It is interesting to note that only phase shear moduli and no Poisson's ratios enter into the bound. Specializing to rigid cylinders, the lower bound is given by

$$\eta\left(1+\frac{2c}{1-c}\right) \leqslant \eta^*, \tag{4.7}$$

while the upper bound becomes infinite. This bound also follows as a special case of cylindrical emulsion bounds given by Hashin [13].

The elastic problem has also been analyzed by Hashin and Rosen [20] using the assemblage model, with composite cylinders instead of composite spheres. This analysis yielded coincident lower and upper bounds, and thus an exact result. The corresponding result for η^* is given by

$$\eta^* = \eta\left(1+\frac{2c}{1-c}\right). \tag{4.8}$$

It is of considerable interest to note that (4.8) is the same as the lower bound (4.7). Therefore (4.7) is the *best lower bound attainable with knowledge of volume fractions only.*

Furthermore, for small c, (4.8) becomes

$$\eta^* = \eta(1+2c). \tag{4.9}$$

This coincides with a dilute suspension formula obtained by Jeffery [21] for very elongated ellipsoidal particles oriented in one direction.

5. CONCLUSION

The problem of the prediction of the effective viscosity coefficient of Newtonian rigid particle suspensions has been treated by variational bounding methods. The bounds for a sphere suspension are in fair agreement with known experimental results. It would be of considerable interest to check experimentally the expression obtained by the coincident bounds for the oriented cylindrical particle suspension.

ACKNOWLEDGMENT

I wish to express my gratitude to Professor M. Reiner for arousing my interest in the subject of viscous flow of suspensions in particular and the theory of heterogeneous media in general, at the time when I was his student. His advice and encouragement have always been invaluable.

REFERENCES

1. A. EINSTEIN, "Eine neue Bestimmung der Moleküldimensionen", *Ann. d. Phys.* **7**, 289 (1906); **34**, 591 (1911).
2. R. RUTGERS, "Relative viscosity and concentration", *Rheologica Acta* **2**, 305 (1962).
3. S. PRAGER, "Diffusion and viscous flow in concentrated suspensions", *Physica* **29**, 129 (1963).

4. H. L. Weissberg and S. Prager, "Viscosity of concentrated suspensions of spherical particles", *Proc. 4th Int. Congress Rheology*, Pt. 2, p. 709, Interscience, New York (1965).
5. H. L. Weissberg and S. Prager, "The viscosity of concentrated suspensions", *Trans. Soc. Rheology* **9,** 1, 321 (1965).
6. Z. Hashin, "Theory of mechanical behavior of heterogeneous media", *Appl. Mech. Rev.* **17,** 1 (1964).
7. S. Prager, "Viscous flow through porous media", *Physics of Fluids* **4,** 1477 (1961).
8. H. Lamb, *Hydrodynamics*, Dover Publications, New York (1945).
9. R. Hill and G. Power, "Extremum principles for slow viscous flow and the approximate calculation of drag", *Quart. J. Mech. Appl. Math.* **9,** 313 (1956).
10. Lord Rayleigh, *Theory of Sound*, vol. II, chap. 19, par. 345, Macmillan, London, (1940).
11. J. N. Goodier, "Slow viscous flow and elastic deformation", *Phil. Mag.* **22,** 678 (1936).
12. Z. Hashin and S. Shtrikman, "On some variational principles in nonhomogeneous and anisotropic elasticity", *J. Mech. Phys. Solids* **10,** 335 (1962).
13. Z. Hashin, "Bounds for viscosity coefficients of fluid mixtures by variational methods", *Proc. Int. Symp. Second Ord. Eff. Elast. Plast. Fluid Dyn.*, Reiner and Abir, Eds., Pergamon Press, New York, p. 434 (1964).
14. B. Paul, "Prediction of elastic constants of multiphase materials", *Trans. AIME* **218,** 36 (1960).
15. Z. Hashin and S. Shtrikman, "A variational approach to the theory of the elastic behaviour of multiphase materials", *J. Mech. Phys. Solids* **11,** 127 (1963).
16. Z. Hashin, "The elastic moduli of heterogeneous materials", *J. Appl. Mech.* **29,** 143 (1962).
17. H. Eilers, "Die Viskosität von Emulsionen hochviskoser Stoffe als Funktion von Konzentration", *Kolloid-Z.* **97,** 313 (1941).
18. V. Vand, "Viscosity of solutions and suspensions. II", *J. Phys. Colloid. Chem.* **52,** 300 (1948).
19. Z. Hashin, "On the elastic behaviour of fibre reinforced materials of arbitrary transverse phase geometry", *J. Mech. Phys. Solids* **13,** 119 (1965).
20. Z. Hashin and B. W. Rosen, "The elastic moduli of fiber reinforced materials", *J. Appl. Mech.* **31,** 223 (1964).
21. G. B. Jeffery, "The motion of ellipsoidal particles immersed in a viscous fluid", *Proc. Roy. Soc.* (London) A **102,** 161 (1923).

ON DECAY OF ELASTIC PROPERTIES IN A VISCOELASTIC MATERIAL UNDER TENSILE STRESS

CHAIM H. LERCHENTHAL

Department of Mechanics, Technion–Israel Institute of Technology, Haifa

ABSTRACT

The gradual loss of resilience in certain viscoelastic materials under tensile stress, as evident in recovery experiments, may be termed "elastic decay".

A theory of this phenomenon is presented, based on the assumption of successive degradation of elastic elements into viscous ones. For mathematical formulation of this concept, "contribution factors" α are introduced, defining the variable relative contribution of elastic and viscous elements to the total deformation. Non-linear behaviour is allowed for by permitting these factors to be time- and stress-dependent, while assuming material parameters to be constant under isothermal conditions. Three such parameters are used: an intrinsic elastic modulus E, an intrinsic coefficient of viscous traction λ, and one probably associated with such influences as network structure, temperature and chemical reactions. The theory yields a simple relation between stress, loading time and maximum elastic deformation (immediate recovery), verified in constant-stress experiments.

INTRODUCTION

Certain viscoelastic materials tend to lose part of their resilience (defined here as the capacity to store elastic energy in a recoverable manner) under tensile stress. This phenomenon is best observed in recovery experiments. If a cylindrical specimen of such a material is stretched under a constant force or stress and released after some time t_l, both the immediate and delayed elastic deformations (ε_{ie}, ε_{de}) on recovery are smaller than those observed during loading (Fig. 1). The longer the load is sustained, the smaller the recovery, while the non-recoverable deformation ε_{irr} increases accordingly. The material in which this effect was studied, though very common, has had little attention outside the circle directly concerned with its technology, namely, baker's dough. The technique of the relevant experiments has been described in principle in earlier publications [1, 2]. Detailed accounts of more recent work are in press [3, 4, 5].

It appears that the phenomenon is common to other soft materials, and most probably to all weakly cross-linked elastomers. The properties of these materials are, however, usually studied in creep and relaxation experiments which, by their very nature, tend to obscure its significance rather than contribute to its elucidation.

The behaviour described in the opening paragraph may be expected from a Burgers body, if the difference between immediate recovery and

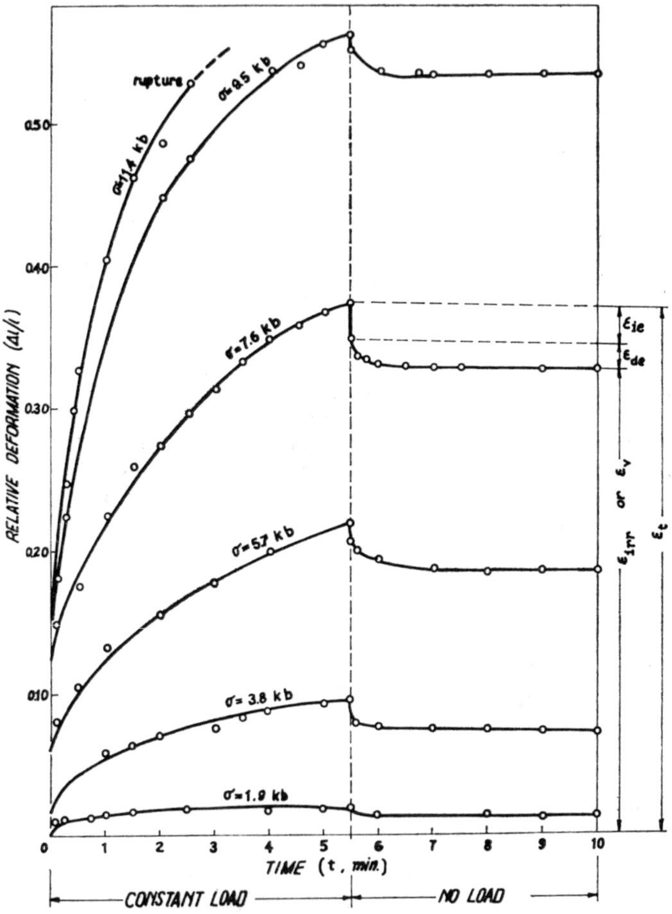

FIG. 1. Recovery after loading at increasing intensity, loading time t_l=const. The indices ie, de, te, irr, v refer to immediate elastic, delayed elastic, total elastic, irrecoverable or viscous deformations. Note that up to $\sigma = 7 \cdot 6$ kb† the elastic recoveries increase with increasing stress, whereas the recovery after loading at higher stresses becomes smaller.

† 1 kb (kilobar) = 1000 dynes/cm² = 1·0 2 g/cm² (technical, i.e. g wt).

initial immediate elastic extension is disregarded. However, when the experiment is repeated under increasing loads—with the specimen replaced for each run—for a specific time t_l, the recovered deformations at first increase sigmoidally with increasing load, but once a certain load level has been exceeded they become smaller, provided the material sustains such loads without rupture or severe necking (Figs. 1 and 2).† This phenomenon cannot be explained by a Burgers-like mechanism, and in fact no array of "classical" models could be devised so as to do justice to this sort of

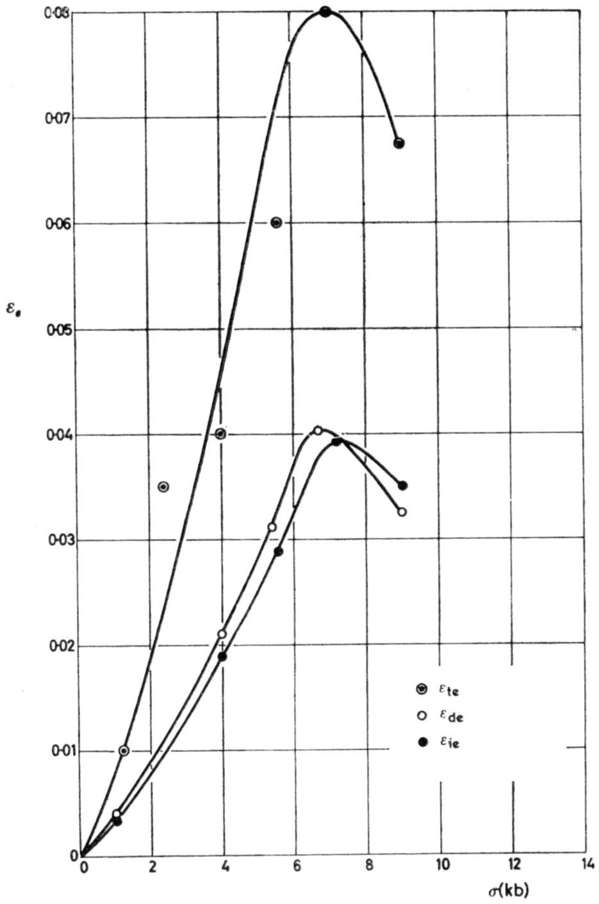

FIG. 2. Total, immediate and delayed recovery versus applied stress; loading time t_l = const. σ_D, ε_D signify decay stress and decay strain, respectively.

† It is important to realize that each point of the curve in Fig. 2 represents an individual experiment. A curve of this shape is, of course, unobtainable in a single run.

behaviour, even if non-linear elements or "spectra" of Maxwell and Kelvin–Voigt models were admitted. It is proposed to call this phenomenon "decay of elasticity" or, for short, "elastic decay", and the peak recoverable energy for a specific set of conditions (stress, loading time, temperature)—"decay point". Although the phenomenon may be associated with molecular relaxation processes, it should not be confused with relaxation in the rheological sense, which is the decay of stress in a material kept under constant deformation. The fact that dough shows decay of resilience, not only under constant extension but also in constant-load and constant-stress experiments, is the very reason why this behaviour defies description by classical models.

Again, although this type of decay is due to internal yield mechanisms, the decay point is not a yield point. The latter is generally defined as transition from elastic to non-elastic (plastic or viscous) deformation in tension or compression experiments, whereas elastic decay manifests itself specifically as the decrease of recovery beyond a specific stress level.

Glucklich and Shelef obtained sigmoid stress–strain curves in an earlier work on dough [1, 2], but did not reach the decay point. Their attempt to attribute the sigmoid shape of the curve to increasing interplay of loosely coupled St. Venant elements is invalidated by short-term experiments. Also, the proposed model does not account for the increase in non-recoverable deformation corresponding to elastic decay.

Attempts have been made previously to attribute certain aspects of non-linear stress–strain behaviour, similar to that described here, to rupture of successive elastic systems, i.e. of Hookean elements in the terminology of models [5, 6] or of molecular networks in that of the chemists [7, 8, 9, 13 and others]. For the model concept this explanation, in its unqualified formulation, is unsatisfactory, because in a constant-stress experiment the remaining Hookean elements should absorb the energy released by the ruptured ones and complete recovery should ensue unless prevented by "one-way" St. Venant elements as long as any Hookean element remains intact. It also seems that this concept has not yielded, so far, a mathematical description consistent with observed facts. The molecular theories by Bueche, Tobolsky and others were developed for amorphous weakly cross-linked polymers, such as soft rubbers, and are based on statistical network geometry and on the thermodynamic laws of rubber elasticity. These theories were successful in explaining certain results of relaxation and rupture experiments, but call for rather intimate knowledge of the structure of the material investigated. Nor do they lend themselves easily to macroscopic resolution of viscous and elastic effects at every stage of an experiment.

THEORY

The dough experiments suggest that no precise or permanent characteristics should be assigned to specific elements, as is the rule in classical rheology. Rather, it appears that certain portions of the material, which at the onset of loading act as elastic (Hookean) elements, undergo gradual degradation into viscous (Newtonian) elements. Model-wise, we can visualize such a mechanism as an array of Hookean elements coupled by shear pins (*SP*) and loosely bridged by Newtonian elements (Fig. 3a).

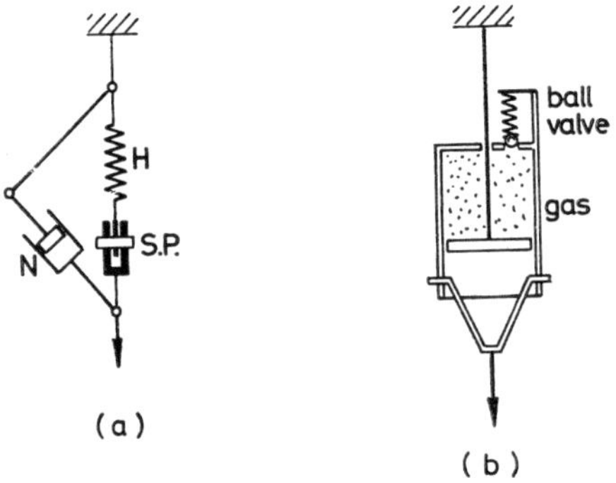

FIG. 3. Decay models. (a) Transformation or decay element composed of conventional units (Hookean spring, shear pin, dashpot). (b) Transformation or decay element composed of "entropy spring" and ball valve.

Once the acting stress has exceeded the tensile strength of the element, represented by rupture of the shear pin, the elastic energy stored in the Hookean element is transformed in part into energy of rupture (formation of new surfaces) and dissipated in part in the deformation of the Newtonian element which replaces the ruptured one. It is proposed to call the scheme in Fig. 3a a "transformation element" (*T*) or "decay element" (*D*)†

$$T = H \overset{\diagup N \diagdown}{-} SP$$

Another possible model representation, which probably permits closer approximation of the physical background of the effect, consists of a gas-filled cylinder representing the "entropy spring" associated with rubber elasticity, and a spring-operated ball valve representing the transformation of elastic into viscous deformation (escape of gas) at a certain pressure level (Fig. 3b).

† The symbol *T* is used, in consistency with previous publications.

At the molecular level, this process is probably best understood as the rupture of cross-links in networks when random-oscillating chain segments are stretched beyond a critical end displacement (Fig. 4). After rupture they recoil, slide or disentangle in a manner visualized macroscopically as viscous flow. To the rheologist, this is equivalent to transformation of an element capable of storing energy (a chain segment firmly knit into its network) into one dissipating energy (a loose segment sliding relative to its neighbours).

To derive mathematical benefit from these concepts, the simplest possible assumptions should serve as starting-point. For this purpose, let us

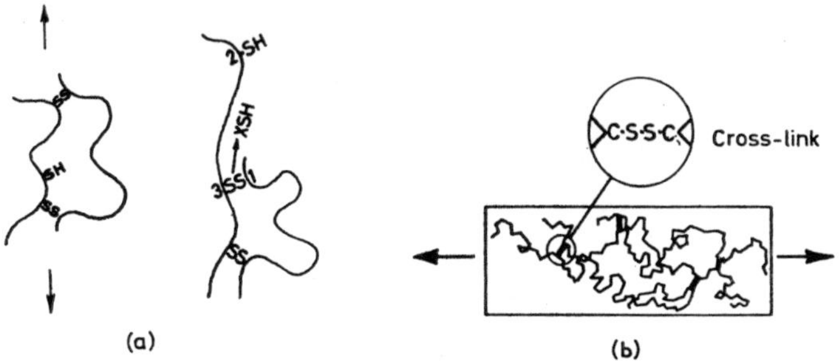

FIG. 4. Molecular network of weakly cross-linked polymer. (a) The first and last stages in the rupture and re-formation of a cross-link in a stretched section of the molecular network of gluten, from ref. [11]. (b) In weakly vulcanized rubbers the S_x cross-links prevent complete disentanglement of chains upon stretching until excessive stress causes successive rupture of bonds; from ref. [12].

first visualize a body consisting of an array of T-elements in series, decaying successively under stress into N-elements until the whole array has been transformed from a pure Hookean body into a pure Newtonian liquid:†

$$T\text{–}T\text{–}T, \ldots, T\text{–}T \rightarrow N\text{–}T\text{–}T, \ldots, T\text{–}T \rightarrow, \ldots, \rightarrow N\text{–}N\text{–}N, \ldots, N\text{–}N.$$

Referring all deformations to the initial unit length $l_0 = 1$, the continuously varying relative contribution of elastic and viscous elements in series to the total deformation may be defined by means of contribution factors α_e and α_v, so that

$$\alpha_e + \alpha_v = 1. \tag{1}$$

The concept underlying this definition is a Maxwell body of unit length, consisting of a Hookean element of length α_e and a dashpot of "length" α_v.

† The model representation of the conditions, required for describing this gradual transformation as a rate process, is not essential for the theoretical solution and is treated in the Appendix.

Since the total viscous flow, after a specific time t, of a Newtonian specimen under tensile stress is proportional to its initial length l, the "length" of a viscous element has a definite physical significance. In classical rheology the nature of an element (either elastic or viscous) is regarded as invariant, hence both the modulus of elasticity E and the coefficient of linear viscous traction $\lambda = 3\eta$ are obtainable by relating the measured deformations (or rates of deformation) to the overall length of the specimen. The departure of the new approach from the classical one lies in carrying the concept of action in series to its logical conclusion. If the mechanical behaviour of a material can be described mathematically as if its total deformation were the sum of those of an elastic element and a viscous one, then it is only reasonable to assign to each of them its own length and calculate the individual contributions accordingly. As long as neither the elastic nor the viscous element changes its nature, the difference between the two approaches is rather academic, but in the light of experimental evidence of elastic decay the principle of "element invariance" is best abandoned and replaced by a new concept: that of variable contributory length, represented by the contribution factors. It will be shown that this approach leads to a simplified theory and possibly also to better insight.

To sum up, we can say that in "classical" rheological bodies, α_e and α_v are constant and their values define specific behaviour: $\alpha_e = 1$, $\alpha_v = 0$—that of a Hookean body; $\alpha_e = 0$, $\alpha_v = 1$ that of a Newtonian liquid; and intermediate constant values ($0 < \alpha_{\text{const}} < 1$) various "grades" of Maxwell liquids. By the new approach, on the other hand, the α factors are variable: they are both time- and stress-dependent.

Because of the simplicity of Eq. (1), we may confine ourselves in most cases to a single factor (say $\alpha_e = \alpha$), with α_v replaced by $(1-\alpha)$. This is the interpretation of α when no subscript (e or v) is attached.

Hence, at any given moment the recoverable elastic strain ε_e will be

$$\varepsilon_e = \frac{\alpha\sigma}{E} \tag{2}$$

and the viscous rate of strain

$$\dot{\varepsilon}_v = \frac{(1-\alpha)\sigma}{\lambda}. \tag{3}$$

Most probably, neither the modulus E nor the coefficient of viscous traction (λ) are constant. However, at first approximation it is assumed that under isothermal conditions they may be considered as virtually invariant or "intrinsic" parameters, variation of the apparent modulus and the apparent viscosity being mainly due to the decay process, manifested in the variability of α. Regarding the latter, observation shows that under a specific load α decreases with time, although at an unknown rate. The simplest law to be proposed is obviously that $\dot{\alpha}$ should at any moment be proportional to the corresponding α, since the smaller the number of elastic elements still intact, the lower the rate of decay. As the successive rupture of bonds requires energy, it is further assumed that the rate of decay at any moment should be proportional to the corresponding available elastic (or stored) energy per unit volume. These assumptions yield

$$\dot{\alpha} = -\alpha \frac{1}{\Phi} \frac{\sigma^2}{2E}, \tag{4}$$

where Φ is a parameter depending on the molecular structure of the material, as well as on ambient conditions such as temperature, chemical agents (oxygen), etc. The physical dimension of Φ must obviously be [dyne·sec/cm²] or [g/cm·sec]. The negative sign is necessary, because in a tensile stress experiment α decreases with time.

The solution of Eq. (4) depends on the nature of the experiment and on the characteristics of Φ and E. As a first step, the equation is solved for the constant-stress (creep) experiment, on the assumption that Φ and E are time-independent. This leads to

$$\alpha = \alpha_0 \, e^{-(\sigma^2 t)/(2\Phi E)}, \tag{5}$$

where α_0 is the elastic contribution factor at the start of the experiment) The expression $2\Phi E/\sigma^2$ has the dimension of time, and we may write

$$\alpha = \alpha_0 e^{-t/t_D}, \tag{5a}$$

where

$$t_D = \frac{2\phi E}{\sigma^2} \tag{5b}$$

is the "decay time" of the stress σ, i.e. the time it takes for $(1/e)$-th of the original number of elastic elements to decay into viscous elements under this stress.

Equation (5) yields the deformation due to a uniform tensile stress in a decay body:

$$\varepsilon = \varepsilon_e + \varepsilon_v = \alpha \frac{\sigma}{E} + \int_0^t (1-\alpha) \frac{\sigma}{\lambda} \, dt \tag{6}$$

$$= \alpha_0 \frac{\sigma}{E} e^{-t/t_D} + \frac{\sigma}{\lambda} \left[t + \alpha_0 t_D (e^{-t/t_D}) - 1\right] \tag{6a}$$

$$= \varepsilon_{e_0} e^{-t/t_D} + \frac{\sigma t}{\lambda}\left[\left(1-\alpha_0 \frac{t_D}{t}\right)+\alpha_0 \frac{t_D}{t}\, e^{-t/t_D}\right]. \tag{6b}$$

Equations (6) represent the law of deformation for a specific experiment under constant-stress conditions. Only the first term represents the recoverable deformation

$$\varepsilon_e = \alpha_0 \frac{\sigma}{E}\, e^{-(\sigma^2 t)/(2E\Phi)}. \tag{6c}$$

Equations (6) were derived for the simplest case, a "decay-Maxwell" body, and the recovery represents here only the immediate elastic behaviour. The present paper concentrates on the consequences of these more elementary assumptions as a first step towards a general solution.

EXPERIMENTAL VERIFICATION

By comparing the recoveries obtained after a specific loading time in a number of experiments under different loads, the decay point for that specific loading time should be obtainable by maximizing the first term in Eq. (6) with respect to the stress.

$$\frac{d\varepsilon_e}{d\sigma} = \frac{d}{d\sigma}\left[\alpha_0 \frac{\sigma}{E}\, e^{-(\sigma^2 t)/(2\Phi E)}\right] = 0$$

)r

$$\frac{\alpha_0}{E}\left(1-\frac{\sigma^2 t}{\Phi E}\right) e^{-(\sigma^2 t)/(2\Phi E)} = 0, \tag{7}$$

Φ being, at first approximation, assumed to be independent of t.

This equation has three solutions. Two or these ($\sigma = \infty$, $t = \infty$) result n $\varepsilon_e = 0$ and confirm that for very high stresses, or very long loading imes, the elastic response should decay completely. The third,

$$\frac{\sigma^2 t_l}{\Phi E} = 1,$$

ields

$$\sigma_D = \sqrt{\left(\frac{E\Phi}{t_l}\right)}. \tag{8}$$

This solution is independent of α_0 and if both E and ϕ are constant (as ssumed for isothermal conditions),

$$\sigma_D \sqrt{t_l} = \sqrt{(\Phi E)} = \text{const.} \tag{8a}$$

'his means that the "decay stresses", i.e. those stresses which produce naximum immediate recovery ε_{ie} for a given loading time t_l, should be nversely proportional to the square root of this time.

The applicability of these results was checked in a large number of immediate-recovery experiments on dough cylinders prepared from a local wheat-flour blend and 55·3% water (600 FU† consistency). As the experiments were carried out at constant load instead of constant stress, the results were corrected for reduced cross-sectional area by means of the known relation

$$\sigma = \frac{P}{A_0}(1+\varepsilon^c), \tag{9}‡$$

where σ is the average stress in the specimen at any moment, P the load, A_0 the original cross-section and $\varepsilon^c (= \Delta l/l_0)$ the Cauchy measure of strain.§

TABLE 1

1 t (sec)	2 P (g)[(d)]	3 $\Delta l_{ie_{max}}$ (mm)	4 $\varepsilon_D = \varepsilon_{ie_{max}}$ [(a)]	5 A_t	6 σ_D (kb)[(b)]	7 $\sigma\sqrt{t} = \sqrt{(E\Phi)}$	8 E (kb)[(b)]
5	26	61	1·22	0·246	106	237	[53]
30	13	40	0·80	0·306	42·5	234	33
180	7·5	17	0·34	0·410	18·2	244	31
330	6·0	12	0·30	0·443	13·4	244	26·7
					Average ≅	240·0	30·2[(c)]

(a) Table 1 is confined to immediate recovery. The total strains (comprising immediate-elastic, delayed-elastic and non-recoverable) measured under the conditions described were 2·36, 2·06, 1·34, 1·15, 1·0 for 5, 30, 90, 180, 330 sec loading time respectively.
(b) See footnote† p. 362.
(c) Average of 30–330 sec loading time.
(d) Technical (g weight).

Table 1 shows the determination of ε_D and σ_D from experimental values of Δl and P for different loading times t_l (Fig. 5) and the calculation of the product $\sigma_D\sqrt{t_l} = \sqrt{(E\Phi)}$ (Table 1, col. 7). This product was practically constant (240 kb sec$^{\frac{1}{2}}\theta$) for $5 < t_l < 330$ sec, so that the points σ_D (experimental) vs. $t_l^{-\frac{1}{2}}$ lie virtually on a straight line as predicted by the

† FU = Brabender Farinograph Unito.

‡ Based on an assumption (valid with satisfactory approximation) of constant specimen volume.

§ In time-dependent processes the results obtained by this correction would not necessarily correspond to those obtained in experiments actualy conducted under constant stress. Hence the approximate character of the experiments themselves (see below).

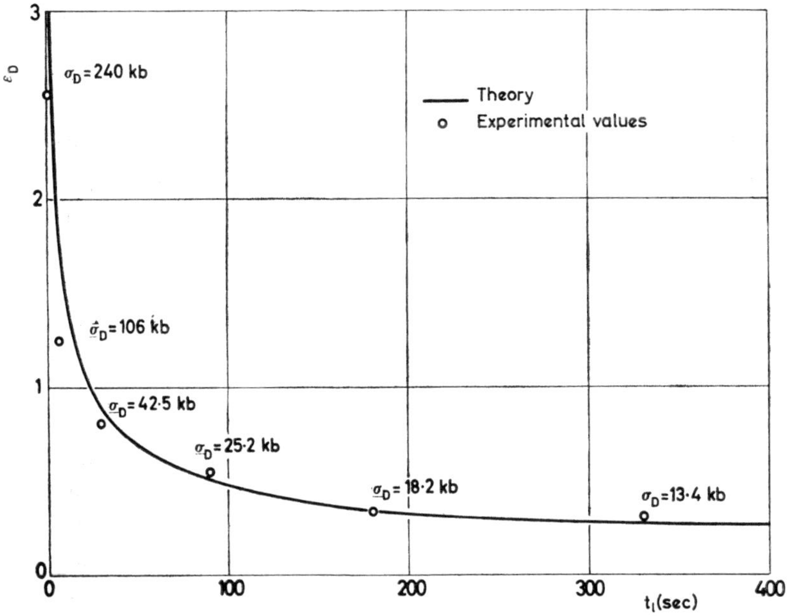

FIG. 5. Decay strain ε_D versus loading time t_l preceding recovery. The corresponding decay stresses are marked at each point.

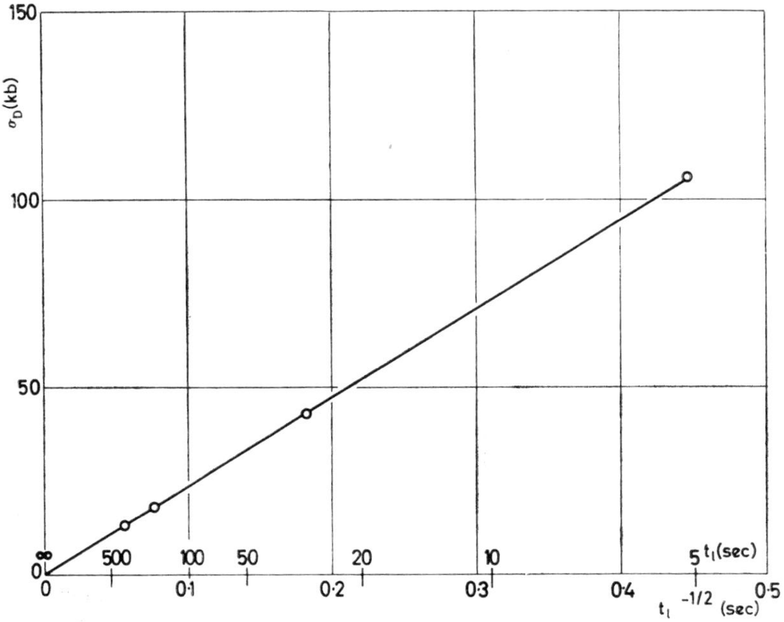

FIG. 6. Decay stress σ_D versus $t_l^{-1/2}$.

theory (Fig. 6). The intrinsic modulus E was then calculated from the relation

$$E = \frac{\sigma_D}{\varepsilon_D\sqrt{e}} = \frac{240}{\varepsilon_D\sqrt{et_l}} \tag{6d}$$

which follows from Eqs. (6c) and (8) for $\alpha_0 = 1$ and $(\sqrt{E\Phi}) = 240$; this calculation, based on experimental values of ε_D, yielded the values entered in col. 8 of Table 1. These values are, in turn, virtually uniform for $30 \leqslant t_l \leqslant 330$, yielding $E_{av} = 30$ kb for this range, while the value for $t = 5$ is considerably larger.

With the aid of the above average $E = 30$ kb, a theoretical ε_D vs. t curve was calculated from Eq. (6d) and superimposed on the experimental values of Fig. 5 showing a good fit for loading times exceeding 5 sec. The deviation in the short time range is attributable to the fact that the values of ε_D derived for loading times equal to and exceeding 30 sec are probably too large (containing a small delayed-recovery component difficult to eliminate under test conditions), in which circumstances the calculated value of E is too small. Within 5 sec no significant amount of delayed recovery can affect the measurements, hence the true value of E is probably closer to the higher level found for $t_l = 5$.

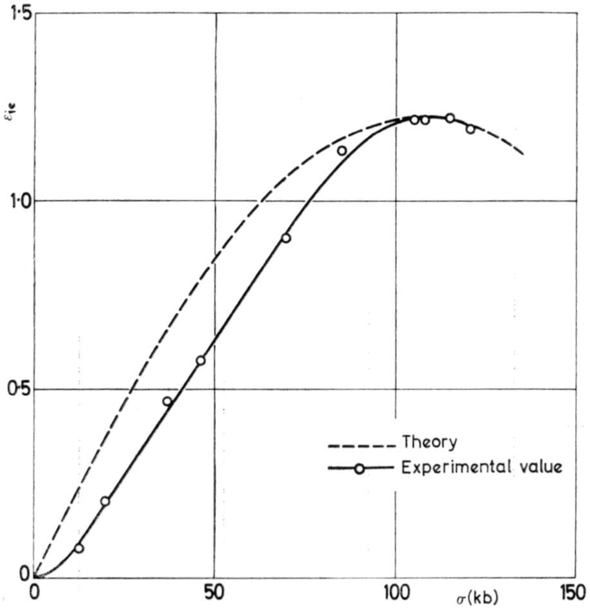

FIG. 7. Immediate elastic recovery versus stress, for constant loading time. Comparison between experimental and theoretical values. Theoretical curve calculated to fit experimental decay point. Loading time $t_l = 30$ sec.

Another reason, of course, may be that the simplifying assumptions, which led to the equations used, do not apply to very short loading times. For this region a more sophisticated approach may be called for, taking into account dynamical effects, difficulty of energy dissipation, and existence of a finite limiting value for σ and ε at loading times approaching zero.

In Fig. 7, the experimental values of the immediate elastic recoveries obtained for $t = 30$ sec are plotted against the acting stress σ, and again the theoretical curve calculated according to Eq. (6c) for $\alpha_0 = 1$, $E = 30$ and $\sqrt{(E\Phi)} = 240$ was superimposed on these values for comparison. Here the agreement with experiment is less satisfactory for lower and medium values of σ and shows the measure of approximation achieved. It should be noted that the experiments also contain an element of approximation: they were performed with constant load, not constant stress as assumed in the mathematical derivation, and in time-dependent processes the results obtained from arithmetic compensation by means of Eq. (9) need not be truly equivalent with results obtained from experiments actually performed under constant stress.

Figure 8 gives a three-dimensional represetation of the immediate elastic recovery in dependence on stress and loading time, the "decay ridge') of the elastic "hill" being marked by a dashed line. The slope beyond that ridge is broken off because it has no practical importance.

DISCUSSION

(*a*) *Specificity of Results*

Equation (8) was derived under a simplifying assumption for a simplified model termed (perhaps not very satisfactorily, for lack of a better term) a "decay-Maxwell" body. Evaluation of the delayed-recovery measurements suggests that the Kelvin–Voigt behaviour of dough calls for a parallel "decay-Kelvin" concept. At the time of writing it is still too early to report on this aspect; the theory developed here could only be checked against the measured immediate component of the recovery. However, the general concept underlying this theory was also tried out in experiments relating tensile strength to rupture time. This application, published elsewhere [4], proved rather promising.

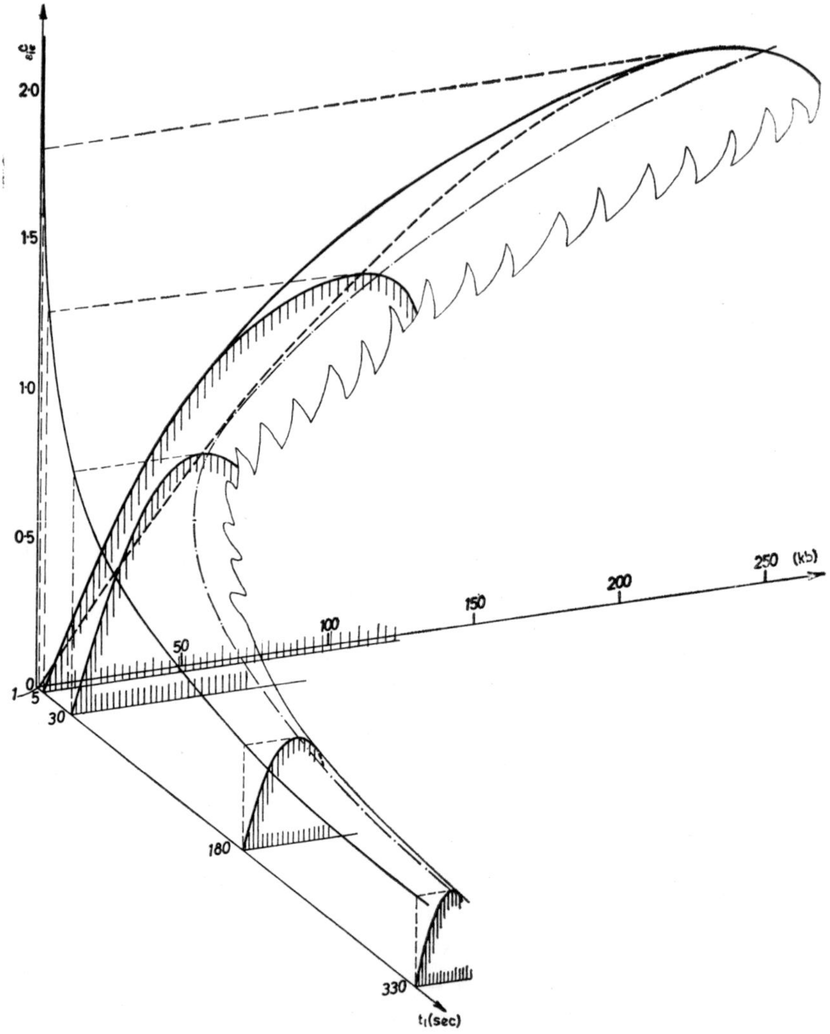

FIG. 8. Axonometrical representation of immediate elastic recovery versus stress and loading time.

(b) *Significance of Parameter Φ*

The parameter Φ was introduced in Eq. (2) as a factor of proportionality related to structure, environment and experimental conditions. Equation (8a) provides food for thoughts regarding its possible physical significance, but these are purely speculative pending verification by additional research, specifically in the field of temperature variation.

To elucidate the physical character of Φ, Eq. (8a) is rewritten as

$$\frac{\sigma_D^2}{2E}\cdot t_l = \frac{\Phi}{2} = \text{const}\left[\frac{(\text{dyne}\cdot\text{cm}\cdot\text{sec}}{\text{cm}^3}\right]. \tag{8b}$$

This means that, under isothermal conditions, the product of loading time and stored energy, corresponding to maximum immediate recovery, should be constant for the type of material treated here. Dimensionally, the product of energy and time is equivalent to angular momentum (in this case both are calculated per unit volume). This equivalence is of interest from the viewpoint of the thermodynamic theory of elastomers: it suggests that in rotating chain segments the component (per unit volume), in the direction of the acting force, of the average angular momentum ("spin") responsible for the elastic behaviour in that direction, is only capable of a critical increment $\Phi_c/2$ if elasticity is to be preserved. Since $\Phi_c/2$ signifies an increase through integration between an initial value Φ_0 and a final value Φ_D, it may be concluded that there exists a specific angular momentum

$$\Phi_D = \Phi_0 + \frac{\Phi_c}{2} \tag{10}$$

responsible for the onset of decay, if Φ_0 is the average angular momentum at zero stress for a given temperature, the term "momentum" being again confined to its component along the stress axis.

The scalar value of the vector representing the angular momentum of a given chain segment depends only on the absolute temperature. Hence, the only way in which the average value of the significant component of this vector can be changed under isothermal conditions is through the average angle between the individual spin vectors and the stress axis, i.e. through the degree of orientation or order; this may provide a connection between the statistical theory of thermodynamics and the proposed hypothesis, which arrived, by purely phenomenological considerations (i.e. without recourse to statistical techniques), at a result (Eq. (6c)), which contains a term representing the standard distribution of the elastic strain, recovered after a certain loading time, in relation to the stress.

CONCLUDING REMARKS

Additional theoretical and experimental work is required to extend this theory to other applications, such as viscous decay and relaxation behaviour, and to confirm it for a wider range of conditions and materials. The present impression is that the very concept of gradual decay through transformation of elastic elements into viscous ones, and the recourse to contribution factors α (which may yield intrinsic parameters of elasticity and viscosity), have proved fruitful in analysing the mechanical behaviour of a material which rheologically is the most complex to date, and which has so far defied attempts at rational description of its special non-linearity.

ACKNOWLEDGEMENT

The theory reported here was developed in connection with a project sponsored by the U.S. Department of Agriculture and headed by Professor M. Reiner. His advice and encouragement are warmly appreciated, as is the devoted collaboration of Miss Carmen B. Funt who conducted and evaluated the experiments, and of the other staff members who participated in this work.

APPENDIX

MODEL DESCRIPTION OF THE RATE PROCESS, AND THE REAL BEHAVIOUR OF DOUGH

To account for the specific behaviour of the material, we have to assume that in the unstressed state the T elements are not aligned axially but connected at different angles by joints exerting delayed-elastic resistance to angular displacement (Fig. 8). These joints may be represented by rotating dashpots (Fig. 9) and denoted in structural diagrams by the symbol  . As long as the cross-link bonds (represented by the shear pins) remain intact, i.e. below a given stress level or the initial yield point, the whole chain of T-elements acts essentially as an elastic or Hookean element (although with non-linear characteristics), as orientational changes affect the apparent modulus of the H-elements but not their resilience, i.e. their capacity to store elastic energy and recover after deformation. When the stress in individual T-elements exceeds this yield point as a result of gradual straightening of the hinges, the bonds begin to rupture, transforming the corresponding T-elements into N-elements as described. This process is supposed to continue at a rate depending on the acting stress until, under a sufficiently large stress and after an adequate time, the whole material is transformed into a Newtonian liquid—unless it ruptures earlier in a tensile creep experiment. This is not the only possible, but perhaps the simplest, model description imaginable to depict behaviour similar to that observed.

The described process is understood to be irreversible, since the energy necessary for rupturing the cross-links is dissipated. However, our work with dough has shown that given sufficient rest time and in the presence of an energy reservoir such as is provided by an isothermal environment or by exothermic chemical processes, some or all of the original resilience may be recovered, indicating restoration of cross-links and entanglements.

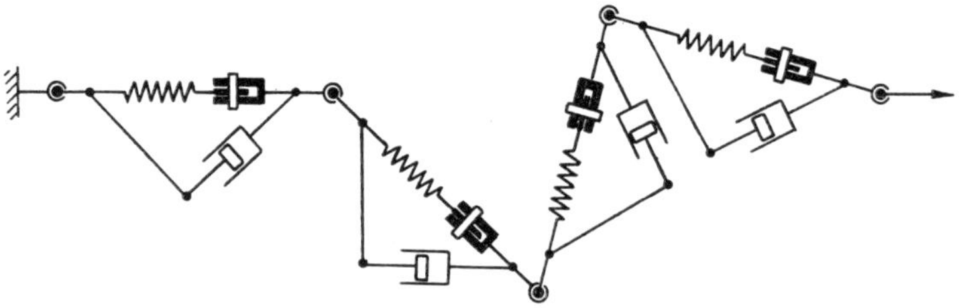

FIG. 9. Array of T-elements connected by rotating dashpots.

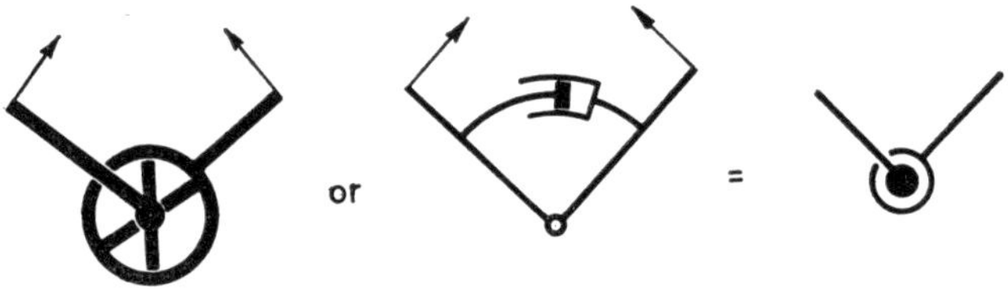

FIG. 10. Rotating dashpots, or viscous hinges, symbol for delayed orientation effects. A mechanism representative of the behaviour described here requires also a spring parallel to the dashpot.

In papers on dough mechanics this process is frequently referred to as "dough relaxation", occasionally as "dough recovery". Since in rheology both relaxation and recovery have other very specific meanings (stress decay under constant strain and reversal of elastic deformation, respectively) it is proposed to use a different term, such as "restoration of elasticity", for describing this process, which is the reversal of decay.

REFERENCES

1. J. GLUCKLICH and L. SHELEF, "An investigation into the rheological properties of flour dough: Studies in shear and compression", *Cereal Chem.* **39,** 242–55 (1962).
2. J. GLUCKLICH and L. SHELEF, "A model representation of the rheological behaviour of wheat-flour dough", *Koll. Z.* **181,** 29–33 (1962).

3. CH. H. LERCHENTHAL and H. G. MULLER, "Research in dough rheology at the Technion Israel Institute of Technology", *Cereal Sci. Today*, **12,** 5, 185–91, (1961).
4. C. B. FUNT, CH. H. LERCHENTHAL and H. G. MULLER, "Influence of additives on dough properties", *Symposium on Rheology and Texture of Foodstuffs*, S.C.I. Monograph No. 27, London, 1967.
5. CH. H. LERCHENTHAL and C. B. FUNT, "The strength of dough in uniaxial tension", *Symposium on Rheology and Texture of Foodstuffs*, S.C.I. Monograph No. 27, London, 1967.
6. R. K. SCHOFIELD and G. W. SCOTT BLAIR, "The relation between viscosity, elasticity and plastic strength of a soft material as illustrated by some mechanical properties of flour dough, IV. The separate contribution of gluten and starch," *Proc. Roy. Soc.* (London), A **160,** No. 900, 87–94 (1937).
7. V. R. REGEL, "Mechanical models of polymers containing breakdown elements", *Polymer Sci. USSR* **6** (3), 437–42 (1965). (Orig.: *Vysokomol Soyed.* **6,** (3) 395–9 (1964).)
8. P. J. FLORY, *Principles of Polymer Chemistry*, Cornell Univ. Press, ch. IX (1953).
9. L. BATEMAN, *The Chemistry and Physics of Rubber-like Substances*, Maclaren, London (1963).
10. F. BUECHE, *Physical Properties of Polymers*, Interscience Publ., New York (1962).
11. J. HLYNKA, *Wheat Chemistry and Technology*, vol. III, p. 506. AACC Minnesota (1964).
12. M. GORDON, *High Polymers*, p. 65, Iliffe, London (1963).
13. F. BUECHE and T. J. DUDEK, "Tensile strength of amorphous gum rubbers", *Rubber Chem. Technology* **36,** 1–10 (1963).
14. AACC (54–21) *Am. Assoc. of Cer. Chem.*, "Physical dough tests, Farinograph method, for flour", Minnesota, U.S.A., pp. 1–6 (1962).

ON THE THEORY OF TORSIONAL RHEOMETRY

RAMESH N. SHROFF† and RAYMOND R. MYERS‡

Lehigh University, Bethlehem, Pa., U.S.A.

ABSTRACT

An alternating series expression relating the complex dynamic viscosity, η^*, to the experimentally measured quantities—the amplitude ratio m and phase angle ϕ—has been derived for bob-in-cup and ring-in-cup geometries. It is similar in form to that derived for a vibrating reed [10]. Both Taylor series and asymptotic expansion of Bessel functions for any one geometry gives similar results. A general condition for the convergence or the useful range of the series expressions is given and the direct analytical expressions are derived for calculating G', G'', and $\tan \delta$ when certain terms in these series can be excluded.

INTRODUCTION

In the rheological investigations carried out in this laboratory in the low-frequency range, two different concentric cylinder geometries are used to study the response of viscoelastic liquids and gels to sinusoidal stresses. In one, a cylindrical bob [1] is suspended coaxially in the fluid sample contained in a cup which is given a fixed angular displacement at various frequencies; in the other, the bob is replaced by a ring [2, 3] so that the sample is sheared between two annuli. The ring-in-cup geometry provides a large torque and shear area and virtually eliminates end effects. Hiller [4] used this geometry for rotational viscometric studies and suggested that the ring dimensions be selected so that the mean rate of shear was equal in both annuli.

For the coaxial cylinder type of elasticoviscometer with bob-in-cup geometry, Markovitz [5] included the moment of inertia of the sample in deriving a relationship between the mechanical properties of the sample and the measured amplitude (m) and phase angle (ϕ) of the motion of the cup relative to that of the suspended bob. The equation of motion

† Current address: Chemplex Company, Rolling Meadows, Illinois, U.S.A.
‡ Current address: Kent State University, Kent, Ohio, U.S.A.

was a Bessel equation and, in the result, the desired mechanical properties were part of the complex argument of Bessel functions. Oldroyd [6] considered the motion of an elasticoviscous liquid characterized by a viscosity coefficient and two relaxation times and arrived at the identical result. Whereas Markovitz used a Taylor series expansion to free the desired quantities from the complex arguments of Bessel functions, Oldroyd employed an asymptotic expansion.

It is the purpose of this article to use both types of expansions of Bessel functions to derive an equation relating the complex dynamic viscosity, η^*, to the experimentally measured quantities (m and ϕ) for ring-in-cup geometry. First, however, it is necessary to extend Oldroy's analysis for bob-in-cup geometry to derive a direct equation relating η^* to m and ϕ.

THEORETICAL ANALYSIS

Bob-in-cup Geometry (asymptotic expansion)

One starts with the Markovitz [5] eq. (54):

$$\frac{\theta_{\text{cup}}}{\theta_{\text{bob}}} = \frac{(1+br^2)[Y_1(ar)J_1(aR)-J_1(ar)Y_1(aR)] - abr^3[J_1(aR)Y_1'(ar)-J_1'(ar)Y_1(aR)]}{abRr^2[J_1'(ar)Y_1(ar)-J_1(ar)Y_1'(ar)]}, \tag{1}$$

where $\theta_{\text{cup}}/\theta_{\text{bob}}$ is the complex ratio. It equals $\cos\phi/m+i\sin\phi/m$, where m is the ratio of the amplitude of the bob to that of the cup and ϕ is the angle by which the bob lags the cup (outer sinusoidally driven cylinder). J_1 and Y_1 are the Bessel functions of the first and second kind, respectively (and of order 1); r and R denote here the radii of the bob and cup, respectively. The values of a and b are given by the formulas: $a = (-i\omega\varrho/\eta^*)^{1/2}$ and $b = -i2\pi h\eta^*\omega/(I\omega^2-k)$, where ω is the circular frequency, h is the depth to which the bob is immersed in the sample, I is the moment of inertia of the bob, k is the restoring constant of the torsion wire, and ϱ is the density of the sample.

It can be shown that Oldroyd's equation [6]

$$\frac{\theta_{\text{cup}}}{\theta_{\text{bob}}} = \frac{\frac{2\pi\varrho r^3h}{\alpha}[J_2(\alpha r)Y_1(\alpha R)-Y_2(\alpha r)J_1(\alpha R)]+\left(\frac{k}{\omega^2}-I\right)[J_1(\alpha r)Y_1(\alpha R)-Y_1(\alpha r)J_1(\alpha R)]}{\frac{2\pi\varrho r^2Rh}{\alpha}[J_2(\alpha r)Y_1(\alpha r)-Y_2(\alpha r)J_1(\alpha r)]} \tag{2}$$

is identical to Eq. (1) when α is replaced by $a\left\{= (-i\omega\varrho/\eta^*)^{\frac{1}{2}}\right\}$. In the above equation $\alpha = [-i\omega\varrho(1+i\omega\lambda_1)/\eta_0(1+i\omega\lambda_2)]^{\frac{1}{2}}$, where η_0, λ_1 and λ_2 are the viscosity coefficient and the two relaxation times which completely characterize the material.

Using the properties of the Wronskian, the asymptotic expansions of the Bessel functions in powers of $(ar)^{-1}$ and the recurrence relations for J and Y, after rearrangement and collection of terms one arrives at

$$\frac{\theta_{\text{cup}}}{\theta_{\text{bob}}} = \frac{1}{a}\left[\frac{1}{(Rr)^{3/2}}\left(\frac{1}{b}+\frac{E}{a^2b}+\frac{E_1}{a^4b}\right)+\left(\frac{r}{R}\right)^{3/2}\left(\frac{D}{a^2}-F\right)\right]\sin(aR-ar)$$
$$+\left[\frac{1}{(Rr)^{3/2}}\left(\frac{S_1}{a^4b}-\frac{S}{a^2b}\right)+\left(\frac{r}{R}\right)^{3/2}\left(1+\frac{F_1}{a^2}\right)\right]\cos(aR-ar), \qquad (3)$$

where $D = 4{\cdot}394\times10^{-2}(7R^2+5r^2)/R^2r^3+0{\cdot}1025(3R^2+r^2)/R^3r^2$,

$E = 0{\cdot}1172(R^2+r^2)/R^2r^2+0{\cdot}1406/rR$,

$E_1 = 3{\cdot}204\times10^{-3}(R^4+r^4)/R^4r^4-3{\cdot}845\times10^{-2}(R^2+r^2)/r^3R^3$,

$F = \frac{3}{8}(r-5R)/rR$,

$F_1 = 0{\cdot}1172(r^2-7R^2)/r^2R^2+0{\cdot}7031/rR$,

$S = \frac{3}{8}(R-r)/rR$

and $S_1 = 4{\cdot}394\times10^{-2}[(3R^2-r^2)/R^2r^3-(3r^2-R^2)/R^3r^2]$.

In the above equation, terms whose denominator contains higher powers than a^2 or a^4b in the large bracket have been neglected [6] for sufficiently large ω to make $|a| > 2\text{ cm}^{-1}$. Equation (3) is similar to Oldroyd's Eq. (27) [6].

The unknown quantity sought, η^*, is buried in the trigonometric functions. If one uses the series expansions for these functions, subsequent rearrangement gives the result

$$\frac{\theta_{\text{cup}}}{\theta_{\text{bob}}} = M+iN\frac{\eta^*}{\omega\varrho}+\sum_{K=1}^{\infty}\left(\frac{i\omega\varrho}{\eta^*}\right)^K[(A'_K+B'_K\varrho)\varrho^{-1}-C'_K\varrho^{-1}\omega^{-2}], \qquad (4)$$

where $M = (r/R)^{3/2}\left[1-\frac{3}{8}(R-r)(r-5R)/rR-\frac{(R-r)^2}{3}\{F_1+D(R-r)/3\}\right]$

$$+\frac{I}{2\pi h(rR)^{3/2}\varrho}\left(1-\frac{k}{I\omega^2}\right)\left[S-E(R-r)+S_1\frac{(R-r)^2}{2!}+E_1\frac{(R-r)^3}{3!}\right], \qquad (5)$$

$$N = \left(\frac{r}{R}\right)^{3/2}[F_1+D(R-r)]-\frac{I}{2\pi h(rR)^{3/2}\varrho}\left(1-\frac{k}{I\omega^2}\right)[S_1+E_1(R-r)], \qquad (6)$$

$$A'_K = \frac{I}{2\pi h(Rr)^{3/2}} \frac{(R-r)^{2K-1}}{(2K-1)!}\left[1+\frac{R-r}{2K}\left\{S-\frac{E(R-r)}{2K+1}+\frac{S_1(R-r)^2}{(2K+1)(2K+2)}\right.\right.$$

$$\left.\left.+\frac{E(R-r)^3}{(2K+1)(2K+2)(2K+3)}\right\}\right], \tag{7}$$

$$B'_K = \left(\frac{r}{R}\right)^{3/2} \frac{(R-r)^{2K}}{(2K)!}\left[1-\frac{3}{8}\frac{R-r}{2K+1}\frac{r-5R}{rR}-\frac{(R-r)^2}{(2K+1)(2K+2)}\right.$$

$$\left.\times\left(F_1+\frac{D(R-r)}{2K+3}\right)\right], \tag{8}$$

and $C'_K = kA'_K/I$.

If terms whose denominator contains higher powers than a^2 or a^4b are retained in Eq. (3), then additional terms in higher powers of $(i\eta^*/\omega\varrho)$ will appear in Eq. (4) along with terms of powers higher than $(R-r)^3$ in the coefficients A'_K. On the other hand, if terms in $1/a^2$ and $1/a^4b$ are also neglected in Eq. (3), then the terms containing D, E_1, F_1 and S_1 in Eqs. (5) through (8) do not appear.

The comparison of the instrument constants A'_1, B'_1, A'_2 and B'_2 with those of Markovitz A_1, B_1, A_2 and B_2 shows a remarkable agreement between two sets of constants except that the constants A'_u differ appreciably from $A_u (u \geqslant 2)$ when gap $(R-r)$ is large (see Table 1). The value of M equals 1 for various combinations of R, r, I and h $(k = 0)$. As can be seen from Table 1, the term containing N can be neglected in comparison with the term $i[(A'_1+B'_1\varrho)\omega - C'_1/\omega]$ for large ω and low η^* in all cases listed in Table 1 except possibly for the apparatus of Markovitz *et al.* [1].

The value of the constant M and the asymptotic instrument constants remain unchanged for the examples given in Table 1 if the terms containing D, E_1, F_1 and S_1 are dropped in the calculations. This suggests that inclusion of additional terms whose denominators contain higher power terms than a^2 or a^4b in the large bracket in Eq. (3) would not affect the values of these constants. In contrast, if the terms containing D, E, E_1, F, F_1, S and S_1 are omitted from the calculations, the asymptotic constants are identical (within 2%) with those of Markovitz. The value of M remains unchanged.

Ring-in-cup Geometry

The differential equation for oscillating motion in cylindrical symmetry is the equation:

$$\frac{d^2\theta}{dz^2}+\frac{3}{z}\frac{d\theta}{dz}-\frac{i\omega\varrho}{\eta^*}\theta = 0, \tag{9}$$

TABLE 1. COMPARISON OF INSTRUMENT CONSTANTS OBTAINED USING ASYMPTOTIC EXPANSION OF BESSEL FUNCTIONS WITH THOSE OF MARKOVITZ FOR BOB-IN-CUP GEOMETRY

R (cm)	r (cm)	I (g·cm²)	h (cm)	Asymptotic constants					Markovitz constants			
				$-N\times10^3$ (cm^{-2})	$A'_1\times10^3$ (g/cm)	$B'_1\times10^4$ (cm²)	$A'_2\times10^5$ (g·cm)	$B'_2\times10^6$ (cm⁴)	$A_1\times10^3$ (g/cm)	$B_1\times10^4$ cm²	$A_2\times10^5$ g·cm	$B_2\times10^6$ cm⁴
0·555	0·395	7·21	4·8[a]	3030	378	93·7	194	18·6	378	93·7	161	18·6
2·1462	2·0962	2677	25·4[b]	1·69	87·9	12·2	3·67	0·25	87·9	12·2	3·66	0·25
2·1207	2·0197	2468	25·4[b]	3·71	176	48·6	30·2	4·09	176	48·6	30·0	4·09
1·745	1·27	258	7·63	58·4	785	842	3100	1480	785	842	2940	1480
1·41	1·27	258	7·63	37·6	315	88	105	14·1	315	88·5	103	14·1
0·857	0·794	7·23	3·25	33·0	39·8	18·4	2·63	0·60	39·8	18·4	2·63	0·60

(a) Markovitz *et al.* [1]. (b) Oldroyd [6]. (c) This work.

where the angular displacement θ is a function of the radial distance z only, and $\eta^*(=\eta'-iG'/\omega)$ is the complex dynamic viscosity.

The solution of Eq. (9) for the outer and inner annulus of ring-in-cup geometry is

$$\theta_0 = [AJ_1(az)+BY_1(az)]/z, \tag{10}$$

and

$$\theta_i = [CJ_1(az)+DY_1(az)]/z, \tag{11}$$

where A, B, C and D are the constants to be determined by the boundary conditions. Here θ_0 and θ_i are the angular displacements of the outer and inner annulus, respectively. The boundary conditions are:

$$\theta_i(R_1) = \theta_0(R_2) = \theta_{\text{cup}}, \tag{12}$$

and

$$\theta_0\,(r_2) = \theta_i(r_1) = \theta_{\text{ring}}, \tag{13}$$

where R_1 and R_2 are the inner and outer radii of the cup, respectively, and r_1 and r_2 are the inner and outer radii of the ring, respectively.

Also, at the ring the following condition is satisfied [7]:

$$T-k\nu_r = I\ddot{\nu}_r, \tag{14}$$

where $\nu_r = \nu_{r_1} = \nu_{r_2}$. Here ν is the instantaneous angular displacement and is equal to $\theta e^{i\omega t}$. T in Eq. (14) is the shearing torque experienced by the ring from the material in the two annuli and is given by the relation:

$$T = 2\pi r_2^3 h\eta^* \left.\frac{d\dot{\nu}}{dz}\right|_{z=r_2} - 2\pi r_1^3 h\eta^* \left.\frac{d\dot{\nu}}{dz}\right|_{z=r_1}. \tag{15}$$

By putting the boundary conditions into Eqs. (10) and (11), one can arrive at the following expression for the complex displacement ratio:

$$\frac{\theta_{\text{cup}}}{\theta_{\text{ring}}} = \frac{\pi}{2}\left\{\frac{[J_1(aR_1)\,Y_1(ar_1)-J_1(ar_1)\,Y_1(aR_1)][J_1(aR_2)Y_1(ar_2)-J_1(ar_2)\,Y_1(aR_2)]}{r_1R_1[J_1(aR_2)\,Y_1(ar_2)-J_1(ar_2)\,Y_1(aR_2)]-r_2R_2[J_1(aR_1)\,Y_1(ar_1)-J_1(ar_1)\,Y_1(aR_1)]}\right\}\times$$

$$\left\{\frac{1}{b}+r_2^2-r_1^2+ar_2^3\frac{Y_1(aR_2)\,J_1'(ar_2)-J_1(aR_2)\,Y_1'(ar_2)}{J_1(aR_2)\,Y_1(ar_2)-Y_1(aR_2)\,J_1(ar_2)}\right.$$

$$\left.-ar_1^3\frac{Y_1(aR_1)\,J_1'(ar_1)-J_1(aR_1)\,Y_1'(ar_1)}{J_1(aR_1)\,Y_1(ar_1)-Y_1(aR_1)\,J_1(ar_1)}\right\}. \tag{16}$$

Markovitz-type Expansion

Treatment of Eq. (16) according to the method of Markovitz gives, after considerable rearrangement and collection of terms, the following expression for the complex displacement ratio:

$$\frac{\theta_{\text{cup}}}{\theta_{\text{ring}}} = 1+\sum_{k=1}^{\infty}\left(\frac{i\omega\varrho}{\eta^*}\right)^K \{(A'_{KK}+B'_{KK}\varrho)\,\varrho^{-1}-C'_{KK}\omega^{-2}\varrho^{-1}\}, \tag{17}$$

where

$$A'_{11} = \frac{I}{4\pi h}\left(\frac{1}{\dfrac{r_1^2R_1^2}{r_1^2-R_1^2}-\dfrac{r_2^2R_2^2}{r_2^2-R_2^2}}\right).$$

$$B'_{11} = \frac{\dfrac{(R_1^2-r_1^2)^2}{8R_1^2}\,\dfrac{1}{BB}-\dfrac{(R_2^2-r_2^2)^2}{8R_2^2}\,\dfrac{1}{AA}}{\left(\dfrac{1}{BB}-\dfrac{1}{AA}\right)}.$$

BB and *AA* in the above expression replace complicated expressions involving the ring-in-cup geometry:

$$BB = 4\ln\frac{r_1}{R_1}+\frac{R_1^2}{r_1^2}-\frac{r_1^2}{R_1^2},$$

$$AA = 4\ln\frac{r_2}{R_2}+\frac{R_2^2}{r_2^2}-\frac{r_2^2}{R_2^2}.$$

The other constants are defined by

$$C'_{11} = kA'_{11}/I,$$

$$A'_{22} = \frac{I}{32\pi h}\bigg/\left(\frac{1}{BB}-\frac{1}{AA}\right),$$

$$B'_{22} = \left(\frac{EE}{DD}-\frac{FF}{CC}\right)\bigg/\left(\frac{1}{DD}-\frac{1}{CC}\right),$$

$$C'_{22} = kA'_{22}/I,$$

$$A'_{33} = \frac{I}{768\pi h}\bigg/\left(\frac{1}{DD}-\frac{1}{CC}\right), \quad \text{etc.,}$$

where

$$EE = 12R_1^2r_1^2\ln\frac{R_1}{r_1}+(R_1^2-r_1^2)(R_1^4-5R_1^2r_1^2-2r_1^4)/R_1^2,$$

$$DD = 12(R_1^2+r_1^2)\ln\frac{R_1}{r_1}-\frac{(R_1^2-r_1^2)(R_1^4+10R_1^2r_1^2+r_1^4)}{R_1^2r_1^2}.$$

The constants *CC* and *FF* are identical with *DD* and *EE* respectively, except that R_1 and r_1 are replaced by R_2 and r_2, respectively.

The various constants are calculated for five different ring and cup systems used in this laboratory and are given in Table 2.

Asymptotic Expansion

Starting with Eq. (16) and using the properties of the Wronskian, the asymptotic expansions of the Bessel functions, and the recurrence relations for J and Y, one gets after rearrangement and collection of terms:

$$\frac{\theta_{\text{cup}}}{\theta_{\text{ring}}} = \frac{\pi}{2}\left\{\frac{1}{\dfrac{\pi a r_1^{\frac{3}{2}} R_1^{\frac{3}{2}}}{2\sin(ar_1 - aR_1)} - \dfrac{\pi a r_2^{\frac{3}{2}} R_2^{\frac{3}{2}}}{2\sin(ar_2 - aR_2)}}\right\} \times \left\{\frac{1}{b} + ar_1^3\frac{\cos(aR_1 - ar_1)}{\sin(ar_1 - aR_1)} - ar_2^3\frac{\cos(aR_2 - ar_2)}{\sin(ar_2 - aR_2)}\right\}. \tag{18}$$

In writing Eq. (18), it was assumed that

$$Y_1(\sigma)\,J_1(\xi) - J_1(\sigma)\,Y_1(\xi) \simeq \sin(\sigma - \xi), \tag{19}$$

TABLE 2. INSTRUMENT CONSTANTS FOR RING-IN-CUP GEOMETRY (Markovitz-type treatment)

R_1 (cm)	r_1 (cm)	r_2 (cm)	(R_2 (cm)	I (g·cm²)	h (cm)	A'_{11} ×10³ (g/cm)	B'_{11} ×10⁴ (cm²)	A'_{22} ×10⁵ (g·cm)	B'_{22} ×10⁶ (cm⁴)
1·595	1·653	1·759	1·832	31·2	1·27	25·55	20·73	1·77	0·654
1·423	1·653	1·759	1·979	31·2	1·27	86·67	246·3	72·0	101·9
1·106	1·653	1·759	2·138	31·2	1·27	165·8	770·4	438·8	967
2·856	3·102	3·175	3·420	325	2·54	80·48	300·1	80·76	150·6
2·706	2·926	3·003	3·321	682	5·08	106·8	324·6	116	156·6

and

$$Y_1(\xi)\,J_2(\sigma) - J_1(\xi)\,Y_2(\sigma) \simeq \cos(\xi - \sigma), \tag{20}$$

where σ and ξ are the arguments. For *bob-in-cup geometry*, such an assumption reduces Eq. (3) to the form

$$\frac{\theta_{\text{cup}}}{\theta_{\text{bob}}} = \frac{1}{R^{\frac{3}{2}} r^{\frac{3}{2}} ab}\sin(aR - ar) + \frac{r^{\frac{3}{2}}}{R^{\frac{3}{2}}}\cos(aR - ar), \tag{21}$$

i.e. the terms containing E, E_1, F, F_1, S, and S_1 drop out. As has been already pointed out in that section, such an assumption is justifiable in the sense that coefficients A'_K and B'_K in Eq. (4) are reduced to those of Markovitz. It should therefore turn out that the asymptotic constants calculated for ring-in-cup geometry using Eq. (18) should be similar in values to those given in Table 2 which used Markovitz-type treatment.

However, Eq. (18) cannot be reduced to a form similar to that of Eq. (17), unless one makes an *assumption that* $R_2-r_2 = r_1-R_1$. *With this assumption, Eq. (18) reduces to*

$$\frac{\theta_{\text{cup}}}{\theta_{\text{ring}}} = \frac{1}{r_1^{\frac{3}{2}}R_1^{\frac{3}{2}}+r_2^{\frac{3}{2}}R_2^{\frac{3}{2}}}\left\{\frac{\sin(aR_2-ar_2)}{ab}+(r_1^3+r_2^3)\cos(aR_2-ar_2)\right\}. \quad (22)$$

If one now makes the expansion for sine and cosine functions and substitutes the expressions for a and b defined in the section on bob-in-cup geometry, then the rearrangement and collection of terms gives the desired result, viz.

$$\frac{\theta_{\text{cup}}}{\theta_{\text{ring}}} = \left(\frac{r_1^3+r_2^3}{r_1^{\frac{3}{2}}R_1^{\frac{3}{2}}+r_2^{\frac{3}{2}}R_2^{\frac{3}{2}}}\right) + \sum_{K=1}^{\infty}\left(\frac{i\omega\varrho}{\eta^*}\right)^K\{(A''_{KK}+B''_{KK}\varrho)\,\varrho^{-1}-C''_{KK}\omega^{-2}\varrho^{-1}\}, \quad (23)$$

where

$$A''_{KK} = \frac{I}{2\pi h}\,\frac{(R_2-r_2)^K}{K!\left(r_1^{\frac{3}{2}}R_1^{\frac{3}{2}}+r_2^{\frac{3}{2}}R_2^{\frac{3}{2}}\right)},$$

$$B''_{KK} = \frac{r_1^3+r_2^3}{r_1^{\frac{3}{2}}R_1^{\frac{3}{2}}+r_2^{\frac{3}{2}}R_2^{\frac{3}{2}}}\,\frac{(R_2-r_2)^{2K}}{(2K)!},$$

$$C''_{KK} = kA''_{KK}/I.$$

The constants calculated from Eq. (23) are then compared with those from Eq. (17) (Markovitz-type treatment). Five sets of geometric constants (R_1, r_1, r_2, R_2, and I, h) were chosen so as to make both gaps equal. These geometric constants were the same as those in Table 2 except the value of R_1 was changed to satisfy the condition

$$R_2-r_2 = r_1-R_1.$$

Table 3 shows that the agreement between the constants using Markovitz-type treatment and those using asymptotic expansion is excellent.

TABLE 3. COMPARISON OF INSTRUMENT CONSTANTS OBTAINED USING MARKOVITZ-TYPE EXPANSION WITH THOSE OBTAINED USING ASYMPTOTIC EXPANSION OF BESSEL FUNCTIONS FOR RING-IN-CUP GEOMETRY

R_1	r_1	r_2	R_2	I	h	Markovitz type expansion				Asymptotic expansion			
						$A'_{11}\times 10^3$	$B'_{11}\times 10^4$	$A'_{22}\times 10^5$	$B'_{22}\times 10^6$	$A''_{11}\times 10^3$	$B''_{11}\times 10^4$	$A''_{22}\times 10^5$	$B''_{22}\times 10^6$
1·58	1·653	1·759	1·832	31·2	1·27	28·53	26·56	2·534	1·183	28·53	26·52	2·534	1·178
1·433	1·653	1·759	1·979	31·2	1·27	85·00	23·91	68·51	96·21	84·83	23·77	68·43	95·86
1·274	1·653	1·759	2·138	31·2	1·27	144·0	700·0	344·0	833·7	143·2	691·1	342·8	827·3
2·857	3·102	3·175	3·420	325	2·54	80·32	299·1	80·33	149·6	80·26	298·6	80·29	149·4
2·608	2·926	3·003	3·321	682	5·08	129·4	502·8	218	423·2	129·2	501·4	217·8	422·5

Useful Range of Eqs. (4), (17) and (23)

To establish the range of validity of Eqs. (4), (17) and (23), it will be sufficient to establish the conditions under which the series on the right-hand side of these equations is convergent. Since it is an alternating power series, its convergence requires that

$$\left|\frac{i\omega\varrho}{\eta^*}\right| \left|\frac{(A_{K+1}+B_{K+1}\varrho)\,\varrho^{-1}-C_{K+1}\varrho^{-1}\omega^{-2}}{(A_K+B_K\varrho)\,\varrho^{-1}-C_K\omega^{-2}\varrho^{-1}}\right| < 1. \quad (24)$$

Oka [8] implies that the Markovitz equation for bob-in-cup geometry is applicable when $\sigma \ll 1$ where $\sigma = (-i\omega\varrho/\eta^*)^{\frac{1}{2}}r$. If that is the case, the series will converge much more rapidly and the estimation of its value can be obtained by considering perhaps only the first two terms in that equation. However, even if $|i\omega\varrho/\eta^*| \gg 1$, the condition set forth in Eq. (24) can be satisfied. In particular, as the gap $(R-r)$ between the two cylinders gets small, the ratio in the second bracket of Eq. (24) decreases. It should also be noted that the gap between the two cylinders can be an appreciable fraction of, but not greater than, the wavelength of the shear wave $[(G'/\varrho)^{\frac{1}{2}}/f]$ propagated in the medium [9].

Calculation of Dynamic Parameters

(a) If only the first order term in $\left(\frac{i\omega\varrho}{\eta^*}\right)$ in Eqs. (4), (17) and (23) is included in the calculations, then

$$J' = (1-\cos\phi/m)/D_1\omega, \quad (25)$$

$$J'' = (\sin\phi/m)/D_1\omega, \quad (26)$$

$$\tan\delta = \sin\phi/(m-\cos\phi), \quad (27)$$

$$\phi = \tan^{-1}[D_1\omega J''/(1-D_1\omega J')], \quad (28)$$

$$m = \frac{1}{[(1-D_1\omega J')^2+(D_1\omega J'')^2]^{\frac{1}{2}}}, \quad (29)$$

where

$$D_1 = (A_1+B_1\varrho)\omega - C_1/\omega,$$

$$J' = G'/(G'^2+G''^2) = 1/G'[1+(\tan\delta)^2], \quad (30)$$

$$J'' = G''/(G'^2+G''^2), \quad (31)$$

and

$$\tan\delta = J''/J' = G''/G'. \quad (32)$$

(b) If the second order term in $(i\omega\varrho/\eta^*)$ is to be included in calculations from Eqs. (4), (17) and (23), J' and J'' can be calculated as follows:

(i) Assume $(D_1^2-4D_2D_4)$ is positive.

Let
$$(A_2+B_2\varrho)\varrho\omega^2-C_2\varrho = D_2,$$
$$1-\cos\phi/m = D_4,$$
$$\sin\phi/m = D_5.$$

Then
$$J' = [D_1+r_A^{\frac{1}{2}}\cos(\theta/2)]/2D_2\omega, \tag{33}$$
$$J'' = -r_A^{\frac{1}{2}}\sin(\theta/2)/2D_2\omega, \tag{34}$$
where
$$r_A = [(D_1^2-4D_2D_5)^2+(4D_2D_5)^2]^{\frac{1}{2}}, \tag{35}$$
and
$$\theta = \tan^{-1}[4D_2D_5/(D_1^2-4D_2D_4)]. \tag{36}$$

If J' and/or J'' are negative, θ is replaced by $\theta+2\pi$ in Eqs. (33) and (34) for J' and J''. The value of θ in Eq. (36) has been restricted in the range $-\pi/2 \leqslant \theta \leqslant \pi/2$ so as to have the values of J' and J'' to be calculated by the digital computer. Values of G' and $G''(=\eta'\omega)$ can then be calculated using Eqs. (30), (31) and (32).

(ii) If $D_1^2-4D_2D_4 < 0$, θ is replaced by $\theta+\pi$ in Eqs. (33) and (34) for J' and J''.

REFERENCES

1. H. Markovitz, P. M. Yavorsky, R. C. Harper, Jr., L. J. Zapas, and T. W. DeWitt, *Rev. Sci. Instr.* **23**, 430 (1952).
2. R. R. Myers, C. J. Knauss and R. D. Hoffman, *J. Appl. Polymer Sci.* **6**, 659 (1962).
3. R. D. Hoffman and R. R. Myers, *Proc. 4th Intern. Congr. Rheology*, Part 2, E. H. Lee, Ed., Interscience, New York, 1965, p. 693.
4. K. H. Hiller, *J. Petrol. Tech.*, p. 779 (1963).
5. H. Markovitz, *J. Appl. Phys.* **23**, 1070 (1952).
6. J. G. Oldroyd, *Quart. J. Mech. and Appl. Math.* **4**, 271 (1951).
7. H. Markovitz, private communication.
8. S. Oka, "Principles of rheometry", in *Rheology, Theory and Applications*, vol. 3, F. R. Eirich, Ed., Academic Press, New York, 1960, p. 50.
9. J. D. Ferry, *Viscoelastic Properties of Polymers*, J. Wiley, New York, 1960, pp. 88–89.
10. R. N. Shroff, to be published in *Frans. Soc. Rheol.*

THE RHEOLOGICAL BEHAVIOUR OF COMPLEX VISCOELASTIC SOLIDS

ZDENĚK SOBOTKA

Czechoslovak Academy of Sciences, Prague

1. INTRODUCTION

The paper deals with the rheological equations and models of complex viscoelastic solids consisting of linear and non-linear elastic and viscous elements.

The general rheological equations representing the shape and volume changes of complex linear viscoelastic bodies, with time-variable rheological parameters such as moduli of elasticity and coefficients of viscosity, have the following form

$$a_m(t)\frac{d^m s_{ij}}{dt^m}+a_{m-1}(t)\frac{d^{m-1}s_{ij}}{dt^{m-1}}+\ldots+a_1(t)\frac{ds_{ij}}{dt}+a_0 t)s_{ij}$$

$$+a_1(t)\int_0^t K_1(t,\tau_1)s_{ij}(\tau_1)\,d\tau_1+a_2(t)\int_0^t\int_0^{\tau_1}K_2(t,\tau_2)s_{ij}(\tau_2)\,d\tau_2 d\tau_1$$

$$+a_{-m}(t)\int_0^t\int_0^{\tau_1}\int_0^{\tau_2}\ldots\int_0^{\tau_{m-2}}\int_0^{\tau_{m-1}}K_m(t,\tau_m)s_{ij}(\tau_m)\,d\tau_m\,d\tau_{m-1}\ldots\,d\tau_3\,d\tau_2\,d\tau_1$$

$$=b_{-n}(t)\int_0^t\int_0^{\tau_1}\int_0^{\tau_2}\ldots\int_0^{\tau_{n-2}}\int_0^{\tau_{n-1}}L_n(t,\tau_n)e_{ij}(\tau_n)\,d\tau_n\,d\tau_{n-1}\ldots\,d\tau_3\,d\tau_2\,d\tau_1$$

$$+\ldots b_2(t)\int_0^t\int_0^{\tau_1}L_2(t,\tau_2)e_{ij}(\tau_2)\,d\tau_2\,d\tau_1+b_{-1}(t)\int_0^t L_1(t,\tau_1)e_{ij}(\tau_1)\,d\tau_1$$

$$+b_0(t)e_{ij}+b_1(t)\frac{de_{ij}}{dt}+b_2(t)\frac{d^2e_{ij}}{dt^2}+\ldots+b_n(t)\frac{d^ne_{ij}}{dt^n}, \qquad (1.1)$$

$$c_m(t)\frac{d^m\sigma_m}{dt^m}+c_{m-1}(t)\frac{d^{m-1}\sigma_m}{dt^{m-1}}+\ldots+c_1\frac{d\sigma_m}{dt}+c_0(t)\sigma_m$$

$$+c_1(t)\int_0^t M_1(t,\tau_1)\sigma_m(\tau_1)d\tau_1+c_2(t)\int_0^t\int_0^{\tau_1}M_2(t,\tau_2)\sigma_m(\tau_2)d\tau_2 d\tau_1$$

$$+\ldots+c_{-m}(t)\int_0^t\int_0^{\tau_1}\int_0^{\tau_2}\int_0^{\tau_{m-2}}\int_0^{\tau_{m-1}}M_m(t,\tau_m)\sigma_m(\tau_m)d\tau_m d\tau_{m-1}\ldots d\tau_3 d\tau_2 d\tau_1$$

$$= k_{-n}(t)\int_0^t\int_0^{\tau_1}\int_0^{\tau_2}\dots\int_0^{\tau_{n-2}}\int_0^{\tau_{n-1}} N_n(t,\tau_n)\varepsilon_m(\tau_n)d\tau_n d\tau_{n-1}\dots, d\tau_3 d\tau_2 d\tau_1$$

$$+\dots+k_2(t)\int_0^t\int_0^{\tau_1} N_2(t,\tau_2)\varepsilon_m(\tau_2)d\tau_2 d\tau_1 + k_1(t)\int_0^t N_1(t,\tau_1)\varepsilon_m(\tau_1)d\tau_1$$

$$+k_0(t)\varepsilon_m + k_1(t)\frac{d\varepsilon_m}{dt} + k_2(t)\frac{d^2\varepsilon_m}{dt^2} + \dots + k_n(t)\frac{d^n\varepsilon_m}{dt^n}, \tag{1.2}$$

where s_{ij} are the components of the stress deviator,
e_{ij} those of the strain deviator,
σ_m the mean normal stress,
ε_m the mean normal strain.

The integrals in Eqs. (1.1) and (1.2) may be eliminated by successive differentiation and the general relationships may be then written in the abbreviated form:

$$a_m(t)\frac{d^m s_{ij}}{dt^m} + a_{m-1}(t)\frac{d^{m-1}s_{ij}}{dt^{m-1}} + \dots + a_1(t)\frac{ds_{ij}}{dt} + a_0(t)s_{ij}$$

$$= b_0(t)e_{ij} + b_1(t)\frac{de_{ij}}{dt} + \dots + b_{n-1}(t)\frac{de_{ij}^{n-1}}{dt^{n-1}} + b_n(t)\frac{d^n e_{ij}}{dt^n} \tag{1.3}$$

$$c_m(t)\frac{d^m\sigma_m}{dt^m} + c_{m-1}(t)\frac{d^{m-1}\sigma_m}{dt^{m-1}} + \dots + c_1(t)\frac{d\sigma_m}{dt} + c_0(t)\sigma_m$$

$$= k_0(t)\varepsilon_m + k_1(t)\frac{d\varepsilon_m}{dt} + \dots + k_{n-1}(t)\frac{d^{n-1}\varepsilon_m}{dt^{n-1}} + k_n(t)\frac{d^n\varepsilon_m}{dt^n}. \tag{1.4}$$

If the rheological parameters do not vary with time, the coefficients a_k, b_k, c_k and d_k of the preceding equations are constant as presented by T. Alfrey [1].

For anisotropic bodies, the author has introduced the following general equation:

$$a_{(m)ij\alpha\beta}(t)\frac{d^m\sigma_{\alpha\beta}}{dt^m} + a_{(m-1)ij\gamma\delta}(t)\frac{d^{m-1}\sigma_{\gamma\delta}}{dt^{m-1}} + \dots + a_{(0)ij\varkappa\lambda}(t)\sigma_{\varkappa\lambda}$$

$$= b_{(0)ij\mu\nu}(t)\varepsilon_{\mu\nu} + \dots + b_{(n-1)ij\varrho\varphi}(t)\frac{d^{n-1}\varepsilon_{\varrho\varphi}}{dt^{n-1}} + b_{(n)ij\psi\omega}(t)\frac{d^n\varepsilon_{\psi\omega}}{dt^n} \tag{1.5}$$

which take the different principal directions of anisotropy of individual stress and strain derivatives into account and where

$\sigma_{\alpha\beta}, \sigma_{\gamma\delta}$ are the components of the stress tensor,
$\varepsilon_{\mu\nu}, \varepsilon_{\varrho\varphi}$ are the components of the strain tensor, and

$a_{(m)ij\alpha\beta}(t)$, $a_{(m-1)ij\gamma\delta}(t)$, $a_{(0)ij\varkappa\lambda}(t)$, $b_{(0)ij\mu\nu}(t),\dots, b_{(m-1)ij\varrho\varphi}(t)$ $b_{(n)ij\psi\omega}(t)$ are the components of various time-dependent fourth-rank tensors of anisotropy.

For particular cases, the author has formulated the concept of equivalent stress and strain, given by

$$\bar{\sigma}_{ij} = a_0(t)\sigma_{ij}+a_1(t)\frac{d\sigma_{ij}}{dt}+a_2(t)\frac{d^2\sigma_{ij}}{dt^2}+\ldots+a_m(t)\frac{d^m\sigma_{ij}}{dt^m}$$

$$+a_{-1}(t)\int_0^t K_1(t,\tau_1)\sigma_{ij}(\tau_1)d\tau_1+a_2(t)\int_0^t\int_0^{\tau_1} K_2(t,\tau_2)\sigma_{ij}(\tau_2)\,d\tau_2 d\tau_1$$

$$+\ldots+a_m(t)\int_0^t\int_0^{\tau_1}\int_0^{\tau_2}\ldots\int_0^{\tau_{m-2}}\int_0^{\tau_{m-1}} K_m(t,\tau_m)\sigma_{ij}(\tau_m)d\tau_m d\tau_{m-1}\ldots\, d\tau_2 d\tau_1 \tag{1.6}$$

$$\bar{\varepsilon}_{ij} = b_0(t)\varepsilon_{ij}+b_1(t)\frac{d\varepsilon_{ij}}{dt}+b_2(t)\frac{d^2\varepsilon_{ij}}{dt^2}+\ldots+b_n(t)\frac{d^n\varepsilon_{ij}}{dt^n}$$

$$+b_{-1}(t)\int_0^t L_1(t,\tau_1)\varepsilon_{ij}(\tau_1)d\tau_1+b_{-2}(t)\int_0^t\int_0^{\tau_1} L_2(t,\tau_2)\varepsilon_{ij}(\tau_2)d\tau_2 d\tau_1$$

$$+\ldots+b_{-n}(t)\int_0^t\int_0^{\tau_1}\int_0^{\tau_2}\ldots\int_0^{\tau_{n-2}}\int_0^{\tau_{n-1}} L_n(t,\tau_n)\varepsilon_{ij}(\tau_n)d\tau_n d\tau_{n-1}\ldots d\tau_2 d\tau_1. \tag{1.7}$$

The equivalent stress and strain have the same dimensions as the actual variables. The equation of linear rheological deformation for anisotropic bodies may be formulated in terms of equivalent stress and strain as follows:

$$\bar{\sigma}_{ij} = A_{ij\varkappa\lambda}\,\bar{\varepsilon}_{\varkappa\lambda}. \tag{1.8}$$

The concept of equivalent stress and strain permits reduction of the complicated rheological equations to simpler expressions relating the above. In the rheological boundary-value problems, it permits integration with respect to the space coordinates separate from that with respect to time.

In introducing the concept of equivalent stress and strain, the author has imposed the following restrictions:

1. The coefficients a_k, b_k in Eqs. (1.6) and (1.7) do not vary with the space coordinates, but may depend on time.
2. The anisotropy tensor does not vary with time, but may depend on the space coordinates.
3. The surface force, or the external loading, vary in quantity only, while its form remains constant. It may be represented by a product of two functions, one dependent on the space coordinates and the other on time.
4. Boundary conditions are either time-constant or else obey the same time law as the external loading.

In this way, the equivalent stress or strain is a function of the actual stress or strain, respectively, of their derivatives and integrals with respect to time, and of time, which varies over the space as a whole.

2. GENERAL EQUIVALENT STRESS–STRAIN RELATIONSHIPS FOR ANISOTROPIC RHEOLOGICAL BODIES AND THE CONCEPT OF TRANSFORMED STRAIN

The general relationship between the equivalent stress and strain components for anisotropic bodies may be expressed in the following form

$$\bar{\sigma}_{ij} = f_{ij}(B_{kl\varkappa\lambda}\bar{\varepsilon}_{\varkappa\lambda}), \tag{2.1}$$

where $B_{kl\varkappa\lambda}$ are the components of the non-dimensional anisotropy tensor of rank 4.

Introducing the equivalent transformed strain tensor of rank 2,

$$\bar{\beta}_{kl} = B_{kl\varkappa\lambda}\bar{\varepsilon}_{\varkappa\lambda}, \tag{2.2}$$

which has the same dimension as the strain tensor, as the anisotropy tensor is non-dimensional, we may consider the general function

$$\bar{\sigma}_{ij} = f_{ij}(\bar{\beta}_{kl}) \tag{2.3}$$

of two coaxial tensors $\bar{\sigma}_{ij}$ and $\bar{\beta}_{kl}$.

The above function may, under certain conditions, be expanded in an absolutely convergent power series as follows

$$\bar{\sigma}_{ij}=A_0\delta_{ij}+A_1\bar{\beta}_{ij}+A_2+\bar{\beta}_{i\alpha}\bar{\beta}_{\alpha j}+A_3\bar{\beta}_{i\alpha}\bar{\beta}_{\alpha\beta}\bar{\beta}_{\beta j}+\ldots, \tag{2.4}$$

where A_0, A_1, A_2, A_3, etc., are scalar coefficients and σ_{ij} is the Kronecker delta.

The left-hand side of Eq. (2.4) being a symmetrical tensor of rank 2, it follows from tensor dimensionality that the absolutely convergent series on the right-hand side is also represented by symmetrical tensors of rank 2, which may be expressed according to the Hamilton–Cayley theorem in terms of three principal tensors:

$$\delta_{ij}, \quad \bar{\beta}_{ij} = B_{ij\varkappa\lambda}\bar{\varepsilon}_{\varkappa\lambda}, \quad \bar{\beta}_{i\alpha}\bar{\beta}_{\alpha j} = B_{i\alpha\varkappa\lambda}B_{\alpha j\mu\nu}\bar{\varepsilon}_{\varkappa\lambda}\bar{\varepsilon}_{\mu\nu}$$

and by functions of the three principal transformed equivalent strain invariants

$$\mathrm{I}_\beta = \bar{\beta}_{ij}\delta_{ij} = B_{ij\varkappa\lambda}\delta_{ij}\bar{\varepsilon}_{\varkappa\lambda}, \tag{2.5}$$

$$\mathrm{II}_\beta = \bar{\beta}_{ij}\bar{\beta}_{ij} = B_{ij\varkappa\lambda}B_{ij\mu\nu}\bar{\varepsilon}_{\varkappa\lambda}\bar{\varepsilon}_{\mu\nu}, \tag{2.6}$$

$$\mathrm{III}_\beta = \bar{\beta}_{ij}\bar{\beta}_{i\alpha}\bar{\beta}_{\alpha j} = B_{ij\varkappa\lambda}B_{i\alpha\mu\nu}B_{\alpha j\varrho\psi}\bar{\varepsilon}_{\varkappa\lambda}\bar{\varepsilon}_{\mu\nu}\bar{\varepsilon}_{\varrho\psi}, \tag{2.7}$$

We then have, for anisotropic bodies, a constitutive stress–strain relationship analogous to that derived by Reiner [11] for isotropic bodies

$$\bar{\sigma}_{ij} = \aleph_0 \delta_{ij} + \aleph_1 B_{ij\varkappa\lambda} \bar{\varepsilon}_{\varkappa\lambda} + \aleph_2 B_{i\alpha\varkappa\lambda} B_{\alpha j\mu\nu} \bar{\varepsilon}_{\varkappa\lambda} \bar{\varepsilon}_{\mu\nu}, \tag{2.8}$$

i.e.

$$\bar{\sigma}_{ij} = \aleph_0 \delta_{ij} + \aleph_1 \bar{\beta}_{ij} + \aleph_2 \bar{\beta}_{i\alpha} \bar{\beta}_{\alpha j}. \tag{2.9}$$

The scalar functions of the invariants $\aleph_0$, $\aleph_1$, $\aleph_2$ follow from three equations analogous to those for isotropic media:

$$\bar{\sigma}_{ij} \delta_{ij} = 3\aleph_0 + \aleph_1 B_{ij\varkappa\lambda} \delta_{ij} + \aleph_2 B_{ij\varkappa\lambda} B_{ij\mu\nu} \bar{\varepsilon}_{\varkappa\lambda} \bar{\varepsilon}_{\mu\nu} \tag{2.10}$$

$$\begin{aligned} \bar{\sigma}_{ij} \bar{\sigma}_{ij} = {} & 3\aleph_0^2 + \aleph_1^2 B_{ij\varkappa\lambda} B_{ij\mu\nu} \bar{\varepsilon}_{\varkappa\lambda} \bar{\varepsilon}_{\mu\nu} + \aleph_2^2 B_{i\alpha\varkappa\lambda} B_{\alpha j\mu\nu} B_{i\beta\xi\varrho} B_{\beta j\varphi\chi} \bar{\varepsilon}_{\varkappa\lambda} \bar{\varepsilon}_{\mu\nu} \bar{\varepsilon}_{\xi\varrho} \bar{\varepsilon}_{\varphi\chi} \\ & + 2\aleph_0 \aleph_1 B_{ij\varkappa\lambda} \delta_{ij} \bar{\varepsilon}_{\varkappa\lambda} + 2\aleph_0 \aleph_2 B_{ij\varkappa\lambda} B_{ij\mu\nu} \bar{\varepsilon}_{\varkappa\lambda} \bar{\varepsilon}_{\mu\nu} + 2\aleph_1 \aleph_2 B_{ij\varkappa\lambda} B_{i\alpha\mu\nu} B_{\alpha j\xi\varrho} \bar{\varepsilon}_{\varkappa\lambda} \bar{\varepsilon}_{\mu\nu} \bar{\varepsilon}_{\xi\varrho}, \end{aligned} \tag{2.11}$$

$$\begin{aligned} \bar{\sigma}_{ij} \bar{\sigma}_{i\alpha} \bar{\sigma}_{\alpha j} = {} & 3\aleph_0^3 + \aleph_1^3 B_{ij\varkappa\lambda} B_{i\alpha\mu\nu} B_{\alpha j\xi\varrho} \bar{\varepsilon}_{\varkappa\lambda} \bar{\varepsilon}_{\mu\nu} \bar{\varepsilon}_{\xi\varrho} \\ & + \aleph_2^3 B_{ij\varkappa\lambda} B_{i\alpha\mu\nu} B_{\alpha\beta\xi\varrho} B_{\beta\gamma\varphi\chi} B_{\gamma\delta\psi\omega} B_{\delta j\eta\theta} \bar{\varepsilon}_{\varkappa\lambda} \bar{\varepsilon}_{\mu\nu} \bar{\varepsilon}_{\xi\varrho} \bar{\varepsilon}_{\varphi\chi} \bar{\varepsilon}_{\psi\omega} \bar{\varepsilon}_{\eta\theta} \\ & + 3\aleph_0^2 \aleph_1 B_{ij\varkappa\lambda} \delta_{ij} \bar{\varepsilon}_{\varkappa\lambda} + 3\aleph_0 \aleph_1^2 B_{ij\varkappa\lambda} B_{ij\mu\nu} \bar{\varepsilon}_{\varkappa\lambda} \bar{\varepsilon}_{\mu\nu} + 3\aleph_0^2 \aleph_2 B_{ij\varkappa\lambda} B_{ij\mu\nu} \bar{\varepsilon}_{\varkappa\lambda} \bar{\varepsilon}_{\mu\nu} \\ & + 3\aleph_0 \aleph_2^2 B_{ij\varkappa\lambda} B_{i\alpha\mu\nu} B_{\alpha\beta\xi\varrho} B_{\beta j\varphi\chi} \bar{\varepsilon}_{\varkappa\lambda} \bar{\varepsilon}_{\mu\nu} \bar{\varepsilon}_{\xi\varrho} \bar{\varepsilon}_{\varphi\chi} \\ & + 3\aleph_1^2 \aleph_2 B_{ij\varkappa\lambda} B_{i\alpha\mu\nu} B_{\alpha\beta\xi\varrho} B_{\beta j\varphi\chi} \bar{\varepsilon}_{\varkappa\lambda} \bar{\varepsilon}_{\mu\nu} \bar{\varepsilon}_{\xi\varrho} \bar{\varepsilon}_{\varphi\chi} \\ & + 3\aleph_1 \aleph_2^2 B_{ij\varkappa\lambda} B_{i\alpha\mu\nu} B_{\alpha\beta\xi\varrho} B_{\beta\gamma\varphi\chi} B_{\gamma j\psi\omega} \bar{\varepsilon}_{\varkappa\lambda} \bar{\varepsilon}_{\mu\nu} \bar{\varepsilon}_{\xi\varrho} \bar{\varepsilon}_{\varphi\chi} \bar{\varepsilon}_{\psi\varrho} \\ & + 6\aleph_0 \aleph_1 \aleph_2 B_{ij\varkappa\lambda} B_{i\alpha\mu\nu} B_{\alpha j\xi\varrho} \bar{\varepsilon}_{\varkappa\lambda} \bar{\varepsilon}_{\mu\nu} \bar{\varepsilon}_{\xi\varrho}. \end{aligned} \tag{2.12}$$

The third term in Eqs. (2.8) and (2.9) represents second-order effects. fn the case of infinitesimal deformation, if the values of the invariant Iunctions $\aleph_0$, $\aleph_1$, $\aleph_2$ are of the same order of magnitude, Eq. (2.8) becomes

$$\bar{\sigma}_{ij} = \aleph_0 \delta_{ij} + \aleph_1 B_{ij\varkappa\lambda} \bar{\varepsilon}_{\varkappa\lambda}. \tag{2.13}$$

The invariant functions are obtainable from

$$\bar{\sigma}_{ij} \delta_{ij} = 3\aleph_0 + \aleph_1 B_{ij\varkappa\lambda} \delta_{ij} \bar{\varepsilon}_{\varkappa\lambda}, \tag{2.14}$$

$$\bar{\sigma}_{ij} \bar{\sigma}_{ij} = 3\aleph_0^2 + 2\aleph_0 \aleph_1 B_{ij\varkappa\lambda} \delta_{ij} \bar{\varepsilon}_{\varkappa\lambda} + \aleph_1^2 B_{i\alpha\varkappa\lambda} B_{\alpha j\mu\nu} \bar{\varepsilon}_{\varkappa\lambda} \bar{\varepsilon}_{\mu\nu} \tag{2.15}$$

after substituting (2.5) and (2.6) as follows:

$$\aleph_0 = \frac{1}{3} \left[\left(I_\sigma - I_\beta \sqrt{\frac{3II_\sigma - I_\sigma^2}{3II_\beta - I_\beta^2}} \right) \right], \tag{2.16}$$

$$\aleph_1 = \sqrt{\left(\frac{3II_\sigma - I_\sigma^2}{3II_\beta - I_\beta^2} \right)}, \tag{2.17}$$

where $I_\sigma = \bar{\sigma}_{ij} \delta_{ij}$, $II_\sigma = \bar{\sigma}_{ij} \bar{\sigma}_{ij}$ are the invariants of the equivalent stress tensor.

After some rearrangements, we obtain the equivalent stress–strain relationships for anisotropic bodies

$$\bar{\sigma}_{ij} - \bar{\sigma}_m \delta_{ij} = \frac{2\bar{\sigma}_i}{3\bar{\beta}_i} (\bar{\beta}_{ij} - \bar{\beta}_m \delta_{ij}), \tag{2.18}$$

where

$$\bar{\sigma}_m = \tfrac{1}{3}(\bar{\sigma}_{11}+\bar{\sigma}_{22}+\bar{\sigma}_{33}) \tag{2.19}$$

is the equivalent mean stress,

$$\bar{\sigma}_i = \frac{1}{\sqrt{2}}\sqrt{[(\bar{\sigma}_{11}-\bar{\sigma}_{22})^2+(\bar{\sigma}_{22}-\bar{\sigma}_{33})^2+(\bar{\sigma}_{33}-\bar{\sigma}_{11})^2+6(\bar{\sigma}_{12}^2+\bar{\sigma}_{23}^2+\bar{\sigma}_{31}^2)]} \tag{2.20}$$

the equivalent effective stress,

$\bar{\beta}_{ij} = B_{ij\varkappa\lambda}\bar{\varepsilon}_{\varkappa\lambda}$ the equivalent transformed strain components,

$$\bar{\beta}_i = \frac{\sqrt{2}}{3}\sqrt{(\bar{\beta}_{11}-\bar{\beta}_{22})^2+(\bar{\beta}_{22}-\bar{\beta}_{33})^2+(\bar{\beta}_{33}-\bar{\beta}_{11})^2+6(\bar{\beta}_{12}^2+\bar{\beta}_{23}^2+\bar{\beta}_{31}^2)]} \tag{2.21}$$

the equivalent transformed effective strain, and

$$\bar{\beta}_m = \tfrac{1}{3}(\bar{\beta}_{11}+\bar{\beta}_{22}+\bar{\beta}_{33}) \tag{2.22}$$

the equivalent transformed mean strain.

Equation (2.18) is analogous to the Hencky relations of the plastic deformation theory for isotropic bodies.

The concept of transformed strain permits formulation of the stress–strain relationships in a manner analogous to that of the isotropic case.

3. COMPLEX VISCOELASTIC SOLID COMPRISING SEVERAL MAXWELL ELEMENTS

The rheological model of a complex viscoelastic solid consisting of one Hookean spring and five Maxwell elements is represented in Fig. 1.

The resultant stress σ in the uniaxial state is the sum of the component

$$\sigma_0 = E_0\varepsilon \tag{3.1}$$

acting in the Hookean spring H_0, and of n components

$$\sigma_k = e^{-E_k t/\lambda_k}\left(E_k\int_0^t \frac{d\varepsilon}{dt}\, e^{E_k t/\lambda_k} - dt + \sigma_{kA}\right) \tag{3.2}$$

acting in the n Maxwell elements.

In Eq. (3.2), σ_{kA} denotes the stress at time $t_0 = 0$, which may be taken equal to zero, E_k the modulus of elasticity in the kth element and λ_k the coefficient of normal viscosity.

Then, the resultant stress is given by

$$\sigma = E_0\varepsilon + \sum_{k=1}^{n} e^{-E_k t/\lambda_k}\left(E_k\int_0^t \frac{d\varepsilon}{dt}\, e^{E_k t/\lambda_k}dt + \sigma_{kA}\right) \tag{3.3}$$

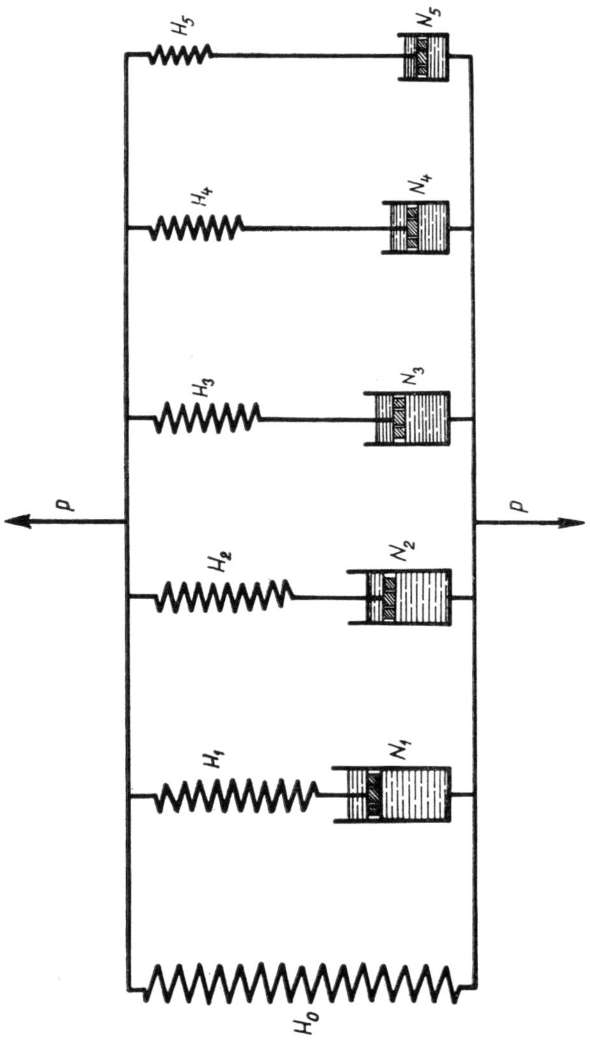

FIG. 1. Rheological model of a complex viscoelastic solid consisting of one Hookean spring and several Maxwell elements.

The integrals in the preceding relationship may be eliminated by successive multiplication by $e^{E_k t/\lambda_k}$ and differentiation. After some rearrangements, the author has obtained for a rheological model comprising one Hookean spring H_0 and five Maxwell elements the following differential stress–strain relation:

$$\frac{d^5\sigma}{dt^5}+\left(\frac{E_1}{\lambda_1}+\frac{E_2}{\lambda_2}+\frac{E_3}{\lambda_3}+\frac{E_4}{\lambda_4}+\frac{E_5}{\lambda_5}\right)\frac{d^4\sigma}{dt^4}$$

$$+\left(\frac{E_1E_2}{\lambda_1\lambda_2}+\frac{E_1E_3}{\lambda_1\lambda_3}+\frac{E_1E_4}{\lambda_1\lambda_4}+\frac{E_1E_5}{\lambda_1\lambda_5}+\frac{E_2E_3}{\lambda_2\lambda_3}\right.$$

$$\left.+\frac{E_2E_4}{\lambda_2\lambda_4}+\frac{E_2E_5}{\lambda_2\lambda_5}+\frac{E_3E_4}{\lambda_3\lambda_4}+\frac{E_3E_5}{\lambda_3\lambda_5}+\frac{E_4E_5}{\lambda_4\lambda_5}\right)\frac{d^3\sigma}{dt^3}$$

$$+\left(\frac{E_1E_2E_3}{\lambda_1\lambda_2\lambda_3}+\frac{E_1E_2E_4}{\lambda_1\lambda_2\lambda_4}+\frac{E_1E_2E_5}{\lambda_1\lambda_2\lambda_5}+\frac{E_1E_3E_4}{\lambda_1\lambda_3\lambda_4}\right.$$

$$+\frac{E_1E_3E_5}{\lambda_1\lambda_3\lambda_5}+\frac{E_1E_4E_5}{\lambda_1\lambda_4\lambda_5}+\frac{E_2E_3E_4}{\lambda_2\lambda_3\lambda_4}+\frac{E_2E_3E_5}{\lambda_2\lambda_3\lambda_5}$$

$$\left.+\frac{E_2E_4E_5}{\lambda_2\lambda_4\lambda_5}+\frac{E_3E_4E_5}{\lambda_3\lambda_4\lambda_5}\right)\frac{d^2\sigma}{dt^2}$$

$$+\left(\frac{E_1E_2E_3E_4}{\lambda_1\lambda_2\lambda_3\lambda_4}+\frac{E_1E_2E_4E_5}{\lambda_1\lambda_2\lambda_4\lambda_5}+\frac{E_1E_2E_4E_5}{\lambda_1\lambda_2\lambda_4\lambda_5}\right.$$

$$\left.+\frac{E_1E_3E_4E_5}{\lambda_1\lambda_3\lambda_4\lambda_5}+\frac{E_2E_3E_4E_5}{\lambda_2\lambda_3\lambda_4\lambda_5}\right)\frac{d\sigma}{dt}+\frac{E_1E_2E_3E_4E_5}{\lambda_1\lambda_2\lambda_3\lambda_4\lambda_5}\sigma=E_0\frac{E_1E_2E_3E_4E_5}{\lambda_1\lambda_2\lambda_3\lambda_4\lambda_5}\varepsilon$$

$$+\left\{E_0\left(\frac{E_1E_2E_3E_4}{\lambda_1\lambda_2\lambda_3\lambda_4}+\frac{E_1E_2E_3E_5}{\lambda_1\lambda_2\lambda_3\lambda_5}+\frac{E_1E_2E_4E_5}{\lambda_1\lambda_2\lambda_4\lambda_5}+\frac{E_1E_3E_4E_5}{\lambda_1\lambda_3\lambda_4\lambda_5}+\frac{E_2E_3E_4E_5}{\lambda_2\lambda_3\lambda_4\lambda_5}\right)\right.$$

$$+\frac{E_2E_3E_4E_5}{\lambda_2\lambda_3\lambda_4\lambda_5}(E_1+E_2+E_3+E_4+E_5)$$

$$+\frac{E_3E_4E_5}{\lambda_3\lambda_4\lambda_5}\left[E_2\left(\frac{E_1}{\lambda_1}-\frac{E_2}{\lambda_2}\right)+E_3\left(\frac{E_1}{\lambda_1}-\frac{E_3}{\lambda_3}\right)+E_4\left(\frac{E_1}{\lambda_1}-\frac{E_4}{\lambda_4}\right)+E_5\left(\frac{E_1}{\lambda_1}-\frac{E_5}{\lambda_5}\right)\right]$$

$$+\frac{E_4E_5}{\lambda_4\lambda_5}\left[E_3\left(\frac{E_1}{\lambda_1}-\frac{E_3}{\lambda_3}\right)\left(\frac{E_2}{\lambda_2}-\frac{E_3}{\lambda_3}\right)+E_4\left(\frac{E_1}{\lambda_1}-\frac{E_4}{\lambda_4}\right)\left(\frac{E_2}{\lambda_2}-\frac{E_4}{\lambda_4}\right)\right.$$

$$\left.+E_5\left(\frac{E_1}{\lambda_1}-\frac{E_5}{\lambda_5}\right)\left(\frac{E_2}{\lambda_2}-\frac{E_5}{\lambda_5}\right)\right]$$

$$+\frac{E_5}{\lambda_5}\left[E_4\left(\frac{E_1}{\lambda_1}-\frac{E_4}{\lambda_4}\right)\left(\frac{E_2}{\lambda_2}-\frac{E_4}{\lambda_4}\right)\left(\frac{E_3}{\lambda_3}-\frac{E_4}{\lambda_4}\right)\right.$$

$$\left.+E_5\left(\frac{E_1}{\lambda_1}-\frac{E_5}{\lambda_5}\right)\left(\frac{E_2}{\lambda_2}-\frac{E_5}{\lambda_5}\right)\left(\frac{E_3}{\lambda_3}-\frac{E_5}{\lambda_5}\right)\right]$$

$$+E_5\left(\frac{E_1}{\lambda_1}-\frac{E_5}{\lambda_5}\right)\left(\frac{E_2}{\lambda_2}-\frac{E_5}{\lambda_5}\right)\left(\frac{E_3}{\lambda_3}-\frac{E_5}{\lambda_5}\right)\left(\frac{E_4}{\lambda_4}-\frac{E_5}{\lambda_5}\right)\Bigg\}\frac{d\varepsilon}{dt}$$

$$+\Bigg\{E_0\left(\frac{E_1E_2E_3}{\lambda_1\lambda_2\lambda_3}+\frac{E_1E_2E_4}{\lambda_1\lambda_2\lambda_4}+\frac{E_1E_2E_5}{\lambda_1\lambda_2\lambda_5}+\frac{E_1E_3E_4}{\lambda_1\lambda_3\lambda_4}\right.$$

$$+\frac{E_1E_3E_5}{\lambda_1\lambda_3\lambda_5}+\frac{E_1E_4E_5}{\lambda_1\lambda_4\lambda_5}+\frac{E_2E_3E_4}{\lambda_2\lambda_3\lambda_4}+\frac{E_2E_3E_5}{\lambda_2\lambda_3\lambda_5}$$

$$\left.+\frac{E_2E_4E_5}{\lambda_2\lambda_4\lambda_5}+\frac{E_3E_4E_5}{\lambda_3\lambda_4\lambda_5}\right)$$

$$+\left(\frac{E_2E_3E_4}{\lambda_2\lambda_3\lambda_4}+\frac{E_2E_3E_5}{\lambda_2\lambda_3\lambda_5}+\frac{E_2E_4E_5}{\lambda_2\lambda_4\lambda_5}+\frac{E_3E_4E_5}{\lambda_3\lambda_4\lambda_5}\right)(E_1+E_2+E_3+E_4+E_5)$$

$$+\left(\frac{E_3E_4}{\lambda_3\lambda_4}+\frac{E_3E_5}{\lambda_3\lambda_5}+\frac{E_4E_5}{\lambda_4\lambda_5}\right)\left[E_2\left(\frac{E_1}{\lambda_1}-\frac{E_2}{\lambda_2}\right)\right.$$

$$\left.+E_3\left(\frac{E_1}{\lambda_1}-\frac{E_3}{\lambda_3}\right)+E_4\left(\frac{E_1}{\lambda_1}-\frac{E_4}{\lambda_4}\right)+E_5\left(\frac{E_1}{\lambda_1}-\frac{E_5}{\lambda_5}\right)\right]$$

$$+\left(\frac{E_4}{\lambda_4}+\frac{E_5}{\lambda_5}\right)\left[E_3\left(\frac{E_1}{\lambda_1}-\frac{E_3}{\lambda_3}\right)\left(\frac{E_2}{\lambda_2}-\frac{E_3}{\lambda_3}\right)\right.$$

$$\left.+E_4\left(\frac{E_1}{\lambda_1}-\frac{E_4}{\lambda_4}\right)\left(\frac{E_2}{\lambda_2}-\frac{E_4}{\lambda_4}\right)+E_5\left(\frac{E_1}{\lambda_1}-\frac{E_5}{\lambda_5}\right)\left(\frac{E_2}{\lambda_2}-\frac{E_5}{\lambda_5}\right)\right]$$

$$+E_4\left(\frac{E_1}{\lambda_1}-\frac{E_4}{\lambda_4}\right)\left(\frac{E_2}{\lambda_2}-\frac{E_4}{\lambda_4}\right)\left(\frac{E_3}{\lambda_3}-\frac{E_4}{\lambda_4}\right)$$

$$+E_5\left(\frac{E_1}{\lambda_1}-\frac{E_5}{\lambda_5}\right)\left(\frac{E_2}{\lambda_2}-\frac{E_5}{\lambda_5}\right)\left(\frac{E_3}{\lambda_3}-\frac{E_5}{\lambda_5}\right)\Bigg\}\frac{d^2\varepsilon}{dt^2}$$

$$+\Bigg\{E_0\left(\frac{E_1E_2}{\lambda_1\lambda_2}+\frac{E_1E_3}{\lambda_1\lambda_3}+\frac{E_1E_4}{\lambda_1\lambda_4}+\frac{E_1E_5}{\lambda_1\lambda_5}\right.$$

$$+\frac{E_2E_3}{\lambda_2\lambda_3}+\frac{E_2E_4}{\lambda_2\lambda_4}+\frac{E_2E_5}{\lambda_2\lambda_5}+\frac{E_3E_4}{\lambda_3\lambda_4}$$

$$\left.+\frac{E_3E_5}{\lambda_3\lambda_5}+\frac{E_4E_5}{\lambda_4\lambda_5}\right)$$

$$+\left(\frac{E_2E_3}{\lambda_2\lambda_3}+\frac{E_2E_4}{\lambda_2\lambda_4}+\frac{E_2E_5}{\lambda_2\lambda_5}+\frac{E_3E_4}{\lambda_3\lambda_4}+\frac{E_3E_5}{\lambda_3\lambda_5}\right.$$

$$\left.+\frac{E_4E_5}{\lambda_4\lambda_5}\right)(E_1+E_2+E_3+E_4+E_5)$$

$$+\left(\frac{E_3}{\lambda_3}+\frac{E_4}{\lambda_4}+\frac{E_5}{\lambda_5}\right)\left[E_2\left(\frac{E_1}{\lambda_1}-\frac{E_2}{\lambda_2}\right)+E_3\left(\frac{E_1}{\lambda_1}-\frac{E_3}{\lambda_3}\right)\right.$$

$$+E_4\left(\frac{E_1}{\lambda_1}-\frac{E_4}{\lambda_4}\right)+E_5\left(\frac{E_1}{\lambda_1}-\frac{E_5}{\lambda_5}\right)$$

$$+E_3\left(\frac{E_1}{\lambda_1}-\frac{E_3}{\lambda_3}\right)\left(\frac{E_2}{\lambda_2}-\frac{E_3}{\lambda_3}\right)+E_4\left(\frac{E_1}{\lambda_1}-\frac{E_4}{\lambda_4}\right)\left(\frac{E_2}{\lambda_2}-\frac{E_4}{\lambda_4}\right)$$

$$\left.+E_5\left(\frac{E_1}{\lambda_1}-\frac{E_5}{\lambda_5}\right)\left(\frac{E_2}{\lambda_2}-\frac{E_5}{\lambda_5}\right)\right\}\frac{d^3\varepsilon}{dt^3}$$

$$+\left[E_0\left(\frac{E_1}{\lambda_1}+\frac{E_2}{\lambda_2}+\frac{E_3}{\lambda_3}+\frac{E_4}{\lambda_4}+\frac{E_5}{\lambda_5}\right)\right.$$

$$+\left(\frac{E_2}{\lambda_2}+\frac{E_3}{\lambda_3}+\frac{E_4}{\lambda_4}+\frac{E_5}{\lambda_5}\right)(E_1+E_2+E_3+E_4+E_5)$$

$$+E_2\left(\frac{E_1}{\lambda_1}-\frac{E_2}{\lambda_2}\right)+E_3\left(\frac{E_1}{\lambda_1}-\frac{E_3}{\lambda_3}\right)$$

$$\left.+E_4\left(\frac{E_1}{\lambda_1}-\frac{E_4}{\lambda_4}\right)+E_5\left(\frac{E_1}{\lambda_1}-\frac{E_5}{\lambda_5}\right)\right]\frac{d^4\varepsilon}{dt^4}$$

$$+(E_0+E_1+E_2+E_3+E_4+E_5)\frac{d^5\varepsilon}{dt^5}. \tag{3.4}$$

For the triaxial state of the isotropic body, we obtain from Eq. (3.3), on replacing σ with s_{ij}, ε with e_{ij}, E with $2G$ and λ with 2η, the relationship for the deviatoric components:

$$s_{ij} = 2G_0e_{ij}+\sum_{k=1}^{n} e^{G_kt/\eta_k}\left(2G_k\int_0^t \frac{de_{ij}}{dt}e^{G_kt/\eta_k}\,dt+e_{ijA}\right) \tag{3.5}$$

where G_k is the shear modulus and η_k the coefficient of the shear viscosity of the kth Maxwell element.

On replacing, in Eq. (3.3), σ with σ_m, ε with ε_m, E with K and λ with ξ, we get the relationship for the hydrostatic components:

$$\sigma_m = K_0\varepsilon_m+\sum_{k=1}^{n} e^{-K_kt/\zeta_k}\left(K_k\int_0^t \frac{d\varepsilon_m}{dt}e^{K_kt/\zeta_k}\,dt+\sigma_{mA}\right) \tag{3.6}$$

where $K_k = \dfrac{E_k}{1-2\mu_k}$ is the bulk modulus,

μ_k Poisson's ratio and

ζ_k the coefficient of volume viscosity of the kth element.

The integrals in Eqs. (3.5) and (3.6) may be eliminated along the same lines as in Eq. (3.3) for the uniaxial state of stress, and we obtain for the deviatoric and hydrostatic components differential relations, analogous to Eq. (3.4), having the form of Eqs. (1.3) and (1.4).

In the case of the anisotropic viscoelastic solids, there are three different types of rheological deformation. The most simple type is that with the same relaxation time in all directions. The stress components in the kth Maxwell element in the model in Fig.1 are in this case given by

$$\frac{d\sigma_{(k)ij}}{dt} + C\sigma_{ij} = B_{(k)ij\varkappa\lambda}\frac{d\varepsilon_{\varkappa\lambda}}{dt}, \tag{3.7}$$

where $B_{(k)ij\varkappa\lambda}$ is the tensor, of rank 4, of moduli of anisotropy, C the common reciprocal relaxation time.

In a more general case, the time of relaxation varies directionally and we have the following equation:

$$\frac{d\sigma_{(k)ij}}{dt} + C_{(k)(ij)}\sigma_{(k)ij} = B_{(k)ij\varkappa\lambda}\frac{d\varepsilon_{\varkappa\lambda}}{dt}, \tag{3.8}$$

where $C_{(k)(ij)}$ are the reciprocal relaxation times, different for each stress component $\sigma_{(k)ij}$.

In the most general case, the $\sigma_{(k)ij}$ rate of rheological deformation in one direction is influenced by all stress and stress-rate components and the principal directions of elastic and viscous anisotropy are non-coincident. This type of deformation is represented by the differential relation

$$\frac{d\sigma_{(k)ij}}{dt} + C_{(k)ij\mu\nu}\sigma_{\mu\nu} = B_{(k)ij\varkappa\lambda}\frac{d\varepsilon_{\varkappa\lambda}}{dt} \tag{3.9}$$

containing the tensor $C_{(k)ij\mu\nu}$, of rank 4, of the reciprocal relaxation times.

The above relationship represents a system of linear differential equations of first order.

Consider first the case represented by Eq. (3.8), of which Eq. (3.7) is merely a particular case.

Equation (3.8) has the following solution:

$$\sigma_{(k)ij} = e^{-C_{(k)(ij)}t}\left(B_{(k)ij\varkappa\lambda}\int_0^t \frac{d\varepsilon_{\varkappa\lambda}}{dt}e^{C_{(k)(ij)}t}\,dt + \sigma_{(kA)ij}\right), \tag{3.10}$$

where $\sigma_{(kA)ij}$ is the initial stress at time $t = 0$.

The resulting stress components of the complex anisotropic solid represented by the rheological model in Fig. 1 is given by

$$\sigma_{ij} = B_{(0)ij\varkappa\lambda}e_{\varkappa\lambda} + \sum_{k=1}^{n} e^{-C_{(k)(ij)}t}\left(\int_0^t B_{(k)ij\varkappa\lambda}\frac{d\varepsilon_{\varkappa\lambda}}{dt}e^{C_{(k)(ij)}t}\,dt + \sigma_{(kA)ij}\right). \tag{3.11}$$

For a solid consisting of one Hookean and three Maxwell elements in parallel, we obtain, on elimination of the integrals in Eq. (3.11), the following differential relationship:

$$\frac{d^3\sigma_{ij}}{dt^3}+(C_{(1)(ij)}+C_{(2)(ij)}+C_{(3)(ij)})\frac{d^2\sigma_{ij}}{dt^2}$$

$$+(C_{(1)(ij)}C_{(2)(ij)}+C_{(1)(ij)}C_{(3)(ij)}+C_{(2)(ij)}C_{(3)(ij)})\frac{d\sigma_{ij}}{dt}$$

$$+C_{(1)(ij)}C_{(2)(ij)}C_{(3)(ij)}\sigma_{ij}$$

$$= B_{(0)ij\varkappa\lambda}C_{(1)(ij)}C_{(2)(ij)}C_{(3)(ij)}\varepsilon_{\varkappa\lambda}$$

$$+\{B_{(0)ij\varkappa\lambda}(C_{(1)(ij)}C_{(2)(ij)}+C_{(1)(ij)}C_{(3)(ij)}+C_{(2)(ij)}C_{(3)(ij)})$$

$$+C_{(2)(ij)}C_{(3)(ij)}(B_{(1)ij\varkappa\lambda}+B_{(2)ij\varkappa\lambda}+B_{(3)ij\varkappa\lambda})$$

$$+C_{(3)(ij)}[B_{(2)ij\varkappa\lambda}(C_{(1)(ij)}-C_{(2)(ij)})+B_{(3)ij\varkappa\lambda}(C_{(1)(ij)}-C_{(3)(ij)})]$$

$$+B_{(3)ij\varkappa\lambda}(C_{(1)(ij)}-C_{(3)(ij)})(C_{(2)(ij)}-C_{(3)(ij)})\}\frac{d\varepsilon_{\varkappa\lambda}}{dt}$$

$$+[B_{(0)ij\varkappa\lambda}(C_{(1)(ij)}+C_{(2)(ij)}+C_{(3)(ij)})+(B_{(1)ij\varkappa\lambda}+B_{(2)ij\varkappa\lambda}+B_{(3)ij\varkappa\lambda})(C_{(2)(ij)}$$

$$+C_{(3)(ij)})+B_{(2)ij\varkappa\lambda}(C_{(1)(ij)}-C_{(2)(ij)})+B_{(3)ij\varkappa\lambda}(C_{(1)(ij)}-C_{(3)(ij)})]\frac{d^2\varepsilon_{\varkappa\lambda}}{dt^2}$$

$$+(B_{(1)ij\varkappa\lambda}+B_{(2)ij\varkappa\lambda}+B_{(3)ij\varkappa\lambda})\frac{d^3\varepsilon_{\varkappa\lambda}}{dt^3}\,. \tag{3.12}$$

The preceding differential relationship permits direct determination of the stress components from the given time-dependent components of the strain tensor.

If, by contrast, the time-dependent stress components are given, Eq. (3.12) yields for $i, j = 1, 2, 3$ and $\varkappa, \lambda = 1, 2, 3$ a system of six linear differential equations of third order for the six components of the symmetrical strain tensor representing the triaxial state of deformation. This is the main rheological difference between such anisotropic bodies and isotropic media.

The System of Simultaneous Differential Equations of Higher Order

Consider, for simplicity, the plane-strain state of the orthotropic Zener body, represented by the rheological model in Fig. 2.

The tensor differential relationship for stress and strain components in this body has the following form,

$$\frac{d\sigma_{ij}}{dt}+C_{(ij)}\sigma_{ij} = B_{(0)ij\varkappa\lambda}C_{(ij)}\varepsilon_{\varkappa\lambda}+(B_{(0)ij\varkappa\lambda}+B_{(1)ij\varkappa\lambda})\frac{d\varepsilon_{\varkappa\lambda}}{dt}, \tag{3.13}$$

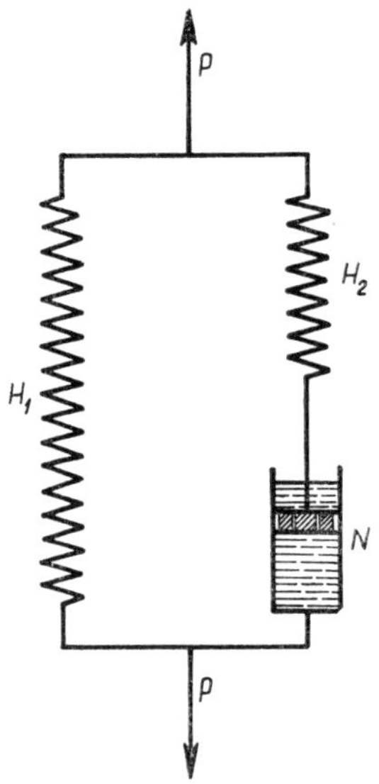

FIG. 2. Rheological model of the Zener body.

which yields three component equations for the plane-strain state:

$$\frac{d\sigma_{11}}{dt}+C_{(11)}\sigma_{11} = B_{(0)1111}C_{(11)}\varepsilon_{11}+B_{(0)1122}C_{(11)}\varepsilon_{22}$$

$$+(B_{(0)1111}+B_{(1)1111})\frac{d\varepsilon_{11}}{dt}+(B_{(0)1122}+B_{(1)1122})\frac{d\varepsilon_{22}}{dt}, \qquad (3.14)$$

$$\frac{d\sigma_{22}}{dt}+C_{(22)}\sigma_{22} = B_{(0)2211}C_{(22)}\varepsilon_{11}+B_{(0)2222}C_{(22)}\varepsilon_{22}$$

$$+(B_{(0)2211}+B_{(1)2211})\frac{d\varepsilon_{11}}{dt}+(B_{(0)2222}+B_{(1)2222})\frac{d\varepsilon_{22}}{dt}, \qquad (3.15)$$

$$\frac{d\sigma_{12}}{dt}+C_{(12)}\sigma_{12} = B_{(0)1212}C_{(12)}\varepsilon_{12}+(B_{(0)1212}+B_{(1)1212})\frac{d\varepsilon_{12}}{dt}. \qquad (3.16)$$

Equation (3.16) permits direct determination of the shear strain ε_{12} from the given time-dependent shear stress σ_{12}.

From Eq. (3.15) we obtain

$$\varepsilon_{22} = \frac{e^{-Lt}}{B_{(0)2222}+B_{(1)2222}}\int_0^t\left[\frac{d\sigma_{22}}{dt}+C_{(22)}\sigma_{22}-B_{(0)2211}C_{(22)}\varepsilon_{11}\right.$$

$$\left.-(B_{(0)2211}+B_{(1)2211})\frac{d\varepsilon_{11}}{dt}\right]e^{Lt}\,dt+\varepsilon_{22(A)}e^{-Lt}, \qquad (3.17)$$

where

$$L = \frac{B_{(0)2222}C_{(22)}}{B_{(0)2222}+B_{(1)2222}}$$

and $\varepsilon_{22(A)}$ is the initial strain at time $t_0 = 0$.

Substitution of Eq. (3.17) in Eq. (3.14) yields, after some rearrangements, the following second-order equation for ε_{11}:

$$[(B_{(0)1111}+B_{(1)1111})(B_{(0)2222}+B_{(1)2222})-(B_{(0)1122}+B_{1(1)1122})^2]\frac{d^2\varepsilon_{11}}{dt^2}$$

$$+[B_{(0)1111}(B_{(0)2222}+B_{(1)2222})C_{(11)}+B_{(0)2222}(B_{(0)1111}+B_{(1)1111})C_{(22)}$$

$$-B_{(0)1122}(B_{(0)1122}+B_{(1)1122})(C_{(11)}+C_{(22)})]\frac{d\varepsilon_{11}}{dt}$$

$$+(B_{(0)1111}B_{(0)2222}-B^2_{(0)1122})C_{(11)}C_{(22)}\varepsilon_{11}$$

$$= B_{(0)2222}C_{(11)}C_{22}\sigma_{11}+[B_{(0)2222}C_{(22)}+(B_{(0)2222}+B_{(1)2222})C_{(11)}]\frac{d\sigma_{11}}{dt}$$

$$+(B_{(0)2222}+B_{(1)2222})\frac{d^2\sigma_{11}}{dt^2}$$

$$-B_{(0)1122}C_{(11)}C_{(22)}\sigma_{22}-[B_{(0)1122}C_{(11)}+(B_{(0)1122}+B_{(1)1122})C_{(22)}]\frac{d\sigma_{22}}{dt}$$

$$-(B_{(0)1122}+B_{(1)1122})\frac{d^2\sigma_{22}}{dt^2}. \tag{3.18}$$

The differential equation for ε_{22} may be derived in quite a similar way. From Eq. (3.14), we have

$$\varepsilon_{11} = \frac{e^{-Rt}}{B_{(0)1111}+B_{(1)1111}}\int_0^t\left[\frac{d\sigma_{11}}{dt}+C_{(11)}\sigma_{11}-B_{(0)1122}C_{(11)}\varepsilon_{22}\right.$$

$$\left.-(B_{(0)1122}+B_{(1)1122})\frac{d\varepsilon_{22}}{dt}\right]e^{Rt}\,dt+\varepsilon_{11(A)}e^{-Rt}, \tag{3.19}$$

where

$$R = \frac{B_{(0)1111}C_{(11)}}{B_{(0)1111}+B_{(1)1111}}.$$

Substituting in Eq. (3.15), we obtain after some rearrangements:

$$[(B_{(0)1111}+B_{(1)1111})(B_{(0)2222}+B_{(1)2222})-(B_{(0)1122}+B_{(1)1122})^2]\frac{d^2\varepsilon_{22}}{dt^2}$$

$$+[B_{(0)2222}(B_{(0)1111}+B_{(1)1111})C_{(22)}+B_{(0)1111}(B_{(0)2222}+B_{(1)2222})C_{(11)}$$

$$-B_{(0)1122}(B_{(0)1122}+B_{(1)1122})(C_{(11)}+C_{(22)})]\frac{d\varepsilon_{22}}{dt}$$

$$+(B_{(0)1111}B_{(0)2222}-B^2_{(0)1122})C_{(11)}C_{(22)}\varepsilon_{22}$$

$$=-B_{(0)1122}C_{(11)}(C_{(22)}\sigma_{11}-[B_{(0)1122}C_{(22)}+(B_{(0)1122}+B_{(1)1122})C_{(11)}]\frac{d\sigma_{11}}{dt}$$

$$-(B_{(0)1122}+B_{(1)1122})\frac{d^2\sigma_{11}}{dt^2}$$

$$+B_{(0)1111}C_{(11)}C_{(22)}\sigma_{22}+[B_{(0)1111}C_{(11)}+(B_{(0)1111}+B_{(1)1111})C_{(22)}]\frac{d\sigma_{22}}{dt}$$

$$+(B_{(0)1111}+B_{(1)1111})\frac{d^2\sigma_{22}}{dt^2}. \tag{3.20}$$

The same result is obtainable directly from Eq. (3.18) by interchanging subscripts 1 and 2.

The equations for the normal strains ε_{11} and ε_{22} in the case represented by Eq. (3.7), with the same relaxation time in all directions, follow from Eqs. (3.19) and (3.20) for $C_{11} = C_{22} = C$.

Consider now the plane-stress state of an orthotropic solid represented by a rheological model consisting of one Hookean and two Maxwell elements in parallel.

According to Eq. (3.9), we may write for the Maxwell elements the following relationships:

$$\frac{d\sigma_{(k)11}}{dt}+C_{(k)1111}\sigma_{(k)11}+C_{(k)1122}\sigma_{(k)22} = B_{(k)11\varkappa\varkappa}\frac{d\varepsilon_{\varkappa\varkappa}}{dt}, \tag{3.21}$$

$$\frac{d\sigma_{(k)22}}{dt}+C_{(k)2211}\sigma_{(k)11}+C_{(k)2222}\sigma_{(k)22} = B_{(k)22\varkappa\varkappa}\frac{d\varepsilon_{\varkappa\varkappa}}{dt}, \tag{3.22}$$

$$\frac{d\sigma_{(k)12}}{dt}+C_{(k)1212}\sigma_{(k)12} = B_{(k)1212}\frac{d\varepsilon_{12}}{dt}. \tag{3.23}$$

From Eq. (3.22), we have

$$\sigma_{(k)22} = e^{-C_{(k)2222}t}\left[\int_0^t\left(B_{(k)22\varkappa\varkappa}\frac{d\varepsilon_{\varkappa\varkappa}}{dt}-C_{(k)2211}\sigma_{11}\right)e^{C_{(k)2222}t}\,dt+\sigma_{(kA)22}\right] \tag{3.24}$$

and substituting it in Eq. (3.21), we obtain, after some rearrangements, the following linear differential equation of second order for the stress $\sigma_{(k)11}$:

$$\frac{d^2\sigma_{(k)11}}{dt^2}+(C_{(k)1111}+C_{(k)2222})\frac{d\sigma_{(k)11}}{dt}+(C_{(k)1111}C_{(k)2222}-C^2_{(k)1122})\sigma_{(k)11}$$

$$=(B_{(k)11\varkappa\varkappa}C_{(k)2222}-B_{(k)22\varkappa\varkappa}C_{(k)1122})\frac{d\varepsilon_{\varkappa\varkappa}}{dt}+B_{(k)11\varkappa\varkappa}\frac{d^2\varepsilon_{\varkappa\varkappa}}{dt^2}. \tag{3.25}$$

The solution to the above equation is

$$\sigma_{(k)11} = e^{-H_{(k)\mathrm{I}}} \int_0^t e^{(H_{(k)\mathrm{I}}-H_{(k)\mathrm{II}})t} \left\{ \int_0^t \left[(B_{(k)11\varkappa\varkappa} C_{(k)2222} - B_{(k)22\varkappa\varkappa} C_{(k)1122}) \frac{d\varepsilon_{\varkappa\varkappa}}{dt} + B_{(k)11\varkappa\varkappa} \frac{d^2\varepsilon_{\varkappa\varkappa}}{dt^2} \right] e^{H_{(k)\mathrm{II}}t}\, dt \right\} dt + c_{(k)\mathrm{I}} e^{-H_{(k)\mathrm{I}}t} + c_{(k)\mathrm{II}} e^{-H_{(k)\mathrm{II}}t}, \tag{3.26}$$

where

$$H_{(k)\mathrm{I},\,\mathrm{II}} = \tfrac{1}{2}(C_{(k)1111} + C_{(k)2222}) \pm \tfrac{1}{2}\sqrt{[(C_{(k)1111} - C_{(k)2222})^2 + 4C^2_{(k)1122}]} \tag{3.27}$$

and $C_{(k)\mathrm{I}}$, $C_{(k)\mathrm{II}}$ are constants of integration.

The resulting stress σ_{11} in the orthotropic viscoelastic solid in question comprises the component $\sigma_{(0)11}$ acting in the Hookean element and the components $\sigma_{(1)11}$ and $\sigma_{(2)11}$ acting in the Maxwell elements, according to the formula

$$\sigma_{11} = B_{(0)11\varkappa\varkappa}\varepsilon_{\varkappa\varkappa} + \sum_{k=1}^{2} \left(e^{-H_{(k)\mathrm{I}}t} \int_0^t e^{(H_{(k)\mathrm{I}}-H_{(k)\mathrm{II}})t} \left\{ \int_0^t \left[(B_{(k)11\varkappa\varkappa} C_{(k)2222} - B_{(k)22\varkappa\varkappa} C_{(k)1122}) \frac{d\varepsilon_{\varkappa\varkappa}}{dt} + B_{(k)11\varkappa\varkappa} \frac{d^2\varepsilon_{\varkappa\varkappa}}{dt^2} \right] e^{H_{(k)\mathrm{II}}t} dt \right\} dt + c_{(k)\mathrm{I}} e^{-H_{(k)\mathrm{I}}t} + c_{(k)\mathrm{II}} e^{-H_{(k)\mathrm{II}}t} \right). \tag{3.28}$$

Eliminating the integrals in Eq. (3.28) by successive differentiation, the author has obtained, after some rearrangements, the following linear differential equation of fourth order for the stress σ_{11}:

$$\begin{aligned}
&\frac{d^4\sigma_{11}}{dt^4} + (H_{(1)\mathrm{I}} + H_{(1)\mathrm{II}} + H_{(2)\mathrm{I}} + H_{(2)\mathrm{II}}) \frac{d^3\sigma_{11}}{dt^3} \\
&+ (H_{(1)\mathrm{I}}H_{(1)\mathrm{II}} + H_{(1)\mathrm{I}}H_{(2)\mathrm{I}} + H_{(1)\mathrm{I}}H_{(2)\mathrm{II}} + H_{(1)\mathrm{II}}H_{(2)\mathrm{I}} \\
&+ H_{(1)\mathrm{II}}H_{(2)\mathrm{II}} + H_{(2)\mathrm{I}}H_{(2)\mathrm{II}}) \frac{d^2\sigma_{11}}{dt^2} \\
&+ (H_{(1)\mathrm{I}}H_{(1)\mathrm{II}}H_{(2)\mathrm{I}} + H_{(1)\mathrm{I}}H_{(1)\mathrm{II}}H_{(2)\mathrm{II}} + H_{(1)\mathrm{I}}H_{(2)\mathrm{I}}H_{(2)\mathrm{II}} \\
&+ H_{(1)\mathrm{II}}H_{(2)\mathrm{I}}H_{(2)\mathrm{II}}) \frac{d\sigma_{11}}{dt} + H_{(1)\mathrm{I}}H_{(1)\mathrm{II}}H_{(2)\mathrm{I}}H_{(2)\mathrm{II}}\sigma_{11} \\
&= H_{(1)\mathrm{I}}H_{(1)\mathrm{II}}H_{(2)\mathrm{I}}H_{(2)\mathrm{II}}B_{(0)11\varkappa\varkappa}\varepsilon_{\varkappa\varkappa} \\
&+ [(H_{(1)\mathrm{I}}H_{(1)\mathrm{II}}H_{(2)\mathrm{I}} + H_{(1)\mathrm{I}}H_{(1)\mathrm{II}}H_{(2)\mathrm{II}} + H_{(1)\mathrm{I}}H_{(2)\mathrm{I}}H_{(2)\mathrm{II}} \\
&+ H_{(1)\mathrm{II}}H_{(2)\mathrm{I}}H_{(2)\mathrm{II}})B_{(0)11\varkappa\varkappa} + H_{(2)\mathrm{I}}H_{(2)\mathrm{II}}(B_{(1)11\varkappa\varkappa}C_{(1)2222} - B_{(1)22\varkappa\varkappa}C_{(1)1122}) \\
&+ (H_{(1)\mathrm{I}}H_{(1)\mathrm{II}} - H_{(2)\mathrm{I}}H_{(2)\mathrm{II}})(B_{(2)11\varkappa\varkappa}C_{(2)2222} - B_{(2)22\varkappa\varkappa}C_{(2)1122})] \frac{d\varepsilon_{\varkappa\varkappa}}{dt} \\
&+ [H_{(2)\mathrm{I}}H_{(2)\mathrm{II}}(B_{(0)11\varkappa\varkappa} + B_{(1)11\varkappa\varkappa} + B_{(2)11\varkappa\varkappa})
\end{aligned}$$

$$+(H_{(1)\mathrm{I}}H_{(1)\mathrm{II}}+H_{(1)\mathrm{I}}H_{(2)\mathrm{I}}+H_{(1)\mathrm{I}}H_{(2)\mathrm{II}}+H_{(1)\mathrm{II}}H_{(2)\mathrm{I}}+H_{(1)\mathrm{II}}H_{(2)\mathrm{II}})B_{(0)11\varkappa\varkappa}$$
$$+H_{(2)\mathrm{II}}(B_{(1)11\varkappa\varkappa}C_{(1)2222}-B_{(1)22\varkappa\varkappa}C_{(1)1122})+(H_{(1)\mathrm{I}}-H_{(2)\mathrm{I}}+H_{(1)\mathrm{II}})$$
$$\times(B_{(2)11\varkappa\varkappa}C_{(2)2222}-B_{(2)11\varkappa\varkappa}C_{(2)1122})+(H_{(1)\mathrm{I}}H_{(1)\mathrm{II}}$$
$$-H_{(2)\mathrm{I}}H_{(2)\mathrm{II}})B_{(2)11\varkappa\varkappa}]\frac{d^2\varepsilon_{\varkappa\varkappa}}{dt^2}$$
$$+[(H_{(2)\mathrm{I}}+H_{(2)\mathrm{II}})(B_{(0)11\varkappa\varkappa}+B_{(1)11\varkappa\varkappa}+B_{(2)11\varkappa\varkappa})$$
$$+(H_{(1)\mathrm{I}}+H_{(1)\mathrm{II}})B_{(0)11\varkappa\varkappa}+B_{(1)11\varkappa\varkappa}C_{(1)2222}-B_{(1)22\varkappa\varkappa}C_{(1)1122}$$
$$+B_{(2)11\varkappa\varkappa}C_{(2)2222}-B_{(2)22\varkappa\varkappa}C_{(2)1122}$$
$$+(H_{(1)\mathrm{I}}+H_{(1)\mathrm{II}}-H_{(2)\mathrm{I}}-H_{(2)\mathrm{II}})B_{(2)11\varkappa\varkappa}]\frac{d^3\varepsilon_{\varkappa\varkappa}}{dt^3}$$
$$+(B_{(0)11\varkappa\varkappa}+B_{(1)11\varkappa\varkappa}+B_{(2)11\varkappa\varkappa})\frac{d^4\varepsilon_{\varkappa\varkappa}}{dt^4}. \tag{3.29}$$

The equation for the stress σ_{22} obtainable by interchanging subscripts 1 and 2 in Eq. (3.29) is as follows:

$$\frac{d^4\sigma_{22}}{dt^4}+(H_{(1)\mathrm{I}}+H_{(1)\mathrm{II}}+H_{(2)\mathrm{I}}+H_{(2)\mathrm{II}})\frac{d^3\sigma}{dt^3}$$
$$+(H_{(1)\mathrm{I}}H_{(1)\mathrm{II}}+H_{(1)\mathrm{I}}H_{(2)\mathrm{I}}+H_{(1)\mathrm{I}}H_{(2)\mathrm{II}}+H_{(1)\mathrm{II}}H_{(2)\mathrm{I}}$$
$$+H_{(1)\mathrm{II}}H_{(2)\mathrm{II}}+H_{(2)\mathrm{I}}H_{(2)\mathrm{II}})\frac{d^2\sigma_{22}}{dt^2}$$
$$+(H_{(1)\mathrm{I}}H_{(1)\mathrm{II}}H_{(2)\mathrm{I}}+H_{(1)\mathrm{I}}H_{(1)\mathrm{II}}H_{(2)\mathrm{II}}+H_{(1)\mathrm{I}}H_{(2)\mathrm{I}}H_{(2)\mathrm{II}}$$
$$+H_{(1)\mathrm{II}}H_{(2)\mathrm{I}}H_{(2)\mathrm{II}})\frac{d\sigma_{22}}{dt}+H_{(1)\mathrm{I}}H_{(1)\mathrm{II}}H_{(2)\mathrm{I}}H_{(2)\mathrm{II}}\sigma_{22}$$
$$=H_{(1)\mathrm{I}}H_{(1)\mathrm{II}}H_{(2)\mathrm{I}}H_{(2)\mathrm{II}}B_{(0)22\varkappa\varkappa}\varepsilon_{\varkappa\varkappa}$$
$$+[(H_{(1)\mathrm{I}}H_{(1)\mathrm{II}}H_{(2)\mathrm{I}}+H_{(1)\mathrm{I}}H_{(1)\mathrm{II}}H_{(2)\mathrm{II}}+H_{(1)\mathrm{I}}H_{(2)\mathrm{I}}H_{(2)\mathrm{II}}$$
$$+H_{(1)\mathrm{II}}H_{(2)\mathrm{I}}H_{(2)\mathrm{II}})B_{(0)22\varkappa\varkappa}$$
$$+H_{(2)\mathrm{I}}H_{(2)\mathrm{II}}(B_{(1)22\varkappa\varkappa}C_{(1)1111}-B_{(1)11\varkappa\varkappa}C_{(1)1122})$$
$$+(H_{(1)\mathrm{I}}H_{(1)\mathrm{II}}-H_{(2)\mathrm{I}}H_{(2)\mathrm{II}})(B_{(2)22\varkappa\varkappa}C_{(2)1111}-B_{(2)11\varkappa\varkappa}C_{(2)1122})]\frac{d\varepsilon_{\varkappa\varkappa}}{dt}$$
$$+[H_{(2)\mathrm{I}}H_{(2)\mathrm{II}}(B_{(0)22\varkappa\varkappa}+B_{(1)22\varkappa\varkappa}+B_{(2)22\varkappa\varkappa})$$
$$+(H_{(1)\mathrm{I}}H_{(1)\mathrm{II}}+H_{(1)\mathrm{I}}H_{(2)\mathrm{I}}+H_{(1)\mathrm{I}}H_{(2)\mathrm{II}}+H_{(1)\mathrm{II}}H_{(2)\mathrm{I}}+H_{(1)\mathrm{II}}H_{(2)\mathrm{II}}$$
$$+H_{(2)\mathrm{I}}H_{(2)\mathrm{II}})B_{(0)22\varkappa\varkappa}+H_{(2)\mathrm{II}}(B_{(1)22\varkappa\varkappa}C_{(1)1111}-B_{(1)22\varkappa\varkappa}C_{(1)1122})$$
$$+(H_{(1)\mathrm{I}}-H_{(2)\mathrm{I}}+H_{(1)\mathrm{II}})(B_{(2)22\varkappa\varkappa}C_{(2)1111}-B_{(2)22\varkappa\varkappa}C_{(2)1122})$$
$$+(H_{(1)\mathrm{I}}H_{(1)\mathrm{II}}-H_{(2)\mathrm{I}}H_{(2)\mathrm{II}})B_{(2)22\varkappa\varkappa}]\frac{d^2\varepsilon_{\varkappa\varkappa}}{dt^2}$$

$$+[(H_{(2)\mathrm{I}}+H_{(2)\mathrm{II}})(B_{(0)22\varkappa\varkappa}+B_{(1)22\varkappa\varkappa}+B_{(2)22\varkappa\varkappa})$$
$$+(H_{(1)\mathrm{I}}+H_{(1)\mathrm{II}})\,B_{(0)22\varkappa\varkappa}+B_{(1)22\varkappa\varkappa}C_{(1)1111}$$
$$-B_{(1)22\varkappa\varkappa}C_{(1)1122}+B_{(2)22\varkappa\varkappa}C_{(2)1111}-B_{(2)11\varkappa\varkappa}C_{(2)1122}$$
$$+(H_{(1)\mathrm{I}}+H_{(1)\mathrm{II}}-H_{(2)\mathrm{I}}-H_{(2)\mathrm{II}})B_{(2)22\varkappa\varkappa}]\,\frac{d^3\varepsilon_{\varkappa\varkappa}}{dt^3}$$
$$+(B_{(0)22\varkappa\varkappa}+B_{(1)22\varkappa\varkappa}+B_{(2)22\varkappa\varkappa})\,\frac{d^4\varepsilon_{\varkappa\varkappa}}{dt^4}\,. \tag{3.30}$$

Equations (3.29) and (3.30) permit determination of stresses σ_{11} and σ_{22} from the given time-dependent strain components $\varepsilon_{\varkappa\varkappa}$. For determination of the strains ε_{11}, ε_{22}, ε_{33}, however, Eqs. (3.29) and (3.30) represent, together with the equation

$$B_{3311}\varepsilon_{11}+B_{3322}\varepsilon_{22}+B_{3333}\varepsilon_{33}=0 \tag{3.31}$$

resulting from the condition of the plane-stress state $\sigma_{33}=0$, a system of two simultaneous linear differential equations of fourth order and one algebraic equation.

Taking Eq. (3.23) into account, we get for the resulting shear stress the expression:

$$\sigma_{12}=B_{(0)1212}\varepsilon_{12}+\sum_{k=1}^{2}e^{-C_{(k)1212}t}\left(\int_0^t B_{(k)1212}\frac{d\varepsilon_{12}}{dt}\,e^{C_{(k)1212}t}dt+\sigma_{(kA)12}\right), \tag{3.32}$$

where $\sigma_{(kA)12}$ are the initial stresses in the Maxwell elements at time $t_0=0$. Eliminating the integrals by successive differentiation, we get the differential relationship

$$\frac{d^2\sigma_{12}}{dt^2}+(C_{(1)1212}+C_{(2)1212})\,\frac{d\sigma_{12}}{dt}+C_{(1)1212}C_{(2)1212}\sigma_{12}$$
$$=B_{(0)1212}C_{(1)212}C_{(2)1212}\varepsilon_{12}$$
$$+[B_{(0)1212}(C_{(1)212}+C_{(2)1212})+B_{(1)1212}C_{(2)1212}+B_{(2)1212}C_{(1)1212}]\,\frac{d\varepsilon_{12}}{dt}$$
$$+(B_{(0)1212}+B_{(1)1212}+B_{(2)1212})\,\frac{d^2\varepsilon_{12}}{dt^2}, \tag{3.33}$$

which permits direct determination of the shear stress from a given time-dependent shear strain, and of the shear strain from a given time-dependent shear stress.

4. RHEOLOGICAL BEHAVIOUR OF A COMPLEX VISCOELASTIC SOLID COMPRISING SEVERAL KELVIN ELEMENTS

Figure 3 represents the rheological model of a complex viscoelastic solid consisting of a Hookean spring and five Kelvin elements.

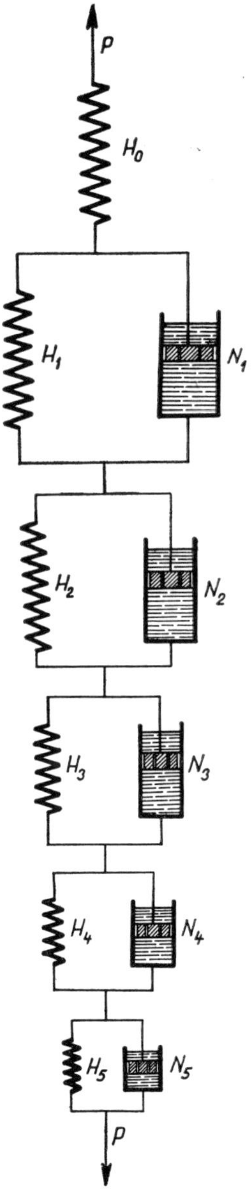

FIG. 3. Rheological model of a complex viscoelastic solid consisting of a Hookean spring and several Kelvin elements.

The resultant strain ε in the uniaxial state is the sum of the component

$$\varepsilon_0 = \frac{\sigma}{E_0} \tag{4.1}$$

acting in the Hookean spring and of n components

$$\varepsilon_k = e^{-(E_k t/\lambda_k)} \left(\int_0^t \frac{\sigma}{\lambda_k} e^{(E_k t/\lambda_k)}\, dt + \varepsilon_{kA} \right) \tag{4.2}$$

acting in the n Kelvin elements, where ε_{kA} denotes the initial strain at the time $t = 0$, which may be taken equal to zero.

Then, the total strain is given by

$$\varepsilon = \frac{\sigma}{E_0} + \sum_{k=1}^{n} e^{-(E_k t/\lambda_k)} \left(\int_0^t \frac{\sigma}{\lambda_k} e^{(E_k t/\lambda_k)}\, dt + \varepsilon_{kA} \right). \tag{4.3}$$

The integrals in the above equation may be eliminated by successive multiplication by $e^{(E_k t/\lambda_k)}$ and differentiation. The author has obtained for such a solid with five Kelvin elements the differential equation of fifth order:

$$\begin{aligned}
&\frac{d^5\varepsilon}{dt^5} + \left(\frac{E_1}{\lambda_1} + \frac{E_2}{\lambda_2} + \frac{E_3}{\lambda_3} + \frac{E_4}{\lambda_4} + \frac{E_5}{\lambda_5}\right) \frac{d^4\varepsilon}{dt^4} \\
&\quad + \left(\frac{E_1E_2}{\lambda_1\lambda_2} + \frac{E_1E_3}{\lambda_1\lambda_3} + \frac{E_1E_4}{\lambda_1\lambda_4} + \frac{E_1E_5}{\lambda_1\lambda_5} + \frac{E_2E_3}{\lambda_2\lambda_3}\right. \\
&\quad \left. + \frac{E_2E_4}{\lambda_2\lambda_4} + \frac{E_2E_5}{\lambda_2\lambda_5} + \frac{E_3E_4}{\lambda_3\lambda_4} + \frac{E_3E_5}{\lambda_3\lambda_5} + \frac{E_4E_5}{\lambda_4\lambda_5}\right) \frac{d^3\varepsilon}{dt^3} \\
&\quad + \left(\frac{E_1E_2E_3}{\lambda_1\lambda_2\lambda_3} + \frac{E_1E_2E_4}{\lambda_1\lambda_2\lambda_4} + \frac{E_1E_2E_5}{\lambda_1\lambda_2\lambda_5} + \frac{E_1E_3E_4}{\lambda_1\lambda_3\lambda_4} + \frac{E_1E_3E_5}{\lambda_1\lambda_3\lambda_5}\right. \\
&\quad \left. + \frac{E_2E_3E_4}{\lambda_2\lambda_3\lambda_4} + \frac{E_2E_3E_5}{\lambda_2\lambda_3\lambda_5} + \frac{E_2E_4E_5}{\lambda_2\lambda_4\lambda_5} + \frac{E_3E_4E_5}{\lambda_3\lambda_4\lambda_5}\right) \frac{d^2\varepsilon}{dt^2} \\
&\quad + \left(\frac{E_1E_2E_3E_4}{\lambda_1\lambda_2\lambda_3\lambda_4} + \frac{E_1E_2E_3E_5}{\lambda_1\lambda_2\lambda_3\lambda_5} + \frac{E_1E_2E_4E_5}{\lambda_1\lambda_2\lambda_4\lambda_5}\right. \\
&\quad \left. + \frac{E_1E_3E_4E_5}{\lambda_1\lambda_3\lambda_4\lambda_5} + \frac{E_2E_3E_4E_5}{\lambda_2\lambda_3\lambda_4\lambda_5}\right) \frac{d\varepsilon}{dt} + \frac{E_1E_2E_3E_4E_5}{\lambda_1\lambda_2\lambda_3\lambda_4\lambda_5}\,\varepsilon \\
&\quad = \left\{ \frac{1}{E_0} \cdot \frac{E_1E_2E_3E_4E_5}{\lambda_1\lambda_2\lambda_3\lambda_4\lambda_5} \right. \\
&\quad + \frac{E_2E_3E_4E_5}{\lambda_2\lambda_3\lambda_4\lambda_5} \left(\frac{1}{\lambda_1} + \frac{1}{\lambda_2} + \frac{1}{\lambda_3} + \frac{1}{\lambda_4} + \frac{1}{\lambda_5}\right) \\
&\quad + \frac{E_3E_4E_5}{\lambda_3\lambda_4\lambda_5} \left[\frac{1}{\lambda_2} \left(\frac{E_1}{\lambda_1} - \frac{E_2}{\lambda_2}\right) + \frac{1}{\lambda_3} \left(\frac{E_1}{\lambda_1} - \frac{E_3}{\lambda_3}\right) \right.
\end{aligned}$$

$$+\frac{1}{\lambda_4}\left(\frac{E_1}{\lambda_1}-\frac{E_4}{\lambda_4}\right)+\frac{1}{\lambda_5}\left(\frac{E_1}{\lambda_1}-\frac{E_5}{\lambda_5}\right)\Bigg]$$

$$+\frac{E_4E_5}{\lambda_4\lambda_5}\left[\frac{1}{\lambda_3}\left(\frac{E_1}{\lambda_1}-\frac{E_3}{\lambda_3}\right)+\frac{1}{\lambda_4}\left(\frac{E_1}{\lambda_1}-\frac{E_4}{\lambda_4}\right)+\frac{1}{\lambda_5}\left(\frac{E_1}{\lambda_1}-\frac{E_5}{\lambda_5}\right)\right]$$

$$+\frac{E_5}{\lambda_5}\left(\frac{E_1}{\lambda_1}-\frac{E_5}{\lambda_5}\right)\left(\frac{E_2}{\lambda_2}-\frac{E_5}{\lambda_5}\right)\left(\frac{E_3}{\lambda_3}-\frac{E_5}{\lambda_5}\right)\left(\frac{E_4}{\lambda_4}-\frac{E_5}{\lambda_5}\right)\Bigg\}\sigma$$

$$+\Bigg\{\frac{1}{E_0}\Bigg(\frac{E_1E_2E_3E_4}{\lambda_1\lambda_2\lambda_3\lambda_4}+\frac{E_1E_2E_3E_5}{\lambda_1\lambda_2\lambda_3\lambda_5}+\frac{E_1E_2E_4E_5}{\lambda_1\lambda_2\lambda_4\lambda_5}$$

$$+\frac{E_1E_3E_4E_5}{\lambda_1\lambda_3\lambda_4\lambda_5}+\frac{E_2E_3E_4E_5}{\lambda_2\lambda_3\lambda_4\lambda_5}\Bigg)+\Bigg(\frac{E_2E_3E_4}{\lambda_2\lambda_3\lambda_4}+\frac{E_2E_3E_5}{\lambda_2\lambda_3\lambda_5}$$

$$+\frac{E_2E_4E_5}{\lambda_2\lambda_4\lambda_5}+\frac{E_3E_4E_5}{\lambda_3\lambda_4\lambda_5}\Bigg)\left(\frac{1}{\lambda_1}+\frac{1}{\lambda_2}+\frac{1}{\lambda_3}+\frac{1}{\lambda_4}+\frac{1}{\lambda_5}\right)$$

$$+\left(\frac{E_3E_4}{\lambda_3\lambda_4}+\frac{E_3E_5}{\lambda_3\lambda_5}+\frac{E_4E_5}{\lambda_4\lambda_5}\right)\Bigg[\frac{1}{\lambda_2}\left(\frac{E_1}{\lambda_1}-\frac{E_2}{\lambda_2}\right)+\frac{1}{\lambda_3}\left(\frac{E_1}{\lambda_1}-\frac{E_3}{\lambda_3}\right)$$

$$+\frac{1}{\lambda_3}\left(\frac{E_1}{\lambda_1}-\frac{E_3}{\lambda_3}\right)+\frac{1}{\lambda_4}\left(\frac{E_1}{\lambda_1}-\frac{E_4}{\lambda_4}\right)+\frac{1}{\lambda_5}\left(\frac{E_1}{\lambda_1}-\frac{E_5}{\lambda_5}\right)\Bigg]$$

$$+\left(\frac{E_4}{\lambda_4}+\frac{E_5}{\lambda_5}\right)\Bigg[\frac{1}{\lambda_3}\left(\frac{E_1}{\lambda_1}-\frac{E_3}{\lambda_3}\right)\left(\frac{E_2}{\lambda_2}-\frac{E_3}{\lambda_3}\right)$$

$$+\frac{1}{\lambda_4}\left(\frac{E_1}{\lambda_1}-\frac{E_4}{\lambda_4}\right)\left(\frac{E_2}{\lambda_2}-\frac{E_4}{\lambda_4}\right)+\frac{1}{\lambda_5}\left(\frac{E_1}{\lambda_1}-\frac{E_5}{\lambda_5}\right)\left(\frac{E_2}{\lambda_2}-\frac{E_5}{\lambda_5}\right)\Bigg]$$

$$+\frac{1}{\lambda_4}\left(\frac{E_1}{\lambda_1}-\frac{E_4}{\lambda_4}\right)\left(\frac{E_2}{\lambda_2}-\frac{E_4}{\lambda_4}\right)\left(\frac{E_3}{\lambda_3}-\frac{E_4}{\lambda_4}\right)$$

$$+\frac{1}{\lambda_5}\left(\frac{E_1}{\lambda_1}-\frac{E_5}{\lambda_5}\right)\left(\frac{E_2}{\lambda_2}-\frac{E_5}{\lambda_5}\right)\left(\frac{E_3}{\lambda_3}-\frac{E_5}{\lambda_5}\right)\Bigg\}\frac{d\sigma}{dt}$$

$$+\Bigg\{\frac{1}{E_0}\Bigg(\frac{E_1E_2E_3}{\lambda_1\lambda_2\lambda_3}+\frac{E_1E_2E_4}{\lambda_1\lambda_2\lambda_4}+\frac{E_1E_2E_5}{\lambda_1\lambda_2\lambda_5}+\frac{E_1E_3E_4}{\lambda_1\lambda_3\lambda_4}$$

$$+\frac{E_1E_3E_5}{\lambda_1\lambda_3\lambda_5}+\frac{E_1E_4E_5}{\lambda_1\lambda_4\lambda_5}+\frac{E_2E_3E_4}{\lambda_2\lambda_3\lambda_4}+\frac{E_2E_3E_5}{\lambda_2\lambda_3\lambda_5}+\frac{E_3E_4E_5}{\lambda_3\lambda_4\lambda_5}\Bigg)$$

$$+\left(\frac{E_2E_3}{\lambda_2\lambda_3}+\frac{E_2E_4}{\lambda_2\lambda_4}+\frac{E_2E_5}{\lambda_2\lambda_5}+\frac{E_3E_4}{\lambda_3\lambda_4}+\frac{E_3E_5}{\lambda_3\lambda_5}+\frac{E_4E_5}{\lambda_4\lambda_5}\right)\Bigg(\frac{1}{\lambda_1}$$

$$+\frac{1}{\lambda_2}+\frac{1}{\lambda_3}+\frac{1}{\lambda_4}+\frac{1}{\lambda_5}\Bigg)+\left(\frac{E_3}{\lambda_3}+\frac{E_4}{\lambda_4}+\frac{E_5}{\lambda_5}\right)\Bigg[\frac{1}{\lambda_2}\left(\frac{E_1}{\lambda_1}-\frac{E_2}{\lambda_2}\right)$$

$$+\frac{1}{\lambda_3}\left(\frac{E_1}{\lambda_1}-\frac{E_3}{\lambda_3}\right)+\frac{1}{\lambda_4}\left(\frac{E_1}{\lambda_1}-\frac{E_4}{\lambda_4}\right)+\frac{1}{\lambda_5}\left(\frac{E_1}{\lambda_1}-\frac{E_5}{\lambda_5}\right)\Bigg]$$

$$+\frac{1}{\lambda_3}\left(\frac{E_1}{\lambda_1}-\frac{E_3}{\lambda_3}\right)\left(\frac{E_2}{\lambda_2}-\frac{E_3}{\lambda_3}\right)+\frac{1}{\lambda_4}\left(\frac{E_1}{\lambda_1}-\frac{E_4}{\lambda_4}\right)\left(\frac{E_2}{\lambda_2}-\frac{E_4}{\lambda_4}\right)$$

$$+\frac{1}{\lambda_5}\left(\frac{E_1}{\lambda_1}-\frac{E_5}{\lambda_5}\right)\left(\frac{E_2}{\lambda_2}-\frac{E_5}{\lambda_5}\right)\Bigg\}\frac{d^2\sigma}{dt^2}$$

$$+\left\{\frac{1}{E_0}\left(\frac{E_1E_2}{\lambda_1\lambda_2}+\frac{E_1E_3}{\lambda_1\lambda_3}+\frac{E_1E_4}{\lambda_1\lambda_4}+\frac{E_1E_5}{\lambda_1\lambda_5}+\frac{E_2E_3}{\lambda_2\lambda_3}\right.\right.$$

$$\left.+\frac{E_2E_4}{\lambda_2\lambda_4}+\frac{E_2E_5}{\lambda_2\lambda_5}+\frac{E_3E_4}{\lambda_3\lambda_4}+\frac{E_3E_5}{\lambda_3\lambda_5}+\frac{E_4E_5}{\lambda_4\lambda_5}\right)$$

$$+\left(\frac{E_2}{\lambda_2}+\frac{E_3}{\lambda_3}+\frac{E_4}{\lambda_4}+\frac{E_5}{\lambda_5}\right)\left(\frac{1}{\lambda_1}+\frac{1}{\lambda_2}+\frac{1}{\lambda_3}+\frac{1}{\lambda_4}+\frac{1}{\lambda_5}\right)$$

$$+\frac{1}{\lambda_2}\left(\frac{E_1}{\lambda_1}-\frac{E_2}{\lambda_2}\right)+\frac{1}{\lambda_3}\left(\frac{E_1}{\lambda_1}-\frac{E_3}{\lambda_3}\right)+\frac{1}{\lambda_4}\left(\frac{E_1}{\lambda_1}-\frac{E_4}{\lambda_4}\right)+\frac{1}{\lambda_5}\left(\frac{E_1}{\lambda_1}-\frac{E_5}{\lambda_5}\right)\Bigg\}\frac{d^3\sigma}{dt^3}$$

$$+\left\{\frac{1}{E_0}\left(\frac{E_1}{\lambda_1}+\frac{E_2}{\lambda_2}+\frac{E_3}{\lambda_3}+\frac{E_4}{\lambda_4}+\frac{E_5}{\lambda_5}\right)\right.$$

$$\left.+\frac{1}{\lambda_1}+\frac{1}{\lambda_2}+\frac{1}{\lambda_3}+\frac{1}{\lambda_4}+\frac{1}{\lambda_5}\right\}\frac{d^4\sigma}{dt^4}+\frac{1}{E_0}\frac{d^5\sigma}{dt^5}. \tag{4.4}$$

For the deviatoric and volumetric strain components at the triaxial state of the isotropic body represented by the rheological model on Fig. 3, we obtain from Eq. (4.3) the following relationships:

$$e_{ij}=\frac{s_{ij}}{2G_0}+\sum_{k=1}^{n}e^{-G_kt/\eta_k}\left(\int_0^t\frac{s_{ij}}{2\eta_k}e^{G_kt/\eta_k}\,dt+e_{(kA)ij}\right), \tag{4.5}$$

$$\varepsilon_m=\frac{\sigma_m}{K_0}+\sum_{k=1}^{n}e^{-K_kt/\zeta_k}\left(\int_0^t\frac{\sigma_m}{\zeta_k}e^{K_kt/\zeta_k}dt+\varepsilon_{(kA)m}\right). \tag{4.6}$$

The integrals in the preceding equations may be eliminated by quite a similar way as in Eq. (4.3) and we obtain, for the deviatoric and volumetric components, two differential relations analogous to Eq. (4.4).

Similarly, as shown in Section 3, there are three different types of rheological deformation of anisotropic bodies consisting of Kelvin elements.

In the first case, the retardation time is the same in all directions, and for a Kelvin element in the rheological model on Fig. 3, the following linear differential equation applies:

$$\frac{d\varepsilon_{(k)ij}}{dt}+D\varepsilon_{ij}=L_{(k)ij\varkappa\lambda}\sigma_{\varkappa\lambda}, \tag{4.7}$$

where $L_{(k)\,ij\varkappa\lambda}$ is the tensor of rank 4 of reciprocal viscosities, D the common reciprocal retardation time.

In a more general case, the time of retardation varies directionally and the rheological deformation of the kth Kelvin element follows from the equation:

$$\frac{d\varepsilon_{(k)ij}}{dt} + D_{(k)(ij)}\varepsilon_{(k)ij} = L_{(k)ij\varkappa\lambda}\sigma_{\varkappa\lambda}, \tag{4.8}$$

where $D_{(k)(ij)}$ are the reciprocal retardation times, different for each strain component $\varepsilon_{(k)ij}$.

Equation (4.8) has the following solution:

$$\varepsilon_{(k)ij} = e^{-D_{(k)(ij)}t}\left(\int_0^t L_{(k)ij\varkappa\lambda}\sigma_{\varkappa\lambda}e^{D_{(k)(ij)}t}dt + \varepsilon_{(kA)ij}\right), \tag{4.9}$$

where $\varepsilon_{(kA)ij}$ is the initial value of the strain $\varepsilon_{(k)ij}$ at time $t_0 = 0$.

For a viscoelastic solid represented by one Hookean and three Kelvin elements series, the resulting deformation is given by

$$\varepsilon_{ij} = A_{(0)ij\varkappa\lambda}\sigma_{\varkappa\lambda} + \sum_{k=1}^{3} e^{-D_{(k)(ij)}t}\left(\int_0^t L_{(k)ij\varkappa\lambda}\sigma_{\varkappa\lambda}e^{D_{(k)(ij)}t}dt + \varepsilon_{(kA)ij}\right), \tag{4.10}$$

where $A_{(0)ij\varkappa\lambda}$ is the tensor, of rank 4, of the reciprocal moduli of elasticity.

After eliminating the integrals by successive differentiation, we obtain from Eq. (4.10), after some rearrangements, the following differential relationship:

$$\begin{aligned}
&\frac{d^3\varepsilon_{ij}}{dt^3} + (D_{(1)(ij)} + D_{(2)(ij)} + D_{(3)(ij)})\frac{d^2\varepsilon_{ij}}{dt^2} \\
&\quad + (D_{(1)(ij)}D_{(2)(ij)} + D_{(1)(ij)}D_{(3)(ij)} + D_{(2)ij}D_{(3)(ij)})\frac{d\varepsilon_{ij}}{dt} \\
&\quad + D_{(1)(ij)}D_{(2)(ij)}D_{(3)(ij)}\varepsilon_{ij} \\
&\quad = \{A_{ij\varkappa\lambda}D_{(1)(ij)}D_{(2)(ij)}D_{(3)(ij)} + L_{(1)ij\varkappa\lambda}D_{(2)(ij)}D_{(3)(ij)} \\
&\quad + D_{(3)(ij)}[(D_{(1)(ij)} - D_{(2)(ij)})L_{(2)ij\varkappa\lambda} + (D_{(1)(ij)} - D_{(3)(ij)})L_{(3)ij\varkappa\lambda}] \\
&\quad + (D_{(1)(ij)} - D_{(3)(ij)})(D_{(2)(ij)} - D_{(3)(ij)})L_{(3)ij\varkappa\lambda}\}\,\sigma_{\varkappa\lambda} \\
&\quad + [A_{ij\varkappa\lambda}(D_{(1)(ij)}D_{(2)(ij)} + D_{(1)ij}D_{(3)ij} + D_{(2)(ij)}D_{(3)(ij)}) \\
&\quad + L_{(1)ij\varkappa\lambda}(D_{(2)(ij)} + D_{(3)(ij)}) + L_{(2)ij\varkappa\lambda}(D_{(1)(ij)} + D_{(3)(ij)}) + L_{(3)ij\varkappa\lambda}(D_{(1)(ij)} \\
&\quad + D_{(2)(ij)})]\frac{d\sigma_{\varkappa\lambda}}{dt} + [A_{ij\varkappa\lambda}(D_{(1)(ij)} + D_{(2)(ij)} + D_{(3)(ij)}) + L_{(1)ij\varkappa\lambda} + L_{(2)ij\varkappa\lambda} \\
&\quad + L_{(3)ij\varkappa\lambda}]\frac{d^2\sigma_{\varkappa\lambda}}{dt^2} + A_{ij\varkappa\lambda}\frac{d^3\sigma_{\varkappa\lambda}}{dt^3}.
\end{aligned} \tag{4.11}$$

This relation permits direct determination of the strain components ε_{ij} from the given time-dependent components of the stress tensor. If, however, the time-dependent strain components are given, Eq. (4.11) yields for $i, j = 1, 2, 3$ and $\varkappa, \lambda = 1, 2, 3$ a system of six linear differential equations of third order for the six components of the symmetrical stress tensor.

The system of simultaneous differential equations may be transformed into a single differential equation of higher order. Consider, for simplicity, the plane-stress problem of the orthotropic Poynting–Thompson body, represented by the rheological model in Fig. 4.

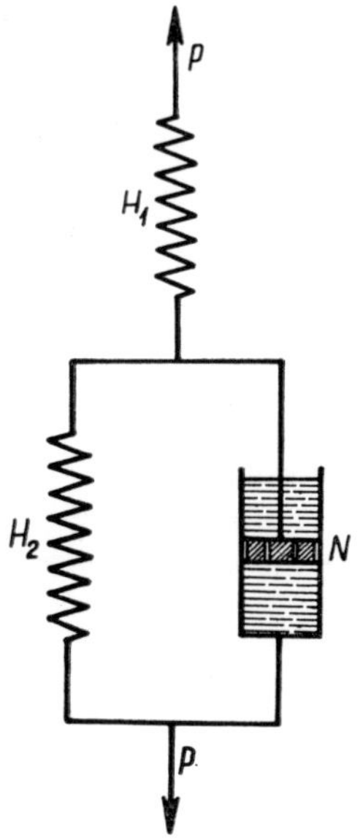

FIG. 4. Rheological model of the Poynting–Thompson body.

The tensor differential equation of this body has the following form:

$$\frac{d\varepsilon_{ij}}{dt}+D_{(ij)}\varepsilon_{ij} = (A_{ij\varkappa\lambda}D_{(ij)}+L_{ij\varkappa\lambda})\sigma_{\varkappa\lambda}+A_{ij\varkappa\lambda}\frac{d\sigma_{\varkappa\lambda}}{dt} \tag{4.12}$$

and yields, for the plane-stress state of the orthotropic body, three equations

$$\frac{d\varepsilon_{11}}{dt}+D_{(11)}\varepsilon_{11} = (A_{1111}D_{(11)}+L_{1111})\sigma_{11}+(A_{1122}D_{(11)}+L_{1122})\sigma_{22}$$

$$+A_{1111}\frac{d\sigma_{11}}{dt}+A_{1122}\frac{d\sigma_{22}}{dt}, \tag{4.13}$$

$$\frac{d\varepsilon_{22}}{dt}+D_{(22)}\varepsilon_{22} = (A_{2211}D_{(22)}+L_{2211})\sigma_{11}+(A_{2222}D_{(22)}+L_{2222})\sigma_{22}$$

$$+A_{2211}\frac{d\sigma_{11}}{dt}+A_{2222}\frac{d\sigma_{22}}{dt}, \tag{4.14}$$

$$\frac{d\varepsilon_{12}}{dt}+D_{(12)}\varepsilon_{12} = (A_{1212}D_{(12)}+L_{1212})\sigma_{12}+A_{1212}\frac{d\sigma_{12}}{dt}. \tag{4.15}$$

Equation (4.15) permits direct determination of the shear stress σ_{12} from the given time-dependent shear strain ε_{12}.

From Eq. (4.14) we obtain

$$\sigma_{22} = e^{-Rt}\left\{\int_0^t \frac{1}{A_{2222}}\left[\frac{d\varepsilon_{22}}{dt}+D_{(22)}\varepsilon_{22}-(A_{2211}D_{(22)}+L_{2211})\sigma_{11}\right.\right.$$

$$\left.\left.-A_{2211}\frac{d\sigma_{11}}{dt}\right] e^{Rt}dt+\sigma_{22(A)}\right\}, \tag{4.16}$$

where $R = D_{(22)}+\dfrac{L_{2222}}{A_{2222}}$

and $\sigma_{22(A)}$ is the initial stress in the Kelvin element at time $t_0 = 0$.

Substitution of Eq. (4.16) in Eq. (4.13) yields, after some rearrangements, the linear differential equation of second order for σ_{11}:

$$(A_{1111}A_{2222}-A^2_{1122})\frac{d^2\sigma_{11}}{dt^2}$$

$$+\{A_{1111}(A_{2222}D_{(22)}+L_{2222})+A_{2222}(A_{1111}D_{(11)}+L_{1111})$$

$$-A_{1122}[A_{1122}(D_{(11)}+D_{(22)})+2L_{1122}]\}\frac{d\sigma_{11}}{dt}$$

$$+[(A_{1111}D_{(11)}+L_{1111})(A_{2222}D_{(22)}+L_{2222})-(A_{1122}D_{(11)}+L_{1122})^2]\sigma_{11}$$

$$= (A_{2222}D_{(22)}+L_{2222})D_{(11)}\varepsilon_{11}$$

$$+[A_{2222}(D_{(11)}+D_{(22)})+L_{2222}]\frac{d\varepsilon_{11}}{dt}+A_{2222}\frac{d^2\varepsilon_{11}}{dt^2}$$

$$-(A_{1122}D_{(11)}+L_{1122})D_{(22)}\varepsilon_{22}$$

$$-[A_{1122}(D_{(11)}+D_{(22)})+L_{1122}]\frac{d\varepsilon_{22}}{dt}-A_{1122}\frac{d^2\varepsilon_{22}}{dt^2}. \tag{4.17}$$

The differential equation for σ_{22} is obtainable from Eq. (4.17) by interchanging subscripts 1 and 2 as follows:

$$(A_{1111}A_{2222}-A^2_{1122})\frac{d^2\sigma_{22}}{dt^2}$$

$$+\{A_{1111}(A_{2222}D_{(22)}+L_{2222})+A_{2222}(A_{1111}D_{(11)}+L_{1111})$$
$$-A_{1122}[A_{1122}(D_{(11)}+D_{(22)})+2L_{1122}]\frac{d\sigma_{22}}{dt}$$
$$+[(A_{1111}D_{(11)}+L_{1111})(A_{2222}D_{(22)}+L_{2222})-(A_{1122}D_{(22)}+L_{1122})^2]\sigma_{22}$$
$$= -(A_{1122}D_{(22)}+L_{1122})D_{(11)}\varepsilon_{11}$$
$$-[A_{1122}(D_{(11)}+D_{(22)})+L_{1122}]\frac{d\varepsilon_{11}}{dt}-A_{1122}\frac{d^2\varepsilon_{11}}{dt^2}$$
$$+(A_{1111}D_{(11)}+L_{1111})D_{(22)}\varepsilon_{22}$$
$$+[A_{1111}(D_{(11)}+D_{(22)})+L_{1111}]\frac{d\varepsilon_{22}}{dt}+A_{1111}\frac{d^2\varepsilon_{22}}{dt^2}. \tag{4.18}$$

As an example of the general case, consider the plane-strain state of the orthotropic viscoelastic solid represented by a rheological model consisting of one Hookean and two Kelvin elements in series.

The general tensor equation for the kth Kelvin element of the complex anisotropic body has the form:

$$\frac{d\varepsilon_{(k)ij}}{dt}+D_{(k)ij\mu\nu}\varepsilon_{(k)\mu\nu} = L_{(k)ij\varkappa\lambda}\sigma_{\varkappa\lambda} \tag{4.19}$$

where $D_{(k)ij\mu\nu}$ is a tensor of rank 4 of the reciprocal retardation times.

Accordingly, for the plane-strain state of the orthotropic body, we have three equations

$$\frac{d\varepsilon_{(k)11}}{dt}+D_{(k)1111}\varepsilon_{(k)11}+D_{(k)1122}\varepsilon_{(k)22} = L_{(k)11\varkappa\varkappa}\sigma_{\varkappa\varkappa}, \tag{4.20}$$

$$\frac{d\varepsilon_{(k)22}}{dt}+D_{(k)2211}\varepsilon_{(k)11}+D_{(k)2222}\varepsilon_{(k)22} = L_{(k)22\varkappa\varkappa}\sigma_{\varkappa\varkappa}, \tag{4.21}$$

$$\frac{d\varepsilon_{(k)12}}{dt}+D_{(k)1212}\varepsilon_{(k)12} = L_{(k)1212}\sigma_{12}. \tag{4.22}$$

From Eq. (4.21) we obtain

$$\varepsilon_{(k)22}= e^{-D_{(k)2222}t}\left[\int_0^t (L_{(k)22\varkappa\varkappa}\sigma_{\varkappa\varkappa}-D_{(k)2211}\varepsilon_{(k)11})e^{D_{(k)2222}t}dt+\varepsilon_{(kA)22}\right], \tag{4.23}$$

where $\varepsilon_{(kA)22}$ is the initial strain at time $t_0 = 0$. Substituting in Eq. (4.20), we obtain for the strain the linear differential equation of second order:

$$\frac{d^2\varepsilon_{(k)11}}{dt^2}+(D_{(k)1111}+D_{(k)2222})\frac{d\varepsilon_{(k)11}}{dt}+(D_{(k)1111}D_{(k)2222}-D^2_{(k)1122})\varepsilon_{(k)11}$$
$$= \left[(L_{(k)11\varkappa\varkappa}D_{(k)2222}-L_{(k)22\varkappa\varkappa}D_{(k)1122})\sigma_{\varkappa\varkappa}+L_{(k)11\varkappa\varkappa}\frac{d\sigma_{\varkappa\varkappa}}{dt}\right]. \tag{4.24}$$

The solution to the above equation is

$$\varepsilon_{(k)11} = \frac{e^{-S_{(k)I}t}}{S_{(k)II}-S_{(k)I}} \int_0^t \Big[(L_{(k)11\varkappa\varkappa}D_{(k)2222}-L_{(k)22\varkappa\varkappa}D_{(k)1122})\sigma_{\varkappa\varkappa}$$

$$+L_{(k)11\varkappa\varkappa}\frac{d\sigma_{\varkappa\varkappa}}{dt}\Big] e^{S_{(k)I}t}dt+\frac{e^{-S_{(k)II}t}}{S_{(k)I}-S_{(k)II}}\int_0^t \Big[(L_{(k)11\varkappa\varkappa}D_{(k)2222}$$

$$-L_{(k)22\varkappa\varkappa}D_{(k)1122})\sigma_{\varkappa\varkappa}+L_{(k)11\varkappa\varkappa}\frac{d\sigma_{\varkappa\varkappa}}{dt}\Big] e^{S_{(k)II}t}dt+c_{(k)I}e^{-S_{(k)I}t}+c_{(k)II}e^{-S_{(k)II}t} \tag{4.25}$$

where

$$S_{(k)\mathrm{I},\,\mathrm{II}} = \tfrac{1}{2}\,[D_{(k)1111}+D_{(k)2222}\pm\sqrt{\{(D_{(k)1111}-D_{(k)2222})^2+4D^2_{(k)1122}\}}] \tag{4.26}$$

and $c_{(k)\mathrm{I}}$, $c_{(k)\mathrm{II}}$ are the constants of integration.

The resultant strain ε_{11} in the orthotropic viscoelastic solid in question consists of the strain $\varepsilon_{(0)11}$ of the Hookean element and $\varepsilon_{(1)11}$ and $\varepsilon_{(2)11}$ of the Kelvin elements:

$$\varepsilon_{11} = A_{11\varkappa\varkappa}\sigma_{\varkappa\varkappa}$$

$$+\sum_{k=1}^{2}\Big\{\frac{e^{-S_{(k)\mathrm{I}}t}}{S_{(k)\mathrm{II}}-S_{(k)\mathrm{I}}}\int_0^t\Big[(L_{(k)11\varkappa\varkappa}D_{(k)2222}-L_{(k)22\varkappa\varkappa}D_{(k)1122})\,\sigma_{\varkappa\varkappa}$$

$$+L_{(k)11\varkappa\varkappa}\frac{d\sigma_{\varkappa\varkappa}}{dt}\Big]e^{S_{(k)\mathrm{I}}t}\,dt+\frac{e^{-S_{(k)\mathrm{II}}t}}{S_{(k)\mathrm{I}}-S_{(k)\mathrm{II}}}\int_0^t\Big[(L_{(k)11\varkappa\varkappa}D_{(k)2222}$$

$$-L_{(k)22\varkappa\varkappa}D_{(k)1122})\,\sigma_{\varkappa\varkappa}$$

$$+L_{(k)11\varkappa\varkappa}\frac{d\sigma_{\varkappa\varkappa}}{dt}\Big]e^{S_{(k)\mathrm{II}}t}\,dt+c_{(k)\mathrm{I}}e^{-S_{(k)\mathrm{I}}t}+c_{(k)\mathrm{I}}e^{-S_{(k)\mathrm{II}}t}\Big\}. \tag{4.27}$$

Eliminating the integrals in the preceding expression by successive differentiation, we obtain, after some rearrangements, for the strain ε_{11} the linear differential equation of fourth order

$$\frac{d^4\varepsilon_{11}}{dt^4}+(S_{(1)\mathrm{I}}+S_{(1)\mathrm{II}}+S_{(2)\mathrm{I}}+S_{(2)\mathrm{II}})\frac{d^3\varepsilon_{11}}{dt^3}$$

$$+(S_{(1)\mathrm{I}}S_{(1)\mathrm{II}}+S_{(1)\mathrm{I}}S_{(2)\mathrm{I}}+S_{(1)\mathrm{I}}S_{(2)\mathrm{II}}+S_{(1)\mathrm{II}}S_{(2)\mathrm{I}}$$

$$+S_{(1)\mathrm{II}}S_{(2)\mathrm{II}}+S_{(2)\mathrm{I}}S_{(2)\mathrm{II}}))\frac{d^2\varepsilon_{11}}{dt^2}$$

$$+(S_{(1)\mathrm{I}}S_{(1)\mathrm{II}}S_{(2)\mathrm{II}}+S_{(1)\mathrm{I}}S_{(1)\mathrm{II}}S_{(2)\mathrm{II}}+S_{(1)\mathrm{I}}S_{(2)\mathrm{I}}S_{(2)\mathrm{II}}$$

$$+S_{(1)\mathrm{II}}S_{(2)\mathrm{I}}S_{(2)\mathrm{II}})\frac{d\varepsilon_{11}}{dt}+S_{(1)\mathrm{I}}S_{(1)\mathrm{II}}S_{(2)\mathrm{I}}S_{(2)}\varepsilon_{11}$$

$$= [S_{(1)\mathrm{I}}S_{(1)\mathrm{II}}S_{(2)\mathrm{I}}S_{(2)\mathrm{II}}A_{11\kappa\kappa}+S_{(2)\mathrm{I}}S_{(2)\mathrm{II}}(L_{(1)11\kappa\kappa}D_{(1)2222}$$
$$-L_{(1)22\kappa\kappa}D_{(1)1122})+(S_{(1)\mathrm{I}}S_{(1)\mathrm{II}}-S_{(2)\mathrm{I}}S_{(2)\mathrm{II}})(L_{(2)11\kappa\kappa}D_{(2)2222}$$
$$-L_{(2)22\kappa\kappa}D_{(2)1122})]\sigma_{\kappa\kappa}$$
$$+ [(S_{(1)\mathrm{I}}S_{(1)\mathrm{II}}S_{(2)\mathrm{I}}+S_{(1)\mathrm{I}}S_{(1)\mathrm{II}}S_{(2)\mathrm{II}}+S_{(1)\mathrm{I}}S_{(2)\mathrm{I}}S_{(2)\mathrm{II}}+S_{(1)\mathrm{II}}S_{(2)\mathrm{I}}S_{(2)\mathrm{II}})A_{11\kappa\kappa}$$
$$+S_{(2)\mathrm{I}}S_{(2)\mathrm{II}}(L_{(1)11\kappa\kappa}+L_{(2)11\kappa\kappa})+S_{(2)\mathrm{II}}(L_{(1)11\kappa\kappa}D_{(1)2222}-L_{(1)22\kappa\kappa}D_{(1)1122})$$
$$+(S_{(1)\mathrm{I}}-S_{(2)\mathrm{I}}+S_{(1)\mathrm{II}})(L_{(2)11\kappa\kappa}D_{(2)2222}-L_{(2)11\kappa\kappa}D_{(2)1122})$$
$$+(S_{(1)\mathrm{I}}S_{((1)\mathrm{II}}-S_{(2)\mathrm{I}}S_{(2)\mathrm{II}})\,L_{(2)11\kappa\kappa}]\,\frac{d\sigma_{\kappa\kappa}}{dt}$$
$$+[(S_{(1)\mathrm{I}}S_{(1)\mathrm{II}}+S_{(1)\mathrm{I}}S_{(2)\mathrm{I}}+S_{(1)\mathrm{I}}S_{(2)\mathrm{II}}+S_{(1)\mathrm{II}}S_{(2)\mathrm{I}}+S_{(1)\mathrm{II}}S_{(2)\mathrm{II}}$$
$$+S_{(2)\mathrm{I}}S_{(2)\mathrm{II}})A_{11\kappa\kappa}+(S_{(2)\mathrm{I}}+S_{(2)\mathrm{II}})(L_{(1)11\kappa\kappa}+L_{(2)11\kappa\kappa})$$
$$+L_{(1)11\kappa\kappa}D_{(1)2222}-L_{(1)22\kappa\kappa}D_{(1)1122}+L_{(2)11\kappa\kappa}D_{(2)2222}-L_{(2)22\kappa\kappa}D_{(2)1122}$$
$$+(S_{(1)\mathrm{I}}+S_{(1)\mathrm{II}}-S_{(2)\mathrm{I}}-S_{(2)\mathrm{II}})\,L_{(2)11\kappa\kappa}]\,\frac{d^2\sigma_{\kappa\kappa}}{dt^2}$$
$$+[(S_{(1)\mathrm{I}}+S_{(1)\mathrm{II}}+S_{(2)\mathrm{I}}+S_{(2)\mathrm{II}})\,A_{11\kappa\kappa}+L_{(1)11\kappa\kappa}+L_{(2)11\kappa\kappa}]\,\frac{d^3\sigma_{\kappa\kappa}}{dt^3}$$
$$+A_{11\kappa\kappa}\frac{d^4\sigma_{\kappa\kappa}}{dt^4}\,. \tag{4.28}$$

The equation for the strain ε_{22} has quite a similar form. It is obtainable from Eq. (4.28) by interchanging subscripts 1 and 2.

The resultant shear strain of the body is given by

$$\varepsilon_{12} = A_{1212}\sigma_{12}+\sum_{k=1}^{2} e^{-D_{1212}t}\left(\int_0^t L_{(k)1212}\sigma_{12}e^{D_{1212}t}\,dt+\varepsilon_{(kA)12}\right), \tag{4.29}$$

where $\varepsilon_{(kA)12}$ are the initial stresses in the Kelvin elements at time $t_0 = 0$. Eliminating the integrals by successive differentiation, we obtain from Eq. (4.29):

$$\frac{d^2\varepsilon_{12}}{dt^2}+(D_{(1)1212}+D_{(2)1212})\frac{d\varepsilon_{12}}{dt}+D_{(1)1212}D_{(2)1212}\varepsilon_{12}$$
$$= (A_{1212}D_{(1)1212}D_{(2)1212}+L_{(1)1212}D_{(2)1212}+L_{(2)1212}D_{(1)1212})\,\sigma_{12}$$
$$+[A_{1212}(D_{(1)1212}+D_{(2)1212})+L_{(1)1212}+L_{(2)1212}]\frac{d\sigma_{12}}{dt}+A_{1212}\frac{d^2\sigma_{12}}{dt^2}. \tag{4.30}$$

This equation permits direct determination of the shear strain from a given time-dependent shear stress, and the shear stress from a given time-dependent shear strain.

The order of the time-dependent stress–strain relationships increases with the number of elements, and with that of tensor components of reciprocal moduli of elasticity, of reciprocal retardation times and of reciprocal coefficients of viscosity.

5. VISCOELASTIC SOLIDS WITH VARIABLE RHEOLOGICAL PARAMETERS

The rheological parameters of many viscoelastic solids vary with time during ageing, because of structural changes due to long-time chemical reactions or to external long-time loading. The increase in moduli of elasticity and coefficients of viscosity of concrete through long-time hydration, or through structural transformation accompanied by volume changes during loading is well known. Other examples of changes in rheological parameters have been observed in plastics.

In order to analyse the variability of the rheological properties of materials, the author [18], [19], [22] introduced rheological models with variable parameters. For such bodies, differential relationships with variable coefficients are obtained.

Consider a viscoelastic solid represented by one Hookean and three Maxwell elements in parallel, with time-variable moduli of elasticity $E_k(t)$ and coefficients of viscosity $\lambda_k(t)$. The rheological equation of the kth Maxwell element has the following form:

$$\varepsilon = \frac{\sigma_k}{E_k} + \int_0^t \frac{\sigma_k}{\lambda_k}\, dt, \tag{5.1}$$

from which we obtain, for the stress σ_k in the kth element, the linear differential equation of first order with variable coefficients:

$$\frac{d}{dt}\left(\frac{\sigma_k}{E_k}\right) + \frac{\sigma_k}{\lambda_k} = \frac{d\varepsilon}{dt} \tag{5.2}$$

which has the solution

$$\frac{\sigma_k}{E_k} = e^{-\int_0^t \frac{E_k}{\lambda_k} dt}\left(\int_0^t \frac{d\varepsilon}{dt} e^{\int_0^t \frac{E_k}{\lambda_k} dt}\, dt + \sigma_{kA}\right), \tag{5.3}$$

where σ_{kA} is the initial stress in the Maxwell element in question.

The resultant stress for the whole viscoelastic solid is

$$\sigma = E_0\varepsilon + \sum_{k=1}^{3} E_k e^{-\int_0^t \frac{E_k}{\lambda_k} dt}\left(\int_0^t \frac{d\varepsilon}{dt} e^{\int_0^t \frac{E_k}{\lambda_k} dt}\, dt + \sigma_{kA}\right). \tag{5.4}$$

Eliminating the integrals by successive differentiation, we obtain the following differential equation of third order with variable coefficients:

$$\frac{1}{A_1 A_3}\frac{d^3}{dt^3}\left(\frac{\sigma}{E_1}\right) + \left[\frac{1}{A_2 A_3}\left(\frac{E_1}{\lambda_1} + \frac{E_2}{\lambda_2} + \frac{E_3}{\lambda_3}\right) + \frac{2}{A_3}\frac{d}{dt}\left(\frac{1}{A_2}\right)\right.$$

$$+\frac{1}{A_2}\frac{d}{dt}\left(\frac{1}{A_3}\right)\Big]\frac{d^2}{dt^2}\left(\frac{\sigma}{E_1}\right)$$

$$+\Big\{\frac{1}{A_2A_3}\left[\frac{E_1E_2}{\lambda_1\lambda_2}+\frac{E_1E_3}{\lambda_1\lambda_3}+\frac{E_2E_3}{\lambda_2\lambda_3}+2\frac{d}{dt}\left(\frac{E_1}{\lambda_1}\right)+\frac{d}{dt}\left(\frac{E_2}{\lambda_2}\right)\right]$$

$$+\frac{1}{A_3}\left(\frac{2E_1}{\lambda_1}+\frac{E_2}{\lambda_2}+\frac{E_3}{\lambda_3}\right)\frac{d}{dt}\left(\frac{1}{A_2}\right)+\frac{1}{A_3}\frac{d^2}{dt^2}\left(\frac{1}{A_2}\right)$$

$$+\left[\frac{1}{A_2}\left(\frac{E_1}{\lambda_1}+\frac{E_2}{\lambda_2}\right)+\frac{d}{dt}\left(\frac{1}{A_2}\right)\right]\frac{d}{dt}\left(\frac{1}{A_3}\right)\Big\}\frac{d}{dt}\left(\frac{\sigma}{E_1}\right)$$

$$+\Big\{\frac{1}{A_2A_3}\Big[\frac{E_1E_2E_3}{\lambda_1\lambda_2\lambda_3}+\frac{E_1}{\lambda_1}\frac{d}{dt}\left(\frac{E_2}{\lambda_2}\right)+\frac{E_2}{\lambda_2}\frac{d}{dt}\left(\frac{E_1}{\lambda_1}\right)+\frac{E_3}{\lambda_3}\frac{d}{dt}\left(\frac{E_1}{\lambda_1}\right)$$

$$+\frac{d^2}{dt^2}\left(\frac{E_1}{\lambda_1}\right)\Big]+\frac{1}{A_3}\frac{d}{dt}\left(\frac{1}{A_2}\right)\left[\frac{E_1E_2}{\lambda_1\lambda_2}+\frac{E_1E_3}{\lambda_1\lambda_3}+2\frac{d}{dt}\left(\frac{E_1}{\lambda_1}\right)\right]$$

$$+\frac{1}{A_2}\frac{d}{dt}\left(\frac{1}{A_3}\right)\left[\frac{E_1E_2}{\lambda_1\lambda_2}+\frac{d}{dt}\left(\frac{E_1}{\lambda_1}\right)\right]+\frac{E_1}{\lambda_1}\frac{d}{dt}\left(\frac{1}{A_2}\right)\frac{d}{dt}\left(\frac{1}{A_3}\right)$$

$$+\frac{1}{A_3}\frac{E_1}{\lambda_1}\frac{d^2}{dt^2}\left(\frac{1}{A_2}\right)\Big\}\frac{\sigma}{E_1}$$

$$=\Big(\Big\{\frac{1}{A_2A_3}\frac{E_2E_3}{\lambda_2\lambda_3}+\frac{1}{A_3}\frac{E_3}{\lambda_3}\frac{d}{dt}\left(\frac{1}{A_2}\right)+\frac{1}{A_3}\Big[\frac{1}{A_2}\frac{d}{dt}\left(\frac{E_2}{\lambda_2}\right)$$

$$+\frac{E_2}{\lambda_2}\frac{d}{dt}\left(\frac{1}{A_2}\right)+\frac{d^2}{dt^2}\left(\frac{1}{A_2}\right)\Big]+\left[\frac{1}{A_2}\frac{E_2}{\lambda_2}+\frac{d}{dt}\left(\frac{1}{A_2}\right)\right]\frac{d}{dt}\left(\frac{1}{A_3}\right)\Big\}\Big[\frac{E_0}{\lambda_1}$$

$$+\frac{d}{dt}\left(\frac{E_0}{E_1}\right)\Big]$$

$$+\left[\frac{1}{A_2A_3}\frac{E_3}{\lambda_3}+\frac{1}{A_2}\frac{d}{dt}\left(\frac{1}{A_3}\right)+\frac{1}{A_3}\frac{d}{dt}\left(\frac{1}{A_2}\right)\right]\Big[\frac{E_1}{\lambda_1}\frac{d}{dt}\left(\frac{E_0}{E_1}\right)$$

$$+\frac{E_0}{E_1}\frac{d}{dt}\left(\frac{E_1}{\lambda_1}\right)+\frac{d^2}{dt^2}\left(\frac{E_0}{E_1}\right)\Big]$$

$$+\frac{1}{A_2A_3}\left[\frac{E_1}{\lambda_1}\frac{d^2}{dt^2}\left(\frac{E_0}{E_1}\right)+2\frac{d}{dt}\left(\frac{E_1}{\lambda_1}\right)\frac{d}{dt}\left(\frac{E_0}{E_1}\right)+\frac{E_0}{E_1}\frac{d^2}{dt^2}\left(\frac{E_1}{\lambda_1}\right)+\frac{d^3}{dt^3}\left(\frac{E_0}{E_1}\right)\right]\Big)\ \varepsilon$$

$$+\Big(1+\frac{1}{A_3}\Big\{\left[\frac{E_2}{A_2\lambda_2}+\frac{d}{dt}\left(\frac{1}{A_2}\right)\right]\left[\frac{E_0}{\lambda_1}+\frac{d}{dt}\left(\frac{E_0}{E_1}\right)\right]+\frac{1}{A_2}\Big[\frac{d^2}{dt^2}\left(\frac{E_0}{E_1}\right)$$

$$+\frac{E_1}{\lambda_1}\frac{d}{dt}\left(\frac{E_0}{E_1}\right)+\frac{E_0}{E_1}\frac{d}{dt}\left(\frac{E_1}{\lambda_1}\right)\Big]\Big\}$$

$$+\Big[\frac{1}{A_2A_3}\frac{E_2}{\lambda_2}\frac{E_3}{\lambda_3}+\frac{3}{A_3}\frac{E_3}{\lambda_3}\frac{d}{dt}\left(\frac{1}{A_2}\right)+\frac{1}{A_2}\frac{E_2}{\lambda_2}\frac{d}{dt}\left(\frac{1}{A_3}\right)+\frac{d}{dt}\left(\frac{1}{A_2}\right)\frac{d}{dt}\Big)\Big(\frac{1}{A_3}$$

$$+\frac{1}{A_3}\frac{d^2}{dt^2}\Big(\frac{1}{A_2}\Big)\Big]\Big(1+\frac{E_0}{E_1}+\frac{E_2}{E_1}+\frac{E_3}{E_1}\Big)+\Big[\frac{1}{A_2A_3}\frac{E_2}{\lambda_2}+\frac{1}{A_3}\frac{d}{dt}\Big(\frac{1}{A_2}\Big)\Big]\frac{d}{dt}\Big(\frac{E_0}{E_1}$$

$$+\frac{E_2}{E_1}+\frac{E_3}{E_1}\Big)+\Big[\frac{1}{A_2A_3}\frac{E_1}{\lambda_1}\frac{E_3}{\lambda_3}+\frac{1}{A_2}\frac{E_1}{\lambda_1}\frac{d}{dt}\Big(\frac{1}{A_3}\Big)$$

$$+\frac{1}{A_3}\frac{E_1}{\lambda_1}\frac{d}{dt}\Big(\frac{1}{A_2}\Big)\Big]\Big(\frac{E_0}{E_1}+\frac{E_3}{E_1}\Big)+\frac{1}{A_2A_3}\frac{d}{dt}\Big[\frac{E_1}{\lambda_1}\Big(\frac{E_0}{E_1}+\frac{E_3}{E_1}\Big)\Big]$$

$$+\Big[\frac{1}{A_2A_3}\frac{E_3}{\lambda_3}+\frac{1}{A_2}\frac{d}{dt}\Big(\frac{1}{A_3}\Big)+\frac{1}{A_3}\frac{E_1}{\lambda_1}\frac{d}{dt}\Big(\frac{1}{A_2}\Big)\Big]\frac{d}{dt}\Big(\frac{2E_0}{E_1}+\frac{E_2}{E_1}+\frac{2E_3}{E_1}\Big)$$

$$+\frac{1}{A_2A_3}\frac{d^2}{dt^2}\Big(\frac{2E_0}{E_1}+\frac{E_2}{E_1}+\frac{2E_3}{E_1}\Big)-\frac{E_3}{E_1}\Big[\frac{1}{A_2A_3}\Big(\frac{E_3}{\lambda_3}\Big)^2+\frac{1}{A_2}\frac{d}{dt}\Big(\frac{1}{A_3}\Big)$$

$$+\frac{1}{A_2}\frac{E_3}{\lambda_3}\frac{d}{dt}\Big(\frac{1}{A_2}\Big)\Big]+\frac{1}{A_2A_3}\frac{d}{dt}\Big(\frac{E_3}{\lambda_3}\Big)-\frac{1}{A_2A_3}\frac{E_3}{\lambda_3}\frac{d}{dt}\Big(\frac{E_3}{E_1}\Big)\Big)\frac{d\varepsilon}{dt}$$

$$+\Big\{\Big[\frac{1}{A_2A_3}\Big(\frac{E_2}{\lambda_2}+\frac{E_3}{\lambda_3}\Big)+\frac{2}{A_3}\frac{d}{dt}\Big(\frac{1}{A_2}\Big)+\frac{1}{A_2}\frac{d}{dt}\Big(\frac{1}{A_3}\Big)\Big]\Big(1+\frac{E_0}{E_1}+\frac{E_2}{E_1}+\frac{E_3}{E_1}\Big)$$

$$+\frac{1}{A_2A_3}\Big[\frac{E_1}{\lambda_1}\Big(\frac{E_0}{E_1}+\frac{E_3}{E_1}\Big)-\frac{E_3}{\lambda_3}\frac{E_3}{E_1}\Big]+\frac{1}{A_3}+\frac{1}{A_2A_3}\frac{d}{dt}\Big(\frac{2E_0}{E_1}+\frac{E_2}{E_1}+\frac{2E_3}{E_1}\Big)\Big\}\frac{d^2\varepsilon}{dt^2}$$

$$+\frac{1}{A_2A_3}\Big(1+\frac{E_0}{E_1}+\frac{E_2}{E_1}+\frac{E_3}{E_1}\Big)\frac{d^3\varepsilon}{dt^3}, \tag{5.5}$$

where

$$A_2=\frac{E_2}{E_1}\Big(\frac{E_1}{\lambda_1}-\frac{E_2}{\lambda_2}\Big)+\frac{d}{dt}\Big(\frac{E_2}{E_1}\Big),$$

$$A_3=\Big[\frac{1}{A_2}\Big(\frac{E_2}{\lambda_2}-\frac{E_3}{\lambda_3}\Big)+\frac{d}{dt}\Big(\frac{1}{A_2}\Big)\Big]\Big[\frac{E_3}{E_1}\Big(\frac{E_1}{\lambda_1}-\frac{E_3}{\lambda_3}\Big)+\frac{d}{dt}\Big(\frac{E_3}{E_1}\Big)\Big]$$

$$+\frac{1}{A_2}\Big\{\Big(\frac{E_1}{\lambda_1}-\frac{E_3}{\lambda_3}\Big)\frac{d}{dt}\Big(\frac{E_3}{E_1}\Big)+\frac{E_3}{E_1}\Big[\frac{d}{dt}\Big(\frac{E_1}{\lambda_1}\Big)-\frac{d}{dt}\Big(\frac{E_3}{\lambda_3}\Big)+\frac{d^2}{dt^2}\Big(\frac{E_3}{E_1}\Big)\Big]\Big\}.$$

In an analogous way, we may derive the differential relations for a body with Kelvin elements in series.

For viscoelastic solids with variable rheological parameters, we always obtain coefficients in the form of the equation presented in Section 1. Because of the variability of the moduli of elasticity and coefficients of viscosity, the complexity of the coefficients in the differential relations increases considerably with the number of rheological elements.

For the anisotropic bodies, we may find, after successive differentiations, relations in the form of Eq. (1.5).

6. TWO-DIMENSIONAL RHEOLOGICAL MODELS OF ORTHOTROPIC VISCOELASTIC BODIES

In order to demonstate the two-dimensional rheological behaviour of the anisotropic bodies in a more comprehensive manner, the author introduced two-dimensional rheological models. The simplest is the two-strip viscoelastic model represented by Fig. 5. It consists of one Hookean (H) and one Newtonian (N) element. Perfect cohesion of the two elements is assumed. Such a model behaves like a Maxwell liquid in the x-direction and like a Kelvin solid in the y-direction.

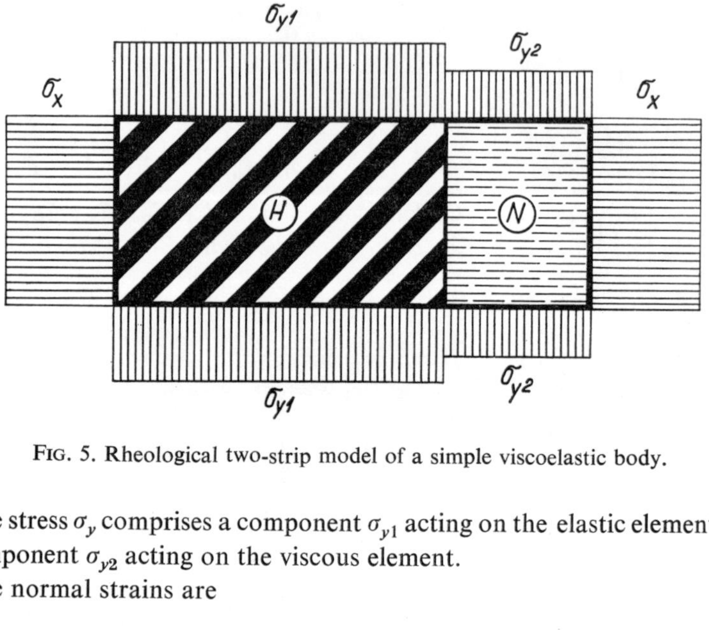

FIG. 5. Rheological two-strip model of a simple viscoelastic body.

The stress σ_y comprises a component σ_{y1} acting on the elastic element and a component σ_{y2} acting on the viscous element.

The normal strains are

$$\varepsilon_x = \frac{1}{E_x}(\sigma_x - \mu_x\sigma_{y1}) + \frac{1}{\lambda_x}\int_0^t (\sigma_x - \nu_x\sigma_{y2})\,dt, \tag{6.1}$$

$$\varepsilon_y = \frac{1}{E_y}(\sigma_{y1} - \mu_y\sigma_x) = \frac{1}{\lambda_y}\int_0^t (\sigma_{y2} - \nu_y\sigma_x)\,dt, \tag{6.2}$$

where μ_x, μ_y are the elastic Poisson ratios and ν_x, ν_y the viscous Poisson ratios.

Equation (6.2) yields a linear differential equation of the first order for σ_{y1}:

$$\frac{d\sigma_{y1}}{dt} + \frac{E_y}{\lambda_y}\sigma_{y1} = \frac{E_y}{\lambda_y}(\sigma_y - \nu_y\sigma_x) + \mu_y\frac{d\sigma_x}{dt} \tag{6.3}$$

which has the solution

$$\sigma_{y1} = e^{-E_y t/\lambda_y}\left\{\int_0^t\left[\frac{E_y}{\lambda_y}(\sigma_y - \nu_y\sigma_x) + \mu_y\frac{d\sigma_x}{dt}\right]e^{E_y t/\lambda_y}\,dt + \sigma_{y1A}\right\} \qquad (6.4)$$

where σ_{y1A} is the initial stress in the elastic element.

Substituting Eq. (6.4) in (6.2), we have

$$\varepsilon_y = e^{-E_y t/\lambda_y}\left[\int_0^t\left(\frac{\sigma_y - \nu_y\sigma_x}{\lambda_y} + \frac{\mu_y}{E_y}\frac{d\sigma_x}{dt}\right)e^{E_y t/\lambda_y}\,dt + \sigma_{y1A}\right] - \frac{\mu_y}{E_y}\sigma_x. \qquad (6.5)$$

After some rearrangements, we obtain

$$\varepsilon_y = e^{-E_y t/\lambda_y}\left\{\int_0^t \frac{i}{\lambda_y}[\sigma_y - (\mu_y+\nu_y)\sigma_x]e^{E_y t/\lambda_y}\,dt + \frac{\sigma_{y1A}}{E_y}\right\}. \qquad (6.6)$$

The expression for ε_y may also be written in the form of a linear differential equation of first order:

$$\lambda_y\frac{d\varepsilon_y}{dt} + E_y\varepsilon_y = \sigma_y - (\mu_y+\nu_y)\sigma_x. \qquad (6.7)$$

To derive the formula for ε_x, we obtain from Eq. (6.2) the partial stress

$$\sigma_{y1} = E_y\varepsilon_y + \mu_y\sigma_x \qquad (6.8)$$

and substitute it in Eq. (6.1), which yields

$$\varepsilon_x = \frac{1-\mu_x\mu_y}{E_x}\sigma_x - \frac{\mu_x E_y}{E_x}\varepsilon_y + \frac{1}{\lambda_x}\int_0^t[(1+\nu_x\mu_y)\sigma_x - \nu_x\sigma_y + \nu_x E_y\varepsilon_y]\,dt, \qquad (6.9)$$

or, after substituting Eq. (6.6):

$$\varepsilon_x = \frac{1-\mu_x\mu_y}{E_x}\sigma_x - \frac{\mu_x E_y}{E_x}e^{-E_y t/\lambda_y}\left\{\int[\sigma_y - (\mu_y+\nu_y)\sigma_x]e^{E_y t/\lambda_y}\,dt + \frac{\sigma_{y1A}}{E_y}\right\}$$
$$+\frac{1}{\lambda_x}\int_0^t\Bigg((1+\nu_x\mu_y)\sigma_x - \nu_x\sigma_y$$
$$+\nu_x E_y e^{-E_y t/\lambda_y}\left\{\int[\sigma_y + (\mu_y+\nu_y)\sigma_x]e^{E_y t/\lambda_y}\,dt + \frac{\sigma_{y1A}}{E_y}\right\}\Bigg)\,dt. \qquad (6.10)$$

The relations for the normal stresses follow from Eqs. (6.7) and (6.9). Obtaining from Eq. (6.7)

$$\sigma_y = E_y\varepsilon_y + \lambda_y\frac{d\varepsilon_y}{dt} + (\mu_y+\nu_y)\sigma_x \qquad (6.11)$$

and substituting it in Eq. (6.9), we have, after differentiation and some rearrangements, the linear differential equation of first order for σ_x:

$$\frac{d\sigma_x}{dt}+\frac{(1-\nu_x\nu_y)E_x}{(1-\mu_x\mu_y)\lambda_x}\sigma_x = \frac{E_x}{1-\mu_x\mu_y}\left[\frac{d\varepsilon_x}{dt}+\left(\frac{\mu_x E_y}{E_x}+\frac{\nu_x\lambda_y}{\lambda_x}\right)\frac{d\varepsilon_y}{dt}\right]. \quad (6.12)$$

For $\mu_x = \nu_x = 0$, this yields the well-known differential equation for the stress in the Maxwell liquid.

The solution of Eq. (6.12) is

$$\sigma_x = e^{-C_x t}\left\{\frac{E_x}{1-\mu_x\mu_y}\int_0^t\left[\frac{d\varepsilon_x}{dt}+\left(\frac{\mu_x E_y}{E_x}+\frac{\nu_x\lambda_y}{\lambda_x}\right)\frac{d\varepsilon_y}{dt}\right]e^{C_x t}\,dt+\sigma_{xA}\right\}, \quad (6.13)$$

where

$$C_x = \frac{(1-\nu_x\nu_y)E_x}{(1-\mu_x\mu_y)\lambda_x}$$

and σ_{xA} is the initial stress.

Substituting Eq. (6.13) in Eq. (6.11), we find

$$\sigma_y = E_y\varepsilon_y+\lambda_y\frac{d\varepsilon_y}{dt}$$

$$+(\mu_y+\nu_y)e^{-C_x t}\left\{\frac{E_x}{1-\mu_x\mu_y}\int_0^t\left[\frac{d\varepsilon_x}{dt}+\left(\frac{\mu_x E_y}{E_x}+\frac{\nu_x\lambda_y}{\lambda_x}\right)\frac{d\varepsilon_y}{dt}\right]e^{C_x t}\,dt+\sigma_{xA}\right\}. \quad (6.14)$$

For $\mu_y = \nu_y = 0$, this yields the well-known formula for the stress in the Kelvin solid.

There are different types of shear deformation of the body represented by the rheological model in Fig. 5. If the boundary of this body is compact and incapable of deflection, which may be represented in Fig. 5 by two rigid plates on the boundaries in the x-direction, the shear strain must be the same for both elements. The body then behaves in shear like a Kelvin solid, according to the equation

$$\tau_{xy} = \tau_{yx} = G\gamma_{xy}+\eta\frac{d\gamma_{xy}}{dt}. \quad (6.15)$$

If, however, the horizontal boundaries are free to deflect, the body behaves in horizontal shear like a Kelvin solid, according to the equation

$$\tau_{yx} = G\gamma_{yx}+\eta\frac{d\gamma_{yx}}{dt}, \quad (6.15a)$$

while in vertical shear the equation of the Maxwell liquid is valid,

$$\frac{d\tau_{xy}}{dt}+\frac{G}{\eta}\tau_{xy} = G\frac{d\gamma_{xy}}{dt}, \quad (6.16)$$

yielding

$$\tau_{xy} = e^{-Gt/\eta}\left(\int_0^t G\,\frac{d\gamma_{xy}}{dt}\,e^{Gt/\eta}dt + \tau_{xyA}\right). \tag{6.17}$$

There are now three different cases to be distinguished.

For the same shear strain $\gamma_{yx} = \gamma_{xy}$ in both directions, we get from Eqs. (6.15) and (6.17) unequal shear stresses $\tau_{yx} \neq \tau_{xy}$.

Because of the inequality of the conjugate shear stresses, the moment stress m_x arises, given by

$$\frac{\partial m_x}{\partial x} + \tau_{xy} - \tau_{yx} = 0, \tag{6.18}$$

involving flexure of the rheological element in Fig. 5 corresponding to the unequal shear displacements of the elastic and viscous portions in the y-direction.

Taking the shear stresses equal, i.e. $\tau_{yx} = \tau_{xy}$, we obtain from Eqs. (6.15a) and (6.16) two unequal shear strains

$$\gamma_{yx} = e^{-Gt/\eta}\left(\int_0^t \frac{\tau_{yx}}{\eta}\,e^{Gt/\eta}dt + \gamma_{yxA}\right), \tag{6.19}$$

$$\gamma_{xy} = \frac{\tau_{xy}}{G} + \int_0^t \frac{\tau_{xy}}{\eta}\,dt. \tag{6.20}$$

Generally, both the shear stress and strain components may be unequal simultaneously, i.e. $\tau_{yx} \neq \tau_{xy}$, $\gamma_{yx} \neq \gamma_{xy}$; they are then given by four equations, namely (6.15a), (6.17), (6.19) and (6.20).

The above relations for non-symmetrical shear apply mainly for micro-stresses in non-homogeneous viscoelastic thin-layered media.

Figure 6 represents a more general two-dimensional orthogonal rheological model corresponding to a unit area of an ideal orthotropic viscoelastic solid. This model behaves, in the axial directions, like the Poynting–Thompson body. This behaviour depends on full cohesion between the elastic and viscous components of the body, entailing equality of strains at the boundaries.

In contradistinction to other rheological models, resolution of the unit area into elastic and viscous components is determined by the relative lengths α_1, α_2, β_1, β_2, which also permits representation of the non-homogeneity of the viscoelastic body.

The resultant strain ε_x in the x-direction consists of two components corresponding to the strains ε_{x1}, ε_{x2} of the right-hand and left-hand parts of the model with relative lengths α_1 and α_2:

$$\varepsilon_x = \alpha_1\varepsilon_{x1} + \alpha_2\varepsilon_{x2}, \tag{6.21}$$

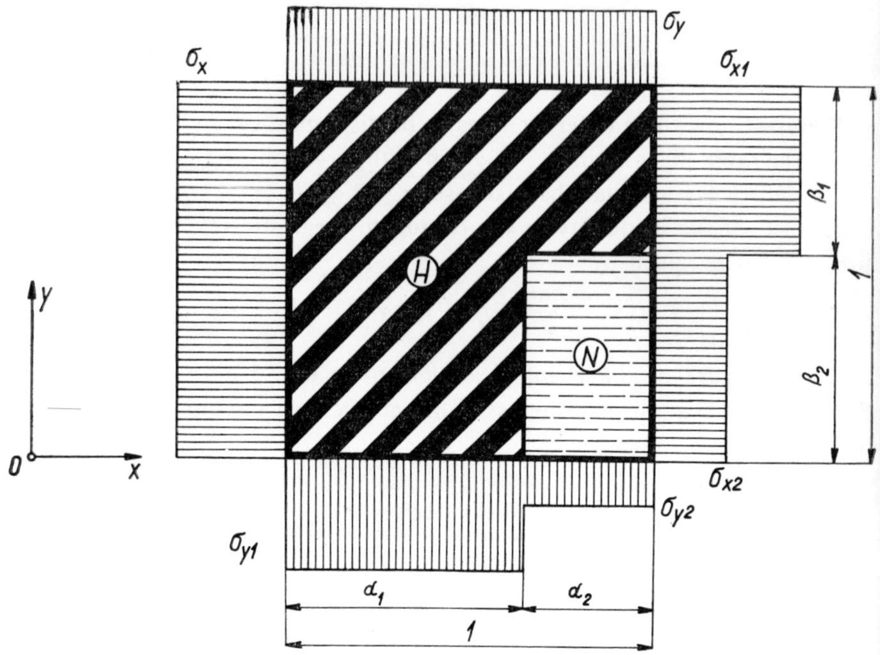

FIG. 6. Two-dimensional orthogonal viscoelastic rheological model.

while in the y-direction it is given by

$$\varepsilon_y = \beta_1\varepsilon_{y1}+\beta_2\varepsilon_{y2}. \tag{6.22}$$

The resultant stresses consist of the components acting on the elastic and viscous elements as shown in Fig. 6 according to the formulae

$$\sigma_x = \beta_1\sigma_{x1}+\beta_2\sigma_{x2}, \tag{6.23}$$

$$\sigma_y = \alpha_1\sigma_{y1}+\alpha_2\sigma_{y2}. \tag{6.24}$$

In the left-hand elastic part of the model, with relative length α_1, the strain ε_{x1} is due to the stress σ_x and to the perpendicular stress which, in the upper part with area $\alpha_1\beta_1$, is equal to σ_y and in the lower part with the area $\alpha_1\beta_2$ to σ_{y1}. Because of the cohesion, the strain in the upper and lower parts must be the same. This produces internal stresses with balanced resultants X_1 and $X_2 = -X_1$.

The strain ε_{x1} in the upper and lower left part of the model on Fig. 6 is then given by

$$\varepsilon_{x1} = \frac{1}{E_x}\left(\sigma_x-\mu_x\sigma_y+\frac{X_1}{\beta_1}\right) = \frac{1}{E_x}\left(\sigma_x-\mu_x\sigma_{y1}+\frac{X_2}{\beta_2}\right) \tag{6.25}$$

whence

$$X_1 = -X_2 = \frac{\mu_x(\sigma_y - \sigma_{y1})}{1/\beta_1 + 1/\beta_2}. \tag{6.26}$$

Substituting Eq. (6.26) in Eq. (6.25), we obtain after, some rearrangements,

$$\varepsilon_{x1} = \frac{1}{E_x}[\sigma_x - \mu_x(\beta_1\sigma_y + \beta_2\sigma_{y1})]. \tag{6.27}$$

The strain ε_{y1} of the upper elastic part is obtainable by quite a similar procedure as follows:

$$\varepsilon_{y1} = \frac{1}{E_y}[\sigma_y - \mu_y(\alpha_1\sigma_x + \alpha_2\sigma_{x1})]. \tag{6.28}$$

The equality of strains ε_{x2} of the right-hand elastic and viscous parts is expressed by

$$\varepsilon_{x2} = \frac{1}{E_x}(\sigma_{x1} - \mu_x\sigma_y) = \frac{1}{\lambda_x}\int_0^t (\sigma_{x2} - \nu_x\sigma_{y2})\,dt, \tag{6.29}$$

whereas for the strain ε_{y2} of the lower elastic and viscous parts we find

$$\varepsilon_{y2} = \frac{1}{E_y}(\sigma_{y1} - \mu_y\sigma_x) = \frac{1}{\lambda_y}\int_0^t (\sigma_{y2} - \nu_y\sigma_{x2})\,dt. \tag{6.30}$$

Obtaining from Eq. (6.24)

$$\sigma_{y2} = \frac{\sigma_y - \alpha_1\sigma_{y1}}{\alpha_2} \tag{6.31}$$

and substituting it in Eq. (6.29), we have, after differentiation, the following relation:

$$\sigma_{y1} = \frac{\alpha_2\lambda_x}{\alpha_1\nu_x E_x}\left(\frac{d\sigma_{x1}}{dt} - \mu_x\frac{d\sigma_y}{dt}\right) + \frac{\sigma_y}{\alpha_1} - \frac{\alpha_2}{\alpha_1\nu_x}\sigma_{x2}. \tag{6.32}$$

Obtaining from Eq. (6.23)

$$\sigma_{x2} = \frac{\sigma_x - \beta_1\sigma_{x1}}{\beta_2} \tag{6.33}$$

and substituting it, together with Eq. (6.32) in Eq. (6.30), we have, after differentiation, a linear differential equation of second order for σ_{x1}:

$$\frac{d^2\sigma_{x1}}{dt^2} + (C_x + C_y)\frac{d\sigma_{x1}}{dt} + (1 - \nu_x\nu_y)C_xC_y\sigma_{x1}$$
$$= \frac{1 - \nu_x\nu_y}{\beta_1}C_xC_y\sigma_x + \frac{E_x}{\lambda_x}\left(\frac{1}{\beta_2} + \mu_y\nu_x\frac{\alpha_1}{\alpha_2}\right)\frac{d\sigma_x}{dt}$$
$$+ \left(\mu_xC_y - \frac{\nu_xE_x}{\alpha_2\lambda_x}\right)\frac{d\sigma_y}{dt} + \mu_x\frac{d^2\sigma_y}{dt^2}, \tag{6.34}$$

where

$$C_x = \frac{\beta_1 E_x}{\beta_2 \lambda_x}, \quad C_y = \frac{\alpha_1 E_y}{\alpha_2 \lambda_y} \tag{6.35}$$

are the reciprocal retardation times. The solution to Eq. (6.34) is

$$\sigma_{x1} = e^{-L_1 t} \int_0^t e^{(L_1-L_2)t} \left\{ \int_0^t \left[\frac{1-\nu_x\nu_y}{\beta_1} C_x C_y \sigma_x \right.\right.$$
$$+ \frac{E_x}{\lambda_x}\left(\frac{1}{\beta_2} + \mu_y\nu_x \frac{\alpha_1}{\alpha_2}\right)\frac{d\sigma_x}{dt} + \left(\mu_x C_y - \frac{\nu_x E_x}{\alpha_2\lambda_x}\right)\frac{d\sigma_y}{dt}$$
$$\left.\left. + \mu_x \frac{d^2\sigma_y}{dt^2}\right] e^{L_2 t} dt \right\} dt + c_1 e^{-L_1 t} + c_2 e^{-L_2 t}, \tag{6.36}$$

where

$$L_{1,2} = \tfrac{1}{2}(C_x + C_y) \pm \tfrac{1}{2}\sqrt{[(C_x - C_y)^2 + 4\nu_x\nu_y C_x C_y]} \tag{6.37}$$

and c_1, c_2 are constants of integration.

Substituting Eqs. (6.32) and (6.36) in Eq. (6.27), Eq. (6.36) in Eq. (6.29), and finally Eqs. (6.27) and (6.29) in Eq. (6.21), we obtain

$$\varepsilon_x = \frac{1}{E_x}\left[\left(\alpha_1 + \alpha_2\frac{\mu_x}{\nu_x}\right)\sigma_x - \mu_x(\alpha_1\beta_1 + \alpha_2 + \beta_2)\sigma_y + \alpha_2\beta_2 \frac{\mu_x^2\lambda_x}{\nu_x E_x}\frac{d\sigma_y}{dt}\right.$$
$$+\alpha_2\left[1 - \frac{\mu_x}{\nu_x}\left(\beta_1 - \beta_2\frac{\lambda_x L_1}{E_x}\right)\right]\left(e^{-L_1 t}\int_0^t e^{(L_1-L_2)t}\left\{\int_0^t\left[\frac{1-\nu_x\nu_y}{\beta_1} C_x C_y \sigma_x\right.\right.\right.$$
$$+\frac{E_x}{\lambda_x}\left(\frac{1}{\beta_2} + \mu_y\nu_x\frac{\alpha_1}{\alpha_2}\right)\frac{d\sigma_x}{dt} + \left(\mu_x C_y - \frac{\nu_x E_x}{\alpha_2\lambda_x}\right)\frac{d\sigma_y}{dt}$$
$$\left.\left.\left. + \mu_x\frac{d^2\sigma_y}{dt^2}\right] e^{L_2 t} dt\right\} dt + c_1 e^{-L_1 t} + c_2 \frac{L_2}{L_1} e^{-L_2 t}\right)$$
$$-\alpha_2\beta_2\frac{\mu_x\lambda_x}{\nu_x E_x} e^{-L_2 t}\int_0^t\left[\frac{1-\nu_x\nu_y}{\beta_1} C_x C_y \sigma_x\right.$$
$$\left.\left. + \frac{E_x}{\lambda_x}\left(\frac{1}{\beta_2} + \mu_y\nu_x\frac{\alpha_1}{\alpha_2}\right)\frac{d\sigma_x}{dt} + \left(\mu_x C_y - \frac{\nu_x E_x}{\alpha_2\lambda_x}\right)\frac{d\sigma_y}{dt} + \mu_x\frac{d^2\sigma_y}{dt^2}\right] e^{L_2 t} dt\right]. \tag{6.38}$$

Substituting Eq. (6.36) in Eq. (6.28), Eqs. (6.32) and (6.36) in Eq. (6.30) and finally Eqs. (6.28) and (6.30) in Eq. (6.22), we obtain

$$\varepsilon_y = \frac{1}{E_y}\left[\left(\beta_1 + \frac{\beta_2}{\alpha_1}\right)\sigma_y - \left[\mu_y(\alpha_1\beta_1 + \beta_2) + \frac{\alpha_2}{\alpha_1\nu_x}\right]\sigma_x\right.$$
$$-\frac{\alpha_2\beta_2\mu_x\lambda_x}{\alpha_1\nu_x E_x}\frac{d\sigma_y}{dt} + \left(\frac{\alpha_2\beta_1}{\alpha_1\nu_x} - \frac{\alpha_2\beta_2\lambda_x}{\alpha_1\nu_x E_x}L_1 - \alpha_2\beta_1\mu_y\right)$$
$$\left(e^{-L_1 t}\cdot\int_0^t e^{(L_1-L_2)t}\left\{\int_0^t\left[\frac{1-\nu_x\nu_y}{\beta_1} C_x C_y \sigma_x + \frac{E_x}{\lambda_x}\left(\frac{1}{\beta_2} + \mu_y\nu_x\frac{\alpha_1}{\alpha_2}\right)\frac{d\sigma_x}{dt}\right.\right.\right.$$

$$+\left(\mu_x C_y - \frac{\nu_x E_x}{\alpha_2 \lambda_x}\right)\frac{d\sigma_y}{dt} + \mu_x \frac{d^2\sigma_y}{dt^2}\Bigg] e^{L_2 t} dt\Bigg\} dt + c_1 e^{-L_1 t} + c_2 \frac{L_2}{L_1} e^{-L_2 t}\Bigg)$$

$$+\frac{\alpha_2\beta_2\lambda_x}{\alpha_1\nu_x E_x} e^{-L_2 t} \int_0^t \Bigg[\frac{1-\nu_x\nu_y}{\beta_1} C_x C_y \sigma_x + \frac{E_x}{\lambda_x}\left(\frac{1}{\beta_2} + \mu_y \nu_x \frac{\alpha_1}{\alpha_2}\right)\frac{d\sigma_x}{dt}$$

$$+\left(\mu_x C_y - \frac{\nu_x E_x}{\alpha_2\lambda_x}\right)\frac{d\sigma_y}{dt} + \mu_x \frac{d^2\sigma_y}{dt^2}\Bigg] e^{L_2 t} dt\Bigg]. \tag{6.39}$$

The normal strains may be also represented in an alternative manner. Substituting from Eq. (6.24) the component

$$\sigma_{y2} = \frac{\sigma_y - \alpha_1 \sigma_{y1}}{\alpha_2} \tag{6.40}$$

in Eq. (6.30) we obtain, after differentiation, a linear differential equation of first order for σ_{y1}:

$$\frac{d\sigma_{y1}}{dt} + \frac{\alpha_1 E_y}{\alpha_2 \lambda_y}\sigma_{y1} = \frac{E_y}{\lambda_y}\left(\frac{\sigma_y}{\alpha_2} - \nu_y \sigma_{x2}\right) + \mu_y \frac{d\sigma_x}{dt}, \tag{6.41}$$

whence

$$\sigma_{y1} = e^{-C_y t}\left\{\int_0^t \left[\frac{E_y}{\lambda_y}\left(\frac{\sigma_y}{\alpha_2} - \nu_y \sigma_{x2}\right) + \mu_y \frac{d\sigma_x}{dt}\right] e^{C_y t} dt + \sigma_{yA}\right\}, \tag{6.42}$$

i.e.

$$\sigma_{y1} = e^{-C_y t}\left\{\int_0^t \frac{E_y}{\lambda_y}\left[\frac{\sigma_y}{\alpha_2} - \left(\mu_y \frac{\alpha_1}{\alpha_2} + \frac{\nu_y}{\beta_2}\right)\sigma_x + \nu_y \frac{\beta_1}{\beta_2}\sigma_{x1}\right] e^{C_y t} dt + \sigma_{yA}\right\}$$
$$+\mu_y \sigma_x, \tag{6.43}$$

where
$$C_y = \frac{\alpha_1 E_y}{\alpha_2 \lambda_y}.$$

Obtaining from Eq. (6.23)

$$\sigma_{x2} = \frac{\sigma_x - \beta_1 \sigma_{x1}}{\beta_2} \tag{6.44}$$

and substituting it together with Eq. (6.42) in Eq. (6.29), we have, after two successive differentiations with respect to σ_{x1}, the linear differential equation (6.34) with the solution (6.36).

Using Eq. (6.43), we obtain, after some rearrangements, the following expressions for the normal strains:

$$\varepsilon_x = \frac{\alpha_1}{E_x}\Bigg\{\sigma_x - \mu_x \Bigg[\beta_1 \sigma_y + \beta_2 e^{-C_y t} \int_0^t \left(\frac{E_y}{\lambda_y}\left(\frac{\sigma_y}{\alpha_2} - \frac{\nu_y \sigma_x}{\beta_2}\right)\right.$$

$$+\mu_y \frac{d\sigma_x}{dt} + \nu_y \frac{\beta_1 E_y}{\beta_2 \lambda_y} e^{-L_1 t} \int_0^t e^{(L_1 - L_2)t} \Bigg\{\int_0^t \Bigg[\frac{1-\nu_x\nu_y}{\beta_1} C_x C_y \sigma_x$$

$$+\frac{E_x}{\lambda_x}\left(\frac{1}{\beta_2} + \frac{\alpha_1}{\alpha_2}\mu_y \nu_x\right)\frac{d\sigma_y}{dt} + \left(\mu_x C_y - \frac{\nu_x E_x}{\alpha_2 \lambda_x}\right)\frac{d\sigma_y}{dt}$$

$$+\mu_x \frac{d^2\sigma_y}{dt^2}\Bigg] e^{L_2 t}dt\Bigg\} dt + \nu_y C_y(c_1 e^{-L_1 t} + c_2 e^{-L_2 t})\Bigg) e^{C_y t}dt + \beta_2\sigma_{yA} e^{-C_y t}\Bigg]\Bigg\}$$

$$+\frac{\alpha_2}{E_x}\left(e^{-L_1 t}\int_0^t e^{(L_1-L_2)t}\left\{\int_0^t\left[\frac{1-\nu_x\nu_y}{\beta_1} C_x C_y \sigma_x\right.\right.\right.$$

$$+\frac{E_x}{\lambda_x}\left(\frac{1}{\beta_2}+\frac{\alpha_1}{\alpha_2}\mu_y\nu_x\right)\frac{d\sigma_x}{dt} + \left(\mu_x C_y - \frac{\nu_x E_x}{\alpha_2\lambda_x}\right)\frac{d\sigma_y}{dt}$$

$$\left.\left.\left.+\mu_x \frac{d^2\sigma_y}{dt^2}\right] e^{L_2 t}dt\right\} dt + c_1 e^{-L_1 t} + c_2 e^{-L_2 t} - \mu_x\sigma_y\right) \tag{6.45}$$

$$\varepsilon_y = \frac{\beta_1}{E_y}\left[\sigma_y - \mu_y\left(\alpha_1\sigma_x + \alpha_2 e^{-L_1 t}\int_0^t e^{(L_1-L_2)t}\left\{\int_0^t\left[\frac{1-\nu_x\nu_y}{\beta_1} C_x C_y \sigma_x\right.\right.\right.\right.$$

$$+\frac{E_x}{\lambda_x}\left(\frac{1}{\beta_2}+\frac{\alpha_1}{\alpha_2}\mu_y\nu_x\right)\frac{d\sigma_x}{dt} + \left(\mu_x C_y - \frac{\nu_x E_x}{\alpha_2\lambda_x}\right)\frac{d\sigma_y}{dt}$$

$$\left.\left.\left.\left.+\mu_x \frac{d^2\sigma_y}{dt^2}\right] e^{L_2 t}\, dt\right\} dt + c_1 e^{-L_1 t} + c_2 e^{-L_2 t}\right)\right]$$

$$+\frac{\beta_2}{E_y}\left[e^{-C_y t}\int_0^t\left(\frac{E_y}{\lambda_y}\left(\frac{\sigma_y}{\alpha_2}-\frac{\nu_y\sigma_x}{\beta_2}\right)+\mu_y\frac{d\sigma_x}{dt}\right.\right.$$

$$+\nu_y\frac{\beta_1 E_y}{\beta_2\lambda_y} e^{-L_1 t}\int_0^t e^{(L_1-L_2)t}\left\{\int_0^t\left[\frac{1-\nu_x\nu_y}{\beta_1} C_x C_y\sigma_x + \frac{E_x}{\lambda_x}\left(\frac{1}{\beta_2}+\frac{\alpha_1}{\alpha_2}\mu_y\nu_x\right)\frac{d\sigma_x}{dt}\right.\right.$$

$$\left.\left.+\left(\mu_x C_y - \frac{\nu_x E_x}{\alpha_2\lambda_x}\right)\frac{d\sigma_y}{dt} + \mu_x\frac{d^2\sigma_y}{dt^2}\right] e^{L_2 t}dt\right\} dt + \nu_y\frac{\beta_1 E_y}{\beta_2\lambda_y}(c_1 e^{-L_1 t} + c_2 e^{-L_2 t})\Bigg)$$

$$\left.\times e^{C_y t}\, dt + \sigma_{yA} e^{-C_y t} - \mu_y\sigma_x\right]. \tag{6.46}$$

The strain ε_z perpendicular to the stress plane is assumed to be the same over the unit area of the model in Fig. 6. This assumption yields

$$\varepsilon_z = \frac{1}{E_z}(\alpha_1\mu_{xz}\sigma_x + \beta_1\mu_{yz}\sigma_y + \alpha_2\beta_1\mu_{xz}\sigma_{x1}$$

$$+\alpha_1\beta_2\mu_{yz}\sigma_{y1}) = \frac{\alpha_2\beta_2}{\lambda_z}\int_0^t (\nu_{xz}\sigma_{x2} + \nu_{yz}\sigma_{y2})\, dt. \tag{6.47}$$

There are again different types of shear deformation of the model in Fig. 6.

In the case of the forced uniform shear strain $\gamma_{xy} = \gamma_{yx}$ occurring over the unit area in Fig. 6, the resultant shear stress consist of the component

$$\tau_{xy1} = \tau_{yx1} = G\gamma_{xy} \tag{6.48}$$

acting on the elastic part with area $1-\alpha_2\beta_2$, and of

$$\tau_{xy2} = \tau_{yx2} = \eta\frac{d\gamma_{xy}}{dt} \tag{6.49}$$

acting on the viscous part with area $\alpha_2\beta_2$. We have

$$\tau_{xy} = \tau_{yx} = (1-\alpha_2\beta_2)\,\tau_{xy1}+\alpha_2\beta_2\tau_{xy2}, \tag{6.50}$$

i.e.

$$\tau_{xy} = (1-\alpha_2\beta_2)\,G\gamma_{xy}+\alpha_2\beta_2\eta\,\frac{d\gamma_{xy}}{dt}, \tag{6.51}$$

whence

$$\gamma_{xy} = e^{-C_{xy}t}\left(\int_0^t \frac{\tau_{xy}}{\alpha_2\beta_2\eta}\,e^{C_{xy}t}\,dt+\gamma_{xyA}\right), \tag{6.52}$$

where

$$C_{xy} = \frac{(1-\alpha_2\beta_2)\,G}{\alpha_2\beta_2\eta} \tag{6.53}$$

is the reciprocal retardation time for the symmetric shear.

Generally, the shear of the orthotropic viscoelastic body represented in Fig. 6 is non-symmetrical, i.e. there is no equality of the conjugate shear stresses and shear strains.

The shear strain γ_{yx}, defined as a horizontal displacement per unit length, is

$$\gamma_{yx} = \beta_1\gamma_{yx1}+\beta_2\gamma_{yx2}, \tag{6.54}$$

while the shear strain in the vertical direction is given by

$$\gamma_{xy} = \alpha_1\gamma_{xy1}+\alpha_2\gamma_{xy2}. \tag{6.55}$$

The shear strain γ_{yx1} occurs in the elastic upper part of the area on Fig. 6 with height β_1, the strain γ_{yx2} in the viscoelastic part with height β_2, the the strain γ_{xy1} in the elastic left part (length α_1) and γ_{xy2} in the viscoelastic right part (length α_2).

The horizontal shear stress τ_{yx} acting on the upper part is resolved into the components τ_{yx1}, acting on the elastic lower part, and τ_{yx2}, acting on the viscous part, according to the relation

$$\tau_{yx} = \alpha_1\tau_{yx1}+\alpha_2\tau_{yx2}. \tag{6.56}$$

Similarly, the vertical shear strain τ_{xy} may be expressed in terms of its components τ_{xy1} and τ_{xy2} as follows:

$$\tau_{xy} = \beta_1\tau_{xy1}+\beta_2\tau_{xy2}. \tag{6.57}$$

The relations between shear stress and strain components have the following form:

$$\tau_{yx} = G\gamma_{yx1}, \quad \tau_{yx1} = G\gamma_{yx2}, \quad \tau_{yx2} = \eta\,\frac{d\gamma_{yx2}}{dt}, \tag{6.58}$$

$$\tau_{xy} = G\gamma_{xy1}, \quad \tau_{xy1} = G\gamma_{xy2}, \quad \tau_{xy2} = \eta\,\frac{d\gamma_{xy2}}{dt}. \tag{6.59}$$

Substitution of Eq. (6.54) in Eq. (6.58) yields

$$\gamma_{yx} = \beta_1 \frac{\tau_{yx}}{G} + \beta_2 \frac{\tau_{yx1}}{G} = \beta_1 \frac{\tau_{yx}}{G} + \beta_2 \int_0^t \frac{\tau_{yx2}}{\eta} dt, \tag{6.60}$$

while on substituting Eqs. (6.59) in Eq. (6.55), we obtain

$$\gamma_{xy} = \alpha_1 \frac{\tau_{xy}}{G} + \alpha_2 \frac{\tau_{xy1}}{G} = \alpha_1 \frac{\tau_{xy}}{G} + \alpha_2 \int_0^t \frac{\tau_{xy2}}{\eta} dt. \tag{6.61}$$

Obtaining from Eq. (6.56)

$$\tau_{yx2} = \frac{\tau_{yx} - \alpha_1 \tau_{yx1}}{\alpha_2} \tag{6.62}$$

and substituting it in Eq. (6.60), we have a linear differential equation of first order for the component τ_{yx1}:

$$\frac{d\tau_{yx1}}{dt} + \frac{\alpha_1 G}{\alpha_2 \eta} \tau_{yx1} = \frac{G}{\alpha_2 \eta} \tau_{yx} \tag{6.63}$$

with the solution

$$\tau_{yx1} = e^{-C_{yx}t} \left(\int_0^t \frac{C_{yx}}{\alpha_1} \tau_{yx} e^{C_{yx}t} dt + \tau_{yx1A} \right), \tag{6.64}$$

where

$$C_{yx} = \frac{\alpha_1 G}{\alpha_2 \eta} \tag{6.65}$$

is the reciprocal relaxation time for the horizontal shear.

Substituting the above expression in Eq. (6.60), we obtain the resultant shear strain in the horizontal direction

$$\gamma_{yx} = \frac{1}{G} \left[\beta_1 \tau_{yx} + \beta_2 e^{-C_{yx}t} \left(\int_0^t \frac{C_{yx}}{\alpha_1} \tau_{yx} e^{C_{yx}t} dt + \tau_{yx1A} \right) \right]. \tag{6.66}$$

Substituting the vertical shear stress component

$$\tau_{xy2} = \frac{\tau_{xy} - \beta_1 \tau_{xy1}}{\beta_2} \tag{6.67}$$

in Eq. (6.61), we have a linear differential equation of first order

$$\frac{d\tau_{xy1}}{dt} + \frac{\beta_1 G}{\beta_2 \eta} \tau_{xy1} = \frac{G}{\beta_2 \eta} \tau_{xy} \tag{6.68}$$

for

$$\tau_{xy1} = e^{-C_{xy}t} \left(\int_0^t \frac{C_{xy}}{\beta_1} \tau_{xy} e^{C_{xy}t} dt + \tau_{xy1A} \right) \tag{6.69}$$

where

$$C_{xy} = \frac{\beta_1 G}{\beta_2 \eta} \tag{6.70}$$

is the reciprocal relaxation time for the vertical shear. Substituting the above expression in Eq. (6.61), we find for the resulting shear strain, in the vertical direction, the following formula:

$$\gamma_{xy} = \frac{1}{G}\left[\alpha_1\tau_{xy} + \alpha_2 e^{-C_{xy}t}\left(\left(\int_0^t \frac{C_{xy}}{\beta_1}\tau_{xy}e^{C_{xy}t}\,dt + \tau_{xyA1}\right)\right)\right]. \quad (6.71)$$

If the conjugate stresses are equal, $\tau_{yx} = \tau_{xy}$, the shear strains γ_{yx} and γ_{xy} are also equal only if the body is isotropic, i.e. for $\alpha_1 = \beta_1$ and $\alpha_2 = \beta_2$.

Substituting Eqs. (6.58) and (6.54) in Eq. (6.56), we obtain, after some rearrangements, a linear differential equation of first order

$$\frac{d\gamma_{yx2}}{dt} + \left(\frac{\beta_2}{\beta_1} + \alpha_1\right)\frac{G}{\alpha_2\eta}\gamma_{yx2} = \frac{G}{\alpha_2\beta_1\eta}\gamma_{yx} \quad (6.72)$$

for

$$\gamma_{yx2} = e^{-D_{yx}t}\left(\int_0^t \frac{G}{\alpha_2\beta_1\eta}\gamma_{yx}e^{D_{yx}t}\,dt + \gamma_{yx2A}\right), \quad (6.73)$$

where

$$D_{yx} = \frac{G}{\alpha_2\eta}\left(\frac{\beta_2}{\beta_1} + \alpha_1\right) \quad (6.74)$$

is the reciprocal retardation time for the horizontal shear.

Substituting Eq. (6.73), together with Eqs. (6.58), in Eq. (6.56), we have the formula for the resultant shear stress in the x-direction:

$$\tau_{yx} = G\left(\frac{\alpha_1}{\eta} - \alpha_2 D_{yx}\right)e^{-D_{yx}t}\left(\int_0^t \frac{G}{\alpha_2\beta_1}\gamma_{yx}e^{D_{yx}t} + \gamma_{yx2A}\right) + \frac{G}{\beta_1}\gamma_{yx}. \quad (6.75)$$

Substituting Eqs. (6.59) in Eq. (6.57), and obtaining γ_{xy1} from Eq. (6.55), we have a linear differential equation of first order

$$\frac{d\gamma_{xy2}}{dt} + \frac{G}{\beta_2\eta}\left(\frac{\alpha_2}{\alpha_1} + \beta_1\right)\gamma_{xy2} = \frac{G}{\alpha_1\beta_2\eta}\gamma_{xy} \quad (6.76)$$

for

$$\gamma_{xy2} = e^{-D_{xy}t}\left(\int_0^t \frac{G}{\alpha_1\beta_2\eta}\gamma_{xy}e^{D_{xy}t}\,dt + \gamma_{xy2A}\right), \quad (6.77)$$

where

$$D_{xy} = \frac{G}{\beta_2\eta}\left(\frac{\alpha_2}{\alpha_1} + \beta_1\right) \quad (6.78)$$

is the reciprocal retardation time for the vertical shear.

Substituting Eq. (6.77), together with Eqs. (6.59), in Eq. (6.57), we find

$$\tau_{xy} = G\left(\frac{\beta_1}{\eta} - \beta_2 D_{xy}\right)e^{-D_{xy}t}\left(\int_0^t \frac{G}{\alpha_1\beta_2}\gamma_{xy}e^{D_{xy}t}\,dt + \gamma_{xy2A}\right) + \frac{G}{\alpha_1}\gamma_{xy}. \quad (6.79)$$

From Eqs. (6.75) and (6.79) we see, that in the case of equal conjugate shear strains $\gamma_{xy} = \gamma_{yx}$, the conjugate shear stresses are also equal $\tau_{yx} = \tau_{xy}$ only if the body has isotropic structure, i.e. if $\alpha_1 = \beta_1, \alpha_2 = \beta_2$.

7. CONCLUSIONS

The rheological behaviour of viscoelastic solids may be represented by differential or integral–differential stress–strain relationships in the form of Eqs. (1.1), (1.2), (1.3) and (1.4) for isotropic bodies, and in the form of Eq. (1.5) for anisotropic bodies. The coefficients of these relations are constant if the rheological parameters are time-independent, but for viscoelastic solids with time-dependent rheological parameters the coefficients are variable.

Under certain restrictions on the rheological behaviour, we may define the deformation by means of the algebraic relationships between the equivalent stress and strain. The concept of equivalent stress and strain permits solution of rheological boundary-value problems with regard to space coordinates separately from that with regard to time.

Introducing the concept of transformed equivalent strain, the author has derived a rheological stress–strain relationship (2.8) analogous to the non-linear tensor equation derived by Reiner for isotropic bodies.

As shown in Sections 3 and 4, anisotropy increases the order of the differential stress–strain relationships for the rheological behaviour of viscoelastic bodies.

A new approach to the definition of the rheological behaviour of the orthotropic viscoelastic bodies is illustrated by two-dimensional rheological models in Section 6.

Starting from the discoveries made by the great founder of modern rheology, Professor Markus Reiner, the author's object was analysis of the rheological behaviour of anisotropic and non-linear complex viscoelastic solids represented by multi-element rheological models.

REFERENCES

1. T. Alfrey, *Mechanical Behavior of High Polymers*, Interscience Publishers, Inc., New York–London, 1948.
2. A. M. Freudenthal, *The Inelastic Behavior of Engineering Materials and Structures*, J. Wiley & Sons, Inc., New York, and Chapman & Hall, Ltd., London, 1950.
3. A. M. Freudenthal and H. Geiringer, "The mathematical theories of inelastic continuum", *Handbuch der Physik* (S. Flügge, ed.), Band VI, Elastizität und Plastizität, Springer-Verlag, Berlin–Göttingen–Heidelberg, 1958.
4. E. Hoffmanová, "Moduli of elasticity of the non-homogeneous two-phase orthotropic materials", *Stavebnický časopis*, **1**, 112–15 (1966).

5. Z. Karni and M. Reiner, "The general measure of deformation", *Second-Order Effects in Elasticity, Plasticity and Fluid Dynamics*, International Symposium of I.U.T.A.M., Haifa, Israel, April 1962 (Edited by M. Reiner and D. Abir). Jerusalem Academic Press, Jerusalem, and Pergamon Press, Oxford–London–New York–Paris, 1964.
6. H. Leaderman, "Viscoelasticity phenomena in amorphous high polymeric systems", *Rheology, Theory and Applications*, vol. II (ed. by. F. R. Eirich), Academic Press, New York, 1956.
7. J. Mandel, "Application du calcul opérationnel à l'étude des corps viscoélastiques", *Cahiers du Groupe Français d'Études de Rhéologie*, tome III, Paris, 1958.
8. W. Nowacki, *Theory of Creep* (in Polish), Arcady, Warsaw, 1963.
9. W. Olszak and P. Perzyna, "Variational theorems in general viscoelasticity", *Ingenieur-Archiv*, **28**, 246–50 (1959).
10. M. Reiner, "A mathematical theory of dilatancy", *American Journal of Mathematics*, **67**, 350–62 (1945).
11. M. Reiner, "Elasticity Beyond the Elastic Limit", *American Journal of Mathematics*, **70**, 433–46 (1948).
12. M. Reiner, *Twelve Lectures on Theoretical Rheology*. North-Holland Publishing Co., Amsterdam, and Interscience Publishers, New York, 1949.
13. M. Reiner, *Deformation and Flow*, H. K. Lewis & Co., London, 1949.
14. M. Reiner, "Theoretical rheology", *Building Materials* (ed. by M. Reiner), North-Holland Publishing Co., Amsterdam, 1954.
15. M. Reiner, "Phenomenological macrorheology", *Rheology, Theory and Applications*, vol. I (ed. by F. R. Eirich), Academic Press, New York, 1956.
16. M. Reiner, *Deformation, Strain and Flow*, H. K. Lewis & Co., London, 1960.
17. Z. Sobotka, *Theory of Plasticity and Limiting States of Engineering Structures*, vol. I (in Czech), Edition of Czechoslovak Academy of Sciences, Prague, 1954.
18. Z. Sobotka, "Rheological laws of the long-time deformation of concrete", *Silicates* **2**, 117–38, Prague, 1958.
19. Z. Sobotka, "Some problems of non-linear rheology", *Second-Order Effects in Elasticity, Plasticity and Fluid Dynamics*, International Symposium of I.U.T.A.M., Haifa, Israel, April 1962 (edited by M. Reiner and D. Abir), Jerusalem Academic Press, Jerusalem, and Pergamon Press, Oxford–London–New York–Paris, 1964.
20. Z. Sobotka, "On rheology of non-linear shell problems", *Non-Classical Shell Problems. Proceedings of I.A.S.S. Symposium, Warsaw, September 1963*, North-Holland Publishing Co., Amsterdam and PWN-Polish Scientific Publishers, Warsaw, 1964.
21. Z. Sobotka, "General stress–strain relationships of anisotropic bodies and the concept of the transformed strain", *Aplikace Matematiky*, **9**, 467–9 (1964).
22. Z. Sobotka, "Rhéologie des problèmes de déformation plane des milieux continus", Symposium International de l'I.U.T.A.M.—*Rhéologie et Mécanique des Sols*, Grenoble, 1964 (ed. by J. Kravtchenko and M. Sirieys), Springer-Verlag, Berlin–Göttingen–Heidelberg.
23. C. Truesdell, "Second-order effects in the mechanics of materials", *Second-Order Effects in Elasticity, Plasticity and Fluid Dynamics*, International Symposium of the I.U.T.A.M., Haifa, Israel, April 1962 (Edited by M. Reiner and D. Abir), Jerusalem Academic Press, Jerusalem, and Pergamon Press, Oxford–London–New York–Paris, 1964.

A RHEOLOGICAL APPROACH TO RESEARCH IN MICROCIRCULATION

K. WEISSENBERG

The Royal Institution, London, U.K.

RESEARCH into circulatory problems has a long history, dating back to Malpighi's discovery in 1686 of circulatory blood flow and to Poiseuille's well-known formula (1835), derived from the classical theory of mechanics, for flow through blood vessels and glass tubes of small diameter. In spite of the valuable pioneer work of the early research workers, little progress was made until, at the turn of the nineteenth century, one became aware of the complexity of circulatory flow. One realized then that every living body is associated with circulatory flow, not only of blood but of a wide variety of different materials and energies. The flow starts from the outside world with intake of food, drink and air, and enters the body through the upper part of the digestive tract and the lungs. It then circulates in the body and interacts there with the blood-stream and with all other tissues, bones, muscles and body fluids, eventually making its exit and returning to the outside world by way of discharge through the lower part of the digestive tract, through the pores of the skin, and by exhalation. One realized, further, that the flow proceeds not only in the form of ducted currents inside the various tubular vessels, but also in the form of ductless cross-currents through the vessel walls. These currents and cross-currents are, in quantity and biological significance, of equal importance for coordination and regulation of all interactions inside the body, and between the body as a whole and its environment, in accordance with metabolic requirements. Finally, one realized that the complexity of the flow phenomena of the various materials and energies could not be accounted for by the theory of classical mechanics, but only by the more modern theory of rheology as discussed below.

One of the founders of modern rheology is M. Reiner who, together with Bingham, studied the flow of materials of various consistencies through glass capillaries, and was the first to develop quantitative equations which accounted for deviations from the classical Poiseuille law. Reiner's work greatly stimulated research into the numerous so-called "anomalous" phenomena of flow, i.e. phenomena not conforming to the classical norms which had at that time been arbitrarily regarded as theoretically established. These phenomena play an important role in the flow of most biological materials, such as blood, serum, synovial fluid, etc., and Reiner's work opened a path to a rheological approach to the problems involved. It would greatly exceed the scope of the present paper to review critically the vast amount of literature available, and it has therefore been decided to review here only some of the phenomena encountered in the flow of biological materials.

Classical mechanics admitted for materials three, and only three, different types of consistency—solid, liquid and gaseous—and assumed for each of them a simple linear law for the relationship between movement and applied force. The three types may be exemplified by steel, water and air respectively, the first having a reversibly-elastic consistency with little extensibility, the second—a viscous and irreversibly plastic consistency with large extensibility in shear and a reversibly-elastic one with little change in volume, and the third—an irreversibly plastic consistency with large change in shear and volume.

In contrast to classical mechanics, one recognizes in modern rheology the existence of materials of the most varied and complex types with generally non-linear laws, which only approximate to linear ones in idealized cases. Here there are innumerable types. Some types of materials (e.g. ceramics, bones, teeth, etc.) are hard, elastic and brittle, while others (e.g. rubber, muscles, thrombi, etc.) are tough and elastically extensible, and still others (e.g. ordinary putty) are soft, plastic and viscous. There are yet other groups of materials (e.g. silicone putty) which exhibit different properties under different applied actions, being soft, plastic and viscous under slow deformation, and hard, brittle and elastic under fast deformation. Finally, mention should be made of a type of materials such as suspensions of varying kinds, which are dilatant and/or thixotropic, with non-linear effects of work-hardening and/or work-softening according to the particular conditions imposed.

Work-hardening may be either irreversible, as seen in stirring cream till it becomes a solid lump of butter, or reversible, as demonstrated in stirring a mixture of borax solution with polyvinyl alcohol. (This fluid solidifies under stirring, but regains its original fluidity after a period of rest.) Blood also belongs to the thixotropic-dilatant type of materials, and its non-linear effects are noticeable in that the flow resistance (apparent viscosity) varies with the rate of shear. For demonstrating this effect, the author arranged a "race" in which one could compare, at various rates of shear, the flow of blood with that of a standard fluid (glycerol–water) obeying a linear law with constant flow resistance. Experiments were made showing that, in outflow from similar tubes, the blood first overtakes the glycerol–water mixture, only to be overtaken by it at a later stage and lose the race.

In flow through the body, one finds that blood and the other materials have consistencies of all the types demonstrated above, and that these consistencies help in regulating the geometry and rates of flow in conjunction with threefold control exercised by various internal organs. First, there is the nervous system, which exercises overall control by generating electric pulses and directing them to all parts of the body. These pulses stimulate muscular activity which enlarges and reduces rhythmically the cavities of the respiratory, digestive and circulatory systems, and these rhythmic movements control the flow of air by periodic bellows action of the lungs, the flow of flood and drink by wavelike peristaltic movement of the digestive tract, and the flow and pressure of the blood delivered to the various organs by pulsating action of the heart and blood-vessels. Second, there are the groups of tubular vessels of the respiratory, digestive and circulatory systems, whose geometry controls the flow patterns in the various ducted currents and the ductless cross-currents in the body. All vessel walls are made up of several coats of semipermeable tubular membrane, of macroscopic dimensions with regard to length and generally also with regard to diameter. In the blood-vessels, however, the tubes are of conical shape and branch out into smaller and still smaller diameters, and eventually into fine-meshed networks of capillaries of microscopically small diameter—microcapillaries for short. Such networks are present in all parts of the body at the key points of all inlets and exits of the various internal organs. Their location at these key positions enables these networks to exercise, throughout the body, the third and final control of the flow, adjusting it everywhere to the metabolic requirements.

It has been found convenient to apply the term *microcirculation* to the flow centred on the microcapillary networks, and the term *macrocirculation* to all remaining flow components centred on the macroscopic portions of the tubular vessels of the respiratory, digestive and circulatory systems. It must, however, be understood that micro- and macrocirculation feed and drain each other, and should be treated as two mutually dependent and complementary components of one and the same flow of matter and energy which preserves the intrinsic unity of the living body. Under such unified treatment, microcirculation is of outstanding biological and medical importance because of its close connection with the metabolism, on which depend health and disease (and ultimately life and death) of the body as a whole and of each of its cells.

The way in which the microcapillary networks operate is highly elaborate. They use their tubular semipermeable membranes and the associated mechanical, electrical and osmotic devices for selective filtration of all flowing material at all inlets and exits of the respiratory, digestive and circulatory systems. The filtered materials are then separated, combined, channelled, divided and subdivided into currents which move as blood inside the tubular ducts of the microcapillaries, and cross-currents of various selected components of the blood which filter through the semipermeable walls into and out of the ducts. The whole process is so adjusted that the microcapillaries draw the essential chemicals from the outside world, and then act as coordinating, sorting and distributing centres. In so doing they ensure that microcirculation maintains the appropriate temperatures and pressures, and satisfies everywhere the local metabolic requirements of chemical interaction, irrigation and drainage. This means that the cross-currents carry into each cell of the body the appropriate amount of oxygen, nutrients, water, and hydrogen and metal ions, and drain off carbon dioxide and toxic waste products of all kinds.

Research in microcirculation can be undertaken along a wide variety of different lines, according to the particular interest of the investigator in mechanics, thermodynamics, optics, physical chemistry, biology, etc. In order to serve the needs of Mankind, it would, however, be necessary to combine all such specialized lines of research and subordinate them to the medical interest with the ultimate aim of a diagnosis and, if necessary, a plan for therapeutic action. This means that one would have to use all available methods to obtain not only a quantitative description of the state of microcirculation in health and disease with respect to metabolic requirements, but also a quantitative account of the changes to be expected by deliberate intervention through dietary measures, climatic conditions, exercise and massage, physio- and chemotherapy, radiation, etc. Here preliminary investigations would be necessary to assess the consistencies of the flowing materials and the threefold controls exercised by the various internal organs, as mentioned earlier.

INSTRUMENTATION

In medically oriented research, the first task is to find suitable instrumentation, i.e. a set of measuring instruments with calibrated scales which would allow quantitative observation of all variables of interest. The required instrumentation is not readily available, and its development involves considerable difficulties. These include difficulties of access (which have to be overcome by surgical techniques and the use of experimental animals) and those due to the variability and complexity of the phenomena involved. Microcirculation and metabolic requirements vary not only with the species and the individual, but also, for the same species and individual, with the particular organ under investigation, and for the same organ—with temperature and pressure, with health and disease, with childhood, youth and old age, with emotion and shock, with motion and rest. Moreover, microcirculation presents in each case an extremely complex system of currents and cross-currents with manifold metabolic interactions between the flowing biological and chemical units (serum, red and white cells, platelets, water, oxygen, carbon dioxide, hydrogen- and metal ions, etc.) and the stationary tissues (capillary walls, bones, muscles, cell membranes, etc.) with which they come into contact. Here it is necessary to measure the flow patterns separately for the various types of biological and chemical units, because the different types have vastly different patterns of flow and interaction with the stationary tissues. For instance, while red cells flow mainly inside the ducts of the blood-vessels, oxygen flows only part of its path with the ducted current of the red cell, while the biologically most important part flows in ductless cross-currents into the blood-stream and later again out of it through the wall membranes of the blood-vessels and those of the red cells. The difference between the two flow patterns cannot be disregarded, because it is through this difference that oxygen is drawn from the air and distributed to every cell in the body so as to keep it alive and healthy. For investigation of the flow patterns in each of the internal organs, there is a whole armoury of biological, chemical and physical instruments and techniques available, but they all have to be miniaturized in order to deal with the small dimensions encountered in microcirculation. One may gain some useful information about the behaviour of the material in *in vivo* flow through the body by a study of *ex vivo* flow under sufficiently simplified conditions, so as to allow individual assessment of each of the phenomena involved. In this way one may, for example, study separately the flow patterns of the blood components and compare them with those of synthetic suspensions. In the preliminary investigation one may introduce further simplifications and even obtain conditions permitting approximation of the behaviour of the biological material by that of a single continuum. Under these conditions an instrument termed *rheogoniometer*,†

† Manufacturers of the instrument are Messrs. Sangamo Controls Ltd., North Bersted, Bognor Regis, Sussex, England.

introduced by the author and specially designed to allow comprehensive testing of the material, can be used. Very small samples of the material suffice for tests, over wide ranges of temperature and rates of shear, for a variety of flow conditions including not only steady and oscillatory motions (with various frequencies and amplitudes), but also the biologically important pulsating processes in which an oscillatory motion is superimposed on a steady one. In each test one can take measurements of the movement and forces present in their variation in time, as well as in their distribution in space about the full solid angle of directions, with special precautions to avoid evaporation and/or skin formation where the material is in contact with air. The tests can then be evaluated to give not only the shear moduli of viscosity and elasticity for the various conditions of flow, but also the relaxation and retardation times, the power consumption, and the characteristics of any thixotropy and/or dilatancy present. There is already a good deal of experimental evidence available, indicating the biological significance of the results of such tests.

THEORY

Once the instrumentation has been established, one can make quantitative observations, and then face the second task. This consists in establishment of "law and order" by incorporating the observations into the mathematical framework of a theory. In order to deal with this task efficiently, one has to review critically the evidence supplied by the observed data. The review reveals that one can no longer accept the classical point of view according to which the system under consideration is observed undisturbed and in isolation, the observed data having the significance of exact and absolute characteristics. Instead, one has to recognize that every quantitative observation is the result of an interaction between the observed system and the observing one at the very moment when the disturbing interaction takes place. Hence, however carefully the experiments may be conducted, the observation inevitably introduces some disturbance, and with it a degree of uncertainty and inaccuracy, which have to be dealt with by statistical methods. In addition, the relation of the observed data to the observing system introduces a considerable amount of arbitrariness, because this system can be chosen in any way one likes with regard to its position in space and time, its units of measurement, and its scale functions indicating how the scales of the measuring instruments are divided, subdivided, and related to the numerical values of the variables. The inevitable interaction of the observed and observing systems, and the attendant uncertainty and relativity of the data, have already been fully recognized as basic conceptions in all modern theories of physics,† and it is important that this recognition should be extended throughout all natural sciences, including chemistry, biology, and last but not least haemorheology of the micro- and macrocirculation, where the interaction of living systems and the instruments used for observation has not always been fully appreciated and may have given entirely misleading results. The question now arises, how can one possibly arrive at a theory, for any given system under observation, from experimental data involving considerable amounts of uncertainty and arbitrariness? It is in answering this question that, under the inspired leadership of Einstein, Heisenberg and Dirac, the most powerful theoretical research tool of modern science was developed, which not only accounted for the many experimental data already observed in various branches of natural science but also opened up vast unexplored regions, whose very existence had not previously even been suspected. This research tool dealt separately with uncertainty and relativity, using the quantum-statistical theory of probability waves for the former, and the group theory of transformation for the latter. The group theory permitted derivation, from all the observed variables considered, of certain parameters and relations which were increasingly free from arbitrariness by being constructed as invariant under ever-wider groups of transformations of the observing system and changes in the material and conditions.‡ Among the invariants so constructed, the quantities of matter and energy play a prominent part, so that development of a useful theoretical scheme for circulatory problems in the living body requires tracing of the movement, through space and time, of each particle of matter and each portion of energy from the intake of food, drink and air to the discharge of excretions and waste products. Another invariant, known as the "geodetic",§ then serves as guide to the path through space and time taken by the matter and energy, however complicated this path may be because of the presence of mechanical, electromagnetic and osmotic forces.

In applying statistical theories to evaluation of experimental results, one has to take into account certain essential differences between the flow en-

† See, for example, W. Heisenberg's "Uncertainty principle in wave and quantum mechanics" (*Naturwiss.* **14,** 989, 1926) and A. Einstein's *Relativity Principle in the Special and General Theories of Relativistic Mechanics* (J. A. Barth, Leipzig, 1916).

‡ Some particularly simple special cases of this method were already known in classical physics, such as the model technique based on parameters which are dimensionless and thus invariant under all changes of the units of measurement used in the observing system.

§ The path along a geodetic line has an extreme length (shortest or longest) compared with all neighbouring pathways.

countered in most technological materials and that in the living body. For the former, one used to apply the continuum theory with statistical averages such that the flow pattern could be approximated by that of a single continuous medium preserving its continuity throughout all its movements. This approximation was justified, because most technological materials (even of very complex structure) are characterized, at the statistical average, by coincidence of all average flow patterns associated with the structural components. However, no such coincidence exists in either macro- or microcirculation. Instead, one observes here distinct separation of the average flow patterns of the blood components, enhanced by the phenomena of aggregation, segregation and sedimentation. Separation is no less pronounced in microcirculation, in which the ducted currents and ductless cross-currents provide different average flow patterns as some blood components are selectively filtered through the vessel walls. To deal with this separation, one requires a more elaborate "multi-continuum" theory in which the flow is approximated by statistical averages taken over a number of different interpenetrating continua in motion relative to one another in the living body. In this theory one would also have to take into account the greater complexity of the boundary conditions, as in technological flow the particles of the flowing material are usually small compared with duct diameters, and the walls are rigid, cylindrical, smooth, impermeable, and of uniform circular cross-section, while in flow through the living body some particles (red cells) are of the same dimensions as the duct diameter, and the walls are elastically deformable, conical, rough, selectively permeable, and of varying elliptical cross-section. Great progress has been made in the development of theoretical and experimental methods for disentanglement of the diverse boundary conditions and flow patterns. However, much further development is required, and this may well be a fruitful field for collaboration between biologist, chemist, physicist and mathematician. It seems likely that a step in the right direction would be made by describing the flow in terms of probability waves (as in wave mechanics) and using the group theory of transformation, to find the invariants in the maze of movements and to trace the geodetic pathways in the flow patterns. In a preliminary investigation, one may gain some useful information about the consistency of the materials involved by studying the various phenomena of flow under *ex vivo* conditions, sufficiently simplified to permit approximation of a biological material by a single continuum. Phenomena so studied often have a paradoxical appearance, but have in fact been predicted from the author's theory of invariants. It should be noted that the consistency of blood and other materials flowing in the living body is such as to exhibit the said phenomena, but for convenience of demonstration the author exemplified them by appropriately chosen technological materials.

The dilatancy phenomena were first noticed by Reynolds when walking on wet sand, i.e. a highly concentrated sand–water suspension. He observed that the sand around his footprints appeared to dry out, indicating that its particles had, on the average, moved further apart from one another, thereby increasing the apparent volume occupied by them and allowing the water to vanish from the surface and seep down into the interstices of the structure. He explained the dilatationary movement of the sand particles on the assumption that prior to putting his foot down the sand particles had been in closest packing, so that any movement could only increase the apparent volume. This explanation was so plausible that it was generally accepted, but it was shown later that dilatancy occurred not only under conditions of closest packing but also when the packing was quite loose and nowhere near the closest. One found, further, a paradoxical effect associated with dilatancy in that, in spite of the solid particles moving further apart, the material work-hardened, solidified, and became more brittle. Moreover, the same material which exhibited positive dilatancy with an increase in apparent volume, work-hardening and solidifying under certain mechanical action, exhibited negative dilatancy under different mechanical actions, with a decrease in apparent volume, work-softening and fluidification.

ACKNOWLEDGEMENT

I gratefully acknowledge my indebtedness to Drs. Knisely, Copley, Harders, Gelin, Scott Blair, Davis and Landau who first introduced me to the problems of haemorheology and microcirculation, and then helped and encouraged my work with many useful discussions and suggestions.